# Lecture Notes in Physics

Edited by J. Ehlers, München, K. Hepp, Zürich
R. Kippenhahn, München, H. A. Weidenmüller, Heidelberg
and J. Zittartz, Köln
Managing Editor: W. Beiglböck, Heidelberg

127

Enrique Sanchez-Palencia



# Non-Homogeneous Media and Vibration Theory

Springer-Verlag
Berlin Heidelberg New York 1980

**Author**

Enrique Sanchez-Palencia
Département de Mécanique, L.A. 229, Université Paris VI,
4 place Jussieu,
F-75230 Paris



ISBN 3-540-10000-8 Springer-Verlag Berlin Heidelberg New York
ISBN 0-387-10000-8 Springer-Verlag New York Heidelberg Berlin



Printing and binding: Beltz Offsetdruck, Hemsbach/Bergstr.
2153/3140-543210

# PREFACE

The present volume deals with perturbation problems in two fields: 1) Homogenization theory in physical phenomena in non-homogeneous media, and 2) Spectral properties of operators with either discrete or continuous spectra, in particular implicit eigenvalue problems, which also applies to the study of scattering frequencies of operators.

The plan is as follows:

Part I (Chapters 1 to 4) contains some elements about boundary value problems. It only contains classical material in a succinct form, without proofs. It should only be useful to the reader not having a sufficient background in this field. On the other hand, it is not complete; in particular, regularity theory for elliptic problems (which is used later) is not given.

Part II (Chapters 5 to 8) deals with the homogenization method for the study of physical phenomena in non-homogeneous media with periodic structure. The presentation emphasizes the role of the homogenized constitutive laws, and is different from that of the current literature.

Part III (Chapters 9 to 14) contains a number of perturbation problems, such as singular perturbations, stiff problems, perturbation of the domain, etc., with emphasis on spectral properties.

Part IV (Chapters 15 to 17) deals with diffraction problems and scattering frequencies. The relation between scattering frequencies and eigenvectors of appropriate operators is shown.

A certain number of results are published for the first time, among them, Chap. 5, Sects.7, 8; Chap. 6, Sects. 6, 7, 8; Chap. 8, Sect. 3; Chap. 9, Sect. 4; Chap. 11, Sect. 6; Chap. 17, Sect. 3 and the Appendix by L. Tartar.

Parts of this work were used in a short course at the Politecnico di Torino (1978) and in a post-graduate course at the Département d'Analyse Numérique de l'Université Paris VI (1979-80). I thank Professors P. Ciarlet, G. Geymonat and P.A. Raviart for their hospitality.

I am very indebted to many colleagues and friends for their aid. In particular, I express my thanks to Professors H. Cabannes, G. Geymonat, T. Lévy and J.L. Lions for useful discussions, comments, correction of proofs and moral support. I am indebted to L. Tartar for allowing me to publish his proof of convergence of the homogenization process in fluid flow as an appendix to this volume. Last but not least, my gratitude to Miss M.F. Couturier for her careful typing of the manuscript.

Paris, December 1979

CONTENTS

# PART I

## NOTIONS ABOUT BOUNDARY VALUE PROBLEMS

This part is an introduction to the problems handled in the sequel. My aim is to give some classical results without proof, but perhaps with explanations and comments.

# CHAPTER 1

## DISTRIBUTIONS AND SOBOLEV SPACES

An extensive theory about this questions may be found in SCHWARTZ [1], NECAS [1], SMIRNOV [1], LIONS and MAGENES [1].

1.- Distributions - Let $\Omega$ be an open set of the N-dimensional euclidean space $R^N$ (possibly $\Omega = R^N$) and $\mathcal{D}(\Omega)$ the set of the infinitely derivable functions with compact support in $\Omega$ (that is to say, identically zero outside of a compact set of $\Omega$). Let us define a topology (or a concept of convergence) on $\mathcal{D}(\Omega)$.

If $\theta^i$ $(i = 1,2 \ldots)$ and $\theta^0$ are functions in $\mathcal{D}(\Omega)$,

$$\theta^i \to \theta^0 \quad \text{in } \mathcal{D}(\Omega)$$

means that the supports of all the $\theta^i$ are contained in a unique compact set of $\Omega$ and $\theta^i$ and all their derivatives tend uniformly to $\theta^0$ and the corresponding derivatives.

Let T be a linear and continuous functional on $\mathcal{D}$, that is to say, a law associating a number (real or complex) $\langle T, \theta \rangle$ to each $\theta \in \mathcal{D}$, and such that this law is linear and

$$\theta^i \to \theta^0 \text{ in } \mathcal{D} \Longrightarrow \langle T, \theta^i \rangle \to \langle T, \theta^0 \rangle$$

Such a functional is called a distribution on $\Omega$ , and the set of such distributions is the space $\mathcal{D}'(\Omega)$. It is possible to define a concept of convergence on $\mathcal{D}'$ : $T^i \to T^0$ in $\mathcal{D}'$ iff

$$\langle T^i , \theta \rangle \to \langle T^0 , \theta \rangle \qquad \forall \theta \in \mathcal{D} .$$

If $f$ is a locally integrable function on $\Omega$ , it is possible to define a distribution $\tilde{f}$ by :

$$\langle \tilde{f} , \theta \rangle = \int_\Omega f(x) \; \theta(x) \, dx$$

which is linear and continuous on $\mathcal{D}$. It is noticeable that if $f^1(x)$ and $f^2(x)$ are equal a.e. (almost everywhere, such is to say, the set $\{x / f^1(x) \neq f^2(x)\}$ is of zero measure), the associated distributions $\tilde{f}^1$ and $\tilde{f}^2$ are the same.

We then see that the distributions generalize the locally integrable functions, but when a function f is considered as a distribution, it is identical with all the functions which one obtains by changing the values of f(x) on a set of measure zero. In fact, the distribution is not associated to a function, but to an equivalence class formed by the functions which are a.e. equal.

If f is continuously differentiable function, by integrating by parts we have (the test function being null in the vicinity of $\partial\Omega$ , frontier of $\Omega$ ) :

$$\left\langle \left(\frac{\partial f}{\partial x_i}\right)^\sim , \theta \right\rangle = \int_\Omega \frac{\partial f}{\partial x_i} \theta \, dx = - \int_\Omega f \frac{\partial \theta}{\partial x_i} dx = \left\langle \tilde{f} , - \frac{\partial \theta}{\partial x_i} \right\rangle$$

this formula is the basis for the definition of distributional derivatives of a distribution. If $T \in \mathcal{D}'$, $\frac{\partial T}{\partial x_i} \in \mathcal{D}'$ is defined by

$$\left\langle \frac{\partial T}{\partial x_i} , \theta \right\rangle = \left\langle T, - \frac{\partial \theta}{\partial x_i} \right\rangle$$

and hence any distribution T has distributional derivatives. In particular, any locally integrable function has distributional derivatives (which in general are not functions). Moreover, it follows from the definition that derivation is a continuous operation in $\mathcal{D}'$, that is to say :

$$T_i \to T \quad \text{in } \mathcal{D}' \Longrightarrow \frac{\partial T_i}{\partial x_k} \to \frac{\partial T}{\partial x_k} \text{ in } \mathcal{D}' \quad .$$

2.- Sobolev spaces - Hereafter, whenever integral is used, we mean the Lebesgue integral ; to ensure that the spaces are complete, this is a fundamental point. An integrable function in the sense of Lebesgue is said to be a summable function. For this integral the following important theorem of Lebesgue holds : (Dominated convergence) :

If $f_n(x)$ is a sequence of summable functions on $\Omega$ such that

$$f_n(x) \to f(x) \quad \text{a.e. in } \Omega \text{ and}$$

$$|f_n(x)| \leq F(x) \quad \text{for a summable function } F,$$

then

$$\int_\Omega f_n(x)\,dx \to \int_\Omega f(x)\,dx .$$

(It also holds for functions with values in a separable,reflexive Banach space)

Let us consider the functions $f(x)$ defined on a domain $\Omega$ whose p-power is summable $(0 < p < \infty)$

$$\int_\Omega |f(x)|^p\,dx < +\infty$$

and we construct the equivalence classes formed by the functions that are equal a.e. The set of such equivalence classes is a vectorial complete space (hence a Banach space) with norm defined by

$$\|f\| = \left(\int_\Omega |f|^p\,dx\right)^{1/p}$$

This space is denoted by $L^p(\Omega)$. It is usual to speak of $L^p$ as the space of functions whose p-powers are summable (with the convention that such a function does not change if its values change on a set of measure zero). In the case $p = +\infty$, $L^\infty(\Omega)$ is the space of functions (classes of functions) which are bounded a.e. It is a Banach space with the norm defined by

$$\|f\| = \operatorname*{ess.\,sup.}_{x \in \Omega} |f(x)|$$

(ess. sup. is the sup. in $\Omega$ up to a set of zero measure).

If $p = 2$, $L^2(\Omega)$ is a Hilbert space, and the norm is associated with the scalar product ( — is for the complex conjugate)

$$(f,g) = \int_\Omega f(x)\,\overline{g(x)}\,dx \qquad \text{(complex functions)}$$

$$(f,g) = \int_\Omega f(x)\,g(x)\,dx \qquad \text{(real functions)}$$

If $\Omega$ is bounded and $f \in L^p$, $g \in L^q$ with $p, q > 1$, $\frac{1}{p} + \frac{1}{q} = 1$, we have the Hölder inequality :

$$\left| \int_\Omega f\,g\,dx \right| \leq \|f\|_{L^p} \, \|g\|_{L^q}$$

It is then easy to prove the (algebraic and topological) embedding

$$(2.1) \qquad p < p' \Rightarrow L^{p'} \subset L^p$$

This result is natural because the larger the exponent p is, the smaller the abmissible singularities of f are. On the other hand, if $\Omega$ is an unbounded domain we do not have (2.1) ; $f \in L^p$ means that the singularities of f are not "too large" and also that f is "sufficiently near to zero in the vicinity of x =∞".

The Sobolev space $W_p^m(\Omega)$ is the space of all distributions such that they are associated (as well as the distributional derivatives of order $\leq m$) with functions belonging to the $L^p$ space. This is a Banach space for the norm

$$\|f\|_{W_p^m} = \left[ \int_\Omega \sum_{0 \leq m_1+\ldots+m_N} \left| \frac{\partial^{m_1+\ldots+m_N} f}{\partial x_1^{m_1} \ldots \partial x_N^{m_N}} \right|^p dx \right]^{1/p}$$

if p = 2, $W_p^m$ is a Hilbert space, usualy denoted by $H^m$, equipped with the evident scalar product, for instance, for m = 1 :

$$(f,g)_{H^1} = (f,g)_{L^2} + \sum_{i=1}^{N} \left(\frac{\partial f}{\partial x_i}, \frac{\partial g}{\partial x_i}\right)_{L^2}$$

It is also possible to define the spaces $W_p^m$ for m real (not integer). In the particular case $\Omega = R^N$, p = 2, the spaces $H^S$ are easily defined by Fourier transform. $H^S$ is the space of the $f \in L^2(R^N)$ such that their Fourier transforms $\hat{f}$ satisfy

$$(1 + \xi^2)^{S/2} \hat{f} \in L^2(R^N)$$

equipped with the (hilbertian) norm

$$\| (1 + \xi^2)^{S/2} \hat{f} \|_{L^2(R^N)}$$

For S integer, this definition coïncides with the preceeding one, thanks to the well known fact that the Fourier transform is an isomorphism in $L^2(R^N)$.

3.- Traces and embedding theorems - The trace is a generalization of the concept of the restriction of a continuous function to a submanifold of its domain of definition (for exemple, $\partial\Omega$ , the boundary of $\Omega$). Nevertheless, it is a deeper and more sophisticated concept.

To fix ideas, let us consider $H^1(\Omega)$, where $\Omega$ has a smooth boundary. According to the definition, an element u of $H^1$ is a distribution associated with a function ; if we consider u as a function, it is in fact still equal to u even after modification on a set of zero measure (for exemple, $\partial\Omega$) and then the restriction of u to $\partial\Omega$ makes no sense.

Nevertheless, the space $C^\infty(\bar{\Omega})$ of infinitely differentiable functions in the closure $\bar{\Omega}$ of $\Omega$ is dense in $H^1(\Omega)$. u can then be considered as the limit (in the $H^1$ topology) of smooth functions $u^i$. For such functions the restriction $u^i|_{\partial\Omega}$ have the usual sense. If it is possible to prove that these functions converge to a (unique !) limit function in an appropriate topology, we shall say that such a limit is the trace $u|_{\partial\Omega}$ of u on $\partial\Omega$.

Let us take a subdomain $\omega$ of $\Omega$ as in the fig. 1 and let us transform it by a diffeomorphism into a cylindrical domain. (In the following inequalities, the local maps are smooth and the constants are modified, but not the essential results).

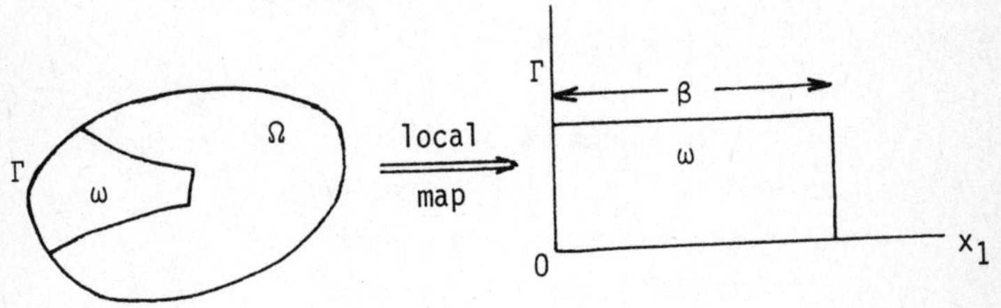


If $v \in C^\infty(\overline{\Omega})$, we have, with $y = (x_2, \dots x_N)$ :

$$v(o,y) = - \int_0^\tau \frac{\partial v}{\partial x_1}(\xi, y)\, d\xi + v(\tau, y)$$

$$|v(0,y)|^2 \leqslant 2 \left| \int_0^\tau \right|^2 + 2\, |v(\tau, y)|^2 \leqslant (\text{Schwarz}) \leqslant 2\beta \int_0^\beta \left| \frac{\partial v}{\partial x_1} \right|^2 d\xi + 2v^2(\tau, y)$$

and by integrating it with respect to $\tau \in (0,\beta)$

$$\beta |v(o,y)|^2 \leqslant 2\beta^2 \int_0^\beta \left| \frac{\partial v}{\partial x_1} \right|^2 d\xi + 2 \int_0^\beta v^2(\tau,y)\, d\tau$$

and by integrating with respect to $y$ over $\Gamma$

$$\beta \| v(o,y) \|^2_{L_2(\Gamma)} \leqslant 2\beta^2 \left\| \frac{\partial v}{\partial x_1} \right\|^2_{L^2(\omega)} + 2 \|v\|^2_{L^2(\omega)} \leqslant C \|v\|^2_{H^1(\Omega)}$$

Then, the restriction operator $v \to v|_\Gamma$ is linear and bounded (and thus continuous) from $H^1(\Omega)$ to $L^2(\Gamma)$ ; and since it is defined on the dense subset $C^\infty(\overline{\Omega})$ of $H^1(\Omega)$, it can be continuously extended (with conservation of the norm) to the whole space $H^1(\Omega)$. <u>This extension is the "trace operator" which is then continuous from $H^1(\Omega)$ into $L^2(\Gamma)$</u>.

A little modification of the preceeding proofs shows that if we "cut" $\omega$ by $x_1 = \alpha$ instead of $x_1 = 0$, the trace $v(\alpha, y)$ (which is of course a function of $\alpha$ with values in $L^2(\Gamma)$) is <u>a continuous function of $\alpha$</u>. This property shows that the trace is also a generalization of the concept of limit value of the function in the vicinity of $x_1 = 0$.

Another important property of the $H^1(\Omega)$ space is that, if $\Omega$ is bounded, the embedding of $H^1(\Omega)$ in $L^2(\Omega)$ is compact (Rellich). This means that the weak convergence in $H^1(\Omega)$ implies the strong convergence in $L^2(\Omega)$ (that is to say, the weak topology of $H^1$ is stronger than the strong topology of $L^2$) and also that a bounded domain in $H^1(\Omega)$ is a precompact set in $L^2(\Omega)$.

The theorems of trace and of Rellich are particular cases of the following embedding theorem of Sobolev and Kondrašov :

Theorem 3.1 - Let $\Omega$ be an open domain of $R^N$, with sufficiently smooth boundary and let $\omega$ be intersection of $\Omega$ and a sufficiently smooth manifold of dimen ion $\nu$ (for exemple, if $\nu = N$,$\omega$ may coïncide with $\Omega$ or be a part of $\Omega$ ; if $\nu = N-1$, $\omega$ may be an intersection of a hyperplan and $\Omega$ , or the boundary of $\Omega$, etc.) If a function $u \in W_p^1(\Omega)$ is given, $u_{|\omega}$ is (if it makes sense) a function defined on $\omega$ , and the operator $u \rightarrow u_{|\omega}$ is called embedding operator (or trace operator, if $\nu < N$). In this conditions, we have the following

a) if $q < \frac{p}{N-p}$ and $\nu > N - p$, the embedding operator is completely continuous (compact) from $W_p^1(\Omega)$ into $L^q$.

b) if $N < p$, $\nu = N$ the embedding operator is completely continuous (compact) from $W_p^1(\Omega)$ into the space $C^0(\overline{\Omega})$ of continuous functions on $\overline{\Omega}$ equipped with the uniform convergence topology.

It is of course possible to apply the preceeding theorem to the function and its derivatives in a reiterative way and obtain properties of the traces of the functions of $W_p^m$ for $m > 0$.

Remark - The elementary proof of the particular case of the trace theorem shows the role of the smoothness of the boundary of $\Omega$ . In fact, the embedding theorem holds for domains bounded by a finite number of smooth surfaces which do not intersect at zero angle. ∎

It is also useful to have embedding and trace theorems for the $H^S(\Omega)$ spaces with real S.

Theorem 3.2 - Let $\Omega$ be a bounded open set of $R^N$ of boundary $\Gamma$ which is a manifold of dimension N-1 and of class $C^\infty$ , and $\Omega$ is on one side of $\Gamma$ . Then,

a) The trace operator is continuous from $H^s(\Omega)$ into $H^{s-1/2}(\Gamma)$ for $s > 1/2$.

b) The embedding $H^s \subset H^r$ for $r < s$ is completely continuous.

It is noticeable that the concept of trace enables us to integrate by parts functions in Sobolev spaces. The proof is immediate by using the density of the smooth functions. Thus, one can integrate for the smooth functions and then pass to the limit in the integrals.

CHAPTER 2

# OPERATORS IN BANACH SPACES

An extensive study of these questions will be found in any text book of functional analysis, for example, KATO [2], SMIRNOV [1], YOSIDA [2], also LIONS [2] and RIESZ - NAGY [1].

The main property of operators and functionals is their continuity, we then will begin by a brief introduction about convergence in Banach spaces.

1.- Strong, weak and weak-star topologies - Let us first recall that a Banach space B is a normed complete space, i.e. a vector space equipped with a norm, and such that every Cauchy sequence has a limit :

$$\|x^i - x^j\| \to 0 \quad i,j \to \infty \Rightarrow \exists\, x \text{ such that } \|x^i - x\| \to 0$$

The norm defines a distance and the associated topology (resp. convergence) is called strong topology (resp. convergence) :

$$x^i \to x^0 \text{ in B strongly} \Leftrightarrow \|x^i - x^0\| \to 0 .$$

Let B' be the dual space of B, i.e. the Banach space of the linear continuous functionals f on B. The value of f for $x \in B$ is denoted by $[f,x]$, the brackets are for the duality product between B' and B. Moreover,

$$\|f\|_{B'} = \sup |[f,x]| \qquad \text{for } \|x\| = 1 .$$

In applications, one must deal with antilinear forms (i.e. forms such that their values in $\lambda x$ are $\bar{\lambda}$ times their values in x, where — is for the complex conjugate). We shall often write "dual" for "antidual" if there is no ambiguity.

Definition 1.1.- A sequence $x^i \in B$ is said to converge weakly (or in the weak topology) to $x^0 \in B$ iff

$$[f,x^i] \to [f,x^0] \quad \forall \; f \in B'$$

and we denote it by

$$x^i \to x^0 \quad \text{in B weakly.}$$

Strong convergence obviously implies weak convergence, but the converse only holds in finite-dimensional spaces.

Proposition 1.1.- If $x^i$ converges to $x^0$ weakly, $\| x^i\|$ is bounded and

$$\| x^0 \| \leq \underline{\lim} \ \|x^i \|$$

where $\underline{\lim}$ is for the lower limit.

Proposition 1.2.- The necessary and sufficient condition for $x^i \in B$ to weakly converge to $x^0$ is that $\|x^i\|$ is bounded and that $[f,x^i]$ converges to $[f,x^0]$ for all f in a dense subset of B'.

Proposition 1.3.- If $f^i$ converges to f in B' strongly and $x^i$ converges to x in B weakly, then

$$[f^i , x^i] \to [f , x] \quad .$$

The new concept of weak-star convergence (or topology) is defined on the dual space B' of B.

Definition 1.2.- A sequence $f^i \in B'$ is said to converge in the weak star topology to $f \in B'$ iff

$$[f^i , x ] \to [f , x] \qquad \forall \ x \in B.$$

It is then evident that weak and weak-star topologies coincide if the space B is reflexive, i.e. if it coincides with the dual of its dual. The spaces $L_p$ and $W_p^m$ with $1 < p < \infty$ are reflexive. By interchanging the roles of x and f we have for weak-star convergence propositions analogous to 1.1 and 1.2.

Definition 1.3.- A Banach space B is said to be separable if there exists a denumerable set of elements which is dense in B.

Proposition 1.4.- All the spaces $L^p$ and $W_p^m$ with $p < \infty$ are separable.

Proposition 1.5.- If B is separable, any bounded set of B' is precompact for the weak star topology.

This is a very useful property in applications. It means that, if $f^i \in B'$ and $\|f^i\| \leq C$, there exists a subsequence $f^\alpha$ and an element f of B' ( $\|f\| \leq C$) such that

$$[f^\alpha , x ] \to [f , x ] \qquad \forall \ x \in B \ .$$

As a consequence, we have

Proposition 1.6.- If B is reflexive and the sequence $x^i \in B$ is such that $\|x^i \| \leq C$, there exists a subsequence $x^\alpha$ and an element $x \in B$ for which

$$[f , x^\alpha] \to [f , x] \qquad \forall f \in B'.$$

Let us recall in this context that if the duality product is formally given by

$$[f , g] = \int_{\Omega} f g \, dx \qquad (\Omega \text{ bounded})$$

the dual space of $L_p(\Omega)$ is $L_q(\Omega)$ $(p^{-1} + q^{-1} = 1)$. These spaces are reflexive for $p \neq 1, \infty$ . Moreover, the dual of $L^1(\Omega)$ is $L^{\infty}$ .

Let us finally recall that a Hilbert space is a Banach space whose norm is associated with a scalar product :

$$\|x\|^2 = (x , x).$$

A Hilbert space may be identified to its dual. Indeed, if H is a Hilbert space and $f \in H'$, there exists a unique $\phi \in H$ such that

$$[f , x] = (x , \phi) \qquad \forall \, x \in H .$$

(Riesz representation theorem)

2.- Operators in Banach spaces - Let X, Y be two Banach spaces. A function A from a part D(A) of X into Y is said to be a linear operator if it preserves linearity. D(A) is called the domain of A. If A is linear, D(A) is a linear manifold of X. The range R(A) is defined as the set of the elements of Y of the form Ax with $x \in D(A)$. If D(A) is dense in X, A is said to be "densely defined". The inverse $A^{-1}x$ is defined iff the map A is one-to-one, i.e. iff $Au = 0 \Rightarrow u = 0$. Then, $D(A^{-1}) = R(A)$, $R(A^{-1}) = D(A)$.

An operator A is said to be continuous iff $x^i \to x \Rightarrow A x^i \to Ax$ (in the strong topologies). The necessary and sufficient condition for A to be continuous is boundedness, i.e.

$$\|Ax\| \leqslant M \|x\| \qquad \forall \, x \in D(A)$$

The smallest number M with this property is called the norm of A and is denoted by $\|A\|$. If A is continuous and densely defined, it may be extended (by continuity) to all X. The space of all continuous operators from X to Y is denoted by $\mathcal{L}(X , Y)$ ; it is a Banach space for the norm $\| A \|$ just defined. A useful property of linear continuous operators is that they transform weakly convergent sequences of X in weakly convergent sequences of Y.

Several kinds of convergence can be defined on $\mathcal{L}(X , Y)$

Definition 2.1.- If $\|A^i - A\| \to 0$ we say that $A^i$ converges in the norm (or uniformly) to A.

If for each $x \in X$

(2.1) $\qquad A^i x \to A x \quad$ in Y strongly

we say that $A^i$ converges <u>strongly</u> to A.

If for each $x \in X$

(2.2) $\qquad A^i x \to Ax \quad$ in Y weakly

we say that $A^i$ converges <u>weakly</u> to A.

It is obvious that weak convergence amounts to

(2.3) $\qquad [y,A^i x] \to [y,Ax] \qquad \forall\, x \in X,\ y \in Y'$ .

On the other hand, uniform convergence implies strong convergence, and this one implies weak convergence. Moreover

<u>Proposition 2.1</u>.- The necessary and sufficient condition for $A^i$ to converge to A is that it converges strongly (resp. weakly) and the convergence in (2.1) (resp. (2.3)) is uniform for $\|x\| \leqslant 1$ (resp. $\|x\| \leqslant 1$, $\|y\| \leqslant 1$).

An operator $A \in \mathcal{L}(X,Y)$ is said to be <u>compact</u> or <u>completely continuous</u> iff the image $Ax^i$ of a bounded sequence $x^i$ contains a Cauchy sequence. Such an operator transforms a weakly convergent sequence into a strongly convergent sequence.

<u>Unbounded linear operators</u> are not defined on the whole space X. They are in many respects very different from bounded linear operators. Nevertheless, there is a class of unbounded operators, the <u>closed linear operators</u> for which many properties of bounded operators still hold. An operator A from X to Y is said to be <u>closed</u> if

$$\left.\begin{array}{l} x^i \to x \\ Ax^i \to y \end{array}\right\} \Rightarrow \left\{\begin{array}{l} x \in D(A) \\ Ax = y \end{array}\right.$$

It is easy to see that this amounts to saying that the graph of A (i.e. the set of the pairs (x,y) of $X \times Y$ such that $x \in D(A)$, $y = Ax$) is closed in $X \times Y$ .

<u>Proposition 2.2 (Closed graph theorem)</u>.- A closed operator whose domain is the whole space X is bounded.

<u>Definition 2.2</u>.- An operator in the Hilbert space H is said to be accretive if

$$\operatorname{Re}(Au\ , u) \geqslant 0 \qquad \forall\, u \in D(A).$$

Definition 2.3.- An operator B is said to be an extension of A if $D(B) \supset D(A)$ and $Bu = Au$ for any $u \in D(A)$. It is a proper extension if D(B) contains elements which are not in D(A).

3.- Self adjoint operators - Let us now consider densely defined operators from a Hilbert space H into itself. Such an operator A is said to be symmetric if

$$(3.1) \qquad (Ax\ ,\ y) = (x\ ,\ Ay) \qquad \forall\ x,y \in D(A)$$

A symmetric operator is said to be semibounded from below if there exists a finite number m such that

$$(3.2) \qquad m = \inf(Ax,x) \qquad \text{for } x \in D(A),\ \|x\| = 1\ .$$

It follows from (3.2) that

$$(3.3) \qquad (Ax,x) \geqslant m\,\|x\|^2 \qquad \forall\ x \in D(A)$$

If $m > 0$, A is said to be positive definite, if $m \geqslant 0$, A is said to be positive.

Let us now introduce the concept of selfadjoint operator. Il $x \in D(A)$ and $y \in H$, we form the scalar product (Ax,y). There exists elements y (for instance $y = 0$) such that

$$(3.4) \qquad (Ax\ ,\ y) = (x\ ,\ y^*) \qquad \forall\ x \in D(A)$$

for a certain $y^*$. In this case, $y^*$ is uniquely defined by y. For, if

$$\left.\begin{aligned} (Ax\ ,\ y) &= (x\ ,\ y_1^*) \\ (Ax\ ,\ y) &= (x\ ,\ y_2^*) \end{aligned}\right\} \qquad \forall\ x \in D(A)$$

we have by substraction

$$(x\ ,\ y_1^* - y_2^*) = 0 \qquad \forall\ x \in D(A)$$

and because D(A) is dense in H we have $y_1^* - y_2^* = 0$. Thus, there exists a well defined operator $A^*$ such that for these y :

$$y^* = A^*\,y\ .$$

Its domain $D(A^*)$ is the set of such y.

An operator is said to be selfadjointiff $A^* = A$ (in particular, $D(A^*) = D(A)$). It is clear that for bounded operators the concepts of symmetry and selfadjointness coincide ; but this is not the case for unbounded operators : selfadjoint operators are a very restricted class of symmetric operators. A symmetric operator is selfadjointiff each element $y \in D(A^*)$ also belongs to D(A). We shall now see a criterion for the selfadjointness of symmetric operators.

Proposition 3.1.- If A is a symmetric operator and its range is the whole space H, then A is selfadjoint.

An important property of selfadjoint operators is

Proposition 3.2. - If A is selfadjoint and positive definite, then $A^{-1}$ exists and is a bounded operator from H to H (and thus defined on the whole H).

The preceeding property shows the great importance of selfadjointness in applications ; if A is selfadjoint, and positive definite, the equation

$$(3.5) \qquad Ax = f$$

with $f \in H$ has a unique solution x. This equation is somewhat nontrivial because A is an unbounded operator. Then, equations involving symmetric unbounded operators often require, in order to have unique solutions, that the operator be selfadjoint. This is the reason why, in applications, "naive" symmetric operators A in an equation of the kind (3.5) must be replaced by a more sofisticated selfadjoint operator $\tilde{A}$ which is an extension of A. The solutions of the equation

$$(3.6) \qquad \tilde{A}x = f$$

are called "generalized" or "weak" solutions.

In this context, the following theorem of Friedrichs is important :

Theorem 3.1 - If A is a symmetric positive definite operator, it can be extended to a selfadjoint operator $\tilde{A}$ which is called the Friedrichs extension of A.

The study of the equation (3.6) for the Friedrichs extension instead of (3.5) is natural in applications if the range R(A) of A is dense in H (as is usually the case). In this case, from the positive definiteness it follows that

$$(3.7) \qquad (Ax^1 - Ax^2, x^1 - x^2) \geqslant m \| x^1 - x^2 \|^2$$

and $x^1 \neq x^2 \Rightarrow Ax^1 \neq Ax^2$. In consequence, $A^{-1}$ is defined from R(A) to H and it follows from (3.7) that

$$\| x^1 - x^2 \| \leqslant m^{-1} \| Ax^1 - Ax^2 \|$$

and thus $A^{-1}$ is bounded and densely defined. Moreover, if $\tilde{A}$ is the Friedrichs extension of A, $\tilde{A}^{-1}$ is, by virtue of proposition 3.2 continuous and defined on the whole space H, and it is of course an extension of $A^{-1}$. It is in fact an extension by continuity of $A^{-1}$. Then, the study of the equation (3.6) instead of (3.5) (which has solution only for $f \in R(A)$) amounts to define as "generalized solution" of (3.5) for $f \notin R(A)$ the limit

$$x^* = \lim x^i$$

where $x^i = A^{-1} f^i$ and $f^i$ is a sequence such that $f^i \to f$.

4.- Resolvent, spectrum and spectral families - Let $A$ be a closed operator in a Banach space $B$ and let $I$ be the identity operator.

Definition 4.1.- A complex number $\lambda$ is said to be an eigenvalue of $A$ iff there exists a non zero element $x \in B$ such that

$$Ax = \lambda x$$

and $x$ is called an eigenvector.

Definition 4.2. - The resolvent domain $\rho(A)$ of $A$ is the set of the complex numbers $\zeta$ such that $A - \zeta I$ (often written $A - \zeta$) has a bounded inverse $(A - \zeta I)^{-1}$. This inverse is defined on the whole space $B$ and is an element of $\mathcal{L}(B , B)$.

The resolvent domain is always an open set (possibly the null set) ; its complement is called the spectrum which is denoted by $\sigma(A)$. The operator $(A - \zeta)^{-1}$ is called the resolvent, which is in fact a holomorphic function of $\zeta$ with values in $\mathcal{L}(B , B)$.

It is clear that if $\lambda$ is an eigenvalue, $\lambda \notin \rho$ (i.e. $\lambda$ belongs to the spectrum). But there are in general points of the complex plane that are not eigenvalues and do not belong to $\rho$. This situation which do not appears in finite-dimensional spaces will be illustrated with examples in the following chapters.

An important property of selfadjoint operators is the following :

Proposition 4.1.- If $A$ is selfadjoint the following properties hold

a) $\sigma(A)$ is contained in the real axis

b) If $(Ax,x) \geqslant C \|x\|^2$ $(C \geqslant 0)$, the portion $\zeta < C$ of the real axis belongs to the resolvent set.

c) If $\zeta$ belongs to the resolvent domain of $A$,

$$\|(A - \zeta)^{-1}\| \leqslant \frac{1}{\operatorname{dist}(\zeta , \sigma(A))}$$

It is clear that part b) of this proposition is a consequence of prop. 3.2 by considering the operator $A - C$ which is also selfadjoint.

We now recall the classical well known properties of compact symmetric (selfadjoint) operators in a Hilbert space $H$.

Theorem 4.1.- Let $A \in \mathcal{L}(H,H)$ be compact and positive (i.e. $(Ax,x) \geqslant 0$) and $Ax \neq 0$ if $x \neq 0$. Then, there exists an infinite sequence $e^i$ of eigenvectors which can be taken to be orthonormal (i.e. $\|e^i\| = 1$, $(e^i,e^j) = 0$ if $i \neq j$)

and which form a basis of H. The numbering i can be made in such a way that the associated eigenvalues $\lambda^i$ (which are real and positive) form a monotone decreasing sequence with zero limit :

(4.1) $$\lambda^1 \geqslant \lambda^2 \geqslant \ldots \geqslant \lambda^n \geqslant \ldots \quad \to 0 \quad .$$

Remark 4.1 - In the sequence (4.1), an eigenvalue may be associated to a finite number of eigenvectors ; in this case, it is customary to repeat it once for each eigenvector. This is the meaning of the signs $\geqslant$ in (4.1). ■

Remark 4.2 - It is clear that, if $\lambda$ is a multiple eigenvalue, the associated eigenvectors $e^i$ are not uniquely determinated. We can choose any orthonormal basis of the (finite dimensional) subspace spanned by them. ■

Remark 4.3 - If we remove in th. 4.1 the hypothesis $Ax \neq 0$ if $x \neq 0$, we have a similar result, but zero may be on eigenvalue of finite or infinite multiplicity. Then, if the $e^i$ are associated to the eigenvalues different from zero, we must join a basis of the space $\{ x \ ; \ Ax = 0 \}$ in order to have a basis of the whole space H. ■

The fact that the $e^i$ form an orthonormal basis of H means that any element $x \in H$ may be represented in a unique way

(4.2) $$x = \sum_1^\infty x^i e^i \qquad ; \qquad x^i = (x \, , e^i)$$

where the series in (4.2) converges in the norm of H. The coefficients $x^i$ are called the components of x in the basis $e^i$. We also have

$$(x,y) = \sum_1^\infty x^i \, \overline{y}^i \qquad ; \quad \|x\|^2 = \sum_i^\infty |x^i|^2 \quad .$$

The action of the operator A has an easy interpretation in the basis $e^i$. From

$$Ae^i = \lambda^i \, e^i$$

and the orthonormality of the $e^i$, we see that the action of A amounts to multiplying every component by $\lambda^i$, i.e. :

(4.3) $$x = \Sigma \, x^i \, e^i \qquad Ax = \Sigma \, \lambda^i \, x^i \, e^i$$

We now write (4 , 3) in another way in order to introduce the concept of spectral family.

Let us now consider the sequence of eigenvalues by counting them only once, disregarding the order of multiplicity. We then write the index in parenthesis, (4.1) gives

(4.4) $$\lambda^{(1)} > \lambda^{(2)} > \lambda^{(3)} > \ldots \to 0$$

Let $P^{(k)}$ be the orthogonal projection of H on the (finite dimensional)

space spanned by the eigenvectors $e^i$ associated with the eigenvalue $\lambda^{(k)}$ i.e.

(4.5) $x = \Sigma x^i e^i \Rightarrow P^{(k)}x = \underset{(k)}{\Sigma} x^i e^i$

where the symbol $\underset{(k)}{\Sigma}$ means that the summation is made over the index i such that $\lambda^i = \lambda^{(k)}$). The identity operator I of H may be written

(4.6) $I = \Sigma P^{(k)}$

and the formula (4.3) becomes

(4.7) $Ax = (\Sigma \lambda^{(k)} P^{(k)})x$

Let us now define the "spectral family" $E(\lambda)$ as a function of the real variable $\lambda$ with values in $\mathcal{L}(H,H)$ defined by

(4.8) $E(\lambda) = \sum_{\lambda^{(k)} \leqslant \lambda} P^{(k)}$

where the sommation is made over the values k such that $\lambda^{(k)}$ is less or equal to $\lambda$. This function is null for $\lambda < 0$, equal to I for $\lambda \geqslant \lambda^{(1)}$ and is piecewise constant, the discontinuities being $P^{(k)}$ at the points $\lambda = \lambda^{(k)}$.

By using the Stieltjes integral, (4.6) and (4.7) become

(4.9) $I = \int_{-\infty}^{+\infty} d\,E(\lambda)$

(4.10) $Ax = \int_{-\infty}^{+\infty} \lambda dE(\lambda)x$

where the integrals are in fact from 0 to $\lambda^{(1)}$. According to the concept of Stieltjes integral, (4.10) is a "limit" of the sum of values $\lambda$ multiplied by the increments of the function $E(\lambda)$ in "short" intervals of $\lambda$.

The expansions((4.7) or (4.10) are useful to define functions of the operator A, for example

$$A^{1/2}\,x = \int_{-\infty}^{+\infty} \lambda^{1/2}\, dE(\lambda)x \equiv (\Sigma(\lambda^{(k)})^{1/2} P^{(k)})\, x$$

The concept of spectral family of an operator is much more general than the particular case of a compact operator. It turns out that every selfadjoint operator has an associated spectral family, which is not in general piecewise constant ; it may be continuous, or piecewise continuous (not constant).

Definition 4.3 - A spectral family $E(\lambda)$ is a function of $\lambda \in R$ with values in $\mathcal{L}(H,H)$ which are orthogonal projectors of the space H with the properties

a) $E(\lambda)$ is nondecreasing : $E(\lambda') \leqslant E(\lambda'')$ for $\lambda' < \lambda''$
(i.e., $E(\lambda)\, E(\mu) = E(\mu)\, E(\lambda) = E(\min(\lambda,\mu))$)

b) $E(\lambda) \underset{\lambda\to-\infty}{\longrightarrow} 0$ ; $E(\lambda) \underset{\lambda\to+\infty}{\longrightarrow} I$ strongly

(i.e., $E(\lambda)x \underset{\lambda\to-\infty}{\longrightarrow} 0$ ; $E(\lambda)x \underset{\lambda\to+\infty}{\longrightarrow} x$ in H strongly $\forall x \in H$)

c) $E(\lambda)$ is strongly continuous on the right

(i.e., $E(\lambda+\varepsilon)x \underset{\varepsilon\searrow 0}{\to} E(\lambda)x$ strongly in H $\forall x \in H$.

<u>Theorem 4.2</u> - If $E(\lambda)$ is a spectral family,

(4.11) $$A = \int_{-\infty}^{+\infty} \lambda \, dE(\lambda)$$

is a selfadjoint operator whose domain D(A) is the set of the $u \in H$ such that the Stieltjes integral

(4.12) $$\int_{-\infty}^{+\infty} \lambda^2 \, d(E(\lambda)u, u) < \infty.$$

<u>Remark 4.4</u> - The parenthesis of (4.12) is evidently the scalar product on H. Moreover, if $u \in D(A)$

$$(Au,v) = \int_{-\infty}^{+\infty} \lambda d(E(\lambda)u,v) \qquad \forall v \in H$$

where the convergence of the integral follows from the estimate

$$|(E(\lambda'')-E(\lambda'))u,v)| = |(E(\lambda'')-E(\lambda'))u,(E(\lambda'')-E(\lambda'))v| \leq$$
$$\leq \|(E(\lambda'')-E(\lambda'))u\| \, \|(E(\lambda'')-E(\lambda'))v\|$$

and from (4.12) and the property b) of $E(\lambda)$.■

Moreover, if $\phi(\lambda)$ is a continuous function, <u>the function $\phi(A)$ of the operator A</u> is defined as the selfadjoint operator given by

(4.13) $$\phi(A) = \int_{-\infty}^{+\infty} \phi(\lambda) \, dE(\lambda)$$

whose domain $D(\phi(A))$ is the set of the $u \in H$ such that the Stieltjes integral

$$\int_{-\infty}^{+\infty} |\phi(\lambda)|^2 \, d(E(\lambda)u, u) < \infty$$

Moreover, <u>the projector $E(\mu)$ corresponds to the (discontinuous !) Heaviside function</u> $H(\mu-\lambda)$, i.e.

(4.14) $$E(\mu) = \int_{-\infty}^{\mu} dE(\lambda) \quad .$$

It is easily seen that <u>an operator commutes with its own functions</u>.

<u>Remark 4.5</u> - As in the previous example of a compact operator, the integrals in (4.11) and (4.13) are in fact extended only to the part of the real axis where the spectral family $E(\lambda)$ is not constant.■ The theorem 4.2 has an important converse which is known as "spectral theorem".

<u>Theorem 4.3</u> - If A is a selfadjoint operator on H, there exists a unique spectral family $E(\lambda)$ such that A admits the representation (4.11). This spectral

family can be calculated by the formula

(4.15) $$(x,[E(b) + E(b-0) - E(a) - E(a-0)]\,y) = \lim_{\sigma \searrow 0} \frac{1}{\pi i} \int_a^b (x,[(A-(\lambda+i\sigma))^{-1} - (A-(\lambda-i\sigma))^{-1}]\,y)\,d\lambda$$

where x, y are arbitrary elements of H.

Remark 4.6 - In (4.15), the symbol E(b - 0)... means the limit of $E(\lambda-\varepsilon)$ as $\varepsilon \searrow 0$ and thus at the points of discontinuity of E, which are the eigenvalues of A, formula (4.15) gives the sum of the left hand and right hand limits of the function E. On the other hand, the integral on the right of (4.15) makes sense because for $\sigma > 0$ the points $\lambda \pm i\sigma$ are contained in the resolvent set of A. ■

Remark 4.7 - It is evident from (4.15) that if the real axis contains a segment belonging to the resolvent set of A, $E(\lambda)$ is constant on it. In particular, for an operator A such that $(Au, u) \geq c\|u\|^2$ $(c \geq 0)$, $E(\lambda)$ is null for $\lambda < c$. Then the integral in (4.11) is extended only to $[c, \infty[$. ■

In the general case, the spectral family $E(\lambda)$ of a selfadjoint operator A is constant on the parts of the real axis which belong to the resolvent domain of A, has discontinuities at the eigenvalues of A and changes continuously elsewhere. If $E(\lambda)$ is constant in each interval between two discontinuities (as in the example of the compact operator) we say that the spectrum of A is discrete. If there are no points of discontinuity of $E(\lambda)$ we say that the spectrum is continuous. We shall see in the theory of scattering some examples of such spectra.

It is noticeable that if $\|A\| \leq M$, then

$$E(\lambda) = 0 \quad \text{for } \lambda < -M$$
$$E(\lambda) = I \quad \text{for } \lambda > M.$$

5.- Sesquilinear forms and associated operators - Let V be a Hilbert space. A function a(u , v) defined on $V \times V$ is said to be a form on V. It is called "sesquilinear" if it is linear (resp. antilinear) with respect to the first (resp. second) argument, i.e.

(5.1) $$a(\lambda u, \mu v) = \lambda\, \bar{\mu}\, a(u, v)$$

where $^{-}$ is for the complex conjugate. It is said to be continuous if there is a constant M such that

(5.2) $$|a(u, v)| \leq M\|u\|\,\|v\|$$

The following theorem is an important tool to prove existence and uniqueness theorems :

Theorem 5.1 (Lax and Milgram) - Let $a(u,v)$ be a sesquilinear form on V, satisfying (5.1), (5.2) and coercive in the sense

(5.3) There exists $c > 0$ such that $|a(u,u)| \geq c \|u\|^2 \quad \forall u \in V$

Moreover, let f be an element of the dual V'.

Then, there exists a unique $u \in V$ such that

(5.4) $$a(v , u) = [f , v] \qquad \forall v \in V.$$

Remark 5.1 - The parenthesis in (5.4) is for the duality product between V' and V. The right hand term is linear with respect to v. By taking the conjugate we have the following result (equivalent to theorem 5.1) :

Under the hypothesis (5.1) - (5.3), if F(v) is a form antilinear and continuous on V, there exists a unique $u \in V$ such that

(5.5) $$a(u , v) = F(v) \qquad \forall v \in V$$

Of course, if the space is real, the two theorems coincide. ■

Remark 5.2 - If a is hermitean, i.e. :

$$a(u,v) = \overline{a(v,u)} \qquad \forall u , v \in V$$

the form is in fact a scalar product on V, equivalent to the natural one, and theorem 5.1 becomes the Riesz representation theorem.

Proof of theorem 5.1 - For $w \in V$ fixed, $a(u , w)$ is a linear and bounded functional on V, and by virtue of the Riesz theorem, there exists a well determinated $Z(w) \in V$ such that

(5.6) $$a(v , w) = (v , Z(w)) \qquad \forall v,w \in V$$

From (5.1) and (5.2) we see that Z is a linear and bounded operator from V to V. If $Z(w) = 0$ it follows that $a(v,w) = 0$ and from (5.3), $w = 0$. Consequently, Z defines a one to one transformation between V and its image Z(V). Let us proove that $Z(V) = V$. Firstly, by taking $v = w$ in (5.6), we have

$$\| w \| \leq \frac{1}{c} \| Z(w) \|$$

consequently, if there exists a sequence $w^i$ such that $Z(w^i)$ converges to a limit x, the sequence $w^i$ is also convergent to an element $w^*$ and by continuity we have $x = Z(w^*)$ ; Z(V) is then closed. As a consequence, if $Z(V) \neq V$, there exists $v \neq 0$ such that

$$(v , Z(w)) = 0 \qquad \forall w \in V$$

and by taking $w = v \Longrightarrow$

$$0 = (v,Z(v)) = a(v,v) \Rightarrow v = 0$$ which is a contradiction. The

operator Z is then a one to one transformation of V. By writting the right hand side of (5.4) under the form (v , F) (where $F \in V$ is given by the Riesz theorem), the solution of (5.4) is given by

$$u = Z^{-1}(F) \quad , \text{Q.E.D.} \blacksquare$$

Let us now consider a situation with two Hilbert spaces which is very useful in applications. We shall give two abstract representation theorems and we shall elaborate on them later in an example.

Let V and H be two Hilbert spaces with $V \subset H$ algebraically and topologically (i.e. the elements of V are also elements of H and there exists a constant $\gamma$ such that

5.7) $$\| u \|_H \leqslant \gamma \| u \|_V \qquad \forall u \in V \quad .$$

Moreover, V is dense in H and H is identified to its dual H'. We then have

$$V \subset H \equiv H' \subset V'$$

the embeddings being dense and continuous. It is clear that if $f \in H$, $v \in V$, the duality product coincides with the scalar product :

$$[ f , v ]_{V'V} = (f,v)_H \quad .$$

Theorem 5.2 (First representation theorem) - Let a(u,v) be a sesquilinear and continuous form on V (i.e. satisfying (5.1) and (5.2)) and coercive in the following sense

(5.8) $$\begin{cases} \text{There exists } c > 0 \ , \ \omega \geqslant 0 \text{ such that} \\ \operatorname{Re} a(u,u) + \omega \|u\|_H^2 \geqslant c \|u\|_V^2 \qquad \forall u \in V \end{cases}$$

Then, we have

1) There exists a well determinated operator $A \in \mathcal{L}(V,V')$ such that

(5.9) $$a(u,v) = [Au,v]_{V'V} \qquad \forall u,v \in V.$$

2) Let us consider the operator $A_H$ which is the restriction of A to H, defined on the domain

(5.10) $$D(A_H) = \{v \ ; \ v \in V, \ Av \in H \} \qquad \text{by :}$$

$$A_H v = Av \quad . \qquad \text{then,}$$

the domain $D(A_H)$ is dense in H and $A_H$ is an unbounded closed operator such that $A_H + \omega I$ is maximal accretive (i.e. it is accretive and it has no proper accretive extension).

Remark 5.3 - It is usual to denote $A_H$ by A ; in general there is no ambiguity : A is a bounded operator from V to V' as well as an unbounded operator (its restriction to H) from H to H. ∎

Remark 5.4 - It is clear that $A + \omega I$ is the operator associated with the form $a(u,v) + \omega(u,v)_H$ : because I is a bounded operator, the domains of A and $A+ \omega I$ are the same, and the term in $\omega$ is unessential.∎

Remark 5.5 - The part 1) of theorem 5.2 is self-evident, as in the first part of the proof of the Lax-Milgram theorem. The part 2) is not trivial and furnish a way to define unbounded maximal accretive operators in a Hilbert space. ∎

Remark 5.6 - It is evident from the definition that

$$D(A) \subset V \quad \text{alebraically}$$

Moreover, it is possible to define on D(A) the "graph norm"

$$\|u\|^2_{D(A)} = \|u\|^2_H + \|Au\|^2_H \tag{5.11}$$

It is easy to prove (by using the closedness of A) that D(A) is a (complete !) Hilbert space for the norm (5.11). We then have

$$D(A) \subset V \subset H \equiv H' \subset V' \subset D(A)'$$

with dense and continuous embeddings.∎

For hermitean forms, we have the sharper theorem :

Theorem 5.3 (Second representation theorem) - Under the hypothesis of theorem 5.2, if

$$a(u,v) = \overline{a(v,u)} \quad \text{and} \tag{5.12}$$

$$a(u,u) \geq 0 \tag{5.13}$$

Then, $A_H$ is selfadjoint, $D(A_H^{1/2}) = V$ and

$$a(u,v) = (A_H^{1/2} u \, , \, A_H^{1/2} v)_H \qquad \forall\, u,v \in V \quad .$$

Remark 5.7 - Let us consider the situation of the first representation theorem, with a(u,v) hermitean. We know that D(A) is dense in H. Moreover, D(A) is dense in V (for the norm of V !). For, by taking into account $B = A + \omega$ , $D(B) = D(A)$, we may consider $\omega = 0$ in (5.8) and the associated form b(u,u) is an equivalent norm on V ; the second representation theorem holds for B. Moreover, the operator $B^{-1/2} = B^{1/2}B^{-1}$ is continuous from H to $D(B^{1/2}) = V$. Let us take $v \in V = D(B^{1/2})$ $B^{1/2}v \in H$ and there exists $v_i \in D(B)$ such that

$$\begin{aligned} &v_i \to B^{1/2} v \quad \text{in } H \implies \\ &B^{-1/2} v_i \to v \quad \text{in } V \end{aligned} \tag{5.13}$$

and $v_i = B^{-1}w_i$ for some $w_i \in H$ ; we have $B^{-1/2}v_i = B^{-1}(B^{1/2}B^{-1}w_i)$ where the paren-parenthesis belongs to H ; then $B^{-1/2}v_i \in D(B)$ and (5.13) is the desired property.∎

6.- Explicit description in a particular case - The particular situation where the embedding of V in H is compact, a(u,v) is hermitean and (5.8) is satisfied with $\omega = 0$ often appears in applications. For the sake of completness we shall repeat here the asumptions.

We have three Hilbert spaces $V \subset H$ with dense and compact embedding (i.e. weak convergence in V implies strong convergence in H) ; H is identified to its dual. We then have

$$V \subset H \subset V'$$

Let a(u,v) be a hermitéan form on V with :

$$(6.1) \qquad \begin{aligned} |a(u,v)| &\leq M \|u\|_V \ \|v\|_V && \forall u,v \in V \\ a(u,u) &\geq C \|u\|_V^2 && \forall u \in V \end{aligned}$$

for positive constants M, C (Let us remark that because of the hermitean condition

$$a(u,v) = \overline{a(v,u)}, \quad \text{i.e.} \quad a(u,u) \text{ is real}).$$

Then, theorems 5.1, 5.2 and 5.3 hold.

Moreover, a(u,v) is a scalar product in V and the associated norm is equivalent to the norm of V. We shall take

$$(6.2) \qquad (u,v)_V = a(u,v)$$

Let us consider A as a bounded operator from V to V'. For $f \in V'$ the equation

$$Au = f \quad \text{is equivalent to} \quad a(u,v) = [f,v]_{V'V} \qquad \forall v \in V \ .$$

which has a unique solution $u \in V$ by the Lax Milgram theorem. Moreover, from (6.1)

$$C \|u\|_V \leq \|f\|_{V'} \quad \text{and then}$$

$$(6.3) \qquad A^{-1} \in \mathcal{L}(V',V)$$

By considering $A^{-1}$ defined only on H, we have

$$(6.4) \qquad A^{-1} \in \mathcal{L}(H,H)$$

moreover, by virtue of (6.3) $A^{-1}$ transform a weakly convergent sequence of H (which is also a weakly convergent sequence of V') in a weakly convergent sequence of V, which is a strongly convergent sequence of H. Consequently $A^{-1}$ is compact in H. Moreover, for $f \in H$ let $u \in V$ be the solution of $Au = f \Leftrightarrow a(u,v) = [f,v]_{V'V} = (f,v)_H \quad \forall v \in V$ and by taking $v = u = A^{-1}f$, we have

$$(f,A^{-1}f)_H = a(u,u) \geq C \|u\|_V^2$$

then, $A^{-1}$ is positive and $A^{-1}f = 0 \Rightarrow f = 0$ and we are in the situation of th. 4.1 We shall denote $\frac{1}{\omega_i^2}$ the eigenvalues of $A^{-1}$ :

$$(6.5) \qquad 0 < \omega_1^2 \leq \omega_2^2 \leq \ldots \leq \omega_i^2 \leq \ \ldots \to \infty$$

and let $e_i$ be the corresponding eigenvectors.

(6.6) $$A^{-1} e_i = \frac{1}{\omega_i^2} e_i \Leftrightarrow Ae_i = \omega_i^2 e_i$$

Then, $\omega_i^2$ are the eigenvalues of A, corresponding to the same eigenvectors, which form an orthonormal basis in H. It is clear from (6.6) that $e_i \in D(A)$

We shall reduce all the spaces to this basis. Firstly,

(6.7) $$v \in H \Leftrightarrow v = \sum_i v_i e_i \text{ with } \sum_i |v_i|^2 < +\infty$$

(6.8) It is then clear from (6.6) that the operator A (resp. $A^{1/2}, A^{-1/2}, A^{-1}$) in the basis $e_i$ operates by multiplying each component by $\omega_i^2$ (resp. $\omega_i$, $\omega_i^{-1}$, $\omega_i^{-2}$).

Incidentaly we see that, in the present case, if $u \in D(A)$, there exists a constant C such that

$$\| u \|_H \leqslant C \| Au \|_H$$

and then, we take

(6.9) $$\| u \|_{D(A)} = \| Au \|_H$$

instead of the graph norm (5.11).

As a consequence of (6.8) we have :

(6.10) $$\sum v_i e_i \in \begin{Bmatrix} D(A) \\ V = D(A^{1/2}) \end{Bmatrix} \Leftrightarrow \begin{Bmatrix} \sum_i (\omega_i^2 |v_i|)^2 < +\infty \\ \sum_i (\omega_i |v_i|)^2 < +\infty \end{Bmatrix}$$

Moreover, in the basis $e_i$, the spaces D(A) and V are easily identified to the space $\ell^2$ of the sequences $\theta_i$ such that $\sum |\theta_i|^2 < +\infty$, which may be identified to its dual by the Riesz theorem. Then, we obtain

(6.11) $$\sum v_i e_i \in \begin{Bmatrix} V' \\ D(A)' \end{Bmatrix} \Leftrightarrow \begin{Bmatrix} \sum (\omega_i^{-1} |v_i|)^2 < +\infty \\ \sum (\omega_i^{-2} |v_i|)^2 < +\infty \end{Bmatrix}$$

and (6.7), (6.10), (6.11) are explicit representations of the different spaces in the basis $e_i$.

We also have

$$(e_i, e_j)_V = a(e_i, e_j) = (Ae_i, e_j)_H = \omega_i^2 (e_i, e_j)$$

and analogous formulae for $A^{1/2}$...We then see that the basis $e_i$ is <u>orthogonal in all the spaces</u> and

$$\| e_i \|^2 \quad \text{in} \quad \begin{Bmatrix} D(A) \\ V \\ H \\ V' \\ D(A)' \end{Bmatrix} = \begin{Bmatrix} \omega_i^4 \\ \omega_i^2 \\ 1 \\ \omega_1^{-2} \\ \omega_1^{-4} \end{Bmatrix}$$

Moreover, the compactness hypothesis reads :

"The set of u such that $\sum (\omega_i |u_i|)^2 < C$ is a precompact set for the norm (square norm !) $\sum |u_i|^2$ "

and by taking $v_i = \omega_i u_i$, we see that "the set of the v such that $\Sigma|v_i|^2 < C$ is a precompact set for the norm (square norm!) $\Sigma(\omega_1^{-1} |v_i|)^2$" such is to say,

(6.12) $$H \subset V' \quad \text{with compact embedding.}$$

Let us now study the equation

(6.13) $$(A - \mu)u = f$$

in the space H (for instance), where $\mu$ is a parameter.

By virtue of (6.8), the equation in the basis $e_i$ is

(6.14) $$(\omega_i^2 - \mu)u_i = f_i \qquad i = 1,2 \ldots.$$

We see that, if $\mu$ is not one of the eigenvalues $\omega_i^2$, the $u_i$ are well determinated ; moreover

(6.15) $$\| u \|^2_{D(A)} = \sum_i \left| \frac{\omega_i^2}{\omega_i^2 - \mu} \right|^2 |f_i|^2$$

and it is easyly seen that there is a constant C such that

$$\left| \frac{\omega_i^2}{\omega_i^2 - \mu} \right| \leqslant C$$

this means that

(6.16) $$(A - \mu)^{-1} \in \mathcal{L}(H , D(A)) \quad ; \quad \mu \neq \omega_i^2 \qquad \forall i$$

On the other hand, if $\mu$ is an eigenvalue of A, $\mu = \omega_k^2$ , the necessary and sufficient condition for u to exist is that $f_i = 0$ for $i \in (k)$ (i.e., for the indexes such that $\omega_i^2 = \omega_k^2$, which are in finite number). This amounts to say that f must be orthogonal to the subspace $N_{(k)}$ spanned by the eigenvectors associated to the eigenvalue $\omega_k^2$. If this condition is satisfied, $u_i$ is well determinated for $i \notin (k)$ and arbitrary for $i \in (k)$. If we search for u in the orthogonal $N^{\perp}_{(k)}$ to $N_{(k)}$, it is well determinated and the operator $f \to u$ is continuous from $N^{\perp}_{(k)}$ (with the topology of H) to $N^{\perp}_{(k)}$ (with the topology of D(A)). We say that $A - \mu$ has a semi-inverse in $N^{\perp}_{(k)}$, bounded from H to D(A). Any other solution of (6.13) is the sum of the preceeding one and any vector in $N_{(k)}$. This situation is a particular case of the Fredholm alternative.

# CHAPTER 3

# EXAMPLES OF BOUNDARY VALUE PROBLEMS

The content of this chapter has been studied extensively. Good references are NECAS [1] or LIONS - MAGENES [1].

1.- The Dirichlet problem for the Laplace equation - Let $\Omega$ be a bounded domain of $R^N$ with regular boundary $\partial\Omega$ ($\Omega$ is on one side of $\partial\Omega$). If $\Delta$ denotes the Laplace operator

$$\Delta = \sum_1^N \frac{\partial^2}{\partial x_i^2} \equiv \frac{\partial}{\partial x_i} \frac{\partial}{\partial x_i}$$

the Dirichlet problem consists of finding a function u such that

$$(1.1) \qquad -\Delta u = f \qquad \text{in } \Omega$$

$$(1.2) \qquad u = 0 \qquad \text{on } \partial\Omega$$

where f is a given function, which may be complex (if f is real, the unknown u is also real, and we can work in the real field).

Let us suppose that such a function u exists. Then, by taking a test function v which satisfies the boundary condition $v = 0$ on $\partial\Omega$, we have by integrating by parts ($n_i$ denote the components of the outer unit normal to $\partial\Omega$) :

$$(1.3) \qquad -\int_\Omega \Delta u.\bar{v}\, dx = -\int_\Omega \frac{\partial}{\partial x_i}\left(\frac{\partial u}{\partial x_i}\bar{v}\right)dx + \int_\Omega \frac{\partial u}{\partial x_i}\frac{\partial \bar{v}}{\partial x_i}\, dx =$$

$$= -\int_{\partial\Omega} n_i\left(\frac{\partial u}{\partial x_i}\bar{v}\right)dS + \int_\Omega \frac{\partial u}{\partial x_i}\frac{\partial \bar{v}}{\partial x_i}\, dx = \int_\Omega \frac{\partial u}{\partial x_i}\frac{\partial \bar{v}}{\partial x_i}\, dx$$

and we then have

$$(1.4) \qquad a(u,v) = {}_\Omega f\,\bar{v}\, dx \qquad \forall v$$

$$(1.5) \quad \text{where} \quad a(u,v) \equiv \int_\Omega \frac{\partial u}{\partial x_i}\frac{\partial \bar{v}}{\partial x_i}\, dx$$

The relation (1.4) suggests to us to apply the Lax-Milgram theorem, because the left-hand side of (1.4) is a sesquilinear form in u,v and the right hand side is an antilinear one in v.

Let us introduce the space

$$(1.6) \qquad H_0^1(\Omega) = \{v \; ; \; v \in H^1(\Omega) \; ; \; v = 0 \text{ on } \partial\Omega\}$$

i.e., the subspace of $H^1(\Omega)$ of the functions which satisfy the boundary condition (1.2). By virtue of the trace theorem, $H^1_0$ is a closed subspace of $H^1$ ; $H^1_0$ is then a Hilbert space, with the standard norm of $H^1$ :

$$\|u\|^2_{H^1_0} = \int_\Omega \left(|u|^2 + \sum_i \left|\frac{\partial u}{\partial x_i}\right|^2\right) dx \tag{1.7}$$

Then, we have

$$|a(u,v)| \leqslant \|u\|_{H^1_0} \|v\|_{H^1_0} \qquad \forall\, u,v \in H^1_0$$

$$\left|\int_\Omega f\ \bar{v}\ dx\right| \leqslant \|f\|_{L^2} \|v\|_{L^2} \leqslant C\|v\|_{H^1_0} \qquad \forall\, v \in H^1_0$$

where C is a constant. Then, a is a sesquilinear continuous form on $H^1_0$ and the right hand side of (1.4) is a continuous antilinear functional on $H^1_0$. Moreover,

Proposition 1.1. - The form $a(u,v)$ is coercive on $H^1_0$. Proof - It suffices to prove the so-called Friedrichs inequality : there exists a constant $\gamma$ such that

$$a(u,u) \geqslant \gamma \|u\|^2_{L^2} \qquad \forall\, u \in H^1_o . \tag{1.8}$$

for, if (1.8) is satisfied, we have

$$a(u,u) \geqslant \inf\left(\frac{1}{2}, \frac{\gamma}{2}\right) \|u\|^2_{H^1_0} .$$

We shall prove (1.8) by contradiction. We may consider $\|u\|_{L^2} = 1$. If (1.8) is not true, there exists a sequence $v^i$ such that

$$\|v^i\|_{L^2} = 1 \quad ; \qquad a(v^i,v^i) \to 0 \tag{1.9}$$

and then there exists a constant C such that

$$\|v^i\|_{H^1_0} \leqslant C$$

i.e., the $v^i$ form a precompact set for the weak topology of $H^1_0$ (c.f. chap. 2, proposition 1.5) and by the Rellich theorem, for the strong topology of $L_2$. Then there exists a subsequence (again denoted by $v^i$) and an element $v^*$ such that

$$v^i \to v^* \quad \text{in } H^1_0 \text{ weakly and in } L \text{ strongly and then} \tag{1.10}$$

$$\|v^*\|_{L^2} = 1 \quad . \tag{1.11}$$

On the other hand, by chap.2, prop. 1.1, we have

$$\|v^*\|^2_{H'} \leqslant \underline{\lim} \|v^i\|^2_{H^1}$$

and by substracting $\|v^*\|^2_{L^2} = \|v^i\|^2_{L^2} = 1$,

$$a(v^*, v^*) \leqslant \underline{\lim}\ a(v^i,v^i)$$

but by (1.9) we then have $a(v^*,v^*) = 0$ which means that, grad $v^* = 0 \Rightarrow v^* = $ cte and $v^* \in H^1_0$ by (1.10), we then have $v^* = 0$, and this is in contradiction with (1.11). ∎

Then, the following abstract problem has one and only one solution.

Abstract (or variational ) Dirichlet problem : Find $u \in H_0^1(\Omega)$ such that

(1.12) $$a(u,v) = \int_\Omega f\,\overline{v}\,dx \qquad \forall v \in H_0^1(\Omega)$$

Moreover, if u is the solution of the abstract problem and it is smooth (in fact of class $C^2(\Omega)$, it satisfies (1.1) and (1.2) ; in fact, (1.2) is a consequence of $u \in H_0^1$ ; as for (1.1), by integrating (1.12) by parts as in (1.3), the integrals over $\partial\Omega$ are null because $v \in H_0^1$, and we have

$$\int_\Omega (f + \Delta u)\overline{v}\,dx = 0 \qquad \forall \overline{v} \in H_0^1$$

since the test function is arbitrary we see that the function in parenthesis is null i.e., (1.1) is satisfied.

Remark 1.1.- If u is not "smooth" by taking $v \in \mathcal{D}(\Omega)$ in (1.12), we see that the equation (1.1) is satisfied in the sense of distributions.■

As for the function f, it is clear that we can take any element of the dual of $H_0^1(\Omega)$ when $L^2(\Omega)$ is identified with its dual (in such a way the duality product is consistent with the integral in (1.12)). This dual space is denoted by $H^{-1}(\Omega)$ and is a space of distributions on $\Omega$ .

In fact, by taking $H = L^2$ ; $V = H_0^1$, (and then $V' = H_0^{1'} \equiv H^{-1}$) and because of the Rellich theorem the embedding of $H_0^1$ in $L_2$ is compact (and dense also). We are in the general framework of section 2.6. By defining

$$A = -\Delta$$

with the boundary condition (1.2), A is a linear continuous operator from $H_0^1$ into $H^{-1}$. Moreover, the operators A and $A^{-1}$ induce an isomorphism between $H_0^{1}$ and $H^{-1}$. Of course, the restriction of A to H is a selfadjoint positive definite operator and there exists a sequence of eigenvalues and eigenfunctions forming a basis of $L_2$ (and $H_0^1$ and $H^{-1}$).

2.- The Neumann problem - As before, $\Omega$ is a bounded domain with regular boundary. We consider the problem

(2.1) $$-\Delta u + u = f \qquad \text{in } \Omega$$

(2.2) $$\frac{\partial u}{\partial \eta} = 0 \qquad \text{on } \partial\Omega$$

where $\eta$ is the outer unit normal to $\partial\Omega$ and f is a given function (for instance in $L^2(\Omega)$).

If u is a solution of (2.1), (2.2), by multiplying (2.1) by an arbitrary test function v (which does not satisfy any boundary condition) and by integrating by parts as in (1.3), we have

(2.3) $$a(u,v) = \int_\Omega f\,\bar{v}\,dx \qquad \forall v$$

where

(2.4) $$a(u,v) \equiv \int_\Omega \left(\frac{\partial u}{\partial x_i}\frac{\partial \bar{v}}{\partial x_i} + u\bar{v}\right) dx = (u,v)_{H^1}$$

Conversely, if u is a function (which does not satisfy any boundary condition for the time being) which satisfies (2.3), by integrating by parts (2.3) we have :

(2.5) $$\int_\Omega (-\Delta u + u - f)\,\bar{v}\,dx = -\int_{\partial\Omega} \frac{\partial u}{\partial n}\,\bar{v}\,dS \qquad \forall v \qquad ..$$

Then, by taking v zero on $\partial\Omega$ but otherwise arbitrary, we see that the expression in parenthesis of the left hand integral is zero, i.e., (2.1) is satisfied. Next, by taking v arbitrary on $\partial\Omega$, we see that u satisfies the boundary condition (2.2).

Then, it is natural to define the

Abstract (or variational) Neumann problem: find $u \in H^1(\Omega)$ such that (2.3) is satisfied for any test function $v \in H^1(\Omega)$.

Of course the form a(u,v) given by (2.4) is continuous and coercive on $H^1$, and the Neumann problem has a unique solution for each f in the dual of $H^1$

Remark 2.1.- The dual of $H^1$ (if $L^2$ is identified to itself) is not easy to describe. In particular, it is not a space of distribution on $\Omega$. Of course, one may take $f \in L^2$ to study the problem. ∎

We are still in the general framework of sect. 2.6 if we take

$$V = H^1(\Omega) \quad ; \quad H = L^2(\Omega).$$

Remark 2.2.- There is a very important difference between the Dirichlet boundary condition (1.2) and the Neumann condition (2.2). The first one was imposed to all the functions of the space $H^1_0$, which is a closed subspace of $H^1$ i.e., a Hilbert space. This is a consequence of the trace theorem, as we know. On the other hand, the Neumann condition is not imposed to all the functions of space $H^1$, it is satisfied only by the solution u of the abstract problem. In fact, if we consider the subspace

$$V^* = \{v \ ; \ v \in H^1(\Omega) \ ; \ \frac{\partial u}{\partial n} = 0 \} \text{ of } H^1$$

this space is not closed (and hence is not a Hilbert space) and the Lax-Milgram theorem does not hold. In fact, the trace theorem does not work for $\frac{\partial u}{\partial n}$ in $H^1$. Conditions of the Dirichlet type are called principal or stable conditions of the Neumann type are called natural or unstable. This terminology is taken from the calculus of variations. ∎

<u>Let us now consider the following problem</u>

(2.6) $$- \Delta u = f \qquad \text{in } \Omega$$

(2.7) $$\frac{\partial u}{\partial n} = 0 \qquad \text{on } \partial\Omega \quad .$$

It is evident that for f = 0 we have the solution u = cst. and we have not uniqueness. Moreover, if there is a solution, by integrating (2.6) on $\Omega$ , and taking into account (2.7) we have

(2.8) $$\int_\Omega f \, dx = -\int_\Omega \frac{\partial}{\partial x_i}\left(\frac{\partial u}{\partial x_i}\right) dx = -\int_{\partial\Omega} n_i \frac{\partial u}{\partial x_i} \, dS = -\int_{\partial\Omega} \frac{\partial u}{\partial n} \, dS = 0$$

i.e., there is a compatibility condition. We then are in a situation different from that of problem (2.1), (2.2) ; this situation recalls the Fredholm alternative.

In fact, if A is the operator associated with the form a(u,v) given by (2.4), the problem (2.6), (2.7) may be written as the operator equation

(2.9) $$(A - 1)u = f$$

in $L_2$(for instance). We are in the situation of (2.6.13), and 1 is an eigenvalue of A. The associated eigenvectors are the functions e satisfying

$$Ae - e = 0$$

and by multiplying it by e in H we have

$$\int_\Omega \left| \frac{\partial e}{\partial x_i} \right|^2 dx = 0$$

i.e. the functions e are constants. In fact, <u>the eigenvalue is simple</u>, and the associated eigenfunction is the constant function. The condition (2.8) is then the compatibility condition : f must be orthogonal to the constant functions in $L_2$. The solution u of (2.6), (2.7) is thus determined up to an additive constant, and it is unique if we impose that

$$\int_\Omega u \, dx = 0$$

i.e., it must be orthogonal to the corresponding eigenvector.

3.- <u>A transmission problem</u> - Let $\Omega$ be a bounded connected domain of $R^N$ with smooth boundary. Let it be divided into two subdomains $\Omega_1$ and $\Omega_2$ by a smooth surface $\Sigma$ (see fig. ). Moreover, the boundary of $\Omega$ is decomposed into two surfaces $\Gamma_1$ and $\Gamma_2$, and $\Gamma_1$ is sufficiently smooth (and then the traces on $\Gamma_1$ are well defined in the general

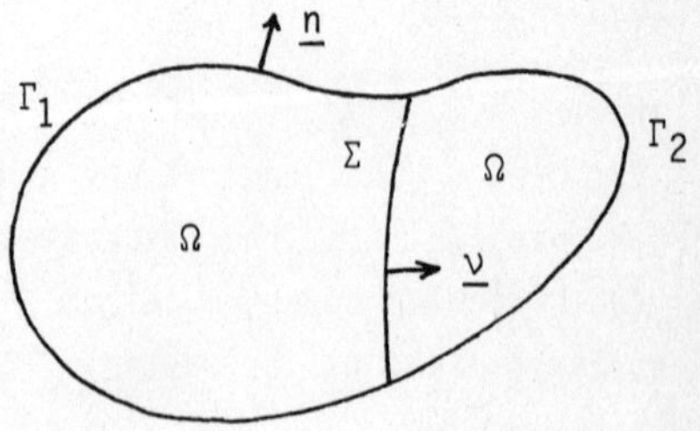

framework of Sobolev spaces). We also assume that the boundary of $\Omega_1$ is formed by $\Gamma_1$ and $\Sigma$. (This fact is unessential).

Let $a_{ij}(x)$ be real (unessential !), piecewise smooth, but may be discontinuous on $\Sigma$ (and of course smooth in $\Omega_1$ and $\Omega_2$). Moreover

(3.1) $$a_{ij}(x) = a_{ji}(x) \quad \text{and}$$

(3.2) $$a_{ij}\,\xi_i\,\xi_j \geqslant \gamma \xi_k\,\xi_k \qquad \forall\, \xi \in R^N \;,\; x \in \Omega\,.$$

then we consider the equations

(3.3) $$-\frac{\partial}{\partial x_i}\Big(a_{ij}(x)\,\frac{\partial u}{\partial x_j}\Big) = f \qquad \text{in } \Omega_1 \text{ and } \Omega_2$$

with the boundary conditions

(3.4) $$u = 0 \quad \text{on } \Gamma_1$$

(3.5) $$n_i\, a_{ij}\,\frac{\partial u}{\partial x_j} = 0 \quad \text{on } \Gamma_2$$

and the transmission conditions

(3.6) $$u_{|1} = u_{|2} \qquad \text{on } \Sigma$$

(3.7) $$\nu_i\, a_{ij}\,\frac{\partial u}{\partial x_j}\Big|_1 = \nu_i\, a_{ij}\,\frac{\partial u}{\partial x_j}\Big|_2 \qquad \text{on } \Sigma\;.$$

where as usual, $n$ is for the unit outer normal to $\Omega$ , $\nu$ is a normal to $\Sigma$ chosen (for instance !) outer to $\Omega_1$.

It is clear that (3.4) is a Dirichlet condition, and (3.5) is a Neumann condition (associated with equation (3.3)) ; by analogy with the preceeding problems, we introduce the space

(3.8) $$V = \{v \mid v \in H^1(\Omega) \quad ; \quad v = 0 \quad \text{on } \Gamma_1\}$$

equipped with the norm of $H^1(\Omega)$. This is a Hilbert space, in view of the trace theorem on $\Gamma_1$ (as $H^1_o$ in section 1).

If $u$ is a solution of (3.1) - (3.7), we multiply (3.3) by $v \in V$ and we integrate over $\Omega$ ; by integrating by parts and taking into account the boundary and transmission conditions, we have

(3.9) $$a(u,v) = \int_\Omega f\,v\,dx \qquad \forall\, v \in V$$

where

(3.10) $$a(u,v) \equiv \int_\Omega a_{ij}(x)\,\frac{\partial u}{\partial x_j}\,\frac{\partial v}{\partial x_i}\,dx$$

Conversely, if $u$ is a function of $V$ (and smooth enough) satisfying (3.9), it is a solution of (3.1) - (3.7). For, if $u \in V$, (3.4) is satisfied ; moreover, the trace theorem on $\Sigma$ shows that (3.6) is satisfied. Then, by integrating by parts (3.9) we have :

$$(3.11) \qquad \int_{\Omega} [f + \frac{\partial}{\partial x_i} (a_{ij} \frac{\partial u}{\partial x_j})]\, v\, dx = \int_{\Gamma_2} n_i\, a_{ij} \frac{\partial u}{\partial x_j}\, v\, dS +$$

$$+ \int_{\Sigma} \nu_i (a_{ij} \frac{\partial u}{\partial x_j}\Big|_1 - a_{ij} \frac{\partial u}{\partial x_j}\Big|_2)\, v\, dS$$

Thus, by taking in particular $v \in \mathcal{D}(\Omega_1)$ or $v \in \mathcal{D}(\Omega_2)$, the right hand side of (3.11) vanishes and we see that (3.3) is satisfed. Note that the left hand side of (3.11) is zero and because v is arbitrary on $\Gamma_2$ and $\Sigma$ we see that (3.5) and (3.7) are also verified.

Then, it is natural to define the

Abstract (variational) formulation - Find $u \in V$ such that (3.9) is satisfied $\forall\, v \in V$.

It is easy to prove (as in section 1) that a(u,v) is a bilinear (and symmetric) form on V, continuous and coercive. (It suffices to remark that $a(u,u) + \|u\|^2_{L^2}$ is a norm[2] equivalent to $\|\cdot\|^2_{H^1}$ and that if $u \in V$ with $a(u,u) = 0$, then $u = \text{cst}$, but since $\Omega$ is connected and u is zero on $\Gamma_1$, the constant is zero).

Then, we are again in the framework of sect. 2.6 with V defined by (3.8) and $H = L^2(\Omega)$.

# CHAPTER 4

## SEMIGROUPS AND LAPLACE TRANSFORM

The reader will find a study of semigroups in Banach spaces and applications in KATO [ 2 ] as well as in HILLE-PHILLIPS and YOSIDA [ 1 ]. The Laplace transform for functions with values in a Banach space may be studied in HILLE-PHILLIPS [ 1 ] GARNIR [ 1 ], SCHWARTZ [ 1 ], [ 2 ], LIONS [ 1 ].

### 1.- Semigroups.- Definitions and generalities

Definition 1.1.- Let B be a Banach space. A semigroup on B is a family of linear, continuous operators G(t) which depends on the parameter $t \geqslant 0$ such that (norms in $\mathcal{L}(B,B)$)

(1.1) $$\| G(t) \| \leqslant M(t)$$

(1.2) $$G(0) = I \qquad (I = \text{identity})$$

(1.3) $$G(t_1 + t_2) = G(t_1)\, G(t_2) \qquad t_1, t_2 \geqslant 0 \quad .$$

Definition 1.2.- The semigroup is said to be strongly continuous (or only continuous) if

(1.4) $$\|G(t)v - v\|_B \to 0 \qquad t \to 0 \qquad \forall v \in B .$$

Proposition 1.1.- If G(t) is a continuous semigroup on B, there exists two real numbers M and $\omega$ such that

(1.4 bis) $$\|G(t)\, v\| \leqslant M\, e^{\omega t} \;\|v\| \qquad \forall v \in B , \quad t \geqslant 0 .$$

We then can take $M(t) = M \exp(\omega t)$ in (1.1)

Proposition 1.2.- If G(t) is a continuous semigroup, for any fixed $v \in B$, the function G(t)v is continuous for the strong topology of B.

This is a direct consequence of (1.1), (1.3) and (1.4).

An important notion in semigroup theory is that of generator. This is (in general is unbounded) an operator A from B to B defined by

(1.5) $$Av = \lim_{t \to 0} \frac{G(t)v - v}{t}$$

whose domain D(A) is the set of the elements v of B for which the right hand side of (1.5) exists.

Proposition 1.3 - D(A) is dense in B ; moreover, if $v \in D(A)$, then $G(t)v \in D(A)$ for $t \geq 0$ and it is a continuously differentiable function. It satisfies

(1.6) $$\frac{d}{dt}(G(t)v) = A(G(t)v) = G(t)Av \qquad \forall t \geq 0 .$$

This proposition shows that if $v \in D(A)$, $G(t)v$ is a solution of the differential equation

(1.7) $$\frac{d}{dt}\, u(t) = Au(t)$$

with the initial condition

(1.8) $$u(0) = v.$$

It is then useful to write $G(t) = \exp(At)$.

In fact, semigroups are a tool for the resolution of equations of the form (1.7), (1.8). In fact, if $G(t)$ is known, $G(t)v$ is defined for $v \in B$, not necessarily $v \in D(A)$. Moreover, because of the density of D(A) in B, for any $v \in B$, there exists a $v^* \in D(A)$ such that $\|v - v^*\|$ is as small as we please and (1.4 bis) shows that $G(t)v - G(t)v^*$ is as small as desired on any finite interval of time. It appears that $G(t)v$ is a limit of solutions of (1.7) with initial value $v^*$ in D(A). It is then natural to define "generalized solution" of (1.7) with $u(0) = v \in B$ by $G(t)v$, even if $v \notin D(A)$. In this connection, it is important to have necessary and sufficient conditions for an operator A to be the generator of a semigroup G.

Theorem 1.1.- (Hille-Yosida).- If $G(t)$ is a continuous semigroup on B, with

(1.9) $$\|G(t)\| \leq M\, e^{\omega t}$$ , then, its generator A satisfies

a) D(A) is dense in B, A is closed

b) The semi-infinite interval $\lambda > \omega$ ($\lambda$ real) belongs to the resolvent set of A and

(1.10) $$\|(\lambda - A)^{-n}\| \leq M(\lambda - \omega)^{-n} \quad , \quad \lambda > \omega \quad , \quad n = 1,2,\ldots$$

Reciprocally, if A is an operator in the Banach space B, satisfying a) and b), it is the generator of a continuous semigroup $G(t)$ which satisfies (1.9).

2.- Contraction semiproups - In physical applications, the space B is often chosen in such a way that the square of the norm is an energy. Then, if there is no input in the system, the energy is either constant or decreasing (if there is a dissipation). In this case, the semigroup is such that

(2.1) $\quad \| G(t) \| \leqslant 1 \qquad \forall t \geqslant 0$

i.e., we can take $M = 1$, $\omega = 0$. Such semigroups are called contraction semigroups.

For such semigroups, the Hille-Yosida's theorem takes a simplified form :

Theorem 2.1.- (Hille-Yosida).- If $G(t)$ is a continuous semigroup of contractions, its generator A satisfies :

a) D(A) is dense in B, A is closed

b) The semi-infinite interval $\lambda > 0$ ($\lambda$ real) belongs to the resolvent set of A and

(2.2) $$\| (\lambda - A)^{-1} \| \leqslant \lambda^{-1} \qquad \lambda > 0 \ ,$$

Reciprocally, if A satisfies a) and b), it is the generator of a contraction semi-group in B.

If the space B is a Hilbert space H, there is another characterization of the contraction semigroups.

Definition 2.1.- An operator A in a Hilbert space H is said to be an accretive (resp. dissipative) operator if

(2.3) $$\mathrm{Re}(Av,v) \geqslant 0 \quad (\text{resp.} \leqslant 0) \quad \forall v \in D(A) \ .$$

Theorem 2.2.- (Lumer-Phillips) - Let A be a linear operator in the Hilbert-space H with domain D(A) dense in H. Then

a) If A is the generator of a contraction semigroup in H, then A is dissipative and the range $R(\lambda - A)$ of $\lambda - A$ is the whole space H for all $\lambda > 0$.

b) If A is dissipative and there exists $\lambda > 0$ such that the range of $\lambda - A$ is the whole space H, then A is the generator of contraction semigroup in H.

Remark 2.1 - The class of operators A considered in theorem 2.2 coincides with the class of the "densely defined maximal dissipative operators" where maximal is understood in the sense that they do not have a dissipative proper extension (c.f. Phillips [1]).∎

In physical processes where there is no dissipation nor energy supply, the norm of the solution of the equation

(2.4) $$\frac{du(t)}{dt} = A\,u(t)$$

is independent of t. In this case, we may hope that solutions u(t) exists for $t < 0$ as well as for $t \geqslant 0$, and we shall deal with a group rather than with a semigroup. In this connection we have the following definitions :

Definition 2.2.- An operator $G \in \mathcal{L}(H,H)$ is said to be unitary if

$$\| G v\| = \| v \| \qquad \forall v \in H .$$

Definition 2.3.- An operator A in a Hilbert space H is said to be skew self-adjoint if its adjoint $A^*$ satisfies

$$A^* = - A$$

which means that iA is selfadjoint.

The characterization of groups of unitary operators is given by the

Theorem 2.3 (Stone) - An operator A in a Hilbert space H generates a group of unitary operators G(t) iff it is skew self-adjoint.

Without giving the proof of this theorem, it is useful to remark that if iA is selfadjoint, A as well as -A, satisfy the conditions of theorem 2.1 (see ch.2, proposition 4.1) and hence the equation (2.4) will have a solution for t either positive or negative.

3.- Miscellaneous properties of semigroups - If A is the generator of the continuous semigroup G(t) in B one may consider the unhomogeneous equation

$$\text{(3.1)} \qquad \begin{cases} \dfrac{du}{dt} = Au + f \\ u(0) = v \end{cases}$$

where f is a continuous function of t with values in B. Then, the (unique) generalizated solution of (3.1) with $v \in B$ for $t \geq 0$ is given by

$$\text{(3.2)} \qquad u(t) = G(t)v + \int_0^t G(t - s)\, f(s)\, ds \qquad t \geq 0$$

which is a continuous function of t with values in B.

Moreover, if $v \in D(A)$ and f(s) is continuously differentiable, u(t) is continuously differentiable with values in B and $u(t) \in D(A)$ for $t \geq 0$. We then have a classical solution of (3.1).

An important class of semigroups are the holomorphic semigroups. They are defined not only for real positive t but also for t in a sector of the complex plane. We have :

Theorem 3.1 - Let A be the generator of a contraction semigroup in the Banach space B. If, in addition to the conditions of theorem 2.1 the resolvent

$(\lambda - A)^{-1}$ is defined in a sector $|\arg\lambda| < \frac{\pi}{2} - \omega$, $\omega > 0$, and is bounded in such a way that for any $\varepsilon > 0$

$$\|(\lambda - A)^{-1}\| \leq \frac{M_\varepsilon}{|\lambda|} \qquad \text{for } |\lambda| < \frac{\pi}{2} + \omega - \varepsilon \tag{3.3}$$

then, the semigroup exp(At) is holomorphic for $|\arg t| < \omega$, uniformly bounded for $|\arg t| < \omega - \varepsilon$ and strongly continuous at $t = 0$.

Remark 3.1 - If exp(At) is holomorphic semigroup, $u(t) = \exp(At)v$ satisfies

$$\frac{du}{dt} = Au \qquad t > 0 \tag{3.4}$$

for any $v \in B$, and in particular, $u \in D(A)$ for $t > 0$. This means that for any v (in particular $v \in D(A)$) the generalized solution of (3.4) belongs to D(A) for $t > 0$, i.e. (3.4) is satisfied in the classical sense for $t > 0$. ∎

A useful criterion for holomorphic semigroups in a Hilbert space H is the following

Theorem 3.2 - If -A is maximal accretive in a Hilbert space H and the complex number

$$(-Av, v)$$

is contained in the sector $|\arg\zeta| \leq \frac{\pi}{2} - \omega$ of the complex plane for any $v \in H$, then, A generates a holomorphic semigroup for $|\arg t| < \omega$ as in theorem 3.1.

An important result about convergence of solutions of evolution equations is the following theorem of Tratter-Kato, which will be proved in chapter 10.

Theorem 3.3 - Let $A_\varepsilon$, $A_0$ be generators of contraction semigroups in the Banach space B, then

a) If

$$(\lambda - A_\varepsilon)^{-1}v \xrightarrow[\varepsilon \to 0]{} (\lambda - A_0)^{-1}v \qquad \forall v \in B \tag{3.5}$$

for some $\lambda$ with $\text{Re}\,\lambda > 0$, then

$$e^{A_\varepsilon t}v \xrightarrow[\varepsilon \to 0]{} e^{A_0 t}v \qquad \forall v \in B \tag{3.6}$$

uniformly in each finite interval of $t \geq 0$.

b) If (3.6) holds for all $t \geq 0$, then (3.5) holds for all $\lambda$ with $\lambda > 0$.

One can say that strong convergence of the resolvent for $\text{Re}\,\lambda > 0$ amounts to strong convergence of the semigroup.

4.- Examples of "parabolic" equations - We consider here a large class of operators which are generators of semigroups, containing as a particular case the heat equation. But in general such equations may not be parabolic.

Proposition 4.1.- Let us consider the Hilbert spaces H and V and the sesquilinear form $a(u , v)$ as in the first representation theorem (theorem 5.2, ch.2) Then, the operator $-A$ (more exactly $-A_H$ restriction to H) is the generator of the continuous semigroup $\exp(-At)$ on H.

Proof : It is easy to prove that $-A$ satisfies a) and b) of the theorem 1.1 (Hille-Yosida). In fact, from (2.5.8) we see that $A + \lambda$ with $\lambda > \omega$ satisfies the conditions of the Lax-Milgram theorem, and thus,

$$(A + \lambda)^{-1} \in \mathcal{L}(H,H)$$

consequently $\lambda > \omega$ belongs to the resolvent set of $-A$. Moreover, if

$$(A + \lambda)u = f \qquad \text{in } H .$$

by scalar multiplication by u in H and using (5.8) of chap. 2, we have

$$C \| u \|_V^2 + (\lambda - \omega) \| u \|_H^2 \leqslant \| f \|_H \| u \|_H \qquad \lambda > \omega$$

and so

$$\| u \|_H \leqslant \frac{\| f \|_H}{\lambda - \omega} \Leftrightarrow \| (A + \lambda)^{-1} \| \leqslant \frac{1}{\lambda - \omega}$$

and the condition b) of the Hille-Yosida's theorem is satisfied with $M = 1$. The condition a) is a direct consequence of the first representation theorem.∎

Remark 4.1 - In particular, if $\omega = 0$ in the coerciveness condition of the first representation theorem, the semigroup $\exp(-At)$ is of contraction in H. Moreover, the conditions of the theorem 3.2 are satisfied because of

$$|\operatorname{Im}(-Av , v)| \leqslant M \| v \|_V^2$$

$$\operatorname{Re}(-Av , v) \geqslant C \| v \|_V^2$$

and so the semigroup is holomorphic.∎

If in addition to the hypothesis of the preceeding remark, we consider that A is selfadjoint and the embedding of V in H is compact, we are in the conditions of chap. 2 sect. 6. It is then easy to give an explicit formula for the semigroup in the basis of eigenvectors $e_i$ (normalized in H). If $v = \sum_i v_i e_i$ then

$$u(t) \equiv e^{-At} v = \sum_i e^{-\lambda_i t} v_i e_i \tag{4.1}$$

the reader may construct explicite examples of semigroups with the operators of chap. 3.

It is easy to obtain "a priori" estimates for the semigroup. Since it is holomorphic, it satisfies in a classical way the equation

$$\frac{du}{dt} + Au = 0 \quad ; \quad u(0) = v$$

and by multiplying this in H by u we have

$$\frac{1}{2}\frac{d}{dt}\|u\|_H^2 + (Au\ ,\ u) = 0 \Rightarrow \frac{1}{2}\frac{d}{dt}\|u\|_H^2 + C\|u\|_V^2 \leq 0$$

and by integrating this from 0 to an arbitrary value T (this is possible because $u \underset{t \to 0}{\to} v$) we have

$$(4.2) \qquad \frac{1}{2}\|u(T)\|_H^2 + C\int_0^T \|u(S)\|_V^2\, dS \leq \frac{1}{2}\|v\|_H^2$$

Analogous formulae can be obtained for the nonhomogeneous equation (3.1).

5.- Examples of "hyperbolic" equations - As in the preceeding section, we consider an abstract framework which contains some typical hyperbolic systems (such as vibrations of membranes and elastic bodies), but also some parabolic systems (such as vibration of plates).

Formally, we are interested in the study of equations of the form :

(5.1) $\dfrac{d^2u}{dt^2} + Au = 0$ and the associated non homogeneous equations. We shall write them in the form of a first order system

$$\begin{Bmatrix} u = u_1, \\ \dfrac{du}{dt} = u_2 \end{Bmatrix} \quad \underline{u} = \begin{pmatrix} u_1 \\ u_2 \end{pmatrix} \quad \text{and (5.1) becomes}$$

$$(5.2) \qquad \frac{d\underline{u}}{dt} = \mathcal{A}\underline{u} \qquad \text{where } \mathcal{A} = \begin{pmatrix} 0 & I \\ -A & 0 \end{pmatrix}$$

we then have :

Proposition 5.1.- Let us consider the spaces H, V and the operator A as in the "first representation theorem" (theorem 2.5.2), with a $(\underline{u},\underline{v})$ hermitian (and then A selfadjoint). Then, the equation (5.2) is associated with a continuous semigroup in the space $V \times H$ ($u_1 \in V$ ; $u_2 \in H$).

Proof - We first remark that by considering $\underline{u} = \underline{v}\, e^{\omega t}$, (5.2) is equivalent to

$$(5.3) \qquad \frac{d\underline{v}}{dt} = (\mathcal{A} - \omega)\underline{v}$$

It then suffices to prove that $\mathcal{A} - \omega$ is the generator of a contraction semigroup in $V \times H$ (the semigroup associated to (5.2) will not be of contractions, in general) where V will be equipped with the scalar product

(5.4) $\qquad (\underline{u},\underline{v})_V = a(\underline{u},\underline{v}) + \omega(\underline{u},\underline{v})_H$, associated with a norm equivalent to the original one. We shall apply the Lumer-Phillips theorem.

We define $D(\mathcal{A} - \omega)$ in such a way that $(\mathcal{A} - \omega)\underline{v} \in V \times H$, i.e.

$$D(\mathcal{A} - \omega) = D(A) \times V$$

which is dense in $V \times H$ (c.f. chap. 2, remark 5.7)

Then, for $\underline{v} \in D(\mathcal{A} - \omega)$ we have

$$\mathrm{Re}\,(\,(-\mathcal{A} + \omega)\underline{v}\,,\,\underline{v})_{V \times H} =$$

$$= \mathrm{Re}\,[\omega \|v_1\|_V^2 - (v_2\,,\,v_1)_V + a(v_1\,,\,v_2) + \omega \|v_2\|_H^2\,] =$$

$$= \mathrm{Re}\,[\omega \|v_1\|_V^2 + \omega \|v_2\|_H^2 - \omega(v_2, v_1)_H\,] \geqslant 0$$

and the operator $-\mathcal{A} + \omega$ is accretive.

Moreover, for sufficiently large $\lambda > 0$, the rang of $\lambda - \mathcal{A} + \omega$ is the whole space $V \times H$ : for, if $\underline{f} \in V \times H$ is given, we must find $\underline{u} \in D(\mathcal{A} - \omega)$ such that

$$(\lambda - \mathcal{A} + \omega)\underline{u} = \underline{f} \quad \text{or equivalently :}$$

(5.5)
$$(\lambda + \omega)^2 u_1 + A\,u_1 = f_2 + (\lambda - \omega) f_1 \in H$$
$$u_2 = (\lambda + \omega)\,u_1 - f_1$$

and for $\lambda$ sufficiently large we can apply the Lax-Milgram theorem to the first equation and find $u_1 \in D(A)$. Then, the second equation gives $u_2 \in V$.

Remark 5.1 - In practical applications to vibrating systems in bounded domains, we must consider the particular case where, in the framework of proposition 5.1, we have $\omega = 0$ (i.e., the coerciveness condition (5.8) of chap. 2 is satisfied with $\omega = 0$). The preceeding proof shows that in this case, the semigroup exp(At) is of contractions for the "energy norm" :

(5.6)
$$\|u\|_E^2 = a(u_1\,,\,u_1) + \|u_2\|_H^2$$

Moreover, the change from t to -t in (5.1) leads to the same system and exp(-At) is also defined and is a contraction semigroup. This shows that exp($\mathcal{A}$t) is in fact a group (defined for all t). In addition, this is a unitary group (i.e., it conserves the energy norm) because

$$\| \exp(\mathcal{A}t)\underline{v} \|_E \leqslant \| \underline{v} \|_E \qquad \text{and}$$

$$\| \underline{v} \|_E = \| \exp(- t) \exp( t)\underline{v} \|_E \leqslant \| \exp(\mathcal{A}t)\underline{v} \|_E$$

Consequently exp(At) is in fact a Stone's group (c.f. th. 2.3). It is also easy in this connection to prove directly that $i\mathcal{A}$ is a selfadjoint operator. In physical applications this is often an example of the conservation of energy. Moreover, if the embedding of V in H is compact, we have, as in (4.1) if

$$u(0) = \sum_i v^i e_i \in V$$

$$\frac{du}{dt}(0) = \Sigma w^i e_i \in H$$

($e_i$ are the eigenvectors, normalized in H, and $\lambda_i$ the eigenvalues)

$$u(t) = \Sigma \; v^i \cos \sqrt{\lambda_i} t + w_i \frac{\sin \sqrt{\lambda_i}\, t}{\sqrt{\lambda_i}}$$

the reader can easyly check that u(t) and u'(t) so defined belong to the spaces V and H. ■

Remark 5.2 - In the application of the proposition 5.1 to vibrating systems in unbounded domains, we often have a situation more general than that of the preceeding remark.

Let us suppose that in proposition 5.1, we have

$$\text{(5.7)} \qquad \begin{cases} a(u,u) \geqslant 0 \\ a(u,u) = 0 \Rightarrow u = 0 \end{cases}$$

(i.e. a(u,u) is a norm on V (in general not equivalent to the norm in (5.4))

Then, for $\underline{w} \in D(\mathcal{A})$, the solution of

$$\text{(5.8)} \qquad \frac{d\underline{u}}{dt} = \mathcal{A}\underline{u} \quad ; \qquad \underline{u}(0) = \underline{w}$$

is a classical solution and (5.8) makes sense. By multiplying (5.8) by $\underline{u}$ in V × H and by taking the real part, we have

$$\frac{1}{2}\frac{d}{dt}(\| u_1 \|_V^2 + \| u_2 \|_H^2) = \text{Re}[\, (u_2,u_1)_V - a(u_1,u_2)]$$

and by using (5.4) we have

$$\frac{d}{dt}(a(u_1 , u_1) + \| u_2 \|_H^2) = 0 \Rightarrow$$

$$\text{(5.9)} \qquad a(u_1,u_1) + \| u_2 \|_H^2 = a(w_1 , w_1) + \| w_2 \|_H^2 \qquad \forall t$$

But for each t, $G(t) = \exp \mathcal{A}t$ is a continuous operator in $V \times H$ ; taking into account that $D(\mathcal{A})$ is dense in $V \times H$, we see that, by continuity, (5.9) holds for any $\underline{w} \in V \times H$, i.e., for any (generalized) solution. Moreover, if we define the energy space $E = \tilde{V} * H$ of norm (5.6), where $\tilde{V}$ is the completion of V for the norm (square) $a(u,u)$, the semigroup may be extented continuously to E and is a unitary group (Stone) in E. It is clear that the generator is an extension of $\mathcal{A}$. ■

6.- Laplace transforms - For the study of the Laplace transformations, we only consider functions or distributions of the real variable t with support in $[0,\infty[$. This means that the functions are zero for $t < 0$, and the distributions are such that their values are zero for test functions which are 0 for $t \geq 0$.

Let v(t) be a locally summable function of the real variable t with values in the reflexive and separable Banach space B. Let us consider, for the complex parameter $p = \xi + i\eta$ the Laplace integral

(6.1) $$\hat{v}(p) = \int_0^\infty v(t)\, e^{-pt}\, dt.$$

we have

$$\| v(t)\, e^{-pt} \| \leq \| v(t) \| e^{-\xi t}$$

and consequently if (6.1) is summable for $\xi = \xi_0$, it is also summable for $\xi \geq \xi_0$, and $\hat{v}(p)$ is a holomorphic function of p for $\xi \geq \xi_0$, with values in B, which is called the Laplace transform of v. Let us define $\xi^*$ as the Inf of the values $\xi_0$ with the preceeding property. $\xi^*$ is called the summability abscissa of the Laplace integral of v. The Laplace transform $\hat{v}(p)$ is then defined for the functions v such that $\xi^*$ exists (is not $+\infty$) and is holomorphic in the half plane $\xi > \xi^*$.

The space $\mathcal{D}'(B)$ of the distributions with values in B is defined (in the same way as the distributions with numerical values) as the space of the linear continuous operators from $\mathcal{D}$ to B.

Let S be the space of the rapidly decreasing functions (i.e., infinitely differentiable functions of $t \in ]-\infty, +\infty[$ such that they decrease, as well as their derivatives of any order more rapidly than any power of $|t|^{-1}$, as $|t| \to \infty$ ). This space is provided with an appropriate topology, but its definition is not often useful in applications. We then define the space S'(B) of the temperated distributions with values in B as the space of the linear continuous operators from S to B. It is a subspace of $\mathcal{D}'(B)$.

Now let v(t) be a distribution with values in B with support in $[0, \infty[$ such that $e^{-\xi t} v(t)$ is a temperated distribution for $\xi \geq \xi_0$. Then, the Laplace transform $\hat{v}(p)$ of the distribution v is defined by

$$(6.2)\qquad \hat{v}(p) = \left\langle e^{-\xi_0 t} v(t),\ e^{-(p-\xi_0)t} \right\rangle = \text{(written)} = \left\langle v(t),\ e^{-pt} \right\rangle$$

which is a holomorphic function of $p = \xi + i\eta$ in the half plane $\xi > \xi^*$ with values in B, (where $\xi^*$ is the Inf of the $\xi_0$ with the preceeding property).

Remark 6.1 - The test function $e^{-(p-\xi_0)t}$ is not an element of S, but it may be extended in $t < 0$ to a function of S, and the value of $\hat{v}(p)$ does not depend on this extension because the support of v is in $[0, \infty[$. ■

We shall denote by $\mathcal{L}(B)$ the class of distributions having a Laplace transform defined in the preceeding way.

A very important theorem in this connection is the following.

Theorem 6.1.- The necessary and sufficient condition for a holomorphic function F of the complex variable $p = \xi + i\eta$ with values in B to be the Laplace transform of a distribution in $\mathcal{L}(B)$ is that it is defined and holomorphic in a halfplane $\xi > \xi^*$ and that it is bounded above in norm by a polynominal in $|p|$

$$(6.2)\qquad \|F(p)\| \leqslant \text{Poly}(\,|p|\,) \quad ; \qquad \xi > \xi^*$$

Remark 6.2.- In fact, the class $\mathcal{L}(B)$ is formed by the distributions which are the (distributional) derivatives of finite order of the continuous functions defined on $]-\infty, +\infty[$ which are zero for $t \leqslant 0$ and such that there exists $\xi^*$ (depending on f) for which

$$(6.3)\qquad \|f(t)\| \leqslant C\, e^{-\xi t} \qquad \text{for } \xi > \xi^*$$

In fact, as we shall see in the sequel, the Laplace image of the distributional derivation is a multiplication by p. It is easy to see that a distribution of the mentioned class has a Laplace transform satisfying (6.2). To prove the converse, by derivating a sufficient number of times, it suffices to prove that if F(p) is a holomorphic function for $\xi > \xi^0$ which satisfies

$$\|F(p)\| \leqslant \frac{C}{|p|^2}$$

then F is the Laplace transform of a continuous function f(t) which is zero for $t \leqslant 0$ and which satisfies (6.3). But this is a well-known result obtained by the classical formula of integration in the p plane for the inverse Laplace transform. ■

Proposition 6.1 - In the situation of th. 6.1, the inverse Laplace transform is unique : i.e. the null distribution is the only one whose Laplace transform is the null function.

The following result of the theory of holomorphic functions is often useful to apply theorem 6.1.

Proposition 6.2 - Let $B_1$, $B_2$ be Banach spaces and let $F(p)$ be a holomorphic function of the complex variable $p$ with values in $\mathcal{L}(B_1, B_2)$. If for $p = p_0$ we have $F^{-1}(p_0) \in \mathcal{L}(B_2, B_1)$ (i.e., the inverse of $F(p_0)$ exists and is continuous from $B_2$ to $B_1$), then for $|p - p_0|$ sufficiently small, $F^{-1}(p)$ exists and is holomorphic with values in $\mathcal{L}(B_2, B_1)$.

The tensorial product of distributions is a generalization of the product of functions. Then, if $T(t_1)$ is a distribution with values in $V$ and $S(t_2)$ is a distribution with values in $\mathcal{L}(V, H)$ (resp. with numerical values), the tensorial product $S \otimes T(t_1, t_2)$ is a distribution of the two variables $t_1$, $t_2$ with values in $H$ (resp. in $C$) defined in such a way that, if $S$ and $T$ are functions we have

$$\langle S \otimes T , \phi \rangle = \iint S(t_1)\, T(t_2)\, \phi(t_1, t_2)\, dt_1\, dt_2$$

The convolution product is then defined by

$$(6.4) \qquad \langle S * T, \phi \rangle = \langle S(t_1) \quad T(t_2) , \phi(t_1 + t_2) \rangle$$

which makes sense if we suppose that $S$ and $T$ are distributions with support in $[0, +\infty[$. In fact, if $\phi$ is in $\mathcal{D}(R)$, $\phi(t_1 + t_2)$ is not in $\mathcal{D}(R^2)$ because it is not of compact support ; nevertheless if $S$ and $T$ have their supports in $[0, \infty[$, only the values of $\phi(t_1 + t_2)$ for $t_1 > 0$, $t_2 > 0$ have an influence in (6.4) and it is possible to modify $\phi(t_1 + t_2)$ in such a way that $\phi(t_1 + t_2) \in \mathcal{D}(R^2)$ without changing the value of the expression (6.4). Then the convolution product of two distributions with support in $[0, \infty[$ is always defined. Moreover, if the distributions are functions, we have

$$\langle S * T , \phi \rangle = \int_0^\infty \int_0^\infty S(t_1)\, T(t_2)\, \phi(t_1 + t_2)\, dt_1\, dt_2 =$$

$$= \int_0^\infty \left( \int_0^\tau S(t)\, T(\tau - t)\, dt \right) \phi(\tau)\, d\tau$$

that is to say

$$(6.5) \qquad S * T(t) = \int_0^t S(\tau)\, T(t - \tau)\, d\tau$$

Moreover, if in (6.4) we take a test function $\phi(\tau)$, zero for $\tau > b$, we see that in (6.4) we can consider only

$$t_1 > 0 \quad , \quad t_2 > 0 \quad , \quad t_1 + t_2 < b$$

this means that the "values" of $S * T(t)$ for $t \leq b$ depend only on the "values" of S and T for $t \leq b$. In physics, if t is the time and S and T are the causes of a result given by $S * T$, we see that this result for $t \in [0,b]$ does not depend on the "values" of S and T for $t > b$. Then, the "result" depend on the past and the present of the causes, but not on the future. We can say that the convolution by distributions with support in $[0,\infty[$ gives a physically admissible causality.

It is now useful to remark that the convolution by the Dirac distribution $\delta$ and its derivatives $\delta'$, $\delta''$ .... give

(6.6) $$\delta * T = T \quad ; \quad \delta' * T = T' \quad , \quad \delta'' * T = T'' \ldots$$

where ' is for the distributional derivative. The Laplace transforms of $\delta$, $\delta'$ ... are

(6.7) $$\hat{\delta} = 1 \quad ; \quad (\delta')^\wedge = p \qquad (\delta'')^\wedge = p^2 \ldots.$$

The image under the Laplace transform of the convolution is the product :

(6.8) $$(S * T)^\wedge = \hat{S}\,\hat{T}$$

As an example, let us prove the formula for the Laplace transform of the contraction semigroups :

Proposition 6.3 - Let exp At be a contraction semigroup in the Banach space B. Then, the resolvent $(p - A)^{-1}$ is defined for $\mathrm{Re}\ p > 0$ and we have

(6.9) $$\int_0^\infty e^{At}\, e^{-pt}\, dt = (p - A)^{-1}.$$

Proof. The relation (6.9) is an identity between operators of $\mathcal{L}(B , B)$ ; it suffices to prove that the two terms operate on $v \in B$ in the same way.

Since $u(t) \equiv e^{At} v$ is continuous and bounded, its Laplace transform

(6.10) $$\hat{u}(p) = \int_0^\infty e^{At}\, v\, e^{-pt}\, dt$$

is well-defined in the classical sense, and is a holomorphic function with values in B for $\xi > 0$ ($p = \xi + i\eta$ ). Moreover, $\| \exp(At)v \| \leq \| v \| \Rightarrow \| \hat{u}(p) \| \leq \| v \| \frac{1}{\xi}$ and then, for p fixed, (6.10) is a continuous operator from B to B. On the other hand, if $v \in D(A)$, we have

$$\frac{du(t)}{dt} = Au(t) \qquad\qquad t \geq 0$$

for the sake of applying the Laplace transform to it, we define u(t) to be zero for $t < 0$ ; then, by writing it with distributional derivatives we have

$$\frac{du(t)}{dt} = Au(t) + v\delta \qquad \text{(distr. deriv.)} \tag{6.11}$$

and we take the Laplace transform of (6.11) by taking into account (6.6), (6.7), (6.8). We then have

$$p\,\hat{u} = A\hat{u} + v$$

so, for p real $> 0$ and by the Hille-Yosida theorem (theorem 2.1) we have $(p + A)^{-1} \in \mathcal{L}(B\ ,\ B)$ and then

$$\hat{u}(p) = (p - A)^{-1}v \quad \text{for } p \text{ real } \quad 0,\ v \in D(A). \tag{6.12}$$

But the two terms of (6.12) are bounded operators in B, and D(A) is dense in B ; then, (6.12) holds for any $v \in B$, and for any real $p > 0$. Moreover, the left hand side of (6.12) is a holomorphic function for complex p in the half plane $\xi > 0$ ; the right hand side of (6.12) is also holomorphic in a neighbourhood of the real axis by virtue of proposition 6.2 ; and by analytic continuation, (6.12) holds in this neighbourhood. In the same way, by analytic continuation, (6.12) holds for any p with $\xi > 0$, $v \in B$, Q.E.D.

# PART II

## HOMOGENIZATION METHOD IN THE PHYSICS OF COMPOSITE MATERIALS

Homogenization deals with the partial differential equations of physics in heterogeneous materials with a periodic structure when $\varepsilon \searrow 0$. ( $\varepsilon$ is the characteristic length of the period). Heuristically, the method is based on the consideration of two length scales associated with the microscopic and macroscopic phenomena.

Chapter 5 has an introductory character. Both formal expansions and proofs of the convergence are given. In chapter 6 to 8 we study some problems in mechanics and electromagnetism. Some results are only formal and the problem of the convergence is open.

# CHAPTER 5

## HOMOGENIZATION OF SECOND ORDER EQUATIONS

1.- Formal expansion - Elliptic equations of the divergence type often appear in physics. They are of the form

$$(1.1) \qquad -\frac{\partial}{\partial x_i}\left(a_{ij}(x)\frac{\partial u}{\partial x_j}\right) = f(x) \quad : \quad a_{ij} = a_{ji}$$

or equivalently

$$(1.2) \qquad -\frac{\partial p_i}{\partial x_i} = f \quad ; \quad p_i = a_{ij}\frac{\partial u}{\partial x_j}$$

Equation (1.1) is, for instance the equation of electrostatics, magnetostatics or time-independent heat transfer. The function u is the electric potential, magnetic potential or temperature respectively, and $\underline{p}$ is the electric displacement, magnetic induction or heat flux, resp. the matrix $a_{ij}(x)$ is the dielectric constant, magnetic permeability or thermal conductivity, resp. and f is a given source term.

The symmetric matrix $a_{ij}(x)$ is a physical property of the material. For a homogeneous matérial, $a_{ij}$ does not depend on x. On the other hand, if the material is not homogeneous, $a_{ij}$ effectively depends on x. For materials with a <u>periodic structure</u>, such as superposition of sheets of different materials, or homogeneous materials with holes filled by another material, $a_{ij}(x)$ is a periodic function of the space variables. In certain cases, the length of the period is very small with respect to the other lengths appearing in the problem. In these cases, one may think that the solution u of (1.1) is approximately the same as the corresponding solution for a "homogenized" matérial with constant matrix $a^h_{ij}$. The homogenization method studies this problem and shows the existence and some properties of the matrix $a^h_{ij}$.

We now study the simplest problem in this connection in a formal way. A proof of the convergence will be given in sect. 4, and some extensions to more complicated problems will be given in the sequel.

Let $\Omega$ be a bounded domain of the space $R^N$ of coordinates $x_i$, fig. 1. We shall consider it as a piece of a heterogeneous material defined as follows.

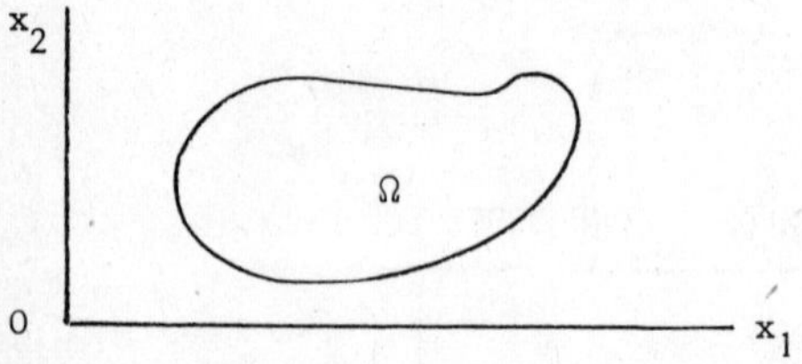


fig. 1

We consider, in the space $R^N$ of coordinates $y_i$, a fixed parallelepiped Y (fig. 2) of edges $y^o_i$, as well as the parallelepipeds obtained by translation of length $n_i y^o_i$ ($n_i$ integer) in the direction of the axis.

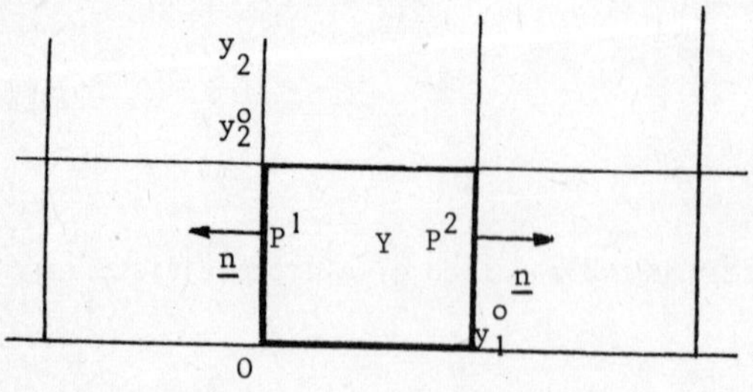


fig. 2

We now consider the Y-periodic, smooth real functions

(1.3) $$a_{ij}(y) = a_{ji}(y)$$

such that there exists $\gamma > 0$ with

(1.4) $$a_{ij}(y)\, \xi_i\, \xi_j \geq \gamma\ \xi_i\, \xi_i \qquad \forall y \in Y\ ,\ \underline{\xi} \in R^N$$

We then define the functions

(1.5) $$a^{\varepsilon}_{ij}(x) = a_{ij}\left(\frac{x}{\varepsilon}\right)$$

where $\varepsilon$ is a real positive parameter, say, $\varepsilon \in ]\ 0, \varepsilon_0]$ . Note that the functions $a^{\varepsilon}_{ij}(x)$ are $\varepsilon Y$-periodic in x; where the period $\varepsilon Y$ is the parallelepiped of edges $\varepsilon\, y^0_i$ .

Then, if f(x) is a given smooth function defined on $\Omega$, we consider the boundary value problem

(1.6) $$-\frac{\partial}{\partial x_i}\left(a^{\varepsilon}_{ij}(x)\, \frac{\partial u^{\varepsilon}}{\partial x_j}\right) = f(x) \qquad \text{in } \Omega$$

(1.7) $$u^{\varepsilon}\Big|_{\partial\Omega} = 0 \qquad \text{on } \partial\Omega$$

Remark 1.1 - For fixed $\varepsilon > 0$, $u^{\varepsilon}$ exists and is unique. This is easily seen, because (1.4) is an ellipticity condition. The existence and uniqueness of $u^{\varepsilon}$ may be studied as in chap. 3 sect. 1. In fact, the abstract formulation of (1.6), (1.7) is : find $u^{\varepsilon} \in H^1_0(\Omega)$ such that

(1.8) $$\int_{\Omega} a^{\varepsilon}_{ij}\ \frac{\partial u^{\varepsilon}}{\partial x_i}\, \frac{\partial v}{\partial x_j}\ dx = \int_{\Omega} fv\ dx \qquad \forall\ v \in H^1_0(\Omega).$$

The coerciveness of the form in the left hand side of (1.8) is evident by (1.4), for in fact,

$$\int_{\Omega} \frac{\partial u}{\partial x_i}\, \frac{\partial u}{\partial x_i}\ dx$$

may be taken as norm$^2$ of $H^1_0(\Omega)$ (see chap. 3, prop. 1.1). Moreover, if f, $a_{ij}$ and $\partial\Omega$ are smooth, standard regularity theory for elliptic equations shows that $u^{\varepsilon}$ is a smooth function. ∎

As in (1.2), we shall define $\underline{p}^{\varepsilon}$ as the vector with components

(1.9) $$p^{\varepsilon}_i(x) \equiv a^{\varepsilon}_{ij}(x)\ \frac{\partial u^{\varepsilon}}{\partial x_j}(x) \quad .$$

Now, we search for an asymptotic expansion of $u^{\varepsilon}(x)$ as a function of $\varepsilon$ for $\varepsilon \searrow 0$. The heuristic device is to suppose that $u^{\varepsilon}$ has a two-scale expansion of the form

(1.10) $$u^{\varepsilon}(x) = u^{0}(x) + \varepsilon\, u^{1}(x\,,\,y) + \varepsilon^{2}\, u^{2}(x\,,\,y) + \ldots \quad ; \quad y = x/\varepsilon$$

where the functions $u^i(x\,,\,y)$ are Y-periodic in the variable y. This means that we postulate the existence of smooth functions $u^i(x\,,\,y)$ defined for $x \in \Omega$ , $y \in R^N$, independent of $\varepsilon$ , Y-periodic in the variable y such that for $y = x/\varepsilon$ , the right hand side of (1.10) is an asymptotic expansion of $u^{\varepsilon}(x)$ (as well as its derivatives).

Remark 1.2 - The function $u^{\varepsilon}(x)$ is defined on $\Omega \times \,]0, \varepsilon_0]$. In fact (1.10) means that there exist $u^i(z\,,\,y)$ defined on $\Omega \times R^N$, Y-periodic in y such that by taking

(1.11) $$\begin{cases} z = x \quad , \quad y = x/\varepsilon \\ \dfrac{\partial}{\partial x_i} = \dfrac{\partial}{\partial z_i} + \dfrac{1}{\varepsilon}\dfrac{\partial}{\partial y_i} \end{cases}$$

the right hand side of (1.10) becomes a uniform expansion of $u^{\varepsilon}(x)$ and its derivatives. In practice, the use of the variable z is a little cumbersome, and we shall use only x and y, with

(1.12) $$\frac{d}{dx_i} = \frac{\partial}{\partial x_i} + \frac{1}{\varepsilon}\frac{\partial}{\partial y_i}$$

where it is understood that the total dependence on x is obtained directly and through the variable y. ■

Now, we explain the physical meaning of the expansion (1.10). Let us consider a term

(1.13) $$u^i(x\,,\,y) \quad , \quad \text{Y-periodic in } y \;, \quad y = \frac{x}{\varepsilon} \;.$$

Let us also consider $u^i(x\,,\,x/\varepsilon)$ with small $\varepsilon$ . It is clear that the dependence with respect to the variable $x/\varepsilon$ is periodic with period $\varepsilon Y$. (see fig. 3). Let us compare the values of $u^i(x\,,\,x/\varepsilon)$ at two points $P^1$ and $P^2$ homologous by periodicity corresponding to two adjacent periods. By periodicity, the dependence in $x/\varepsilon$ is the same, and the dependence on x is almost the same because the distance $P^1P^2$ is small and $u^i$ is a smooth function. On the other hand, let $P^3$ be a point homologous to $P^1$ by periodicity, but located far from $P^1$. The dependence of $u^i$ on y is the same, but the dependence on x is very different because $P^1$ and $P^2$ are not near each other. Finally, we compare the values of $u^i$ at two different points $P^1$ and $P^4$ of the same period. The dependence on x is almost the same because $P^1$ and $P^2$ are near from one another, but the dependence on y is very different because $P^1$ and $P^4$ are not homologous by periodicity (in fact, the

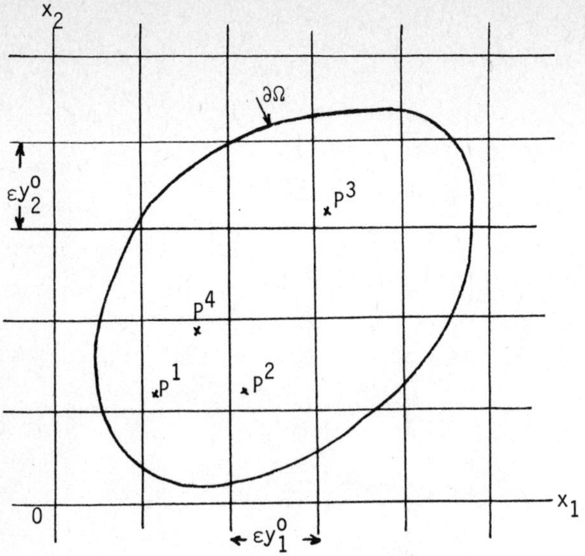


fig. 3

distance $P^1P^4$ is "large" when measured with the variable y !). Consequently, (1.13) means that the $u^i$ takes values that are almost the same in neighbouring periods, but very different in distant periods. Such functions will be called "locally periodic".

Remark 1.3 - The preceeding considerations show that it is natural to search for $u^\varepsilon$ in the form of the expansion (1.10). In fact, $u^\varepsilon$ depends on the (periodic) coefficients $a^\varepsilon_{ij}$ and on $\partial\Omega$ and f. It is natural to search for $u^\varepsilon$ depending on x in two different forms. Firstly, periodically of period $\varepsilon Y$, secondly in an aperiodic fashion. In any case, the expansion (1.10) is heuristic and it fails near the boundary $\partial\Omega$ , where aperiodic phenomena are preponderant and a boundary layer arises.■

Remark 1.4 - In (1.10) we postulate that $u^0$ depends only on x. This is a particular case of $u^0(x , y)$ (constant with respect to y). If we start with $u^0(x , y)$, the subsequent treatment of the problem shows that $u^0$ is constant with respect to to y. In fact (1.10) means that $u^\varepsilon$ is the smooth function $u^0(x)$ plus a little highly oscillating term (fig. 4). It is not difficult to see that $u^\varepsilon$ must have this form (by considering the one-dimensional case, for instance)■

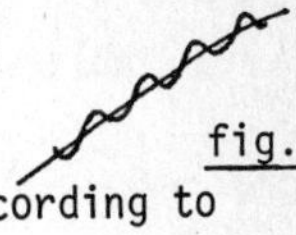

fig. 4

The following step of our study is to expand grad $u^\varepsilon$ and $p^\varepsilon$ according to

(1.10) and (1.12). We have

$$\frac{du^\varepsilon}{dx_i} = \left(\frac{\partial u^o}{\partial x_i} + \frac{\partial u^1}{\partial y_i}\right) + \varepsilon\left(\frac{\partial u^1}{\partial x_i} + \frac{\partial u^2}{\partial y_i}\right) + \varepsilon^2 \ldots. \tag{1.14}$$

$$p_i^\varepsilon(x) = p_i^o(x\,,\,y) + \varepsilon\, p_i^1(x\,,\,y) + \varepsilon^2 \ldots \equiv a_{ij}(y)\frac{du^\varepsilon}{dx_j} \tag{1.15}$$

$$\begin{cases} p_i^o(x\,,\,y) = a_{ij}(y)\left(\dfrac{\partial u^o}{\partial x_j} + \dfrac{\partial u^1}{\partial y_j}\right) \\ p_i^1(x\,,\,y) = a_{ij}(y)\left(\dfrac{\partial u^1}{\partial x_j} + \dfrac{\partial u^2}{\partial y_j}\right) \\ \ldots\ldots\ldots\ldots\ldots\ldots\ldots\ldots\ldots \end{cases} \tag{1.16}$$

and all the terms are Y-periodic in y and the expansions hold for $y = x/\varepsilon$.

Next, we write equation (1.6) in the form

$$-\frac{d}{dx_i}\, p_i^\varepsilon(x) = f \Leftrightarrow \left(-\frac{\partial}{\partial x_i} - \frac{1}{\varepsilon}\frac{\partial}{\partial y_i}\right)(p_i^o + \varepsilon\, p_i^1 + \ldots) = f(x) \tag{1.17}$$

and we expand it in powers of $\varepsilon$ . We must have an identity for any small $\varepsilon$, and in in consequence, the coefficients of the successive powers of $\varepsilon$ must be zero, i.e.

$$\varepsilon^{-1} \Rightarrow \frac{\partial p_i^o}{\partial y_i} = 0 \Leftrightarrow \frac{\partial}{\partial y_i}\left[a_{ij}(y)\left(\frac{\partial u^o}{\partial x_j}(x) + \frac{\partial u^1}{\partial y_j}(x\,,\,y)\right)\right] = 0 \tag{1.18}$$

$$\varepsilon^o \Leftrightarrow -\frac{\partial p_i^o}{\partial x_i} - \frac{\partial p_i^1}{\partial y_i} = f \tag{1.19}$$

and so on. As we shall see soon, (1.18) and (1.19) are equations for the microscopic (or local) and macroscopic (or homogenized) behaviour of $u^\varepsilon$ respectively. It is worthwhile in this connection to recall that, according to (1.10), x and y are two independent variables.

Let us begin by studying (1.19). We consider the operator "average" or "mean" defined on any Y-periodic function $\phi(y)$ by

$$\tilde{\phi} = \frac{1}{|Y|}\int_Y \phi(y)\, dy \tag{1.20}$$

where $|Y|$ is the measure of Y. It is clear that $\tilde{\phi}$ does not depend on y ; moreover, if $\phi$ depends also on the variable x, <u>the mean operator commutes with differenciation with respect to x</u>.

By applying the operator $\sim$ to (1.19), we have

$$-\frac{\partial \tilde{p}_i^o}{\partial x_i} - \left(\frac{\partial p_i^1}{\partial y_i}\right)^\sim = \tilde{f} = f \tag{1.21}$$

(Note that f(x) is a function of x only, it may be considered as function of x and y, constant with respect to y, and we have the second equality in (1.21).

On the other hand, by using the divergence theorem we have

$$(1.22)\qquad \left(\frac{\partial p_i^1}{\partial y_i}(x\,,\,y)\right)^{\sim} = \frac{1}{|Y|}\int_Y \frac{\partial p^1}{\partial y_i}\,dy = \frac{1}{|Y|}\int_{\partial Y} n_i\, p_i(x\,,\,y)\,dS$$

where n is the outer unit normal to the boundary of Y, $\partial Y$. But $p_i(x\,,\,y)$ is Y-periodic in y, in consequence, the integral on two opposite faces of $\partial Y$ takes opposite values because $\underline{p}(x\,,\,y)$ (resp. $\underline{n}(y)$) takes the same (resp. opposite) values in homologous points such as $P^1$, $P^2$ (see fig. 2). We then have

$$(1.22)\qquad \left(\frac{\partial p_i^1}{\partial y_i}\right)^{\sim} = 0$$

and (1.21) becomes

$$(1.23)\qquad -\frac{\partial \tilde{p}_i^0}{\partial x_i}(x) = f(x)$$

Note that $p_i^0$ is a function of x and y but its mean value depends only on x. In consequence, (1.23) is a macroscopic equation in x. Moreover, (1.23) shows that $\tilde{\underline{p}}^0$ (mean value of the first term of the expansion (1.15)) satisfies an equation analogous to (1.2).

The next step is to search for a relation between $\tilde{\underline{p}}^0$ and $u^0$ in order to obtain an equation of the type (1.1). This will be made by using the local equation (1.18).

2.- Study of the local problem - We write the local equation (1.18) in the form

$$(2.1)\qquad -\frac{\partial}{\partial y_i}\left(a_{ij}(y)\,\frac{\partial u^1}{\partial y_j}\right) = \frac{\partial u^0}{\partial x_j}\,\frac{\partial a_{ij}}{\partial y_i}(y) \qquad\qquad \text{Y-periodic}$$

and we consider it as an equation in the unknown $u^1(y)$ ; $u^0$ is considered for the time being as known. Of course, $u^0$ and $u^1$ depend on the parameter x. The right hand side of (2. 1 ) is then known, and (2. 1 ) is in fact an elliptic equation in $u^1(y)$ ; moreover, $u^1(y)$ must be Y-periodic, and this condition plays the role of boundary conditions. We shall see that $u^1(y)$ exists and is unique (up to an additive constant).

In order to study this problem, we introduce the two spaces of Y-periodic functions:

$$H_Y = \{ u \in L^2_{loc}(R^N) \ ; \ u \text{ is Y-periodic} \}$$

$$V_Y = \{ u \in H^1_{loc}(R^N) \ ; \ u \text{ is Y-periodic} \}$$

which are Hilbert spaces for the scalar products.:

$$(2.2) \qquad (u , v)_{H_Y} = \int_Y u\, v\, dy$$

$$(u , v)_{V_Y} = \int_Y \left(\frac{\partial u}{\partial y_i} \frac{\partial v}{\partial y_i} + u\, v\right) dy$$

Remark 2.1 - A Y-periodic function defined on $R^N$ does not belong to $L^2(R^N)$ because its "norm" is infinite. Then, $H_Y$, $V_Y$ are spaces of functions whose restrictions to any bounded domain belong to $L^2$ or $H^1$ of this domain. This is the meaning of $L^2_{loc}$ , $H^1_{loc}$. It is immediate to see that $H_Y$, $V_Y$ are complete (and thus Hilbert spaces) for the norms associated to (2.2). In fact, Y-periodic functions may be considered as functions defined on a period Y only. In this case $H_Y$ may be identified with $L^2(Y)$ and $V_Y$ may be identified with the space of the functions of $H^1(Y)$ such that the traces on the opposite faces of $\partial Y$ are the same.■

As a consequence of the Rellich theorem, and chap. 2, (6.12), we have :

Proposition 2.1 - If $H_Y$ is identified with its dual,

$$V_Y \subset H_Y \subset V'_Y$$

with dense and compact embeddings.

The variational formulation of (2.1) is

Find $u^1 \in V_Y$ such that

$$(2.3) \qquad \int_Y a_{ij}(y) \frac{\partial u^1}{\partial y_j} \frac{\partial v}{\partial y_i} dy = \frac{\partial u^0}{\partial x_j} \int_Y \frac{\partial a_{ij}}{\partial y_i}(y)\, v\, dy \qquad \forall v \in V_Y$$

The equivalence of (2.3) and (2.1) is easily proved. If we multiply (2.1) by a test function $v \in V_Y$ and we integrate in Y (note that this is the scalar product in $H_Y$) we obtain (2.3) by using the following formula of integration by parts:

$$(2.4) \quad \int_Y \frac{\partial}{\partial y_i}\left(a_{ij} \frac{\partial u^1}{\partial y_i}\right) v\, dy + \int_Y a_{ij} \frac{\partial u^1}{\partial y_i} \frac{\partial v}{\partial y_i} dy = \int_{\partial Y} n_i\, a_{ij} \frac{\partial u^1}{\partial y_j} v\, dS = 0$$

where the surface integral in the right hand side vanishes by periodicity, as in (1.22). Conversely, if $u^1$ satisfies (2.3), we may use (2.4) again (note that by interior regularity theory for elliptic equations $u^1 \in H^2_{loc}$) to obtain

$$\int_Y \left[\frac{\partial}{\partial y_i}\left(a_{ij}\frac{\partial u^1}{\partial y_j}\right) + \frac{\partial u^o}{\partial x_j}\frac{\partial a_{ij}}{\partial y_i}\right] v \, dy = 0 \qquad \forall v \in V_Y$$

which implies (2.1).

To study (2.3), we note that the form in the left hand side is not coercive on $V_Y$, but the form

(2.5) $$b(u^1 , v) \equiv \int_Y a_{ij}\frac{\partial u^1}{\partial y_j}\frac{\partial v}{\partial y_i} dy + \int_Y u^1 v \, dy$$

is coercive on $V_Y$ by virtue of the ellipticity condition (1.4). This problem recalls the Neumann problem (2.6) of chap. 3 and will be solved in an analogous manner. If B is the operator associated with the form b according to the first representation theorem, we evidently have

(2.6) $$B = A + I \qquad ; \qquad A = -\frac{\partial}{\partial y_i}(a_{ij}\frac{\partial}{\partial x_j})$$

on $H_Y$ (the periodicity conditions are included in $H_Y$ and they are not written explicitly). Moreover, by writing (2.1) in the form

(2.7) $$(B - I)\, u^1 = \frac{\partial u^o}{\partial x_j}\frac{\partial a_{ij}}{\partial y_i}$$

we are in the framework of chap. 2, (6.13). The necessary and sufficent condition for (2.7) to be solvable is that the right hand side is orthogonal in $H_Y$ to the solutions of

(2.8) $$(B - I)w = 0$$

But if w is a solution of (2.8), by multiplying (2.8) by w we have

$$0 = b(w , w) - (w , w)_{H_Y} = \int_Y a_{ij}(y)\frac{\partial w}{\partial y_i}\frac{\partial w}{\partial y_j} dy$$

which implies w = cost. by (1.4). Consequently, the compatibility condition for equation (2.7) is

$$\frac{\partial u^o}{\partial x_j}\int_Y \frac{\partial a_{ij}}{\partial y_i}(y)\, dy = 0$$

which is in fact satisfied by the periodicity of $a_{ij}$:

$$\int_Y \frac{\partial a_{ij}}{\partial y_i} dy = \int_{\partial Y} n_i \, a_{ij} \, dS = 0$$

Consequently, $u^1(y)$ (x is always a parameter) is determined up to an additive constant. Of course, $u^1$ is uniquely determined if we impose that its mean value must be zero. Moreover, let $w^k$ be the unique solution of

$$(2.9)\quad \begin{cases} \text{Find } w^k \in V_y \quad ; \quad \text{with } \tilde{w}^k = 0 \\ \displaystyle\int_Y a_{ij} \frac{\partial w^k}{\partial y_j} \frac{\partial v}{\partial y_i}\, dy = \int_Y \frac{\partial a_{ik}}{\partial y_i}\, v\, dy \qquad \forall\, v \in V_k \end{cases}$$

Then, by virtue of the linearity of problem (2.3), we have :

$$(2.10)\qquad u^1(x\ ,\ y) = \frac{\partial u^o(x)}{\partial x_k}\, w^k(y) + C(x)$$

where the term C(x) is the additive constant (function of the parameter x). As a result, we have

Proposition 2.1 - If $u^o(x)$ is known, $u^1(x\ ,\ y)$ is determined up to an additive function of x. It has the form (2.10), where $w^k$ is uniquely defined by (2.9).

3.- Formulae for the homogenized coefficients and their properties - We go back to equation (1.23). Our aim is to find a relation between $\tilde{p}^o(x)$ and $u^o(x)$. This is easily obtained from (1.16) and proposition 2.1 :

$$(3.1)\qquad p_i^o(x\ ,\ y) = a_{ij}(y)\left(\frac{\partial u^o(x)}{\partial x_j} + \frac{\partial u^1(x\ ,\ y)}{\partial y_j}\right) =$$

$$= a_{ij}(y)\left[\frac{\partial u^o(x)}{\partial x_j} + \frac{\partial u^o(x)}{\partial x_k}\frac{\partial w^k(y)}{\partial y_j}\right] = \left[a_{ik}(y) + a_{ij}(y)\frac{\partial w^k(y)}{\partial y_j}\right]\frac{\partial u^o(x)}{\partial x_k}$$

and by applying the average operator defined by (1.20)

$$(3.2)\qquad \tilde{p}_i^o(x) = \left[a_{ik}(y) + a_{ij}(y)\,\frac{\partial w^k(y)}{\partial y_j}\right]^{\sim} \frac{\partial u^o(x)}{\partial x_k}\ .$$

We then have :

Proposition 3.1 - $\tilde{\underline{p}}^o(x)$ is related to $u^o(x)$ by

$$(3.3)\qquad \tilde{p}_i^o(x) = a_{ik}^h \frac{\partial u^o(x)}{\partial x_k}$$

where the constant coefficients $a_{ij}^h$ (the "homogenized "coefficients) are

$$(3.4)\qquad a_{ik}^h = \left[a_{ik}(y) + a_{ij}(y)\,\frac{\partial w^k(y)}{\partial y_j}\right]^{\sim} \equiv \left[a_{ij}(y)\left(\delta_{jk} + \frac{\partial w^k(y)}{\partial y_j}\right)\right]^{\sim}$$

where $\sim$ is the mean, defined by (1.20). They are well determined constants which only depend on the local coefficients $a_{ij}(y)$.

Equation (1.23) then becomes an equation in $u^o(x)$ :

(3.5) $$-\frac{\partial}{\partial x_i}\left(a^h_{ik}\ \frac{\partial u^o(x)}{\partial x_k}\right) = f \quad .$$

We now search for symmetry and positivity relations for the homogenized coefficients.

If in (2.9) we integrate by parts the right hand side, we see that it is equal to

$$\int_{\partial Y} n_i\ a_{ik}\ v\ dS - \int_Y a_{ik}\ \frac{\partial v}{\partial y_i}\ dy$$

and the integral over $\partial Y$ vanishes by periodicity, and we have, $\forall\, v \in V_Y$ :

$$0 = \int_Y a_{ij}\left(\frac{\partial w^k}{\partial y_j} + \delta_{jk}\right)\frac{\partial v}{\partial y_i}\ dy = \int_Y a_{mj}\ \frac{\partial (w^k + y^k)}{\partial y_j}\ \frac{\partial\, v}{\partial y_m}\ dy$$

and by taking $v = w^i$ and adding and substracting the same quantity, we have

$$\int_Y a_{mj}\frac{\partial (w^k + y^k)}{\partial y_i}\ \frac{\partial (w^i + y^i)}{\partial y_m}\ dy = \int_Y a_{mj}\ \frac{\partial (w^k + y^k)}{\partial y_j}\delta_{im}\ dy =$$

$$= \int_Y a_{ij}\left(\frac{\partial w^k}{\partial y_j} + \delta_{kj}\right) dy$$

and by comparing this with (3.4) we have :

(3.6) $$a^h_{ik} = \frac{1}{|Y|}\int_Y a_{mj}(y)\ \frac{\partial (w^k(y) + y^k)}{\partial y_j}\ \frac{\partial (w^i(y) + y^i)}{\partial y_m}\ dy.$$

It is then evident by virtue of (1.3) that

(3.7) $$a^h_{ik} = a^h_{ki}$$

Now we derive another expression for the homogenized coefficients which will be generalized to some nonlinear problems in the next chapter. If x is a parameter we may define the following bilinear form of $\underline{\text{grad}}_x u^o$ :

(3.8) $$W\left(\frac{\partial u^o}{\partial x_i}\right) \equiv \frac{1}{2}\, a^h_{ik}\ \frac{\partial u^o}{\partial x_i}\ \frac{\partial u^o}{\partial x_k} = \text{(by (3.6) and (2.10))} =$$

$$= \frac{1}{2\ |Y|}\int_Y a_{mj}(y)\left(\frac{\partial u^1}{\partial y_j} + \frac{\partial u^o}{\partial x_j}\right)\left(\frac{\partial u^1}{\partial y_m} + \frac{\partial u^o}{\partial x_m}\right) dy$$

On the other hand, (3.3) becomes

(3.9) $$\tilde{p}^o_i(x) = \frac{\partial W}{\partial\left(\frac{\partial u^o}{\partial x_i}\right)}$$

We then have a relation between $\underline{\text{grad}}\ u^o$ and $\tilde{\underline{p}}^o$ given by (3.8), (3.9) ; it contains the homogenized coefficients only in an implicit form, and it may be generalized to nonlinear problems.

Remark 3.1 - It is clear that x is only a parameter in (3.8). The function W is a function of the N components $q_i$ of the vector $\underline{\text{grad}}_x u^0$ :

$$q_i \equiv \frac{\partial u^0}{\partial x_i} \qquad \blacksquare$$

Moreover, the coefficients $a^h_{ik}$ satisfy an ellipticity condition analogous to (1.4).

(3.10) $$a^h_{ik}\, \xi_i\, \xi_k \geqslant \delta\, \xi_i\, \xi_i \quad , \quad \delta > 0 \quad , \quad \forall \underline{\xi} \in R^N$$

(in fact relation (3.10) holds with $\delta = \gamma$ , the same constant as in (1.4), but this will not be proved here (see Bensoussan, Lions, Papanicolaou [2], chap. 1, sect. 3.4). To obtain (3.10) it suffices to prove that its left hand side is positive for any $\xi \neq 0$ ; this amounts to saying that $\underline{\text{grad}}\, u^0 \neq 0 \Rightarrow W \neq 0$, and this is true, because, by (1.4)

$$W = 0 \Rightarrow 0 = \int_Y \sum_s \left(\frac{\partial u^1}{\partial y_s} + \frac{\partial u^0}{\partial x_s}\right)^2 dy$$

i.e., for s = 1 ... N we have

$$\frac{\partial u^1}{\partial y_s} + \frac{\partial u^0}{\partial x_s} = 0 \Rightarrow 0 = \int_Y \frac{\partial u^1}{\partial y_s} dy = - \frac{\partial u^0}{\partial x_s} |Y| \Rightarrow \underline{\text{grad}}_x u^0 = 0 \quad .$$

Finally to obtain a well posed problem for $u^0$, we only need a boundary condition for $u^0$. From (1.7) and (1.10) we obtain

(3.11) $$u^0(x)\Big|_{\partial\Omega} = 0 \qquad \text{on } \partial\Omega$$

Note that this relation is formal, (see Remark 1.3) but it will be rigorously proved in next section.

We summarize the results of the preceeding considerations :

Proposition 3.2 - If we postulate an expansion of the form (1.10), the first term $u^0(x)$ is determined as a solution of the elliptic equation (3.5) with the boundary condition (3.11). The coefficients $a^h_{ik}$ are given by any of the formulae (3.4), (3.6) or (3.8) and (3.9). They satisfy the symmetry and positivity relations (3.7) and (3.10).

The only formal point in the preceeding considerations is the form of the expansion (1.10). In fact, it is possible to prove that $u^\varepsilon$ converges to the function $u^0$ given by proposition 3.2. An elementary proof, based on the maximum principle is given in Bensoussan–Lions–Papanicolaou [2] , chap. 1, sect 2.4. Unfortunately such a proof does not hold for other problems (such as Neumann conditions, elliptic systems ...). In the next section we shall give a proof due to Tartar which holds with minor modifications for many other problems.

4.- Proof of the convergence - This section is devoted to the proof of the following theorem of De Giorgi and Spagnolo (Tartar's proof) :

Theorem 4.1 - Under the hypothesis of the preceeding sections, if $u^\varepsilon$ (resp. $u^0$) is the solution of (1.6), (1.7) (resp. (3.5), (3.11)), we have

$$u^\varepsilon \to u^0 \quad \text{in } H_0^1(\Omega) \text{ weakly} \tag{4.1}$$

We begin by a lemma which helps us to understand the sense of convergence in $L^2(\Omega)$ weakly.

Lemma 4.1 - Let $\phi \in L^2(Y)$. If we extend it periodically to $R^N$, we have

$$\phi(\tfrac{x}{\varepsilon}) \to \tilde{\phi} \qquad \text{in } L^2(\Omega) \text{ weakly} \tag{4.2}$$

[ (4.2) means that $\phi(x/\varepsilon)$ tends to the function defined on $\Omega$ which is equal to the constant $\tilde{\phi}$ given by (1.20)]

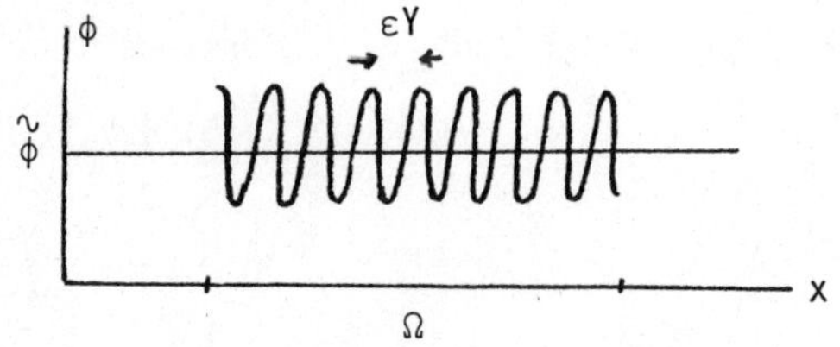


Proof - It is clear that $\phi(x/\varepsilon)$ is bounded in the $L^2(\Omega)$ norm as $\varepsilon \to 0$. Because $\mathcal{D}(\Omega)$ is dense in $L^2(\Omega)$ it suffices to prove (see chap. 2, prop. 1.2) that

$$\int_\Omega \phi(x/\varepsilon)\,\theta(x)\,dx \longrightarrow \tilde{\phi}\int_\Omega \theta(x)\,dx \qquad \forall\,\theta \in \mathcal{D}(\Omega) \tag{4.3}$$

But this is immediate. Let $\theta_\varepsilon^*$ be the function which takes in each $\varepsilon Y$-period a constant value equal to $\theta$ at the center of the period. Because $\theta$ is smooth, we clearly have

$$\int_\Omega \phi(\tfrac{x}{\varepsilon})(\theta - \theta_\varepsilon^*)\,dx \underset{\varepsilon\to 0}{\longrightarrow} 0$$

$$\int_\Omega \phi(\tfrac{x}{\varepsilon})\,\theta_\varepsilon^*(x)\,dx = \tilde{\phi}\int_\Omega \theta_\varepsilon^*(x)\,dx \underset{\varepsilon\to 0}{\longrightarrow} \tilde{\phi}\int_\Omega \theta(x)\,dx \quad . \blacksquare$$

Now, we prove theorem 4.1. The function $u^\varepsilon$ (resp. $u^0$) is the element of $H_0^1(\Omega)$ which satisfy (1.8) (resp. an analogous relation with $a_{ij}^h$ , $u^0$ instead of

$a^{\varepsilon}_{ij}$ , $u^{\varepsilon}$).

By taking $v = u^{\varepsilon}$ in (1.8) and taking into account (1. 4) we have

$$\gamma \int_{\Omega} \frac{\partial u^{\varepsilon}}{\partial x_i} \frac{\partial \bar{u}^{\varepsilon}}{\partial x_i} dx \leq \int_{\Omega} a^{\varepsilon}_{ij} \frac{\partial u^{\varepsilon}}{\partial x_i} \frac{\partial u^{\varepsilon}}{\partial x_j} dx \leq C' \|u^{\varepsilon}\|_{L^2} \leq C'' \|u^{\varepsilon}\|_{H^1_0}$$

But the left hand side may be taken as the norm$^2$ in $H^1_0$ (see chap. 3, prop. 1.1) and thus

(4.4) $$\|u^{\varepsilon}\|_{H^1_0} \leq C$$

where C is some constant independent of $\varepsilon$ . This means that $u^{\varepsilon}$ remains in a bounded set of $H^1_0$, i.e., in a precompact set for the weak topology of $H^1_0$ (see chap. 2, prop. 1.6). Consequently, we can extract a sequence $\varepsilon \to 0$ such that

(4.5) $$u^{\varepsilon} \to u^* \qquad (u^* \in H^1_0) \quad \text{in } H^1_0(\Omega) \quad \text{weakly}$$

the theorem will be proved if we show that for any sequence as (4.5), we have $u^* = u^0$. From (4.5), the partial derivatives of $u^{\varepsilon}$ are bounded in $L^2(\Omega)$ ; by multiplying them by $a_{ij}(x/\varepsilon)$, which are smooth bounded functions, we see that

(4.6) $$\|p^{\varepsilon}_i(x)\|_{L^2} \equiv \|a^{\varepsilon}_{ij}(x) \frac{\partial u^{\varepsilon}(x)}{\partial x_j}\|_{L^2} \leq C$$

and by extracting a sequence $\varepsilon \to 0$ from the preceeding one, we have

(4.7) $$p^{\varepsilon}_i \to p^*_i \qquad (p^*_i \in L^2(\Omega)) \quad \text{in } L^2(\Omega) \text{ weakly} \quad .$$

Now, by writing (1.8) in the form

$$\int_{\Omega} p^{\varepsilon}_i \frac{\partial v}{\partial x_i} dx = \int_{\Omega} f\ v\ dx \qquad \forall\ v \in H^1_0$$

we can pass to the limit for any fixed $v \in H^1_0$ and get

(4.8) $$\int_{\Omega} p^*_i \frac{\partial v}{\partial x_i} dv = \int_{\Omega} f\ v\ dx \qquad \forall\ v \in H^1_0$$

Let us suppose that

(4.9) $$p^*_i(x) = a^h_{ij} \frac{\partial u^*}{\partial x_j}(x) \qquad \text{in } \Omega$$

Then, (4.8) shows that $u^* \in H^1_0$ satisfies the variational formulation of the problem $u^0$ ; by uniqueness, $u^* = u^0$. Consequently, we only have to prove (4.9). To show this, we shall take test functions of a special suitable form.

If $w^k(y)$ is the function defined by (2.9), we write

(4.10) $$w_{\varepsilon}(x) \equiv x_k + \varepsilon\, w^k(x/\varepsilon)$$

(note that this function is in fact the sum of the two first terms of the expansion, $u^0(x) + \varepsilon u^1(x/\varepsilon)$ for $u^0 = x_k$). It is clear that

(4.11) $$w_\varepsilon(x) \to x_k \quad \text{in } L^2(\Omega) \text{ strongly.}$$

Moreover, $w_\varepsilon$ satisfies the equation

(4.12) $$-\frac{\partial}{\partial x_j}\left(a_{ij}\left(\frac{x}{\varepsilon}\right) \frac{\partial w_\varepsilon(x)}{\partial x_i}\right) = 0 \qquad \text{in } R^N$$

for, (2.9) gives, in the sense of distributions :

$$-\frac{\partial}{\partial y_j}\left(a_{ij} \frac{\partial w^k}{\partial y_i}\right) = \frac{\partial a_{ik}}{\partial y_i}$$

which is equivalent to (4.12). Then, by multiplying (4.12) by any $v \in H^1_o$ we have :

(4.13) $$\int_\Omega a_{ij}\left(\frac{x}{\varepsilon}\right) \frac{\partial w_\varepsilon}{\partial x_j} \frac{\partial v}{\partial x_i}\, dx = 0 \qquad \forall v \in H^1_o(\Omega)$$

Now, to avoid difficulties with the boundary condition, we take a function $\phi \in \mathcal{D}(\Omega)$ and we write (1.8) with $v = \phi\, w_\varepsilon$ and (4.13) with $v = \phi\, u^\varepsilon$ . By substracting and taking into account that $a_{ij} = a_{ji}$ we have

(4.14) $$\int_\Omega a_{ij}\left(\frac{x}{\varepsilon}\right)\left[\frac{\partial u^\varepsilon}{\partial x_j} \frac{\partial \phi}{\partial x_i} w_\varepsilon - \frac{\partial w_\varepsilon}{\partial x_i} \frac{\partial \phi}{\partial x_j} u^\varepsilon\right] dx = \int_\Omega f\, \phi\, w_\varepsilon\, dx$$

Now we can pass to the limit $\varepsilon \to 0$ in (4.14) because each term is the scalar product in $L^2$ of an element which converges weakly and another which converges strongly (chap. 2 , prop. 1.3). Indeed :

$p_i^\varepsilon(x) \equiv a_{ij}\left(\frac{x}{\varepsilon}\right) \frac{\partial u^\varepsilon}{\partial x_j}$ converges in $L^2(\Omega)$ weakly by (4.7).

$\frac{\partial \phi}{\partial x_i} w_\varepsilon$ converges to $\frac{\partial \phi}{\partial x_i} x_k$ in $L^2$ strongly by (4.11) (note that $\phi$ is smooth and fixed).

$a_{ij}\left(\frac{x}{\varepsilon}\right) \frac{\partial w_\varepsilon}{\partial x_i}$ is $\varepsilon Y$-periodic and tends in $L^2$ weakly to its mean value :

$$\left[a_{ij}(y) \left(\delta_{ik} + \frac{\partial w^k(y)}{\partial y_i}\right)\right]^\sim = \text{by (3.4)} = a^h_{jk} \quad .$$

Finally $\frac{\partial \phi}{\partial x_j} u^\varepsilon$ converges to $\frac{\partial \phi}{\partial x_j} u^*$ in $L^2$ strongly by (4.5) and the Rellich theorem.

We obtain

(4.15) $$\int_\Omega (p^*_j\, x_k - a^h_{jk}\, u^*) \frac{\partial \phi}{\partial x_j}\, dx = \int_\Omega f\, \phi\, x_k\, dx$$

Moreover, by (4.8) with $v = \phi x_k$, the right hand side of (4.15) is

$$(4.16) \qquad = \int_\Omega p_j^* \frac{\partial(\phi x_k)}{\partial x_j} dx$$

The relation (4.15) (with the right hand side (4.16)) holds for any $\phi \in \mathcal{D}(\Omega)$ ; this means that, in the sense of distributions on $\Omega$, we have :

$$-\frac{\partial}{\partial x_j}(p_j^* x_k - a_{jk}^h u^*) = -\frac{\partial p_j^*}{\partial x_j} x_k \Leftrightarrow p_k^* = a_{jk}^h \frac{\partial u^*}{\partial x_j}$$

which is the desired relation (4.9). ∎

## 5.- Generalization to other elliptic problems and convergence of the resolvents -

The considerations of the sections 1 to 4 apply with minor modifications to a great variety of problems.

Let us consider, instead of (1.6), (1.7) a Neumann problem of the type :

$$(5.1) \qquad -\frac{\partial}{\partial x_i}\left(a_{ij}^\varepsilon(x) \frac{\partial u^\varepsilon}{\partial x_j}\right) + a_o^\varepsilon(x)\, u^\varepsilon = f^\varepsilon(x) \quad \text{in } \Omega$$

$$(5.2) \qquad a_{ij}^\varepsilon(x) \frac{\partial u^\varepsilon}{\partial x_j} n_i = 0 \quad \text{on } \partial\Omega$$

where n is the outer unit normal to $\partial\Omega$ and $a_o^\varepsilon(x) = a_o(\frac{x}{\varepsilon})$ and $a_o(y)$ is a Y-periodic smooth function with

$$(5.3) \qquad a_o(y) \geq \gamma \qquad y \in R^N$$

and $f^\varepsilon(x)$ is a sequence of functions such that

$$(5.4) \qquad f^\varepsilon(x) \to f^* \qquad \text{in } L^2(\Omega) \text{ weakly.}$$

Then we have :

<u>Theorem 5.1</u> - $u^\varepsilon \to u^o$ in $H^1(\Omega)$ weakly where $u^o$ is the (unique) solution of

$$(5.5) \qquad -\frac{\partial}{\partial x_i}\left(a_{ij}^h \frac{\partial u^o}{\partial x_j}\right) + \tilde{a}_o\, u^o = f^* \qquad \text{in } \Omega$$

$$(5.6) \qquad a_{ij}^h \frac{\partial u^o}{\partial x_j} n_i = 0 \qquad \text{on } \partial\Omega \quad .$$

<u>Remark 5.1</u> - Note that the homogenized coefficients $a_{ij}^h$ are given by the same formulas as in the Dirichlet problem, (3.4). The new term $a_o^\varepsilon$ gives in the homogenized equation its mean value : $a_o^h \equiv \tilde{a}_o$ (This fact also holds for Dirichlet conditions !). On the other hand, the Neumann condition (5.6) is associated with the homogenized coefficients ; while (5.2) is associated with the given coefficients (which depend on $\varepsilon$). ∎

The proof of Th. 5.1 is almost the same as that of Th. 4.1. The variational formulation of (5.1), (5.2) is : (as in chap. 3, (2.1), (2.2)) : Find $u^\varepsilon \in H^1(\Omega)$ such that

$$(5.7)\qquad \int_\Omega (a^\varepsilon_{ij} \frac{\partial u^\varepsilon}{\partial x_i} \frac{\partial v}{\partial x_j} + a^\varepsilon_o \; u^\varepsilon \; v)dx = \int_\Omega f^\varepsilon v \; dx \qquad \forall v \in H^1(\Omega)$$

The convergence in (4.5) now holds in $H^1$ weakly and by the Rellich theorem, in $L^2(\Omega)$ strongly. Moreover, by lemma 4.1 :

$$a^\varepsilon_o \to \tilde{a} \qquad \text{in } L^2(\Omega) \text{ weakly}$$

and then, we obtain instead of (4.8) :

$$(5.8)\qquad \int_\Omega (p^*_i \frac{\partial v}{\partial x_i} + \tilde{a}_o \; u^* v) \; dx = \int_\Omega f^* v \; dx \qquad \forall v \in H^1$$

and this shows that $p^*$ and $u^*$ satisfy a certain equation and <u>an associated boundary condition as in the classical Neumann problems</u>(chap. 3, sect. 2). Indeed, by integrating (5.8) by parts we have

$$\int_\Omega \left(- \frac{\partial p^*_i}{\partial x_i} + \tilde{a}_o \; u^* - f^*\right) v \; dx + \int_{\partial\Omega} (n_i \; p^*_i) v \; dx = 0 \quad \forall v \in H^1$$

which implies that the two functions in parenthesis are zero. By proving (4.9) we have (5.5) and (5.6), as desired. The proof of (4.9) is of course the same as in th. 4.1 because it is a local property independent of the boundary conditions.

It is also possible to consider transmission problems with coefficients $a_{ij}(y)$ which are not smooth. To fix ideas, we take $a_{ij}$ to be piecewise constant, (or piecewise smooth). We may consider the period Y divided in two regions $Y^1$, $Y^2$ separated by the smooth surface $\Gamma$ , and $a_{ij}(y)$ takes the constant values $a^1_{ij}$ and

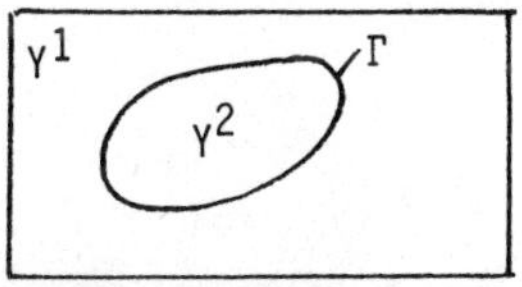


fig. 5

$a^2_{ij}$ in $Y^1$ and $Y^2$ respectively. Then, equation (1.6) (or (5.1), because the term $a_o$ is irrelevant) must be interpretated in the sense of distributions ;

$$(5.9)\qquad [\,u^\varepsilon\,] = 0 \qquad \text{on } \Gamma$$

$$(5.10)\qquad [\,a_{ij}(y) \frac{\partial u^\varepsilon}{\partial x_i} n_j] = 0 \qquad \text{on } \Gamma$$

where the brakets mean, as usual, the discontinuity across $\Gamma$ . In the usual weak sense, the variational formulation of the problem for $u^\varepsilon$ is (1.8) (or (5.7)) ;

this problem is analogous to that of chap. 3, sect. 3 ; condition (5.9) is contained in $u^\varepsilon \in H_0^1(\Omega)$, and (5.10) is the natural condition associated with the variational formulation.

Theorems 4.1 or 5.1 holds in the present case (the proofs are exactely the same but the formal asymptotic expansion and the expressions for the homogenized coefficients are slightly modified, in the following way :

In the formal expansion, we note that $a_{ij}(y)$ is piecewise constant, but we must include the transmission conditions on $\Gamma$ .

Relation (1.18) becomes (5.11), (5.12) :

$$\frac{\partial}{\partial y_i}\left(a_{ij}(y)\,\frac{\partial u^1(x\,,\,y)}{\partial y_j}\right) = 0 \qquad \text{in } Y^1 \text{ and } Y^2 \tag{5.11}$$

$$\left[a_{ij}(y)\left(\frac{\partial u^o}{\partial x_j}+\frac{\partial u^1}{\partial y_j}\right)n_j\right] = 0 \qquad \text{on } \Gamma \tag{5.12}$$

where (5.12) is the first term of the expansion of (5.10) with (1.10). To find $u^1$ we also must include the Y-periodicity condition.

It is easily seen that (5.11) and (5.12) may be written as

$$\frac{\partial}{\partial y_i}\left(a_{ij}(y)\left(\frac{\partial u^o}{\partial x_j}+\frac{\partial u^1}{\partial y_j}\right)\right) = 0 \qquad \text{on } Y$$

in the sense of distributions. Then, the variational formulation of the local problem is (instead of (2.3)) :

(5.13)

$$\left|\begin{array}{l}\text{Find } u^1 \in V_Y \text{ such that} \\ \displaystyle\int_Y a_{ij}(y)\left(\frac{\partial u^o}{\partial x_j}+\frac{\partial u^1}{\partial y_j}\right)\frac{\partial v}{\partial y_i}\,dy = 0 \qquad \forall\, v \in V_Y\end{array}\right.$$

with an analogous modification for (2.9). Note that (2.3) in sect. 2 may also be written in the form (5.13) because, by integrating by parts the right hand side of (2.3) we have, by virtue of the periodicity :

$$\int_Y \frac{\partial a_{ij}}{\partial y_i}\, v\, dy = -\int_Y a_{ij}\,\frac{\partial v}{\partial y_i}\, dy$$

Consequently, <u>(5.13) is more general than (2.3), because (5.13) holds for continuous as well as discontinuous coefficients.</u>

Of course, it is also possible to consider the case where $a_{ij}$ is dicontinuous on $\Gamma$ and smoothly variable in $Y^1$ and $Y^2$. Formula (5.13) also holds in this case.

It is also possible to consider elliptic equations with complex coefficients. In this case, we consider (to use the preceeding calculations) only the symmetric case (1.3). It is clear that the operator (1.6) is no longer selfadjoint, and in fact, the homogenized operator is not symmetric in general.

To fix ideas, we consider the problem of sections 1 to 4 with the following modifications. Instead of (1.4), the ellipticity condition is ($^-$ is for the complex conjugate)

(5.14) $$\left| a_{ij}(y)\, \xi_i\, \bar{\xi}_j \right| \geqslant \gamma\, |\xi|^2 \qquad \forall \xi \quad C^N$$

and (1.8) becomes

(5.15) $$\int_\Omega a_{ij}(y) \frac{\partial u^\varepsilon}{\partial x_i} \frac{\partial \bar{v}}{\partial x_j}\, dx = \int_\Omega f\, \bar{v}\, dx \qquad \forall v \in H_0^1(\Omega)$$

Existence and uniqueness follow from the Lax Milgram theorem (see chap. 2, formula (5.5))

In (2.3) and (2.9), we must write $\bar{v}$ instead of v. The expression (3.4) for the homogenized coefficients holds in this case. Moreover, instead of (3.6), we obtain (we take $v = \bar{w}^i$ instead of $v = w^i$ in the proof) :

$$a^h_{ik} = \frac{1}{|Y|} \int_Y a_{mj}(y) \frac{\partial(w^k + y^k)}{\partial y_j} \frac{\partial(\bar{w}^i + y^i)}{\partial y_m}\, dy$$

which is not in general symmetric. But we have an ellipticity property of the homogenized problem

(5.16) $$\left| a^h_{ij}\, \xi_i\, \bar{\xi}_j \right| \geqslant \gamma\, |\xi|^2 \qquad \forall \xi \in C^N \ .$$

This can be proved as in sect. 3, because (5.16) amounts to proving that grad $u^0 \neq 0$ implies that the right hand side of (3.8) (with the complex conjugates in the second factors) is not zero.

Moreover, theorem 4.1 holds in the case of complex coefficients. The proof is the same as in sect. 4, but in (4.8) and (4.13), we have $\bar{v}$ instead of v. Moreover, to obtain (4.14) we must write $v = \bar{\phi}\, \bar{w}_\varepsilon$ in (1.8) and $v = \bar{\phi}\, \bar{u}^\varepsilon$ in (1.13) ; this gives (4.14) without modification, and the proof follows.

To finish this section, we give a result about the convergence of the resolvents at the point zero, which is useful to obtain, in a simple way, results about the convergence of eigenvectors and eigenvalues (without hypothesis of sel-adjointness, in particular in the framework of complex coefficients) (see chap.11, sect 3).

Let $A_\varepsilon$ and $A_h$ be the elliptic operators in (1.6), and (3.5), with the Dirichlet boundary conditions, or the operators in (5.1), (5.5) with the Neumann boundary conditions. According to the first representation theorem, we consider $A_\varepsilon$, $A_h$ as unbounded maximal accretive operators in $L^2(\Omega)$ ($\Omega$ is bounded). Then, we have the following theorem which applies to spectral properties of the homogenization process (see chap. 11, sect. 3) :

Theorem 5.2 - Under the preceeding conditions

(5.17) $$\| A_\varepsilon^{-1} - A_h^{-1} \|_{\mathcal{L}(L^2,L^2)} \to 0 \qquad \text{as} \quad \varepsilon \to 0$$

The proof is exactly analogous to that of chapter 9, formula (6.7), and will not be given here. It is based on the consideration of $A_\varepsilon$, $A_h$ as bounded operators from V into V' where

$$V \subset H \subset V'$$

with dense and compact embedding. Here, we shall take $H = L^2(\Omega)$, $V = H_0^1(\Omega)$ for the Dirichlet problem and $V = H^1(\Omega)$ for the Neumann problem.

6.- Homogenization of evolution equations - Now we consider parabolic and hyperbolic equations analogous to (1.6), (1.7) and the generalizations of sect. 5. To fix ideas, we consider real smooth and symmetric coefficients $a_{ij}(y)$ as in sect. 1. Moreover, let $\rho(y)$ be a real smooth Y-periodic function with

(6.1) $$\rho(y) \geq \gamma \qquad \gamma > 0 \qquad \forall\, y \in Y$$

(It is of course possible to take the same $\gamma$ as in (1.4). We also consider the corresponding homogenized coefficients $a_{ij}^h$ of (3.4). Let $A_\varepsilon$ , $A_h$ be the operators

(6.2) $$A_\varepsilon = - \frac{\partial}{\partial x_i} \left(a_{ij}\left(\frac{x}{\varepsilon}\right) \frac{\partial}{\partial x_j}\right)$$
$$A_h = - \frac{\partial}{\partial x_i}\left(a_{ij}^h \frac{\partial}{\partial x_j}\right)$$

with Dirichlet boundary condition (1.7) on the boundary $\partial\Omega$ of a bounded domain $\Omega$ of $R^N$.

Definition 6.1 - Let $u^\varepsilon$ be the solution of the parabolic problem

(6.3) $$\rho\left(\frac{x}{\varepsilon}\right) \frac{\partial u^\varepsilon}{\partial t} + A_\varepsilon u^\varepsilon = 0 \qquad ; \qquad (u^\varepsilon|_{\partial\Omega} = 0)$$

(6.4) $$u^\varepsilon(0) = u_0 \qquad L^2(\Omega)$$

for $x \in \Omega$ , $t \in [\,0,\infty[$ . Let also $u^h$ be the solution of

(6.5) $$\tilde{\rho} \frac{\partial u^h}{\partial t} + A_h u^h = 0 \qquad (u^h|_{\partial\Omega} = 0)$$

(6.6) $$u^h(0) = u_0$$

for $x \in \Omega$ , $t \in [0,\infty[$ .

Remark 6.1 - (6.4) and (6.6) are initial conditions. Equations (6.3) and (6.5) are written for functions with values in $L^2(\Omega)$ ; then the boundary conditions on $\partial\Omega$ are included in $A_\varepsilon$ and $A_h$ . This is the reason why these boundary conditions are written in parentheses. On the other hand, the initial value $u_0$ is the same in (6.4) and (6.6). The right hand sides of (6.3) and (6.5) are zero, but it is also possible to consider non homogeneous problems (see remark 6.3 later). Note also that the homogenized form of $\rho(x/\varepsilon)$ is its mean value $\tilde{\rho}$. ∎

Theorem 6.1 - If $\rho(y)$ is independent of y, we have

(6.7) $$u^\varepsilon(t) \to u^h(t) \qquad \text{in } L^2(\Omega) \text{ strongly}$$

for any fixed t. (see def. 6.1).

Proof - This is an immediate consequence of the Trotter-Kato theorem (see chap. 4 Th. 3.3 and chap. 10). Indeed, if $\rho$ is a constant we may take it equal to 1. We consider the semigroups associated with (6.3) and (6.5) in the framework of chapter 4, sect. 4 and by the Trotter-Kato theorem, (6.7) is equivalent to

(6.8) $$(I + A_\varepsilon)^{-1} v \to (I + A_\varepsilon)^{-1} v \quad \text{in } L^2(\Omega) \text{ strongly}$$

for any test function $v \in L^2(\Omega)$. But this amounts to proving that if $w^\varepsilon$, $w^h$ are defined by

(6.9) $$w^\varepsilon - \frac{\partial}{\partial x_i}\left(a_{ij}\left(\frac{x}{\varepsilon}\right) \frac{\partial w^\varepsilon}{\partial x_j}\right) = v \qquad ; \qquad w^\varepsilon\big|_{\partial\Omega} = 0$$

and an analogous equation with $w^h$ and $a^h_{ij}$, we have

(6.10) $$w^\varepsilon \to w^h \text{ in } L^2(\Omega) \text{ strongly.}$$

But from theorem 4.1 (or rather theorem 5.1 with Dirichlet boundary condition, due to the term $\rho$ ) $w^\varepsilon$ converges to $w^h$ in $H^1_0$ weakly and by the Rellich theorem we have (6.10).

Let us now consider the case where $\rho$ is variable.

Remark 6.2 - Let us consider (6.3), (6.4) with fixed $\varepsilon$ . It is possible to obtain the existence and uniqueness of $u^\varepsilon$ from semigroup theory. If we take the $L^2$-scalar product of (6.3) with $v \in H^1_0$, we see that (6.3) is equivalent to

(6.11) $$b^\varepsilon\left(\frac{\partial u^\varepsilon}{\partial t}, v\right) + a^\varepsilon(u^\varepsilon, v) = 0 \qquad \forall\, v \in H^1_0$$

where

(6.12) $$b^{\varepsilon}(u\ ,\ v) \equiv \int_{\Omega} \rho(\frac{x}{\varepsilon})\ u\ v\ dx$$

(6.13) $$a^{\varepsilon}(u\ ,\ v) \equiv \int_{\Omega} a_{ij}(\frac{x}{\varepsilon})\ \frac{\partial u}{\partial x_i}\ \frac{\partial v}{\partial x_j}\ dx$$

Now, from (6.1), we see that $b^{\varepsilon}$ is a scalar product on $L^2$ with norm equivalent to the standard one. Let $H^{\varepsilon}$ be the space $L^2$ equipped with the scalar product $b^{\varepsilon}$ . On the other hand, $a^{\varepsilon}$ is a symmetric bounded and coercive form on $H_o^1$, which is a Hilbert space densely embedded into $H^{\varepsilon}$ as well as in $L^2$. Then, (6.11) is equivalent to

(6.14) $$\frac{\partial u^{\varepsilon}}{\partial t} + A_{\varepsilon}^{*}\ u^{\varepsilon} = 0$$

where $A_{\varepsilon}^{*}$ is the operator associated with the form $a^{\varepsilon}$ in the framework of the first representation theorem <u>in $H^{\varepsilon}$</u> (in the same way as $A_{\varepsilon}$ is associated with $a^{\varepsilon}$ <u>in $L^2$</u>). The operator $A_{\varepsilon}^{*}$ is not to be confused with $\rho^{-1}A_{\varepsilon}$. Semigroup theory then applies to (6.14) (see chap. 4, sect. 4). It is then immediate to come back to (6.3).

<u>Theorem 6.2</u> - In the framework of def. 6.1, we have

$$u^{\varepsilon} \to u^h \qquad \text{as } \varepsilon \searrow 0$$

in $L^{\infty}(0\ ,\ \infty\ ;\ L^2(\Omega))$ weakly $*$ and in $L^2(0\ ,\infty\ ;\ H_o^1(\Omega))$ weakly.

<u>Proof</u> - The classical a priori estimate (see remark 6.2 and chap. 4, (4.2)) give

$$\begin{cases} \|u^{\varepsilon}(t)\|_{H^{\varepsilon}} \leqslant \|u_o\|_{H^{\varepsilon}} \leqslant C & \forall\ t \geqslant 0 \\ \int_0^{\infty} \|u^{\varepsilon}(t)\|^2_{H_o^1}\ dt \leqslant \|u_o\|_{H^{\varepsilon}} \leqslant C & \end{cases}$$

and by extracting subsequences, we have

(6.15) $$u^{\varepsilon} \xrightarrow[\varepsilon \to 0]{} u^{*} \quad \begin{cases} \text{in } L^{\infty}(0\ ,\infty\ ;\ L^2) \text{ weakly } * \\ \text{in } L^2(0\ ,\infty\ ;\ H_o^1) \text{ weakly} \end{cases}$$

The theorem will be proved if we show that $u^{*} = u^h$ for any subsequence.

By taking the Laplace transform of (6.15),

(6.16) $$\hat{u}^{\varepsilon} \to \hat{u}^{*} \quad \begin{cases} \text{in } L^2 \text{ weakly} \\ \text{in } H_o^1 \text{ weakly} \end{cases}$$

for any p with Rep $>$ 0. (This is obtained by taking $e^{-pt}v$ as test function in (6.15)).

On the other hand, $\hat{u}^\varepsilon$ can be written as (see remark 6.2 and chap. 4, prop. 6.3) :

$$\hat{u}^\varepsilon = (p + A_\varepsilon)^{-1} u_0 \quad \text{in} \quad H^\varepsilon, \text{ for } \operatorname{Re} p > 0 .$$

and this amounts to saying that

$$p\rho(\tfrac{x}{\varepsilon})\, \hat{u}^\varepsilon + A_\varepsilon \hat{u}^\varepsilon = u_0 \quad \text{in } L^2 , \ \operatorname{Re} p > 0 \tag{6.17}$$

and in the same way

$$p\, \tilde{\rho}\, \hat{u}^h + A_h \hat{u}^h = u_0 \quad \text{in } L^2 , \operatorname{Re} p > 0 . \tag{6.18}$$

Now we consider (6.17) and (6.18) with real p. By theorem 5.1 (with Dirichlet boundary condition !), $\hat{u}^\varepsilon$ converges to $\hat{u}^h$ in $H^1_o$ weakly, and by (6.16), $\hat{u}^h = \hat{u}^*$ for real p. Moreover, the Laplace transforms are holomorphic functions in the half plane $\operatorname{Re} p > 0$ ; thus $\hat{u}^h = \hat{u}^*$ for any p, and this implies $u^h = u^*$ .■

Now we study hyperbolic problems.

Definition 6.2 - $u^\varepsilon$ is the solution of the hyperbolic problem

$$\rho(\tfrac{x}{\varepsilon}) \frac{\partial^2 u^\varepsilon}{\partial t^2} + A_\varepsilon u^\varepsilon = 0 \qquad (u^\varepsilon|_{\partial\Omega} = 0) \tag{6.19}$$

$$u^\varepsilon(0) = u_0 \in H^1_o(\Omega) \quad ; \quad u^{\varepsilon\prime}(0) = u_1 \in L^2(\Omega) \tag{6.20}$$

for $x \in \Omega$ , $t \in [0 , \infty[$ . Let also $u^h$ be the solution of

$$\tilde{\rho} \frac{\partial^2 u^h}{\partial t^2} + A_h u^h = 0 \tag{6.21}$$

$$u^h(0) = u_0 \quad ; \quad u^{h\prime}(0) = u_1 \tag{6.22}$$

for $x \in \Omega$ , $t \in [0,\infty[$ .

Theorem 6.3 - In the framework of def. 6.2, we have

$$u^\varepsilon \rightarrow u^h \qquad \text{in } L^\infty(0 , \infty ; H^1_o) \text{ weakly } *$$

$$u^{\varepsilon\prime} \rightarrow u^{h\prime} \qquad \text{in } L^\infty(0 , \infty ; L^2) \text{ weakly } * .$$

Proof - The proof is analogous to that of th. 6.2. By remark 6.2, (6.19) is equivalent to (we define $V^\varepsilon$ as $H^1_o$ with the scalar product $(A^\varepsilon u , v)$)

$$\frac{\partial^2 u^\varepsilon}{\partial t^2} + A^*_\varepsilon u = 0 \qquad \text{in } H^\varepsilon$$

and for fixed $\varepsilon$, the problem is in the framework of chap. 4, remark 5.1 in $V^\varepsilon \times H^\varepsilon$. In this space, $u^\varepsilon$, $u^{\varepsilon\prime}$ are associated with a unitary group and thus

(6.23) $$\| u^\varepsilon \|^2_{V^\varepsilon} + \| u^{\varepsilon\prime} \|^2_{H^\varepsilon} = \| u_0 \|^2_{V^\varepsilon} + \| u_1 \|^2_{H^\varepsilon} \leqslant C$$

for any t. By extraction of subsequences,

$$u^\varepsilon \to u^* \quad \text{in } L^\infty(0\ ,\ \infty\ ;\ H^1_0) \text{ weakly } *$$

$$u^{\varepsilon\prime} \to u^{*\prime} \quad \text{in } L^\infty(0\ ,\ \infty,\ \ L^2) \text{ weakly } *$$

and by taking the Laplace transform :

(6.24) $$\hat{u}^\varepsilon \to \hat{u}^* \quad \text{in} \quad H^1_0 \text{ weakly} \quad .$$

By taking the Laplace transform of the semigroup we have

$$\left.\begin{array}{l} p\ \hat{u}^\varepsilon - (u^{\varepsilon\prime})^\wedge = u_0 \\ p(u^{\varepsilon\prime})^\wedge + A^*_\varepsilon\ \hat{u}^\varepsilon = u_1 \end{array}\right\} \Rightarrow p^2\ \hat{u}^\varepsilon + A^*_\varepsilon\ \hat{u}_\varepsilon = u_1 + pu_0 \quad .$$

and this is equivalent to

$$p^2\ b(\tfrac{x}{\varepsilon})\ \hat{u}^\varepsilon + A_\varepsilon\ \hat{u}_\varepsilon = b(\tfrac{x}{\varepsilon})\ (u_1 + p\ u_0)$$

for any p with Rep $> 0$. We then finish the proof as in th. 6.2.

Remark 6.3 - It is not difficult to obtain theorems analogous to th. 6.1 to 6.3 if the right hand side of the equations is a fixed function f instead of zero. If f has a Laplace transform, the proofs are analogous to the preceeding ones. If f has not a Laplace transform (for example $f = e^{t^2}$),one study $t \in [0\ ,\ T]$ for any fixed T it is then possible to take f = 0 for $t > T$, and estimates (6.15) and (6.23) hold.

## 7.- Homogenization of a boundary in heat transfer theory. Formal expansion -

Now we study the heat equation in a bounded domain whose boundary is a waved surface of very small period, with the boundary condition

(7.1) $$\frac{\partial u}{\partial n} + \lambda\ u = 0$$

where $\lambda$ is a real positive constant and n is the outer unit normal to the domain. The boundary condition (7.1) is classical in some problems of heat transfer. Physically, if k is the conductivity of the medium, $-k\ \partial u/\partial n$ is the flux of heat at the surface of the body ; (7.1) means that this flux is proportional to the temperature near the boundary. This is the case if the body is cooled by a flow of fluid at temperature 0. The more the temperature u of the body is, the more the heat flux towards the fluid is, but this one is not heated because it is in motion and the heated particles are replaced by cool particules.

In the basis of the preceeding physical considerations, it is natural to expect that the waved surface shall radiate more heat than a smooth (homogenized !)

surface. The boundary condition for the homogenized problem shall be different from (7.1). This is the reason why the radiators are waved !

We shall study the problem in the two dimensional case, but the three-dimensional case is obtained in the same way (see remark 7.2 later).

Let $\Omega_0$ be a bounded open domain of $R^2$ with smooth boundary $\partial\Omega_0$ and with outer unit normal N. Let s be the curvilinear abscissa of the curve $\partial\Omega_0$. In a neighbourhood of $\partial\Omega_0$, s and N are curvilinear coordinates of the plane.

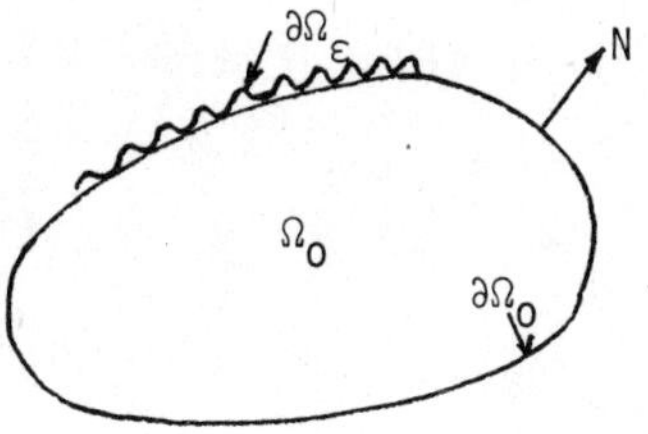


fig. 6

Moreover, in the rectangular coordinates $y_1$, $y_2$, we consider a smooth periodic function $y_2 = F(y_1)$ of period 1 (fig. 7). We then define the boundary $\partial\Omega_\varepsilon$ of the domain $\Omega_\varepsilon$ as the curve defined in the coordinates s, N by $N = \varepsilon F(s/\varepsilon)$ (see fig. 6) where $\varepsilon$ is a real positive small parameter.

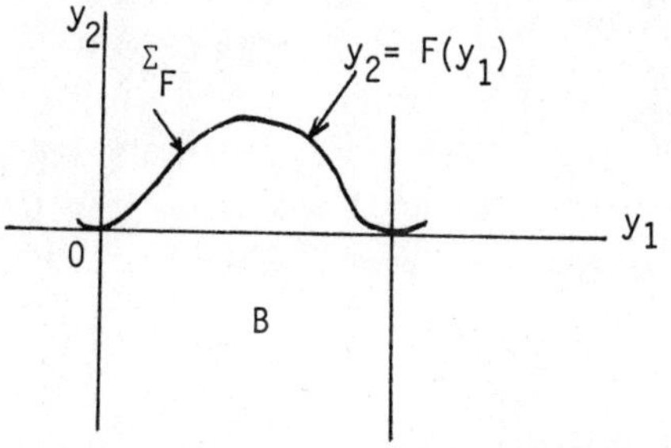


fig. 7

Note that the waves of $\partial\Omega_\varepsilon$ are "almost" homothetic to the curve of fig. 7, "almost" because there is a little distortion due to the curvilinear coordinates, but this distortion tends to zero as $\varepsilon \searrow 0$. We then consider the domain $\Omega_\varepsilon$ enclosed by $\partial\Omega_\varepsilon$ .

Now we consider the heat equation in $\Omega_\varepsilon$ with fixed initial value $u_0(x)$ (which is a smooth function defined on a neighbourhood of $\Omega_0$, and thus on any $\Omega_\varepsilon$). $u^\varepsilon(x , t)$, $x \in \Omega_\varepsilon$ , $t \in [ 0 , \infty [$ is defined by

(7.2) $$\frac{\partial u^\varepsilon}{\partial t} - \Delta\ u^\varepsilon = 0 \qquad \text{for } x \in \Omega_\varepsilon \ , \ t \in ]0 , \infty[$$

(7.3) $$\frac{\partial u^\varepsilon}{\partial n} + \lambda\, u^\varepsilon = 0 \qquad \text{on } \partial\Omega_\varepsilon$$

(7.4) $$u^\varepsilon(x\ ,\ 0) = u_o(x) \qquad \text{on } \Omega_\varepsilon$$

This problem has a unique solution, as we shall see in the next section. (It is of course also possible to consider the non homogeneous problem with f instead of 0 at the right hand side of (7.2)).

We now postulate an asymptotic expansion for $u^\varepsilon$ . The parameter $\varepsilon$ only occurs in the boundary. It is then natural to hope that the expansion of $u^\varepsilon$ will have a boundary layer term depending on the x and y variables, for $y_1 = s/\varepsilon$ , $y_2 = N/\varepsilon$ , and that this term will be periodic of period 1 in the $y_1$ variable. Consequently, we shall write :

(7.5) $$u^\varepsilon(x\ ,\ t) = u^o(x\ ,\ t) + \varepsilon\, u^1(x\ ,\ y\ ,\ t) + \varepsilon^2 \ \ldots.$$
$$\text{for } y_1 = \frac{s}{\varepsilon}\ ,\quad y_2 = \frac{N}{\varepsilon}\ ,\quad u^1 \text{ is 1-periodic in } y_1$$

(7.6) $$\underline{\text{grad}}_y\, u^1 \xrightarrow[y_2 \to -\infty]{} 0$$

Condition (7.6) means that the "waving" introduced by $u^1$ in (7.5) tends to zero far the boundary, i.e. $u^1$ is a boundary layer term. In fig. 7, the region

(7.7) $$B = \{y_1\ ,\ y_2\ ;\ y_1 \in ]\ 0\ ,\ 1[\quad ;\quad y_2 < F(y_1)\}$$

is in fact the period of $u^1$ if x and t are parameters.

In order to replace (7.5) into (7.2), we remark that for small $\varepsilon$ , the coordinates y are asymptotically orthogonal and we can handle them as orthogonal coordinates for the study of the first term $u^1$. Then

(7.8) $$\frac{d}{dx_i} = \frac{\partial}{\partial x_i} + \frac{1}{\varepsilon}\frac{\partial}{\partial y_i}$$
$$\frac{du^\varepsilon}{dx_i} = \frac{\partial u^o}{\partial x_i} + \frac{\partial u^1}{\partial y_i} + \varepsilon \ldots.$$

(7.9) $$\Delta\, u^\varepsilon = \frac{d}{dx_i}\frac{du^\varepsilon}{dx_i} = \frac{1}{\varepsilon}\frac{\partial}{\partial y_i}\frac{\partial u^1}{\partial y_i} + \ldots$$

In the same way, (7.3) gives

$$\frac{du^\varepsilon}{dn} = n_i\,\frac{du^\varepsilon}{dx_i} = n_i\left(\frac{\partial u^o}{\partial x_i} + \frac{\partial u^1}{\partial y_i}\right) + \varepsilon \ \ldots$$

Consequently, taking the terms $\varepsilon^{-1}$ of (7.2) and $\varepsilon^{0}$ of (7.3) we obtain the following local problem for $u^1(y)$ (where x, t are parameters)

$$(7.10) \qquad \begin{cases} \Delta_y u^1 = 0 & \text{in } B \\ \dfrac{\partial u^1}{\partial n} = -\left(\dfrac{\partial u^0}{\partial n} + \lambda u^0\right) & \text{for } y_2 = F(y_1) \\ u^1 \text{ is B-periodic and satisfies (7.6)} \end{cases}$$

We shall see later that this problem has a solution which is unique up to an additive constant if a compatibility condition is satisfied by $u^0$. In order to obtain such a condition, let us suppose that $u^1$ exists. Then, by integrating by parts, with the periodicity conditions, we have ($\Sigma_F$ is the curve $y_2 = F(y_1)$).:

$$(7.11) \qquad 0 = \int_B \Delta u^1 \, dy = \int_{\partial B} \frac{\partial u^1}{\partial n} d\sigma = \int_{\Sigma_F} \frac{\partial u^1}{\partial n} d\sigma = -\int_{\Sigma_F} \left(\frac{\partial u^0}{\partial n} + \lambda u^0\right) d\sigma =$$

$$= -\frac{\partial u^0}{\partial x_i} \int_{\Sigma_F} n_i \, d\sigma - \lambda u^0 \int_{\Sigma_F} d\sigma$$

but

$$(7.12) \qquad \int_{\Sigma_F} d\sigma = |\Sigma_F| \quad ; \quad \int_{\Sigma_F} n_i \, d\sigma = \delta_{i2}$$

where $\Sigma_F$ denotes the measure (length) of the arc $\Sigma_F$. $\delta_{i2}$ is the Kronecker symbol (if the period of F in the direction $y_1$ is not one, the second relation (7.12) must be multiplied by the period). Let us define the "waving coefficient" $\Gamma$ of the boundary by

$$\Gamma = |\Sigma_F|$$

this is evidently the ratio of the lenght of $\partial\Omega_\varepsilon$ to the lenght of $\partial\Omega_0$.

Note that (7.11) is written in the basis associated with $y_1, y_2$ ; consequently we have $x_1 = s$, $x_2 = N$ ; (7.11) becomes

$$(7.13) \qquad \frac{\partial u^0}{\partial N} + \lambda \Gamma u^0 = 0$$

which is the desired compatibility condition.

Consequently, the "limit problem" for $u^0(x , t)$ is

$$(7.14)\quad \begin{cases} \dfrac{\partial u^o}{\partial t} = \Delta\, u^o & \text{for } x \in \Omega_o \,,\ t \in \,]\,0\,,\infty\,[ \\ u^o(x\,,0) = u_o(x) & \text{for } x \in \Omega_o \\ (7.13) & \text{for } x \in \partial\Omega_o \,,\ t \in \,]\,0\,,\infty\,[\,. \end{cases}$$

Note that the first equation of (7.14) is immediately obtained from (7.5) out of the boundary layer (see (7.6)).

<u>Proposition 7.1</u> - If we postulate an expansion of the type (7.5), the first term $u^o(x\,,t)$ (i.e., the limit of $u^\varepsilon$ as $\varepsilon \searrow 0$) is uniquely determined by (7.14). This is a problem analogous to (7.2) - (7.4) but in $\Omega_o$ instead of $\Omega_\varepsilon$ and with the boundary condition (7.13) instead of (7.3). Note that the coefficient $\lambda$ is multiplied by the waving coefficient $\Gamma$ of $\partial\Omega_\varepsilon$.

It is worthwhile to prove the existence of $u^1$ (i.e. of the boundary layer).

<u>Theorem 7.1</u> - If $u^o$ is the solution of (7.14) the local problem (7.10) for $u^1(y)$ has a solution, which is unique up to an additive constant.

The remainder of this section is devoted to the proof of this theorem. The exact definition of "solution" will be clear later.

Let us construct a B-periodic function $a(y)$ satisfying

$$(7.15)\qquad \frac{\partial a}{\partial n} = -\,\frac{\partial u^o}{\partial x_i}\,n_i \;-\; \lambda u^o \qquad \text{on } \Sigma_F$$

and identically zero for sufficiently large $-y_2$. It is clear that for the study of the local problem, x and t are parameters ; moreover $x_1 = s$ , $x_2 = N$. The function evidently exists and is smooth (because F does). Moreover

$$(7.17)\qquad \int_B \Delta\, a\; dy = 0$$

For, by integrating by parts, (7.17) is equal to

$$\int_{\partial B} \frac{\partial a}{\partial n}\, d\sigma$$

this integral vanishes on $\Sigma_F$ by (7.15) and (7.11) (which is equivalent to (7.13)) ; on the remainder of $\partial B$ the integral vanishes by the periodicity and nullity conditions.

Now, we take the new unknown $v = u^1 - a$. The problem for v is

$$(7.18)\qquad \Delta\, v = -\,\Delta\, a$$

(7.19) $$\left.\frac{\partial v}{\partial n}\right|_{\Sigma_F} = 0 \quad ; \quad \underline{\text{grad}}\ v \xrightarrow[y_2 \to -\infty]{} 0$$

(7.20) v is B-periodic

Let us define the set V of the functions w which are B-periodic and smooth, constant for sufficiently large $-y_2$. We immediatly obtain

(7.21) $$\int_B \Delta\, v\, w\, dy = \int_{\partial B} \frac{\partial v}{\partial n}\, w\, d\sigma - \int_B \frac{\partial v}{\partial y_i} \frac{\partial w}{\partial y_i}\, dy \qquad \forall\, w \in V$$

and the term $\int_{\partial B}$ vanishes by periodicity and (7.19). We then have

(7.22) $$\int_B \frac{\partial v}{\partial y_i} \frac{\partial w}{\partial y_i}\, dy = \int_B \Delta\, a\, w\, dy \qquad \forall\, w \in V$$

Moreover, if v satisfies (7.20), the second relation of (7.19) and (7.22), by using (7.21) we see that (7.18) and the first relation of (7.19) are satisfied. It is then easy to obtain a variational formulation of (7.18) - (7.20) up to an additive constant.

We consider the equivalence class obtained by identifying the elements of V difference of which is a constant. We introduce the scalar product

(7.23) $$(\hat{v}\ ,\ \hat{w})_{\hat{V}} = \int_B \frac{\partial v}{\partial y_i} \frac{\partial w}{\partial y_i}\, dy$$

in the space of the equivalence class, where v, (or w) is any element of the equivalence class $\hat{v}$ (or $\hat{w}$). Note that $(\hat{v}\ ,\ \hat{v}) = 0 \Rightarrow v = \text{const.} \Rightarrow \hat{v} = 0$. We then define $\hat{V}$ as the Hilbert space obtained by completion of the equivalence class space with the norm associated with (7.23).

The variational formulation of (7.18) - (7.20) is :

(7.24) Find $\hat{v} \in \hat{V}$ such that
$$(\hat{v}\ ,\ \hat{w})_{\hat{V}} = \int_B \Delta\, a\, \hat{w}\, dy \qquad \forall\, \hat{w} \in \hat{V}$$

where the right hand side is for the right hand side of (7.22) with any $w \in \hat{w}$ (note that by virtue of (7.17), this value is independent of the particular w chosen, it only depends on the equivalence class $\hat{w}$).

The existence and uniqueness of $\hat{v}$ will be proved if we show that the right hand side of (7.24) is a bounded functional on $\hat{V}$. To this end, we note that $\Delta a$ is zero for sufficiently large $-y_2$ ; consequently the domain of integration is in fact a bounded set, denoted by Bd, where the Poincaré's inequality (see Mikhlin [1] p. 337) :

$$(7.25) \qquad \int_{Bd} w^2\, dy \leqslant C\left[\int_{Bd} \frac{\partial w}{\partial y_i}\frac{\partial w}{\partial y_i}\, dy + \left(\int_{Bd} w\, dy\right)^2\right]$$

holds. Moreover, for a given equivalence class $\hat{w}$, we may choose $w \in \hat{w}$ in such a way that the mean value of w on Bd is zero. We then have :

$$\left|\int_B \Delta\, a\, \hat{w}\, dy\right| \leqslant C\left(\int_{Bd} w^2\, dy\right)^{1/2} \leqslant C'\left(\int_{Bd} \frac{\partial w}{\partial y_i}\frac{\partial w}{\partial y_i}\, dy\right)^{1/2} = C'\,\|\hat{w}\|_{\hat{V}}$$

and the functional is bounded. The theorem is proved.

<u>Remark 7.1</u> - The local problem does not hold in a bounded domain and the Rellich compactness theorem does not hold. Consequently it is not easy to see if the Fredholm's alternative holds or not. This is the reason why we introduced the space of the equivalence class. ■

<u>Remark 7.2</u> - If we consider the three-dimensional case, all the results hold but $\Gamma$ is defined as the ratio of the areas of the surfaces $\partial\Omega_\varepsilon$ and $\partial\Omega_0$. ■

8.- <u>Proof of the convergence</u> - This section is devoted to the proof of the convergence $u^\varepsilon \to u^0$ in the problem of the preceeding section. We begin with the associated stationary problem. For spectral properties, see chap. 11, sect. 6.

Let $\Omega_\varepsilon$ and $\Omega_0$ be defined as in the preceeding section. If $\Gamma$ is the waving coefficient and $\lambda, \mu$ are two real positive constants, we consider

$$(8.1) \qquad (-\Delta + \mu)\, u^\varepsilon = f \qquad \text{in } \Omega_\varepsilon$$

$$(8.2) \qquad \frac{\partial u^\varepsilon}{\partial n} + \lambda\, u^\varepsilon = 0 \qquad \text{on } \partial\Omega_\varepsilon$$

and

$$(8.3) \qquad (-\Delta + \mu)\, u^0 = f \qquad \text{in } \Omega_0$$

$$(8.4) \qquad \frac{\partial u^0}{\partial n} + \lambda\,\Gamma\, u^0 = 0 \qquad \text{on } \partial\Omega_0$$

where f is a given function of $L^2(R^2)$ (and thus of $L^2(\Omega_\varepsilon)$ for any $\varepsilon$ ).

It is easily seen that $u^\varepsilon$ and $u^0$ are uniquely determined. In fact, the variational formulations of (8.1), (8.2) (proof of which is immediate) is :

Find $u^\varepsilon \in H^1(\Omega_\varepsilon)$ such that

$$(8.5) \qquad \int_{\Omega_\varepsilon} \frac{\partial u^\varepsilon}{\partial x_i}\frac{\partial v}{\partial x_i}\, dx + \mu\int_{\Omega_\varepsilon} u^\varepsilon v\, dx + \lambda\int_{\partial\Omega_\varepsilon} u^\varepsilon\, v\, d\sigma = \int_{\Omega_\varepsilon} f\, v\, dx \quad \forall v \in H^1(\Omega_\varepsilon)$$

and an analogous one for $u^0$.

<u>Theorem 8.1</u> - If $u^\varepsilon$, $u^0$ are defined by (8.1) - (8.4),

(8.6) $$u^\varepsilon\big|_{\Omega_0} \xrightarrow[\varepsilon \searrow 0]{} u^0 \qquad \text{in } H^1(\Omega_0) \text{ weakly} .$$

If we take $v = u^\varepsilon$ in (8.5), we immediately have

(8.7) $$\| u^\varepsilon \|_{H^1(\Omega_\varepsilon)} \leqslant C \qquad (\text{indep. of } \varepsilon)$$

Then, by extracting a sequence $\varepsilon \to 0$ :

(8.8) $$u^\varepsilon\big|_{\Omega_0} \longrightarrow u^* \qquad \text{in } H^1(\Omega_0) \text{ weakly}$$

and it suffices to prove that $u^* = u^0$. Because $\Omega_0$ has a smooth boundary it suffices to prove that

(8.9) $$\int_{\Omega_0} \frac{\partial u^*}{\partial x_i} \frac{\partial v}{\partial x_i} dx + \mu \int_{\Omega_0} u^* v \, dx + \lambda \Gamma \int_{\partial\Omega_0} u^* v \, d\sigma = \int_{\Omega_0} f v \, dx$$

for any smooth function v. It is clear that, by (8.8), the integrals over $\Omega_\varepsilon$ in (8.5) converge to the corresponding integrals over $\Omega_0$ in (8.9) (Note that the integrals over $\Omega_\varepsilon - \Omega_0$ are bounded above by

(8.10) $$\Big| \int_{\Omega_\varepsilon - \Omega_0} u^\varepsilon v \, dx \Big| \leqslant \| u^\varepsilon \|_{L^2(\Omega_\varepsilon - \Omega_0)} \| v \|_{L^2(\Omega_\varepsilon - \Omega_0)}$$

and an analogous one for the derivatives. The right hand side of (8.10) tends to zero by (8.7) and $|\Omega_\varepsilon - \Omega_0| \to 0$).

Consequently, theorem 8.1 will be proved if we prove the following lemma.

<u>Lemma 8.1</u> -

(8.11) $$\int_{\partial\Omega_\varepsilon} u^\varepsilon v \, d\sigma \longrightarrow \Gamma \int_{\partial\Omega_0} u^* v \, d\sigma$$

for any smooth function v.

<u>Proof</u> - We shall write (8.11) in the curvilinear coordinates s, N.

(8.12) $$\partial\Omega_0 \text{ is } N = 0 \quad ; \quad d\sigma = ds$$
$$\partial\Omega_\varepsilon \text{ is } N = \varepsilon\, F_\varepsilon(s) \text{ where } F_\varepsilon(s) \equiv F(\tfrac{s}{\varepsilon}) \quad .$$

Then, if $g_{ij}$ are the components of the metric tensor,

$$d\sigma^2 = g_{11}\, ds^2 + g_{22}\, dN^2 + 2\, g_{12}\, ds\, dN \quad ; \quad dN = F'_\varepsilon(s)\, ds$$

But it is clear that, because the curvilinear coordinates are smooth,

$$g_{11} \to 1 \quad ; \quad g_{12} \to 0 \quad ; \quad g_{22} \to 1$$

uniformly as $N \to 0$. Thus, we have

$$(8.13) \qquad d\sigma = [\,1 + F_\varepsilon'(s)^2\,]^{1/2}\, ds \;\; (1 + \delta(\varepsilon)) \quad \text{on } \partial\Omega_\varepsilon$$

where $\delta(\varepsilon)$ tends to zero as $\varepsilon \searrow 0$ (uniformly).

Consequently, the left hand side of (8.11) is :

$$(8.14) \qquad \int_{\partial\Omega_0} u^\varepsilon(s\,,\varepsilon\, F_\varepsilon(s))v\,(s\,,\,\varepsilon F_\varepsilon(s)) \sqrt{1 + F_\varepsilon'(s)^2}\;(1 + \delta(\varepsilon))\, ds$$

On the other hand, by the trace theorem, from (8.8) we have

$$(8.15) \qquad u^\varepsilon \big|_{\partial\Omega_0} \to u^*\big|_{\partial\Omega_0} \quad \text{in } L^2(\partial\Omega_0) \;\; \text{strongly}$$

Moreover, using (8.7) and a calculation analogous to that of the trace theorem, chap. 1, sect. 3, we have, by using (8.7) :

$$|u^\varepsilon(s,\varepsilon F_\varepsilon(s)) - u^\varepsilon(s,0)|^2 = \left|\int_0^{\varepsilon F_\varepsilon} \frac{\partial u^\varepsilon}{\partial N}(s,\xi)\, d\xi\right|^2 \leqslant \varepsilon F_\varepsilon \int_0^{\varepsilon F_\varepsilon} \left|\frac{\partial u^\varepsilon}{\partial N}\right|^2 d\xi \;\Rightarrow$$

$$(8.16) \qquad \int_{\partial\Omega_0} |u^\varepsilon(s,\, \varepsilon F_\varepsilon) - u^\varepsilon(s,0)|^2\, ds \leqslant C\varepsilon \int_{\Omega_\varepsilon - \Omega_0} |\text{grad } u^\varepsilon|^2\, dx \leqslant C\varepsilon$$

This, with (8.15) implies that :

$$(8.17) \qquad u^\varepsilon(s,\varepsilon F_\varepsilon(s)) \;\; \text{is bounded in } L^2(\partial\Omega_0)$$

As a consequence, we can neglect the term $\delta(\varepsilon)$ in (8.14) because it tends to zero. For the same reason, we may write $v(s,0)$ instead of $v(s,\varepsilon F_\varepsilon)$ in (8.14). Thus, it suffices to prove that

$$(8.18) \qquad \int_{\partial\Omega_0} u^\varepsilon(s,\varepsilon F_\varepsilon)\; v(s,0) \sqrt{1 + F_\varepsilon'^2}\; ds \to \Gamma \int_{\partial\Omega_0} u^*(s,0)\, ds \quad .$$

But from (8.15) and (8.16), $u^\varepsilon(s,\varepsilon F_\varepsilon)$ converges to $u^*(s,0)$ in $L^2(\partial\Omega_0)$ strongly. Moreover, $\sqrt{1 + F_\varepsilon'^2}$ is a $\varepsilon$-periodic function, thus by lemma 4.1 it tends in $L^2(\partial\Omega_0)$ weakly to its mean value which is $\Gamma$, and (8.18) follows. Lemma 8.1 (and thus theorem 8.1) is proved.∎

Now we consider the parabolic problem of the preceeding section.

<u>Theorem 8.2</u> - If $u^\varepsilon$ (resp. $u^0$) are the solutions of (7.2) - (7.4) (resp. (7.14)),

(8.19) $$u^\varepsilon\Big|_{\Omega_0} \xrightarrow[\varepsilon \searrow 0]{} u^0$$

$u^\varepsilon$ and $\frac{\partial u^\varepsilon}{\partial x_k}$ converge in $L^\infty(0, \infty; L^2(\Omega^0))$ weakly *

Proof - First, taking into account that the Laplace operator with the boundary condition (8.2) is associated with the form in (8.5), we see that $u^\varepsilon$ and $u^0$ are well determined in the framework of semigroup theory. The proof is then analogous to that of sect. 6. From (7.2) - (7.4) we see that $u^\varepsilon$ is bounded in $L^\infty(0,\infty;L^2(\Omega_\varepsilon))$ and $\frac{\partial u^\varepsilon}{\partial x_k}$ in $L^2(0, \infty; L^2(\Omega_\varepsilon))$ and by extracting a subsequence

(8.20) $$u^\varepsilon\Big|_{\Omega_0} \to u^*$$

in the topologies indicated in theorem 8.2. By taking the Laplace transform, (which commutes with $|_{\Omega_0}$ ) :

(8.21) $$\hat{u}^\varepsilon\Big|_{\Omega_0} \to \hat{u}^* \qquad \text{in } H^1(\Omega_0) \text{ weakly}$$

but $\hat{u}^\varepsilon$ satisfies

$$A_\varepsilon \hat{u}^\varepsilon + p\,\hat{u}^\varepsilon = u_0 \quad \text{for Re } p > 0$$

where $A_\varepsilon$ is the Laplace operator in $\Omega_\varepsilon$ with the boundary condition (7.3). We have an analogous relation for $\hat{u}^0$. Theorem 8.1 with (8.21) then shows that $\hat{u}^0 = \hat{u}^*$ for real p (and by analytic continuation, for Re $p > 0$). Then $u^* = u^0$ and (8.20), the desired result, follows ■

9.- Asymptotic expansion of an integral identity - We consider here another method of expansion to find again the results of sections 1 - 4. This method is based on the variational formulation (1.8) instead of the classical formulation (1.6), (1.7). The advantage of the new method is that it deals only with first order derivatives and the calculations are shorter than with the classical method.

Let us begin with a formal calculation. Let $F(x , y)$ be a function of x and y, Y-periodic in y in the framework of (1.10). Let $\Phi$ be the function defined by

(9.1) $$\Phi(\varepsilon) = \int_\Omega F(x , \frac{x}{\varepsilon})\, dx$$

We then have

(9.2) $$\lim_{\varepsilon \to 0} \Phi(\varepsilon) = \int_\Omega \tilde{F}(x)\, dx$$

where, as usual, $\tilde{F}(x) = \frac{1}{|Y|} \int_Y F(x , y)\, dy$ .

To obtain (9.2), we consider (9.1) as the sum of the integrals on the periods $\varepsilon Y$.

(9.3) $$\Phi(\varepsilon) = \sum_{\text{periods}} |\varepsilon Y| \left( \frac{1}{|\varepsilon Y|} \int_{\varepsilon Y} F(x , \frac{x}{\varepsilon})\, dx \right)$$

and in each period, we may consider x constant in the first argument because the corresponding variation of F tends to zero. Then, by writing each integral in the variable y, we have

$$\Phi(\varepsilon) \cong \sum_{\text{periods}} |\varepsilon Y| \left( \frac{1}{|Y|} \int_{Y} F(x , y)\, dy \right)$$

because the jacobian of the transformation is $|\varepsilon Y| / |Y|$, and (9.2) follows.

Now, <u>we consider the Dirichlet problem with periodic coefficients under the form (1.8), i.e.</u> :

(9.4) $$\begin{cases} \text{Find } u^\varepsilon \in H_0^1(\Omega) \text{ such that} \\ \displaystyle\int_\Omega a_{ij}(\frac{x}{\varepsilon}) \frac{\partial u^\varepsilon}{\partial x_i} \frac{\partial v}{\partial x_j}\, dx - \int_\Omega f\, v\, dx = 0 \qquad \forall\, v \in H_0^1(\Omega) \end{cases}$$

We then introduce the asymptotic expansion (1.10) for the solution, i.e. :

(9.5) $$u^\varepsilon(x) = u^0(x) + \varepsilon u^1(x,y) + \varepsilon^2 \ldots \quad , \qquad y = \frac{x}{\varepsilon} \quad , \quad Y\text{-periodic in } y$$

Moreover, (9.4) is satisfied for any $v \in H_0^1$. In particular, for small $\varepsilon$ we may take test functions of the form (see also remark 9.1 later)

(9.6) $$v = v^\varepsilon = v^0(x) + \varepsilon\, v^1(x , y) + \ldots. \quad ; \quad y = \frac{x}{\varepsilon} \quad ; \quad Y\text{-periodic in } y.$$

As usual, we have

$$\frac{\partial u^\varepsilon}{\partial x_i} = \frac{\partial u^0}{\partial x_i} + \frac{\partial u^1}{\partial y_i}$$

and an analogous relation for v. The integral identity (9.4) becomes <u>at the first order</u> :

(9.7) $$\int_\Omega a_{ij} \frac{\partial u^0}{\partial x_i} \frac{\partial v^0}{\partial x_j}\, dx + \int_\Omega a_{ij} \frac{\partial u^1}{\partial y_i} \frac{\partial v^0}{\partial x_j}\, dx + \int_\Omega a_{ij} \frac{\partial u^0}{\partial x_i} \frac{\partial v^1}{\partial y_j}\, dx +$$
$$+ \int_\Omega a_{ij} \frac{\partial u^1}{\partial y_i} \frac{\partial v^1}{\partial y_j}\, dx - \int_\Omega f\, v^0\, dx = 0$$

for any $v^0$, $v^1$ (Y-periodic in y). If in particular we take $v^1 = 0$, $v^0$ arbitrary, we have

(9.8) $$\int_\Omega a_{ij}(y) (\frac{\partial u^0}{\partial x_i} + \frac{\partial u^1}{\partial y_i}) \frac{\partial v^0}{\partial x_j}\, dx - \int_\Omega f\, v^0\, dx = 0$$

and by taking the limit as $\varepsilon \to 0$ according to (9.1), (9.2) :

$$\int_\Omega \tilde{p}_j^o \frac{\partial v^0}{\partial x_j}\, dx - \int_\Omega f\, v^0\, dx = 0 \qquad \text{where} \tag{9.9}$$

$$p_j^o \equiv a_{ij}\left(\frac{\partial u^0}{\partial x_i} + \frac{\partial u^1}{\partial y_i}\right) \tag{9.10}$$

and (9.10) shows that (in the sense of distributions) :

$$- \frac{\partial \tilde{p}_j^o}{\partial x_j} = f \quad . \tag{9.11}$$

Moreover, by replacing (9.8) into (9.7), only the terms in $v^1$ remain ; and by taking the limit value according to (9.1), (9.2),

$$\int_\Omega \left[ a_{ij}\left(\frac{\partial u^0}{\partial x_i} + \frac{\partial u^1}{\partial y_i}\right) \frac{\partial v^1}{\partial y_j} \right]^\sim dx = 0 \qquad \forall\, v^1 \ \text{Y-periodic} \tag{9.12}$$

This relation may be considered as an equation to find $u^1(x\,,\, y)$ (defined on $\Omega \times Y$) if $u^0(x)$ is given. In fact, it is easier to take

$$v^1 = \theta(x)\, w(y) \qquad ; \qquad \theta \in \mathcal{D}(\Omega) \qquad ; \qquad w \in V_Y \tag{9.13}$$

where $V_Y$ is the space of periodic functions defined in sect. 2. We then have

$$\int_\Omega \left[ a_{ij}\left(\frac{\partial u^0}{\partial x_i} + \frac{\partial u^1}{\partial y_i}\right) \frac{\partial w}{\partial y_j} \right]^\sim \theta\, dx = 0 \qquad \forall\, \theta\,,\, w$$

and this implies

$$\left[ a_{ij}\left(\frac{\partial u^0}{\partial x_i} + \frac{\partial u^1}{\partial y_i}\right) \frac{\partial w}{\partial y_j} \right]^\sim = 0 \qquad \forall\, w \in V_Y \tag{9.14}$$

which is exactly the equation for the local behaviour, written under the form (5.13).

Consequently, the homogenized problem is obtained under the form (9.9) (or 9.11), (9.14). Of course, $u^0 \in H_o^1(\Omega)$ gives the boundary condition on $\partial\Omega$ .

Remark 9.1 - We have only used test functions $v^1$ of the form (9.13). In fact, we may consider instead of (9.6) :

$$\begin{cases} v = v^0(x) + \theta(x)\, w(y) & ; \\ v^0 \in H_o^1(\Omega) \quad ; \quad \theta \in \mathcal{D}(\Omega) \quad ; \quad w \in V_Y \end{cases} \tag{9.15}$$

and the expansion (9.5) of the unknown may be assumed only out of a boundary layer near $\partial\Omega$ (see remark 1.3). The preceeding considerations furnish in this case the behaviour out of a neighbourhood of $\partial\Omega$ . ■

10.- <u>Method of the conservation law</u> - In physics it is customary to consider an equation of the type (1.1), (1.2) as equivalent to the conservation law

$$(10.1) \qquad \int_{\partial D} p_i \, n_i \, ds + \int_D f \, dx = 0 \qquad \forall D$$

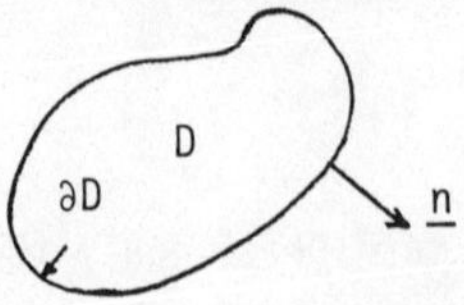


where D is any subdomain of the domain of definition of u. If u satisfies (1.2), we obtain (10.1) by integration of (1.2) on D and utilisation of the divergence theorem. Conversely, if (10.1) is satisfied, for any D, by the divergence theorem, we have

$$(10.2) \qquad \int_D \left( \frac{\partial p_i}{\partial x_i} + f \right) dx = 0$$

and, if the integrand is continuous (which is a hypothesis of regularity for u), this shows that it is zero. (For, if it is positive or negative at a point, it is also positive or negative in a neighbourhood of this point, and (10.2) should not hold by taking D equal to that neighbourhood).

This remark is the basis of another method to obtain the macroscopic equations. This method is often useful in physics because it shows the physical meaning of the equation (conservation of mass, momentum, energy ...).

If in (1.6) we take only the first term of the expansion, we have (1.18) and consequently the study of the local behaviour and the homogenized coefficients. The macroscopic equation may be obtained in the following way (without using equation (1.19)).

We consider a domain D whose dimensions are of order O(1) (independent of $\varepsilon$), made of whole periods, but otherwise arbitrary. From equations (1.6) and (1.9), the remark at the begining of this section gives :

$$\int_{\partial D} p_i^{\varepsilon} \, n_i \, dS + \int_D f \, dx = 0$$

and thus, at the first order,

$$(10.3) \qquad \int_{\partial D} p_i^{0} \, n_i \, dS + \int_D f \, dx = 0$$

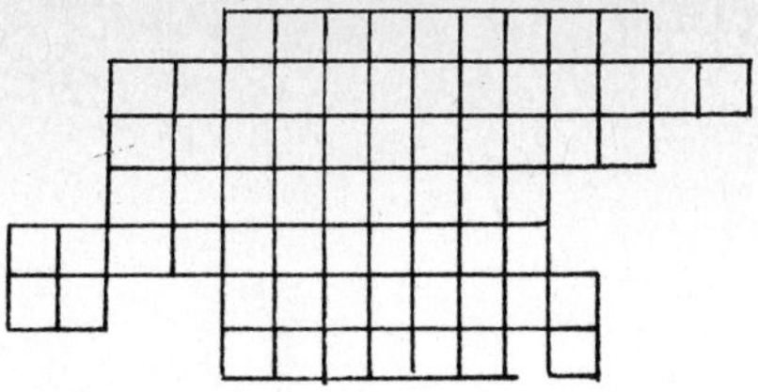

Let us now suppose (this will be proved later) that the integral of $p_i^0 n_i$ on each face of a period is the same as that of its mean value $\tilde{p}_i^0 n_i$. We then have

$$\int_{\partial D} \tilde{p}_i^0 n_i \, dS + \int_D f \, dx = 0 \tag{10.4}$$

which is a conservation law for functions of the macroscopic variable x. This implies

$$\frac{\partial \tilde{p}_i^0}{\partial x_i} + f = 0 \tag{10.5}$$

which is the desired relation (1.23). (Note that the integrand of (10.4) is independent of $\varepsilon$ and the reasoning leading to (10.5) holds for small $\varepsilon$ as for arbitrary D).

Now, we prove the assumed property on the integrals of $p_i^0 n_i$ and $\tilde{p}_i^0 n_i$.

<u>Proposition 10.1</u> - The integral of $p_i^0(y) n_i$ on a face of a period is the same as the integral of its mean value $p_i^0 n_i$.

<u>Proof</u> - We consider a period (x is a parameter). By integrating (1.18) on a region R (see figure) we obtain

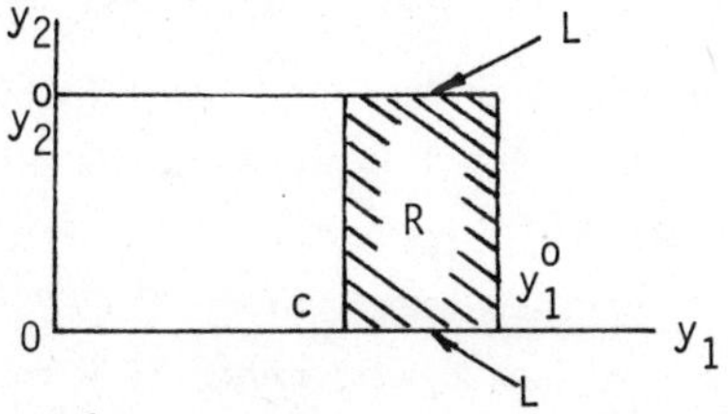


$$\int_{\partial R} p_i^0 n_i \, dS = 0 \tag{10.6}$$

and by Y-periodicity, the surface integral on the lateral surface L vanishes, and we see that the integral

$$\Phi(c) = \int_0^{y_2^0} \int_0^{y_3^0} p_1^0(c, y_2, y_3) \, dy_2 \, dy_3 \tag{10.7}$$

is equal to

$$\int_0^{y_2^0}\int_0^{y_3^0} p_i^0(y_1^0, y_2 , y_3)\, dy_2\, dy_3$$

i.e., $\Phi$ is independent of c. We then have

$$(10.8) \qquad \boxed{\Phi = \frac{1}{y_1^0}\int_0^{y_1^0} \Phi\, dc = \frac{y_2^0\, y_3^0}{|Y|}\int_Y p_1^0\, dy = y_2^0\, y_3^0\, \tilde{p}_1^0}$$

and the equality of the right hand sides of (10.7) and (10.8) is the desired relation.

Remark 10.1 - Proposition 10.1 may be stated by saying that "The volume and surface means coincide". It is clear that this fact is not general ; it holds for vectors of zero divergence in the variable y (see (1.18), which implies (10.6)). ∎

11.- Comments and bibliographical notes - The homogenization method for the study of composite materials is based on the study of periodic solutions of partial differential equations, and their asymptotic behaviour as the period tends to zero. It is fitted for the study of composite materials with periodic structure, and it may also be taken as a model of other composite materials. The hypothesis of periodic structure permits a rigorous treatment of the problem, which is useful in very controversed problem, such that mixtures of solids and fluids, fluid flow in porous media and so on (see the following chapters). Generally speaking, we shall not give comparaisons and references for other methods ; this is made in the papers cited in the litterature. Nevertheless, we point out Sendeckyj [ 1 ] as a general reference for other methods.

First papers on homogenization were Sanchez-Palencia [ 1 ], [ 2 ], [ 3 ], where only a formal study relevant of asymptotic methods (such as in Cole [ 1 ] or Van Dyke [ 1 ]) was given. De Giorgi and Spagnolo [ 1 ] gave the first proof of the convergence of the method, based on the G-convergence theory of partial differential equations (see Spagnolo [ 1 ] , [ 2 ] ). For recent developments of G-convergence, including non linear problems, see for instance Biroli [ 1 ] , Boccardo et Marcellini [ 1 ], Carbone [ 1 ] , Marcellini and Sbordone [ 1 ] , Murat [ 2 ] , Tartar [ 3 ] and the recent review paper of De Giorgy [ 1 ] . Tartar [ 3 ], [ 4 ] gave a simplified proof of convergence (sect. 4) which applies to other problems. Bensoussan, Lions and Papanicolaou [ 2 ] studied a great number of homogenization problems ; this is the general reference in homogenization theory. It includes time-varying coefficients, probabilistic methods and wave propagation for small wave length. An abridged version of some parts of this book may be seen in Lions [ 5 ] .

The convergence of the resolvents in the norm (sect. 5) is suggested by Boccardo and Marcellini [1] and gives in a natural way the perturbation of eigenvalues and eigenvectors (see chapter 11, sect. 3) without a hypothesis of self-adjointness (compare with Kesavan [1] ). The homogenization of a boundary (sect. 7 and 8) seems new (see Brizzi et Chalot for other related problems). The method of section 9 is fitted the asymptotic study of problems in variational form, and is systematically used in the following chapters, including variational inequalities (chap. 6, sect. 6 and 7). See also, in this convection, Bensoussan, Lions et Papanicolaou [2] , remarks at the end of chapter 1, sect. 18.

For other homogenization problems, see Babuska [1] , Bakhvalov [1], [2], Bensoussan, Lions and Papanicolaou [3] , Desgraupes [1] ; for some non linear problems, see Artola et Duvaut [2] , Damlamian [1] . The problems of homogenization of media with holes (see also the following chapters and the appendix by L. Tartar) were studied by Tartar [3] , [4] , Cioranescu and Saint Jean Paulin [1] and Vanninathan [1] , [2] . Scattering problems were studied by Codegone [1]. Numerical results and comparaison with exact solutions were given by Bourgat [1] , Bourgat et Dervieux [1] and Bourgat et Lanchon [1] .

# CHAPTER 6

# HOMOGENIZATION IN ELASTICITY

# AND ELECTROMAGNETISM

In this chapter we apply the homogenization method to the study of elastic and electromagnetic phenomena in composite materials with periodic structure (as the period tends to zero). An interesting feature is that, in some cases, the homogenized equations are integro-differential althought the basic equations are differential. Then, the macroscopic phenomena exhibit memory effets (see sect. 4, 8 and 9).

1.- A model problem in elastostatics - The partial differential system of elasticity is elliptic, and several boundary value problems may be solved in much the same way that the examples of chap. 3 for elliptic equations. We introduce here a model problem analogous to that of chap. 3, sect. 3 which will be handled in the sequel. Other boundary value problems may be studied by analogous methods.

Let $\Omega$ be an open connected domain of $R^3$ with smooth boundary $\partial\Omega$ . The boundary is made of two parts, $\partial_1\Omega$, $\partial_2\Omega$ which are portions of regular surfaces

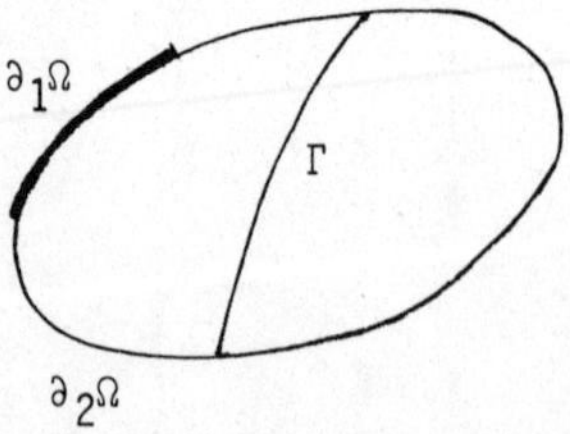


(their common boundary is smooth). Moreover, $\Omega$ may be divided into two parts $\Omega_1$, $\Omega_2$ by a smooth surface $\Gamma$ . (Note that the fig. 1 is made in the two-dimensional case).

In the framework of linear elasticity, let $\underline{u}(x)$ be the displacement and $e_{ij}(u)$ the components of the deformation tensor :

(1.1) $$e_{ij}(\underline{u}) = \frac{1}{2}\left(\frac{\partial u_i}{\partial x_j} + \frac{\partial u_j}{\partial x_i}\right)$$

and $\sigma_{ij}$ the strain tensor, given by

(1.2) $$\sigma_{ij}(u) = a_{ijkh}\, e_{kh}(\underline{u})$$

where $a_{ijkh}$ are the elastic constants. We shall admit that they satisfy the symetry and positivity properties :

(1.3) $$a_{ijkh} = a_{jikh} = a_{ijhk} = a_{khij}$$

(1.4) $$a_{ijkh}\, e_{ij}\, e_{kh} \geqslant \alpha\, e_{ij}\, e_{ij} \quad ; \quad \alpha > 0 \qquad \forall\, e_{ij} \text{ (symmetric)}$$

If the material of the body is heterogeneous, the coefficients $a_{ijkh}$ are functions of x. We shall admit that $a_{ijkh}$ are piecewise smooth functions, having discontinuities on the surface $\Gamma$ .

Moreover, if $f_i$ (resp. $F_i$) are the components of the given body forces (resp. surface forces, defined on $\partial_2\Omega$ ), and if $\Omega$ is an elastic body clamped by $\partial_1\Omega$), the equations and boundary conditions are

(1.5) $$0 = \frac{\partial \sigma_{ij}}{\partial x_j} + f_i \qquad \text{in } \Omega$$

(1.6) $$u_i = 0 \qquad \text{on } \partial_1\Omega$$

(1.7) $$\sigma_{ij}\, n_j = F_i \qquad \text{on } \partial_2\Omega$$

where n is the unit outer normal to $\partial\Omega$ .

It is clear that (1.5) is understood in the sense of distributions. On $\Gamma$ we have

(1.8) $$[\underline{u}] = 0 \qquad \text{on } \Gamma$$

(1.9) $$[\sigma_{ij}\, n_j] = 0 \qquad \text{on } \Gamma$$

In order to establish the variational formulation of the problem (1.5) - (1.7), we introduce the space

(1.10) $$V = \{\underline{u} \; ; \; u_i \quad H^1(\Omega) \quad ; \quad u_i\big|_{\partial_1\Omega} = 0\}$$

which is a Hilbert space for the norm of $(H^1(\Omega))^3$ (note that V is a closed subspace of $(H^1)^3$ by virtue of the trace theorem).

By virtue of (1.3),

$$a(\underline{u}\ ,\ \underline{v}) \equiv \int_\Omega a_{ijkh}\, e_{ij}(\underline{u})\, e_{kh}(\underline{v})\, dx \equiv \int_\Omega a_{ijkh} \frac{\partial u_i}{\partial x_j} \frac{\partial v_k}{\partial x_h}\, dx \tag{1.11}$$

is a bounded symmetric form on V.

Definition 1.1 - The variational formulation of the problem (1.5)-(1.7) is : Find $\underline{u} \in V$ such that

$$a(\underline{u}\ ,\ \underline{v}) = \int_\Omega f_i\, v_i\, dx + \int_{\partial_2\Omega} F_i\, v_i\, dS \qquad \forall\ \underline{v} \in V\ . \tag{1.12}$$

It is not difficult to obtain the equivalence between (1.12) and (1.5)-(1.7) if $\underline{u}$ is a smooth function (except on $\Gamma$ ). If $\underline{u}$ satisfies (1.5)-(1.7), by multiplying (1.5) by $v_i$ and by integrating by parts,

$$\int_\Omega f_i\, v_i\, dx = -\int_{\partial\Omega} n_j\, \sigma_{ij}\, v_i\, dS + \int_\Omega \sigma_{ij} \frac{\partial v_i}{\partial x_j}\, dx = \tag{1.13}$$

$$= -\int_{\partial_2\Omega} F_i\, v_i\, dS + \int_\Omega a_{ijkh}\, e_{kh}(\underline{u})\, e_{ij}(\underline{v})\, dx$$

where (1.2), (1.3), (1.7) and the fact that $\underline{v}$ is zero on $\partial_1\Omega$ have been used. We see that $\underline{u}$ satisfies (1.12). Conversely, if $\underline{u} \in V$ satisfies (1.12), by integrating by parts we obtain the first equality (1.13). Then, by taking $\underline{v} \in (\mathcal{D}(\Omega))^3$ we have

$$\int_\Omega f_i\, v_i\, dx = \int_\Omega \sigma_{ij} \frac{\partial v_i}{\partial x_j}\, dx = -\int_\Omega \frac{\partial \sigma_{ij}}{\partial x_j}\, v_i\, dx$$

and (1.5) is satisfied. The first relation (1.13) then becomes

$$\int_{\partial\Omega} (n_j\, \sigma_{ij} - F_i)\, v_i\, dS = 0 \qquad \forall\ \underline{v} \in V$$

and (1.7) follows. As for (1.6), it is a consequence of $\underline{u} \in V$.

In the preceeding proof of the equivalence of the classical and variational formulations of the problem we considered $a_{ijkh}$ continuous. If they are (as we said) discontinuous on $\Gamma$ , we have a supplementary term

$$\int_\Gamma [\sigma_{ij}\, n_j\, ]\, v_i\, dS$$

in (1.13). In the first part of the proof, it vanishes by (1.9) ; in the second part, it shows that (1.9) is satisfied. Note also that (1.8) is merely a consequence of $\underline{u} \in V$.

In order to prove the existence and uniqueness of $\underline{u}$ satisfying (1.12), we shall apply the Lax-Milgram theorem. To this end, we have :

Lemma 1.1 - (Korn's inequality) - If $\Omega$ is a bounded domain with smooth boundary, there exists $\gamma > 0$ such that

(1.14) $$\int_\Omega e_{ij}(\underline{v})e_{ij}(\underline{v})dx + \int_\Omega v_i\, v_i\, dx \geq \gamma\, \|\underline{u}\|^2_{(H^1)^3} \quad \forall \underline{u} \in (H^1)^3$$

The proof may be seen, for instance, in Duvaut - Lions [1 ] , chap. 3, sect sect. 3.3.

Lemma 1.2 - The form $a(\underline{u}\,,\,\underline{v})$ is coercive on $V$ (i.e., there exists $\delta > 0$ such that

$$a(\underline{v}\,,\,\underline{v}) \geq \delta \,\|\,\underline{v}\,\|^2_{(H^1)^3} \qquad \forall\, \underline{v} \in V \quad .$$

Proof - It is analogous to that of the Friedrichs inequality (chap. 3, prop. 1.1). Note that by (1.14), the left hand side of (1.14) may be taken as norm$^2$ in $V$ . We shall do this in the sequel. On the other hand, by (1.4),

$$a(\underline{v}\,,\,\underline{v}) \geq \alpha \int_\Omega e_{ij}(\underline{v})\; e_{ij}(\underline{v})\; dx \qquad \forall\, \underline{v} \in V$$

Thus, it suffices to prove that there exists $C > 0$ such that

(1.15) $$\int_\Omega v_i\, v_i\, dx \leq C \int_\Omega e_{ij}(\underline{v})\; e_{ij}(\underline{v})\; dx \qquad \forall\, \underline{v} \in V\ .$$

If (1.15) does not hold, by taking into account (1.14) we see that there exists a sequence $\underline{v}^k$ such that

$$\|\underline{v}^k\|_{(L^2)^3} = 1 \quad ; \quad \int_\Omega e_{ij}(\underline{v}^k)\; e_{ij}(\underline{v}^k) dx \to 0 \qquad k \to \infty$$

$$\underline{v}^k \to \underline{v}^* \quad \text{in } V \text{ weakly and } (L^2)^3 \text{ strongly} \Longrightarrow$$

(1.16) $$\|\,\underline{v}^*\|_{(L^2)^3} = 1 \quad ; \quad \|\,v^k\|_V \leq C$$

and by chap. 2, prop. 1.1,

$$\|\underline{v}^*\|^2_V \leq \underline{\lim}\, \|\underline{v}^k\|^2_V \Longrightarrow$$

$$\int_\Omega e_{ij}(\underline{v}^*)\; e_{ij}(\underline{v}^*) dx \leq \underline{\lim} \int_\Omega e_{ij}(\underline{v}^k)\; e_{ij}(\underline{v}^k)\; dx = 0$$

and this implies that $\underline{v}^*$ is a displacement field of solid body, i.e. $\underline{v}^* = \underline{a} + \underline{b} \wedge \underline{x}$, (see for instance Germain [1] , p. 52) and $\underline{v}^*|_{\partial_1\Omega} = 0 \Rightarrow \underline{v}^* = 0$, which is in contradiction with (1.16). ∎

By virtue of the trace theorem, the right hand side of (1.12) is bounded by

$$\left|\int_\Omega f_i\, v_i\, dx + \int_{\partial_2\Omega} F_i\, v_i\, dS\right| \leqslant C\left( \|\underline{v}\|_{(L^2)^3} + \|\underline{v}\|_{L^2(\partial_2\Omega)}\right) \leqslant C'\, \|\underline{v}\|_V$$

and is then a linear and bounded functional on V. By the Lax-Milgram theorem we have :

<u>Theorem 1.1</u> - The solution of the problem of definition 1.1 exists and is unique.

<u>Remark 1.1</u> - It is clear that the embedding of V in $H = (L^2(\Omega))^3$ is compact. Consequently, the preceeding problem is in the framework of chap. 2, sect. 6 and the eigenvectors and eigenvalues exist and form a basis.■

2.- <u>Homogenization in elasticity</u> - Now we consider a problem of homogenization analogous to that of chap. 5, sect. 1-5 but with the elasticity system. The process is about the same as in the preceeding chapter and certain details will not be given.

The coefficients $a_{ijkh}(y)$ are Y-periodic and satisfy (1.3), (1.4). Moreover, we shall take them continuous (if they are piecewise continuous, the same results are obtained, as in chap. 5, sect. 5, transmission problem). Then, we study the elasticity problem with the coefficients

(2.1) $$a^\varepsilon_{ijkh}(x) \equiv a_{ijkh}\left(\frac{x}{\varepsilon}\right)$$

For $\Omega$ given as in sect. 1, we consider

(2.2) $$0 = \frac{\partial \sigma^\varepsilon_{ij}(x)}{\partial x_j} + f_i(x) \qquad \text{in } \Omega$$

(2.3) $$\underline{u}^\varepsilon = 0 \qquad \text{on } \partial_1\Omega$$

(2.4) $$\sigma^\varepsilon_{ij}\, n_j = F_i(x) \qquad \text{on } \partial_2\Omega$$

(2.5) $$\sigma^\varepsilon_{ij}(x) = a^\varepsilon_{ijkh}(x)\, e_{kh}(\underline{u}^\varepsilon)$$

Note that the given forces $\underline{f}$, $\underline{F}$ are independent of $\varepsilon$. We search for expansions :

(2.6) $$\underline{u}^\varepsilon(x) = \underline{u}^0(x) + \varepsilon\, \underline{u}^1(x , y) + \varepsilon^2 \ldots$$

(2.7) $$e^\varepsilon_{ij} \equiv e_{ij}(\underline{u}^\varepsilon) = e^0_{ij}(x , y) + \varepsilon\, e^1_{ij}(x , y) + \varepsilon^2 \ldots$$

(2.8) $$\sigma^\varepsilon_{ij} \equiv a_{ijkh}(y)\, e^\varepsilon_{kh} = \sigma^0_{ij}(x , y) + \varepsilon \sigma^1_{ij}(x , y) + \varepsilon^2 \ldots$$

Y-periodic in y.

With a self-evident notation, we have

(2.9) $$\frac{d}{dx_i} = \frac{\partial}{\partial x_i} + \frac{1}{\varepsilon}\frac{\partial}{\partial y_i} \quad ; \quad e_{ij}(v) = e_{ijx}(v) + \frac{1}{\varepsilon} e_{ijy}(v)$$

(2.10) $$e^o_{ij}(x\ ,\ y) = e_{ijx}(u^0) + e_{ijy}(u^1)$$

By introducing (2.6) - (2.9) into (2.2), we have, at the orders $\varepsilon^{-1}$ and $\varepsilon^0$ respectively :

(2.11) $$\frac{\partial}{\partial y_i}\sigma^o_{ij} = 0$$

(2.12) $$\frac{\partial \sigma^1_{ij}}{\partial y_i} + \frac{\partial \sigma^o_{ij}}{\partial x_i} + f_j = 0$$

In order to obtain the macroscopic equation, we apply the operator "mean"

$$\tilde{\cdot} = \frac{1}{|Y|}\int_Y \cdot\ dy$$

to (2.12) ; by taking into account that, by Y-periodicity,

(2.13) $$\left(\frac{\partial \sigma^1_{ij}}{\partial y_i}\right)^{\sim} = \frac{1}{|Y|}\int_Y \frac{\partial \sigma^1_{ij}}{\partial y_i}\, dy = \frac{1}{|Y|}\int_{\partial Y} \sigma^1_{ij}\, n_i\, dS = 0\ ,$$

we obtain the equation for the macroscopic behaviour :

(2.14) $$\frac{\partial \tilde{\sigma}^o_{ij}}{\partial x_i} + f_j = 0$$

On the other hand, the homogenized elastic coefficients will be obtained from (2.11). By (2.10), it becomes

(2.15) $$\begin{cases} -\dfrac{\partial}{\partial y_i}\left[\, a_{ijkh}(y)\, e_{khy}(\underline{u}^1)\right] = e_{khx}(\underline{u}^0)\, \dfrac{\partial a_{ijkh}(y)}{\partial y_i} \\ Y\text{ - periodic in } y. \end{cases}$$

In order to study (2.15) we introduce the spaces (2.16) which are Hilbert spaces for the scalar products (2.17)

(2.16) $$\begin{cases} H_Y = \{\ \underline{u}\ ;\ u_i \quad L^2_{loc}(R^3)\ ,\ Y\text{-periodic}\ \} \\ V_Y = \{\ \underline{u}\ ;\ u_i \quad H^1_{loc}(R^3)\ ,\ Y\text{-periodic}\ \} \end{cases}$$

(2.17) $$\begin{cases} (\underline{u}\ ,\ \underline{v})_{H_Y} = \displaystyle\int_Y u_i\, v_i\, dy \\ (\underline{u}\ ,\ \underline{v})_{V_y} = (\underline{u}\ ,\ \underline{v})_{H_Y} + \displaystyle\sum_i \left(\frac{\partial \underline{u}}{\partial y_i}\ ,\ \frac{\partial \underline{v}}{\partial y_i}\right)_{H_Y} \end{cases}$$

The variational formulation of (2.15) is

$$\text{Find } \underline{u}^1 \in V_Y \text{ such that } \forall \underline{v} \in V_Y$$

(2.18) $$\int_Y a_{ijkh}(y)\, e_{khy}(\underline{u}^1)\, e_{ijy}(\underline{v})\, dy = e_{khx}(\underline{u}^0) \int_Y \frac{\partial a_{ijkh}}{\partial y_i} v_j\, dy$$

which is analogous to (1.12) and is obtained in the same way. It is clear that the form in the left hand side of (2.18) is not coercive on $V_Y$, but if we add the form $(\underline{u}, \underline{v})_{H_Y}$ it is coercive on $V_Y$ by (1.4) and (1.14) (Note that Y has not a smooth boundary; nevertheless, the Korn's inequality (1.14) holds for $\underline{v} \in V_Y$ because by the periodicity condition, all the points of Y may be considered as interior points). We are in a situation analogous to that of (2.6)-(2.8) of chapter 5. The eigenvectors associated to the eigenvalue zero are such that

$$\int_Y a_{ijkh}\, e_{khy}(\underline{v})\, e_{ij}(\underline{v})\, dy = 0 \implies e_{ij}(\underline{v}) = 0$$

and $\underline{v}$ is a displacement of solid body. By Y-periodicity, $\underline{v} = \underline{c} = \underline{\text{cost}}$ ; Then, the compatibility condition for (2.18) is

$$e_{khx}(\underline{u}^0) \int_Y \frac{\partial a_{ijkh}}{\partial y_i} c_j\, dy = 0$$

which is satisfied by integrating by parts and using the Y-periodicity. Then, <u>for given $\underline{u}^0$, $u^1$ exists and is unique up to an additive constant vector</u>.

We then may write

(2.19) $$\underline{u}^1 = e_{khx}(\underline{u}^0)\, \underline{w}^{kh} + \underline{cte}$$

where <u>$w^{kh}$ is defined by</u> : $\underline{w}^{kh} \in V_Y$ and $\forall \underline{v} \in V_Y$ :

(2.20) $$\int_Y a_{ij\ell m}\, e_{\ell my}(\underline{w}^{kh})\, e_{ijy}(\underline{v})\, dy = \int_Y \frac{\partial a_{ijkh}}{\partial y_i} v_j\, dy$$

It is then easy to obtain the homogenized coefficients. From (2.8), (2.10) and (2.19), we have

$$\sigma^o_{ij} = a_{ij\ell m}\, [e_{\ell mx}(\underline{u}^0) + e_{\ell my}(\underline{u}^1)] =$$

$$= a_{ij\ell m}[\delta_{\ell k}\, \delta_{mh} + e_{\ell my}(\underline{w}^{kh})]\, e_{khx}(\underline{u}^0)$$

and the homogenized strain-stress relation is

(2.21) $$\tilde{\sigma}^o_{ij} = a^h_{ijkh}\, e_{khx}(\underline{u}^0) = a^h_{ijkh}\, \tilde{e}^o_{kh} \quad \text{where}$$

(2.22) $$a^h_{ijkh} \equiv \{a_{ij\ell m}[\delta_{\ell k}\, \delta_{mh} + e_{\ell my}(\underline{w}^{kh})]\}^{\sim}$$

Note that the second equality (2.21) is a consequence of the periodicity of $\underline{u}^1$ (as in (2.13)) :

(2.23) $$\tilde{e}^{o}_{kh} = [\, e_{khx}(\underline{u}^{0}) + e_{khy}(\underline{u}^{1})]^{\sim} = e_{khx}(\underline{u}^{0})$$

Moreover, as in chap. 5,(3.9), the homogenized coefficients are implicitely given by the relation

(2.24) $$\tilde{\sigma}^{o}_{ij} = \frac{\partial W}{\partial e_{ijx}(\underline{u}^{0})}$$ where **W** is the function defined by

(2.25) $$W(e_{ijx}(\underline{u}^{0})) \equiv \frac{1}{2|Y|} \int_Y a_{kh\ell m}\, [\, e_{\ell mx}(\underline{u}^{0}) + e_{\ell my}(\underline{u}^{1})][e_{khx}(\underline{u}^{0}) + e_{khy}(\underline{u}^{1})]\, dy$$

and it is immediate that the homogenized coefficients $a^{h}_{ijkh}$ satisfy the symmetry relations (1.3). Moreover, they also satisfy a positivity condition of the type (1.4). This is equivalent to the fact that if the tensor $e_{ijx}(u^{0})$ is not null, W is not zero.

Consequently, <u>the homogenized coefficients are associated to a linear classical elasticity</u>

Now we formulate the limit problem satisfied by $\underline{u}^{0}(x)$ and we shall prove the convergence of solutions afterwards. <u>The homogenized problem is</u> :

(2.26) $$0 = \frac{\partial \sigma_{ij}}{\partial x_j} + f_i \qquad \text{in } \Omega$$

(2.27) $$\underline{u}^{0} = 0 \qquad \text{on } \partial_1\Omega$$

(2.28) $$\sigma_{ij}\, n_j = F_i \qquad \text{on } \partial_2\Omega$$

(2.29) $$\sigma_{ij} = a^{h}_{ijkh}\, e_{kh}(\underline{u}^{0})$$

<u>Theorem 2.1</u> - If $\underline{u}^{\varepsilon}$ (resp. $\underline{u}^{0}$) is the solution of (2.2) - (2.5) (resp. (2.26) - (2.29)), we have

(2.30) $$\underline{u}^{\varepsilon} \to \underline{u}^{0} \quad \text{in } V \text{ weakly}$$

(V is defined in (1.10)). Moreover, if $A^{\varepsilon}$ is the sefadjoint operator of H ($H = (L^2(\Omega))^3$) defined by

(2.31) $$\begin{cases} (A_\varepsilon\, \underline{v})_i = \dfrac{\partial \sigma_{ij}(\underline{v})}{\partial x_j} & ; \quad \sigma_{ij}(v) = a^{\varepsilon}_{ijkh}\, e_{kh}(\underline{v}) \\ \underline{v}\big|_{\partial_1\Omega} = 0 & ; \quad \sigma_{ij}\, n_j\big|_{\partial_2\Omega} = 0 \end{cases}$$

and $A_h$ is the operator analogous to (2.31) but defined with the homogenized coefficients, then

(2.32) $$\| A_\varepsilon^{-1} - A_h^{-1} \|_{\mathcal{L}(H,H)} \longrightarrow 0 \qquad \text{as } \varepsilon \to 0 .$$

Spectral properties of the homogenization follows from (2.32) ; see chap. 11, sect 3.

Proof of th. 2.1 - As in theorem 5.2, the property (2.32) is a consequence of the compact embedding of V in H (Rellich). We shall prove (2.30) as in theorem 4.1. The variational formulation of (2.2) - (2.5) is : $\underline{u}^\varepsilon \in V$ and

(2.33) $$\int_\Omega a^\varepsilon_{ijkh}\, e_{kh}(\underline{u}^\varepsilon) e_{ij}(\underline{v})\,dx = \int_\Omega f_i\, v_i\,dx + \int_{\partial_2\Omega} F_i\, v_i\, dS \quad \forall\, \underline{v} \in V \;.$$

We immediately see that $\|\underline{u}^\varepsilon\|_V$ is bounded independtly of $\varepsilon$, and, as in chap. 5, (4.5), (4.7) :

(2.34) $$\underline{u}^\varepsilon \to \underline{u}^* \qquad \text{in V weakly, and H strongly}$$

(2.35) $$\sigma^\varepsilon_{ij} \equiv a^\varepsilon_{ijkh}\, e_{kh}(\underline{u}^\varepsilon) \to \sigma^*_{ij} \quad \text{in } L^2(\Omega) \text{ weakly.}$$

We pass to the limit in (2.33) with fixed $\underline{v}$ :

(2.36) $$\int_\Omega \sigma^*_{ij}\, e_{ij}(\underline{v})\,dx = \int_\Omega f_i\, v_i\, dx + \int_{\partial_2\Omega} F_i\, v_i\, dS \qquad \forall\, \underline{v} \in V$$

and we see that $\underline{u}^*$ is the solution of (2.26) - (2.29) if we prove that

(2.37) $$\sigma^*_{kh} = a^h_{khij}\, e_{kh}(\underline{u}^*)$$

To this end, for fixed k,h, we consider the vector field with component $i = \delta_{ik}\, x_h$ (note that this field has a constant $e_{\ell m}$ matrix with all components zero except $e_{kh}$ and $e_{hk}$ which are unity). We then construct the vector field $\underline{w}_\varepsilon$ :

(2.38) $$w_{\varepsilon i} = \delta_{ik} x_h + \varepsilon\, w_i^{kh}\left(\frac{x}{\varepsilon}\right)$$

which satisfies the equation

$$-\frac{\partial}{\partial x_i}\left(a^\varepsilon_{ij\ell m}\, e_{\ell m}(\underline{w}_\varepsilon)\right) = 0 \qquad \text{in } R^3$$

as is easily seen from (2.15) and (2.20). As in chap. 5, (4.11) and (4.13) :

(2.39) $$w_{\varepsilon i} \to \delta_{ik}\, x_h \qquad \text{in } L^2(\Omega) \text{ strongly}$$

(2.40) $$\int_\Omega a^\varepsilon_{ij\ell m} \frac{\partial w_{\varepsilon \ell}}{\partial x_m} \frac{\partial v_i}{\partial x_j}\, dx = 0 \qquad \forall\, \underline{v} \in (H^1_0(\Omega))^3$$

we write (2.33) with $\underline{v} = \phi\, \underline{w}_\varepsilon$ (we bear in mind the second relation (1.11)) and (2.40) with $\underline{v} = \phi\, \underline{u}^\varepsilon$ (where $\phi$ is any element of $\mathcal{D}(\Omega)$) ; by substraction, with the last relation (1.3), we obtain :

$$\int_\Omega a_{ij\ell m}\left[\frac{\partial u^\varepsilon_\ell}{\partial x_m}\, w_{\varepsilon i} \frac{\partial \phi}{\partial x_j} - \frac{\partial w_{\varepsilon i}}{\partial x_j}\, u^\varepsilon_\ell \frac{\partial \phi}{\partial x_m}\right] dx = \int_\Omega f_i\, \phi\, w_{\varepsilon i}\, dx + \int_{\partial_2\Omega} F_i \phi\, w_{\varepsilon i}\, dS$$

which is analogous to chap. 5,(4.14). We then pass to the limit by using (2.34), (2.39) and chap. 5, Lemma 4.1, noting that

$$\widetilde{[a^{\varepsilon}_{ij\ell m} \frac{\partial w_{\varepsilon i}}{\partial x_j}]} = a^h_{kh\ell m}$$

and we obtain

$$(2.41) \qquad \int_\Omega [\sigma^*_{kj} x_h \frac{\partial\phi}{\partial x_j} - a^h_{kh\ell m} u^*_\ell \frac{\partial\phi}{\partial x_m}] dx = \int_\Omega f_k \phi x_h dx + \int_{\partial_2\Omega} F_k \phi x_h dS$$

The right hand side, by virtue of (2.36) becomes

$$\int_\Omega \sigma^*_{kj} ( \delta_{jh} \phi + x_h \frac{\partial\phi}{\partial x_j}) dx$$

Several terms of (2.41) cancel and by integrating by parts (in the sense of distributions) we have :

$$\int_\Omega (a^h_{\ell mkh} \frac{\partial u^*_\ell}{\partial x_m} - \sigma^*_{kh}) \phi dx = 0$$

which is equivalent to (2.37) and the proof is finished.∎

Remark 2.1 - The preceeding considerations apply without modification to the case of discontinuous coefficients (and hence a composite material with periodic structure). The boundary conditions analogous to chap. 5, (5.9), (5.10) are

$$(2.42) \qquad [\underline{u}^\varepsilon] = 0 \qquad \text{on } \Gamma$$

$$(2.43) \qquad [a^\varepsilon_{ijkh} e_{kh}(\underline{u}^\varepsilon) n_j] = 0 \qquad \text{on } \Gamma$$

and the homogenized solution is associated to a homogeneous material with constant coefficients given by (2.22), where $\underline{w}^{kh}$ is defined by

$$(2.44) \qquad \int_Y a_{ij\ell m} [ e_{\ell my}(\underline{w}^{kh}) + \delta_{\ell k} \delta_{mh}] e_{ijy}(\underline{v}) dy = 0 \qquad \forall \underline{v} \in V_Y$$

instead of (2.20). (Note that for continuous coefficients, (2.44) is equivalent to (2.20)).

Moreover, as in chap. 5, sect. 5, the preceeding considerations hold for complex coefficients.∎

As in chap. 5, sect. 6, it is immediate to obtain homogenization properties for the dynamic problem

$$(2.45) \qquad \rho^\varepsilon(x) \frac{\partial^2 u^\varepsilon_i}{\partial t^2} - \frac{\partial \sigma^\varepsilon_{ij}}{\partial x_j} = f_i \quad ; \qquad u^\varepsilon_i|_{\partial_1\Omega} = \sigma_{ij} n_j|_{\partial_2\Omega} = 0$$

$$(2.46) \qquad \underline{u}^\varepsilon(0) = \underline{u}^{\varepsilon\prime}(0) = 0 \qquad (\,' \equiv \frac{d}{dt})$$

(The fact that the initial values are zero is of course not essential), where

(2.47) $\rho^{\varepsilon}(x) \equiv \rho(\frac{x}{\varepsilon})$ and $\rho(y)$ is Y-periodic

and positive, and the tensor $\sigma_{ij}^{\varepsilon}$ is related to $\underline{u}^{\varepsilon}$ by (2.5). As in chap. 5, sect. 6, the homogenized coefficients $a_{ijkh}^{h}$ are the same as in the static case and

$$\rho^{h} = \tilde{\rho} \equiv \frac{1}{|Y|} \int_{Y} \rho(y) \, dy \quad . \tag{2.48}$$

3.- A problem with couple-wise applied forces -

Homogenization techniques are useful in other problems with rapidly varying functions. As an example we consider here an elastic homogeneous medium acted upon by given forces which are periodic in the space variables, with zero mean value on each period. The periods tend to zero and the intensity of the forces is such that the couple of the given forces by unit volume tends to a non zero limit.

In fact, we define the Y-periodic functions $F_i(y)$, $i = 1,2,3$ such that

$$\tilde{F}_i \equiv \frac{1}{|Y|} \int_{Y} F_i(y) \, dy = 0 \tag{3.1}$$

and the functions $\Phi(x)$ defined on $\Omega$ and $f_i(x , y)$ , $i = 1, 2, 3$ defined on $\Omega \times Y$, Y-periodic in y. Moreover, we consider the constant coefficients $a_{ijkh}$ satisfying (1.3), (1.4) and we consider the problem

$$0 = \frac{\partial \sigma_{ij}^{\varepsilon}}{\partial x_j} + \frac{\Phi(x)}{\varepsilon} F_i(\frac{x}{\varepsilon}) + f_i(x , \frac{x}{\varepsilon}) \quad \text{in } \Omega \tag{3.2}$$

$$\sigma_{ij}^{\varepsilon} = a_{ijkh} \, e_{kh}(\underline{u}^{\varepsilon}) \tag{3.3}$$

$$u^{\varepsilon}\Big|_{\partial\Omega} = 0 \quad \text{on } \partial\Omega \quad . \tag{3.4}$$

It is clear that $\Phi(x)$ is a measure of the intensity of the couples, which are associated to the forces with components $\Phi F_i \, \varepsilon^{-1}$. On the other hand, there are "ordinary" forces $f_i$.

We search for an expansion in the form (2.6) - (2.8). We of course have (2.9), (2.10). We introduce this into (3.2) and obtain at orders $\varepsilon^{-1}$ and $\varepsilon^{0}$ respectively

$$0 = \frac{\partial \sigma_{ij}^{0}}{\partial y_j} + \Phi(x) \, F_i(y) \tag{3.5}$$

$$0 = \frac{\partial \sigma_{ij}^{0}}{\partial x_j} + \frac{\partial \sigma_{ij}^{1}}{\partial y_j} + f_i(x , y) \tag{3.6}$$

By applying the mean operator $\sim$ to (3.6) and using the periodicity of $\sigma^1$ (see (2.13)) we have

$$(3.7) \qquad 0 = \frac{\partial \tilde{\sigma}^0_{ij}}{\partial x_j} + \tilde{f}_i(x)$$

which is an equation for the macroscopic behaviour. Now, we search for the microscopic behaviour. Equation (3.5) becomes (note that the coefficients are constant) :

$$(3.8) \qquad 0 = \frac{\partial}{\partial y_j} [a_{ijkh}\, e_{kh}(\underline{u}^1)] + \Phi(x)\, F_i(y)$$

which is an equation for $\underline{u}^1$ (which is also Y-periodic) analogous to (2.15). Only the right hand side differs from (2.15). We consider the spaces $V_Y$, $H_Y$ defined by (2.16) ; the variational formulation of the problem is

$$(3.9) \qquad \begin{cases} \text{Find } \underline{u}^1 \in V_Y \text{ such that, } \quad \forall \underline{v} \in V_Y \quad , \\ \int_Y a_{ijkh}\, e_{kh}(\underline{u}^1)\, e_{ij}(\underline{v})\, dy = \Phi \int_Y F_i\, v_i\, dy \end{cases}$$

The necessary and sufficient condition for the existence of solution is, as in sect. 2 :

$$(3.10) \qquad \Phi \int_Y F_i(y)\, dy = 0 \qquad\qquad i = 1, 2, 3 \ .$$

(Note that the necessity is immediately obtained by integration of (3.8) on Y). This condition is satisfied by hypothesis (3.1). Consequently, as in (2.19), we have

$$(3.11) \qquad \underline{u}^1 = \underline{w}(y)\, \Phi(x) + \underline{c}(x)$$

where $\underline{w}(y)$ is the solution of (3.9) with $\Phi = 1$ ; it is unique if we impose $\tilde{\underline{w}} = 0$. Of course the vector $\underline{c}(x)$ is not determinated .

Bearing in mind (2.10), (3.11) and the fact that $\underline{c}$ is independent of y, we have

$$(3.12) \qquad \sigma^0_{ij} = a_{ijkh}\, [e_{khx}(\underline{u}^0) + \Phi\, e_{khy}(\underline{w})]$$

We then apply the operator $\sim$ and we note that

$$\tilde{e}_{khy}(\underline{w}) = \frac{1}{2|Y|} \int_Y \left( \frac{\partial w_k}{\partial y_h} + \frac{\partial w_h}{\partial y_k} \right) dy = \frac{1}{2|Y|} \int_{\partial Y} (n_h w_k + n_k w_h)\, dS = 0$$

We then obtain

$$(3.13) \qquad \tilde{\sigma}^0_{ij} = a_{ijkh}\, e_{khx}(\underline{u}^0)$$

Equation (3.7) (with the boundary condition (3.4) at the first order) shows that $u^0(x)$ is uniquely determinated by :

$$(3.14)\qquad \frac{\partial}{\partial x_j}[a_{ijkh}\, e_{khx}(\underline{u}^0)] + f_i = 0$$

$$(3.15)\qquad u^0 = 0 \qquad \text{on } \partial\Omega$$

which is the ordinary elasticity problem without the terms associated with $F_i$.

Remark 3.1 - As a result, the first term $\underline{u}^0$ of the expansion (2.6) is independent of the forces $F_i$. Oppositely, the first term $\sigma^0_{ij}$ of (2.8) is given by (3.12) and it depends on $F_i$. The mean value of $\sigma^0_{ij}$ on a period is given by (3.13) : it is thus independent of $F_i$ and satisfies the symmetry condition

$$(3.16)\qquad \tilde{\sigma}^0_{ij} = \tilde{\sigma}^0_{ji}$$

but this is a relation between the mean values on a period. In fact, the local values are such that their values on the boundary of a period equilibrate the couple of $\underline{F}$. If $\gamma_{ijk}$ is the tensor equal to one (resp. -1) for ijk in cyclic (resp. anti-cyclic) progression, we have, by (3.5) and (3.16) :

$$(3.17)\qquad \boxed{\int_{\partial Y} \gamma_{ijk}\, y_j\, \sigma^0_{k\ell} n_\ell\, dS} = \int_Y \gamma_{ijk} \frac{\partial}{\partial y_\ell}(y_j\, \sigma^0_{k\ell})\, dy =$$

$$= \delta_{j\ell}\gamma_{ijk}\, \tilde{\sigma}^0_{k\ell}|Y| + \int_Y \gamma_{ijk}\, y_j \frac{\partial \sigma^0_{k\ell}}{\partial y_\ell}\, dy =$$

$$\boxed{= - \Phi \int_Y \gamma_{ijk}\, y_j\, F_k\, dy}$$

which is not zero in general∎

Remark 3.2 - The elastic energy by unit volume is not the same as in the case $\underline{F} = 0$. In fact, the density of energy is, at the first order :

$$(3.18)\qquad a_{ijkh}\, e_{kh}(\underline{u}^\varepsilon)\, e_{ij}(\underline{u}^\varepsilon) \cong a_{ijkh}\, e^0_{kh}\, e^0_{ij} =$$

$$= a_{ijkh}\, [e_{khx}(\underline{u}^0) + e_{khy}(\underline{u}^1)]\, [e_{ijx}(\underline{u}^0) + e_{ijy}(\underline{u}^1)]$$

But as in the calculation between (3.12) and (3.13), we have

$$\frac{1}{|Y|}\int_Y a_{ijkh}\, e_{khx}(\underline{u}^0)\, e_{ijy}(\underline{u}^1)\, dy = a_{ijkh}\, e_{khx}(\underline{u}^0)\, \tilde{e}_{ijy}(\underline{u}^1) = 0$$

and of course

$$(3.19)\qquad \frac{1}{|Y|}\int_Y a_{ijkh}\, e_{khy}(\underline{u}^1)\, e_{ijy}(\underline{u}^1)\, dy = \beta\ \Phi^2$$

where $\beta > 0$ is defined as the left hand side of (3.19) with $\underline{w}$ instead of $\underline{u}^1$. Consequently, the asymptotic density of energy is given by

$$a_{ijkh}\, e_{khx}(\underline{u}^0) e_{ijx}(\underline{u}^0) + \beta\, \Phi^2$$

and we see that the energies associated to $\underline{u}^0$ and F are uncoupled.

4.- Homogenization in viscoelasticity

We consider a problem of viscoelasticity with instantaneous memory (and thus associate with differential equations). We shall see that the homogenized equations are integro-differential and they are associated with a viscoelasticity with non instantaneous memory. We have the following equations and initial and boundary conditions :

$$\rho^{\varepsilon}(x)\frac{\partial^2 u_i^{\varepsilon}}{\partial t^2} - \frac{\partial \sigma_{ij}^{\varepsilon}}{\partial x_j} = f_i \tag{4.1}$$

$$\underline{u}^{\varepsilon}(0) = \underline{u}^{\varepsilon\prime}(0) = 0 \qquad\qquad (\prime = \frac{d}{dt}) \tag{4.2}$$

$$u_i^{\varepsilon}\Big|_{\partial_1\Omega} = \sigma_{ij}^{\varepsilon}\, n_j\Big|_{\partial_2\Omega} = 0 \tag{4.3}$$

(the fact that the initial values are zero is not essential), where

$$\rho^{\varepsilon}(x) = \rho(\tfrac{x}{\varepsilon}) \quad \text{and} \quad \rho(y) \quad \text{is Y-periodic and positive.} \tag{4.4}$$

Moreover, we define the two systems of Y-periodic coefficients $a^o_{ijkh}$, $a^1_{ijkh}$, both satisfying the symmetry and positivity conditions (1.3) and (1.4). The strain-stress relation is :

$$\sigma_{ij}^{\varepsilon} = a^{o\varepsilon}_{ijk\ell}\, e_{k\ell}(\underline{u}^{\varepsilon}) + a^{1\varepsilon}_{ijk\ell}(\frac{\partial \underline{u}^{\varepsilon}}{\partial t}) \tag{4.5}$$

where $\quad a^{m\varepsilon}_{ijk\ell}(x) = a^{m}_{ijk\ell}(\frac{x}{\varepsilon})$

Remark 4.1 - If we formally take the Laplace transform of (4.5), it appears that the strain-stress relation is

$$\hat{\sigma}^{\varepsilon}_{ij} = (a^o_{ijk\ell} + p\, a^1_{ijk\ell})\; e_{k\ell}\,(\hat{\underline{u}}^{\varepsilon})$$

and the homogenized coefficients $(a^o_{ijk\ell} + p\, a^1_{ijk\ell})^h$ depend on p. This means that the homogenized behaviour will be given by a convolution product (associated to a non-instantaneous memory).∎

Theorem 4.1 - If $H = (L^2(\Omega))^3$ and V is defined by (1.10), for any continuous function $\underline{f}$ with values in H, the solution $\underline{u}^{\varepsilon}$ of problem (4.1) - (4.5) exists and is unique in the classical framework of semigroup theory (see the proof for details).

Proof - It is immediate to see that (4.1), (4.3) is formally equivalent to

$$(\underline{u}^{\varepsilon\prime\prime}, \underline{v})_{H_{\varepsilon}} + a^0(\underline{u}^{\varepsilon}, \underline{v}) + a^1(\underline{u}^{\varepsilon\prime}, \underline{v}) = \int_{\Omega} f\, v\, dx \qquad \forall v \in V \tag{4.6}$$

where

$$(4.7) \qquad (\underline{u}\ ,\ \underline{v})_{H_\varepsilon} \equiv \int_\Omega \rho^\varepsilon\, u_i\, v_i\, dx$$

$$(4.8) \qquad a^m(\underline{u}\ ,\ \underline{v}) \equiv \int_\Omega a^{m\varepsilon}_{ijm\ell}\, e_{m\ell}(\underline{u})\, e_{ij}(\underline{v})\, dx \quad ; \quad m = 0\ ,1\ .$$

The relation (4.6) will be taken as basis for the definition of the solution. Let us define $H_\varepsilon$ (resp. $V_\varepsilon$) as the space H (resp. V) equipped with the scalar product defined by (4.7) (resp. $a^0(u\ ,\ v)$ in (4.8). Let $A^0$, $A^1$ be the self-adjoint operators in $H_\varepsilon$ associated (according to the second representation theorem) with the forms $a^0$, $a^1$ in $H_\varepsilon$ . The relation (4.6) is equivalent to

$$(4.9) \qquad \underline{u}'' + A^0\,\underline{u} + A^1\,\underline{u}' = \underline{F} \quad ; \quad (F\ ,\ v)_{H_\varepsilon} \equiv \int_\Omega \rho^\varepsilon\, f_i\, v_i\, dx$$

Note that the boundary condition $\sigma_{ij}\, n_j = 0$ on $\partial_2\Omega$ is implicite in the formulation (4.9). Equation (4.9) with the initial condition (4.2) is equivalent to the first order system

$$(4.10) \qquad \frac{dv}{dt} + \mathcal{A}\, v = \phi \quad ; \quad v = \begin{pmatrix} v_1 \\ v_2 \end{pmatrix} \quad ; \quad \phi = \begin{pmatrix} 0 \\ F \end{pmatrix} \quad ; \quad v(0) = 0$$

where $\mathcal{A}$ is defined in $\mathcal{H}_\varepsilon = V_\varepsilon \times H_\varepsilon$ by

$$(4.11) \qquad \mathcal{A} = \begin{pmatrix} 0 & -I \\ A^0 & A^1 \end{pmatrix} \quad ; \quad D(\mathcal{A}) = \{ v \in \mathcal{H}_\varepsilon \ ;\ A^0 v_1 + A^1 v_2 \in H_\varepsilon \}$$

The Lumer-Phillips theorem shows that $-\mathcal{A}$ is the generator of a contraction semigroup in $\mathcal{H}_\varepsilon$ . For ,

1) $D(\mathcal{A})$ is dense in $\mathcal{H}_\varepsilon$ because it contains $D(A^0) \times D(A^1)$ which is dense in $V_\varepsilon \times H_\varepsilon$ (see chap. 2, remark 5.7)

2) $\mathcal{A}$ is accretive because

$$\mathrm{Re}(\mathcal{A}\, v\ ,\ v)_{\mathcal{H}_\varepsilon} = a^{1\varepsilon}(v_2\ ,\ v_2) \geq 0\ .$$

3) The range of $I + \mathcal{A}$ is the whole space $\mathcal{H}_\varepsilon$ . This amounts to say that for any $\psi_1 \in V_\varepsilon$ , $\psi_2 \in H_\varepsilon$, there exists $v_1$ , $v_2$ belonging to $D(\mathcal{A})$ such that

$$\left.\begin{array}{l} -\,v_2 + v_1 = \psi_1 \\ A^0\, v_1 + A^1 v_2 + v_2 = \psi_2 \end{array}\right\} \Leftrightarrow \left\{\begin{array}{l} v_2 = v_1 - \psi_1 \\ (A^0 + A^1 + I)v_1 = \psi_2 + A^1\,\psi_1 + \psi_1 \end{array}\right.$$

which is easily solved under its second form by the Lax-Milgram theorem (note that the right hand side $\psi_2 + A^1\,\psi_1$ is an element of $V'_\varepsilon$ , the dual of $V_\varepsilon$ when $H_\varepsilon$ is identified to its dual.

The proof is complete. ∎

Now, we study the asymptotic behaviour of $\underline{u}^{\varepsilon}$ as $\varepsilon \searrow 0$. Let us consider a formal expansion of the form ($y = x/\varepsilon$ ; all functions are Y-periodic in y) :

(4.12) $\underline{u}^{\varepsilon}(x , t) = \underline{u}^{0}(x , t) + \varepsilon\, \underline{u}^{1}(x , y , t) + \varepsilon^2 \dots \qquad y = \frac{x}{\varepsilon}$

(4.13) $e^{\varepsilon}_{ij}(x , t) = e^{o}_{ij}(x , y , t) + \varepsilon\, e^{1}_{ij}(x , y , t) + \varepsilon^2$

(4.14) $\sigma^{\varepsilon}_{ij}(x , t) = \sigma^{o}_{ij}(x , y , t) + \varepsilon\, \sigma^{1}_{ij}(x , y , t) + \dots =$

$$= (a^{o}_{ijkh}(y)\, e^{o}_{kh} + a^{1}_{ijkh} \frac{\partial}{\partial t} e^{o}_{kh}) + \varepsilon \dots$$

(4.15) $e^{o}_{ij}(x , y) = e_{ijx}(\underline{u}^{0}) + e_{ijy}(\underline{u}^{1})$

We introduce these expansions in (4.1) and we have, at the orders $\varepsilon^{-1}$ and $\varepsilon^{0}$:

(4.16) $$\frac{\partial \sigma^{o}_{ij}}{\partial y_j} = 0$$

(4.17) $$\rho^{\varepsilon} \frac{\partial^2 u^{o}_{i}}{\partial t^2} - \frac{\partial \sigma^{o}_{ij}}{\partial x_j} - \frac{\partial \sigma^{1}_{ij}}{\partial y_j} = f$$

and we apply the operator $\sim$ (mean value on Y) we obtain, by virtue of the Y-periodicity, as in (2.13) :

(4.18) $$\tilde{\rho} \frac{\partial^2 u^{o}_{i}}{\partial t^2} - \frac{\partial \tilde{\sigma}^{o}_{ij}}{\partial x_j} = f$$

which is the homogenized equation ; we may adjoin the initial and boundary conditions formally obtained from (4.2), (4.3) (this will be justified in the sequel) :

(4.19) $\underline{u}^{0}(x , 0) = \dfrac{\partial \underline{u}^{0}(x , 0)}{\partial t} = 0 \quad ; \quad \underline{u}^{0}\Big|_{\partial_1 \Omega} = 0 \quad \tilde{\sigma}^{o}_{ij}\, n_j\Big|_{\partial_2 \Omega} = 0$

We only have to find a relation between $\underline{u}^{0}$ and $\tilde{\sigma}^{o}_{ij}$. This will be obtained from (4.16). As in (2.18), if $V_Y$ is defined by (2.16), equation (4.16) with the periodicity condition is equivalent to

(4.20) $$\int_Y \sigma^{o}_{ij} \frac{\partial v_i}{\partial y_j}\, dy = 0 \qquad \forall\, v \in V_Y$$

where, by virtue of (4.14) and (4.15) :

(4.21) $$\sigma^{o}_{ij} = (a^{o}_{ijk\ell} + a^{1}_{ijk\ell} \frac{\partial}{\partial t})(e_{k\ell x}(\underline{u}^{0}) + e_{k\ell y}(\underline{u}^{1}))$$

Remark 4.2 - The local equation (4.20) is thus an evolution equation in time. This is a consequence of the fact that the strain-stress relation (4.5) contains derivatives in time. ■

In order to avoid difficulties associated to the fact that $\underline{u}^1$ is only determinated up to an additive constant, we introduce the space $\tilde{V}_Y$ of the functions of $V_Y$ with zero mean value :

(4.22) $$\tilde{V}_Y = \{\underline{v} \in V_Y \quad ; \quad \tilde{\underline{v}} = 0\}$$

and we introduce the scalar product

(4.23) $$(\underline{u}\ ,\ \underline{v})_{\tilde{V}_Y} = \int_Y a^1_{ijk\ell}(y)\ \frac{\partial u_k}{\partial y_\ell}\ \frac{\partial v_i}{\partial y_j}\ dy$$

which is associated to a norm equivalent to the natural norm of $(H^1)^3$ (this is proved as Lemma 1.2).

Now, (4.20) with (4.21) is equivalent to :

(4.24) $$\begin{cases} \text{Find } \underline{u}^1(t) \text{ with values in } \tilde{V}_Y \text{ such that} \\ \displaystyle\int_Y [a^0_{ijk\ell}(y) + a^1_{ijk\ell}(y)\frac{\partial}{\partial t}]\left(\frac{\partial u^0_k}{\partial x_\ell} + \frac{\partial u^1_k}{\partial y_\ell}\right)\frac{\partial v_i}{\partial y_j}\ dy = 0 \quad \forall \underline{v} \in \tilde{V}_Y \end{cases}$$

(We use the last equality (1.11) and we bear in mind that (4.20), (4.21) is independent of a constant added to $\underline{u}^1$ (for each t) ). Moreover, if $B \in \mathcal{L}(\tilde{V}_Y\ ,\ \tilde{V}_Y)$, $f^m_{k\ell} \in \tilde{V}_Y$ are defined by

(4.25) $$(B\ \underline{u}\ ,\ \underline{v})_{\tilde{V}_Y} = \int_Y a^0_{ijk\ell}\ \frac{\partial u_k}{\partial y_\ell}\ \frac{\partial v_i}{\partial y_j}\ dy$$

(4.26) $$(\underline{f}^m_{k\ell}\ ,\ \underline{v})_{\tilde{V}_Y} = \int_Y a^m_{ijk\ell}\ \frac{\partial v_i}{\partial y_j}\ dy \qquad\qquad m = 0\ ,\ 1$$

the relation (4.24) is equivalent to

(4.27) $$\left(B\ \underline{u}^1 + \frac{\partial \underline{u}^1}{\partial t} + \underline{f}^0_{k\ell}\ \frac{\partial u^0_k}{\partial x_\ell} + \underline{f}^1_{k\ell}\ \frac{\partial}{\partial t}\ \frac{\partial u^0_k}{\partial x_\ell}\ ,\ \underline{v}\right)_{\tilde{V}_Y} = 0 \quad ;\ \forall \underline{v} \in \tilde{V}_Y$$

and thus the first factor in (4.27) must be zero. Moreover, from (4.2), all the terms in the expansion (4.12) are zero for t = 0. We then have the following equation to find $\underline{u}^1$ when $\underline{u}^0$ is known :

(4.28) $$\begin{cases} \dfrac{\partial \underline{u}^1}{\partial t} + B\ \underline{u}^1 = -\ \underline{f}^0_{k\ell}\ \dfrac{\partial u^0_k}{\partial x_\ell} - f^1_{k\ell}\ \dfrac{\partial}{\partial t}\ \dfrac{\partial u^0_k}{\partial x_\ell} \\ \underline{u}^1(0) = 0 \end{cases}$$

which may be solved by standard semigroup theory (note that B is bounded and in fact it is generator of a group). We write

(4.29) $$\underline{w} = \underline{u}^1 + \underline{f}^1_{ij}\ \frac{\partial u^0_i}{\partial x_j} \qquad ; \qquad \underline{f}^2_{ij} = B\ \underline{f}^1_{ij} - \underline{f}^0_{ij} \implies$$

(4.30) $$\frac{\partial \underline{w}}{\partial t} + B\,\underline{w} = f^2_{ij}\,\frac{\partial u^o_i}{\partial x_j} \quad ; \quad \underline{w}\,(0) = 0$$

and by chap. 4, (3.2) :

(4.31) $$\underline{w}(t) = \int_0^t e^{-(t-s)B}\,\underline{f}^2_{ij}\,\frac{\partial u^o_i}{\partial x_j}(s)\,ds = \underline{u}^1(t) + \underline{f}^1_{ij}\,\frac{\partial u^o_i}{\partial x_j}$$

it is clear that $\underline{u}^o$ determine $\underline{u}^1$ and that $\underline{u}^1$ at time t depends on the values of $u^o(s)$ for $s \leq t$.

By differenciating (4.31) we have

(4.32) $$\frac{d\underline{u}^1}{dt}(t) = \underline{f}^2_{ij}\,\frac{\partial u^o_i}{\partial x_j}(t) - \underline{f}^1_{ij}\,\frac{\partial}{\partial t}\,\frac{\partial u^o_i}{\partial x_j} - \int_0^t B\,e^{-(t-s)B}\,\underline{f}^2_{ij}\,\frac{\partial u^o_i}{\partial x_j}(s)\,ds$$

and now we may write $\sigma^o_{ij}$ as a functional of $u^o(s)$ $(s < t)$. First, we note that by virtue of (4.14), (4.15) and the symmetry of the coefficients, we have :

(4.33) $$\sigma^o_{ij} = [a^o_{ijk\ell} + a^1_{ijk\ell}\,\frac{\partial}{\partial t}]\left(\frac{\partial u^o_k}{\partial x_\ell} + \frac{\partial u^1_k}{\partial y_\ell}\right)$$

Thus, by (4.31) and (4.32) and taking the mean value, we obtain :

(4.34) $$\boxed{\tilde{\sigma}^o_{ij}(t) = \beta^o_{ijmn}\,\frac{\partial u^o_m}{\partial x_n}(t) + \beta^1_{ijmn}\,\frac{\partial}{\partial t}\,\frac{\partial u^o_m}{\partial x_n}(t) + \int_0^t g_{ijmn}(t-s)\,\frac{\partial u^o_m}{\partial x_n}(s)\,ds}$$

where $\beta^o$ , $\beta^1$ (resp. g) are coefficients (resp. functions) defined by

(4.35) $$\begin{cases} \beta^o_{ijmn} = [a^o_{ijmn} - a^o_{ijk\ell}\,\frac{\partial}{\partial y_\ell}\,(\underline{f}^1_{mn})_k + a^1_{ijk\ell}\,\frac{\partial}{\partial y_\ell}(\underline{f}^2_{mn})_k\,]^\sim \\ \beta^1_{ijmn} = \left[a^1_{ijmn} - a^1_{ijk\ell}\,\frac{\partial}{\partial y_\ell}\,(\underline{f}^1_{mn})_k\right]^\sim \\ g_{ijmn}\,(\xi) = \left[a^o_{ijk\ell}\,\frac{\partial}{\partial y_\ell}(e^{-\xi B}\,\underline{f}^2_{mn})_k - a^1_{ijk\ell}\,\frac{\partial}{\partial y_\ell}(B\,e^{-\xi B}\,\underline{f}^2_{mn}\,)_k\right]^\sim \end{cases}$$

It is clear that the strain-stress law (4.34) contains an elastic term $(\beta^o)$, a viscoelastic term with instantaneous memory $(\beta^1)$ and a term with long memory (g). Consequently, (4.34) is a law of a different kind than (4.5).

Because B is a positive defined operator, $g(\xi)$ decays exponentially for $\xi \to +\infty$ and consequently the memory vanishes exponentially

(4.36) The homogenized problem is (4.18) with the initial and boundary conditions (4.19) and the strain-stress relation (4.34)

Remark 4.3 - In particular cases, the memory term in g is null. This is the case if the coefficients $a^0$ and $a^1$ are proportional. In this case, the operator B in (4.25) is of the form $\lambda I$ (I = identity) and $\underline{f}^2$ in (4.29) is zero. ■

Now, we study some properties of the relation (4.34). We take the Laplace transform of (4.34) where $\underline{u}^0(t)$ and $g_{ijmn}(t)$ are continuated with zero values for $t < 0$) and we obtain

$$(4.37) \qquad \hat{\tilde{\sigma}}^o_{ij}(p) = (\beta^o_{ijmn} + p\,\beta^1_{ijmn} + \hat{g}_{ijmn}(p))\,\frac{\partial \hat{u}^o_m}{\partial x_n}(p)$$

we define the coefficients $\alpha(p)$ by

$$(4.38) \qquad p\,\alpha_{ijmn}(p) \equiv \beta^o_{ijmn} + p\,\beta^1_{ijmn} + \hat{g}_{ijmn}$$

which is defined for $\mathrm{Re}\,p > 0$. On the other hand, the Laplace transform of (4.24), where $\underline{u}^1$ is a function of t with values in $\tilde{V}_Y$ and x is a parameter is

$$(4.39) \qquad \int_Y (a^o_{ijk\ell} + p\,a^1_{ijk\ell})\left(\frac{\partial \hat{u}^o_k}{\partial x_\ell} + \frac{\partial \hat{u}^1_k}{\partial y_\ell}\right)\frac{\partial v_i}{\partial y_j}\,dy = 0 \qquad \forall \underline{v} \in \hat{V}_Y$$

and from (4.33) :

$$(4.40) \qquad \hat{\sigma}^o_{ij} = (a^o_{ijmn} + p\,a^1_{ijmn})\left(\frac{\partial \hat{u}^o_k}{\partial x_\ell} + \frac{\partial \hat{u}^1}{\partial y_\ell}\right)$$

Note that we are searching for properties of the homogenized law (4.34) (i.e., of $\beta^0$, $\beta^1$, g) which is independent of the "given" function $\underline{u}^0(x)$. Consequently, we may suppose that $\underline{u}^0(x\,,\,t)$ has a Laplace transform, defined for instance for $\mathrm{Re}\,p > C$.

We see that (4.39) is analogous to the equation (2.18) of the local behaviour in elastostatics with the coefficients

$$(4.41) \qquad a^o_{ijk\ell} + p\,a^1_{ijk\ell}$$

By writting them under the form

$$p\left(\frac{a^0}{p} + a^1\right)$$

we see that for sufficiently large $|p|$, the sesquilinear form in (4.39) is coercive on $\hat{V}_Y$. Consequently, $\hat{\underline{u}}^1$ is well defined and the mean values may be calculated as in (2.20), (2.21) (for sufficiently large $|p|$). The homogenized coefficients are those associated to (4.41), i.e.

$$(4.42) \qquad p\,\alpha_{ijmn}(p) = (a^o_{ijmn} + p\,a^1_{ijmn})^h$$

Moreover, these coefficients are holomorphic functions of p for large $|p|$. This is easily seen from (2.20) ; if the coefficients are holomorphic functions of p (and moreover, of $p^{-1}$ for small $|p^{-1}|$), the operator $B(p^{-1})$ associated with the form in the left hand side of (2.20) is holomorphic (see later chap. 11, prop. 4.3) as well as $B^{-1}(p^{-1})$. Relation (2.20) becomes

$$B(p^{-1})w^{kh} = F(p^{-1})$$

where F is a holomorphic function of $p^{-1}$. Consequently, $\underline{w}^{kh}$ are holomorphic functions of $p^{-1}$ and by (2.22) the homogenized coefficients are too. It appears that $\alpha_{ijmn}(p)$ are holomorphic functions of p (and of $p^{-1}$ for small $|p^{-1}|$ ) and they take, for $p = \infty$ the values $a^{1h}_{ijmn}$ (i.e., the homogenized coefficients associated to $a^{1}_{ijmn}(y)$. Thus, we have proved the following proposition :

<u>Proposition 4.1</u> - The functions $\alpha_{ijmn}(p)$ defined by (4.38) are holomorphic functions of p for Rep $> 0$. Moreover, they are holomorphic functions of $p^{-1}$ for $|p^{-1}|$ sufficiently small and they take for $p = \infty$ the values

$$\alpha_{ijmn}(\infty) = a^{1h}_{ijmn}$$

where $a^{1h}_{ijmn}$ are the homogenized coefficients associated with $a^{1}_{ijmn}(y)$.

Now <u>we study the existence and uniqueness of the function</u> $\underline{u}^{0}$. We shall do this by Laplace transform. We recall that $\mathcal{L}(B)$ is the class of the functions having a Laplace transform in the framework of chapter 4, sect. 6 (where some properties of the Laplace transform are given).

<u>Theorem 4.2</u> - Let $\underline{f} \in \mathcal{L}(H)$. There exists one and only one $\underline{u}^{0} \in \mathcal{L}(V)$ which satisfies (in a generalized sense that will be seen in the proof) equation (4.18) with the initial and boundary conditions (4.19) and the stress-train relation (4.34).

<u>Proof</u> - We consider the Laplace transform of (4.18), (4.19), (4.34) and we take into account (4.37), (4.38) ; we have

$$\text{(4.42)} \qquad p^2 \rho\, \hat{u}^{o}_{i} - \frac{\partial \hat{\tilde{\sigma}}^{o}_{ij}}{\partial x_j} = \hat{f}_i$$

$$\text{(4.43)} \qquad \hat{u}^{o}_{i}\Big|_{\partial_1\Omega} = 0 \quad ; \quad \hat{\tilde{\sigma}}^{o}_{ij}\, n_j\Big|_{\partial_2\Omega} = 0$$

$$\text{(4.44)} \qquad \hat{\tilde{\sigma}}^{o}_{ij} = p\, \alpha_{ijmn}(p)\, \frac{\partial \hat{u}^{o}_{m}}{\partial x_n}$$

we shall prove that, for sufficiently large Rep, (4.42) - (4.44) define a unique $\hat{\underline{u}}^{o}$ having an inverse Laplace transform, which is the solution of the problem.

We define the form

$$a(p\ ;\ \underline{u}\ ,\ \underline{v}) = \int_\Omega \alpha_{ijmn}(p)\ \frac{\partial u_m}{\partial x_n}\ \frac{\partial \overline{v}_i}{\partial x_j}\ dx$$

which is sesquilinear on V. Moreover, for $p = \infty$ , it is coercive on V (see proposition 4.1 and Lemma 1.2 ; note that the coefficients $a^{1h}_{ijkh}$ satisfy a positivity condition analogous to (1.4) as a consequence of the hypothesis on $a^{1}_{ijkh}(y)$). Moreover, the coefficients $\alpha_{ijkh}(p)$ are holomorphic in $p^{-1}$ and thus, for p in a neighbourhood of infinity, the form a is coercive. In particular we have

Lemma 4.1 - There exists $c > 0$ and $\gamma > 0$ such that if $p = \xi + i\eta$ , and $\xi > c$, we have

(4.45) $\quad \mathrm{Re}\,a(p\ ,\ \underline{v}\ ,\ \underline{v}) \geqslant \gamma \|\underline{v}\|^2_V \qquad \forall\, v \in V$ .

On the other hand, the variational formulation of (4.42) - (4.44) is (after division by p) :

(4.46) $\quad$ Find $\hat{\underline{u}}^o \in V$ such that
$p\,\beta(\hat{\underline{u}}^o\ ,\ \underline{v})_H + a(p\ ;\ \hat{\underline{u}}^o\ ,\ \underline{v}) = \frac{1}{p}(\hat{f}\ ,\ v)_H \qquad \forall\, \underline{v} \in V$

The left hand side of (4.46) is a sesquilinear and continuous form on V ; moreover, for $\xi > c$, its real part is more or equal than $\gamma \|v\|^2$ ; consequently it is coercive and by the Lax-Milgram theorem $\hat{u}^o(p)$ exists and is unique for sufficiently large $\xi$ (because $\hat{f}$ is a Laplace transform and thus a holomorphic function of p for sufficiently large $\xi$). Moreover, $\hat{\underline{u}}^o(p)$ is a holomorphic function of p (this is easily seen as in prop. 4.1); if $A(p)$ is the operator of $\mathcal{L}(V\ ,\ V')$ associated to the form in the left hand side of (4.46), this equation becomes

$$A(p)\hat{\underline{u}}^o = \frac{1}{p}\,\hat{\underline{f}} \Rightarrow \hat{\underline{u}}^o = A(p)^{-1}\,\frac{1}{p}\,\hat{\underline{f}}$$

and $A(p)^{-1}$ is a holomorphic family of operators). Moreover, by taking $\underline{v} = \hat{\underline{u}}^o$ in (4.46), for sufficiently large $\xi$( $\xi > d$, say), we have :

$$\gamma \|\underline{v}\|^2_V \leqslant \mathrm{Re}\,\{\tfrac{1}{p}(\hat{\underline{f}}\ ,\ \underline{v})_H\} \leqslant \frac{1}{|p|}\left|(\hat{\underline{f}}\ ,\ \underline{v})_H\right| \leqslant \frac{c}{d}\,\|\hat{\underline{f}}\|_H\,\|\underline{v}\|_V$$

and we see that $\|\underline{v}\|_V$ (as well as $\|\hat{f}\|_H$) is bounded by a polynomial in $|p|$ . Consequently, $\hat{\underline{u}}^o$ is the Laplace transform of a well determinated distribution with values in V (see chap. 4, theorem 6.1).■

Consequently, $\underline{u}^o$ is a well-determinated function. Now it is easy to prove the convergence of the functions $\underline{u}^\varepsilon$ (given by theorem 4.1) to $\underline{u}^o$ (given by theorem 4.2). In fact, we have :

<u>Theorem 4.3</u> - Let $f \in L^2(0, T; H)$. Then :

(4.47) $\underline{u}^\varepsilon \to \underline{u}^0$ in $L^\infty(0, T; V)$ weakly *

(4.48) $\underline{u}^{\varepsilon\prime} \to \underline{u}^{0\prime}$ in $L^\infty(0, T; H)$ weakly *

where $\underline{u}^\varepsilon$ and $\underline{u}^0$ are given by theorems 4.1 and 4.2 respectively.

<u>Proof</u> - We continuate the function $\underline{f}$ with value 0 for $t > T$. Then, $\underline{u}^\varepsilon$ and $\underline{u}^0$ exist in the framework of theorems 4.1 and 4.2.

We search for an a priori estimate for $\underline{u}^\varepsilon$. We take $\underline{v} = \underline{u}^{\varepsilon\prime}$ in (4.6) and we neglect the positive term $a^1$ ; then

(4.49) $$\frac{1}{2}\frac{d}{dt} Z^\varepsilon(t) \leq \| \underline{f} \|_{H_\varepsilon} \|\underline{u}^{\varepsilon\prime}\|_{H_\varepsilon} \qquad \text{where}$$

$$Z^\varepsilon(t) = \|\underline{u}^{\varepsilon\prime}\|^2_{H_\varepsilon} + a^0(\underline{u}^\varepsilon, \underline{u}^\varepsilon)$$

From (4.49) we have

$$\frac{dZ^\varepsilon}{dt} \leq 2 \|\underline{f}\| (Z^\varepsilon)^{1/2} \leq \|\underline{f}\|^2 + Z^\varepsilon$$

$$\frac{d}{dt}(e^{-t} Z^\varepsilon(t)) \leq \| \underline{f}(t) \|^2 \quad ;$$

and by integrating from 0 to t and taking into account that $Z^\varepsilon(0) = 0$, we have

(4.50) $$Z^\varepsilon(t) \leq e^t \int_0^t \|\underline{f}(S)\|^2_{H_\varepsilon} e^{-S}\, dS$$

and using (4.50) with $t \in [0, T]$ and then (4.49) with $t \geq T$ (where $f = 0$) we see that $Z^\varepsilon(t)$ is bounded by a constant independent of $\varepsilon$ and t. Moreover, the norms of $H_\varepsilon$ and H are equivalent, and then

$$\|\underline{u}^\varepsilon\|_V \leq C \quad ; \quad \|\underline{u}^{\varepsilon\prime}\|_H \leq C$$

where C is a constant independent of $\varepsilon$ and t. By weak-star compactness, we can extract subsequences such that

(4.51) $$\begin{cases} \underline{u}^\varepsilon \to \underline{u}^* & \text{in } L^\infty(0, \infty; V) \text{ weakly } * \\ \underline{u}^{\varepsilon\prime} \to u^{*\prime} & \text{in } L^\infty(0, \infty; H) \text{ weakly } * \end{cases}$$

The theorem will be proved if we show that for any subsequence of the type (4.51), $\underline{u}^* = \underline{u}^0$. By taking the Laplace transform of (4.51), we have, for Re $p > 0$ :

(4.52) $$\hat{\underline{u}}^\varepsilon \to \hat{u}^* \qquad \text{in } V \text{ weakly}$$

On the other hand, $\hat{\underline{u}}^\varepsilon$ satisfies

(4.53) $$p^2 \rho^\varepsilon \hat{\underline{u}}^\varepsilon - \frac{\partial \hat{\sigma}^\varepsilon_{ij}}{\partial x_j} = \hat{f}_i$$

(4.54) $\underline{\hat{u}}^{\varepsilon} \Big|_{\partial_1\Omega} = 0 \qquad ; \qquad \hat{\sigma}^{1\varepsilon}_{ij}\, n_j \Big|_{\partial_2\Omega} = 0$

(4.55) $\hat{\sigma}^{\varepsilon}_{ij} = (a^{o\varepsilon}_{ijk\ell} + p\, a^{1\varepsilon}_{ijk\ell})\, e_{k\ell}(\underline{u}^{\varepsilon})$

and by taking into account (4.42). We can pass to the limit for real positive p as in the problem (4.1), (4.2) and we see that for real positive p, $\underline{\hat{u}}^{*}$ satisfies (4.42) - (4.44). By the uniqueness of this problem, we have $\underline{\hat{u}}^{*} = \underline{\hat{u}}^{0}$ for real positive p and thus, by analytic continuation, for any p with positive real part. By the uniqueness of the inverse Laplace transform, we have $\underline{u}^{*} = \underline{u}^{0}$; Q.E.D.

## 5.- Fissured elastic body. Generalities

We introduce here the boundary value problem for a fissured body subject to one-side constraint (the two lips of the fissure may be open, but they cannot overlap). This leads to a variational inequality. In the following sections, we consider the homogenized behaviour of a body with many small, periodically distributed fissures.

We assume that the fissure is without friction : when it is closed at a point, the force acting on each lip at that point is normal to the fissure.

Before setting the problem, we recall two theorems about variational inequalities which will be used in the sequel. Proofs may be seen in Lions [4 ], (see also Lions [3 ]).

Let V be a real Hilbert space, L, a linear bounded functional on V, a, a bilinear continuous form on V, and K a convex closed set of V. Then :

**Theorem 5.1** - There exists a unique $u \in K$ such that

(5.1) $$a(u\,,\, v - u) \geqslant L(v - u) \qquad \forall v \in K.$$

Moreover, let H be a Hilbert space (identified to its dual) such that

$$V \subset H \subset V'$$

with continuous and dense embedding, and let

$$f\,,\ \frac{\partial f}{\partial t} \in L^2(0\,,\,T\,;\,V') \quad ; \quad f(0) \in H \quad \text{and} \quad u_0 \in K \cap D(A)$$

where A is the operator associated to the form a according to the first representation theorem. Then, we have :

Theorem 5.2 - There exists a unique u such that $u(t) \in K$ a. e. in t and

$$u, \frac{\partial u}{\partial t} \in L^2(0, T ; V) \quad ; \quad u(0) = u_0 \quad ;$$

(5.2) $$\left(\frac{\partial u}{\partial t}, v - u\right)_H + a(u(t), v - u) \geq (f, v - u)_{V'V} \qquad \forall v \in K \ .$$

Now, let us consider an elastic body $\Omega$ , fixed by its boundary $\partial\Omega$ . It is assumed to be homogeneous, the elastic coefficients $a_{ijkh}$ are constant and satisfy the symmetry and positivity conditions (1.3) and (1.4). Let F (the fissure) be a piece of a smooth surface. It may be not connected (i.e., we may be have several fissures) but the domain

(5.3) $$\Omega_F = \Omega \setminus F$$

is assumed to be connected (i.e., F do not decompose the body into several pieces).

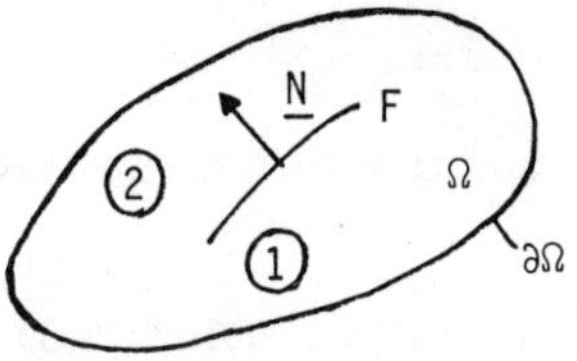


We assume that F contains its boundary (i.e., F is closed) which is a smooth curve. Then, $\Omega_F$ is open. It is clear that $\Omega_F$ has not a smooth boundary : nevertheless, the Korn's inequality holds for it (it suffices to decompose $\Omega_F$ into two smooth domains). Moreover, the trace theorem, under the form of chap. 1, th. 3.1 holds for each side of F. Moreover, let $\underline{N}$ be a unit normal to F ($\underline{N}$ is taken outer to a side of F, noted 1; the other side is noted 2 (see fig.)). We use the classical notation for the discontinuity of a function across F :

(5.4) $$[\phi] = \phi_2 - \phi_1$$

The classical formulation of the problem, with the notations of sect. 3, is :

(5.5) $$0 = \frac{\partial \sigma_{ij}}{\partial x_j} + f_i \qquad \text{in } \Omega_F$$

(5.6) $$u_i = 0 \qquad \text{on } \partial\Omega$$

(5.7) $$[u_i \ N_i] \geq 0 \qquad \text{on } F$$

(5.8) $$\sigma_{ij} n_j \Big|_1 = \sigma_{NN} \Big|_1 N_i \ ; \ \sigma_{ij} n_j \Big|_2 = - \sigma_{NN} \Big|_1 N_i \ ; \ \sigma_{NN} \Big|_1 \leq 0 \text{ on } F$$

(5.9) $$\text{if } (5.7) > 0 \Rightarrow \sigma_{NN} = 0 \ .$$

where $\underline{n}$ is of course exterior to the open domain $\Omega - F$.

The meaning of the preceeding boundary conditions is the following : if $\sigma_{NN}$ is the stress in the direction of N for a section normal to N, the two first relations (5.8) express that the force on F is normal and the action is opposite to the reaction. The third relation (5.8) means that there is compression (but not traction !) on F. Relation (5.9) means that, if the fissure is open at a point, the force is zero at this point.

In order to state the variational formulation of this problem, we introduce the space $V_F$ and the convex set $K_F$ defined by

(5.10) $$V_F = \{\underline{u} \ ; \ u_i \in H^1(\Omega_F) \ ; \ u_i\big|_{\partial\Omega} = 0\}$$

(5.11) $$K_F = \{\underline{u} \ ; \ \underline{u} \in V_F \ ; \ [u_i N_i] \geqslant 0 \quad \text{a.e. on } F\}$$

Lemma 5.1 - $V_F$ equipped with the classical norm of $(H^1(\Omega_F))^3$ is a Hilbert space. $K_F$ is a closed convex set of $V_F$.

This Lemma is evident by taking into account that $L^2$ convergence implies a.e. convergence.

The variational formulation of the problem (5.5) - (5.9) is the following

(5.12)
Find $\underline{u} \in K_F$ such that

$$a(\underline{u}, \underline{v} - \underline{u}) \geqslant \int_{\Omega_F} f_i (v_i - u_i)\, dx \qquad \forall v \in K_F$$

where a is the bilinear form on $V_F$ given by (1.11).

It is not hard to prove the formal equivalence of (5.5) - (5.9) and (5.12). If $\underline{u}$ satisfies (5.5) - (5.9), we multiply (5.5) by $\underline{v} \in K_F$ and by integrating by parts on $\Omega_F$, we obtain (5.12). On the other hand, if $\underline{u} \in K_F$ satisfies the virtual power identity (5.12), we see that :

a) by taking $\underline{v} - \underline{u}$ with compact support in $\Omega_F$, (5.5) is satisfied.

b) by taking $\underline{v}$ such that $[\underline{v}-\underline{u}]\big|_F = 0$, we see that $\sigma_{ij} n_j$ is continuous through F ; moreover, by taking $\underline{v}$ with arbitrary tangential components on each side of F, we obtain the two first relations (5.8).

c) by taking $\underline{v}-\underline{u}$ zero on a side of F and with normal component $\geqslant 0$ on the other side, we have the third relation (5.8). Finally, if in a point $[u_i n_i] > 0$, we may take $\underline{v}$ arbitrary on the both sides of F in a neighbourhood of that point and we have (5.9). The equivalence is proved.

By virtue of the preceeding considerations about the validity of the Korn's inequality, the left (resp. right) hand side of (5.12) is a bilinear continuous and coercive (resp. linear continuous) form on $V_F$. Theorem 5.1 holds for problem (5.12) and consequently $\underline{u}$ exists and is unique.

## 6.- Homogenization of an elastic body with small, periodic fissures. Formal expansion

Now we consider a sequence of problems of the type of the preceeding section depending on a small parameter $\varepsilon$ . In the classical framework of homogenization, we consider the period Y in $R^3$ with a smooth fissure F which do not intersects the boundary of Y. Then, $Y_F$ denotes the open domain $Y \setminus F$. Moreover, we

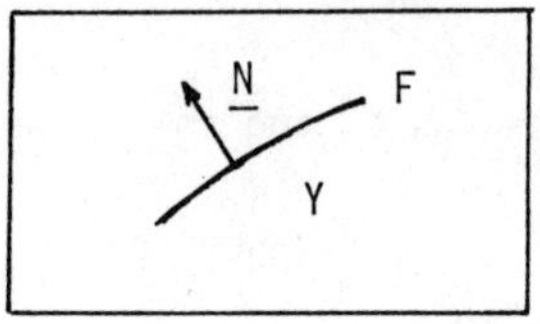


consider the bounded domain $\Omega$ (the body) as in the preceeding section, and we denote denote by $\Omega_{F\varepsilon}$ the fissured body, i.e. :

(6.1) $$\Omega_{F\varepsilon} = \Omega \cap \{ \underline{x} \ ; \frac{x}{\varepsilon} \in Y_F \}$$

Remark 6.1 - The domain $\Omega_{F\varepsilon}$ is open if the fissures $\varepsilon F$ do not intersect $\partial\Omega$ . We shall assume in the sequel that this is the case. ■

The space $V_{F\varepsilon}$ and the convex set $K_{F\varepsilon}$ are defined as in (5.10), (5.11) but with $\Omega_{F\varepsilon}$ instead of $\Omega_F$.

The solution $\underline{u}^\varepsilon(x)$ of the elasticity problem in $\Omega_{F\varepsilon}$ is the (unique) functions $\underline{u}^\varepsilon$ satisfying :

(6.2) $$\begin{cases} \underline{u}^\varepsilon \in K_{F\varepsilon} \\ \displaystyle\int_\Omega a_{ijkh} \frac{\partial u_k^\varepsilon}{\partial x_h} \frac{\partial (v_i - u_i^\varepsilon)}{\partial x_j} dx \geqslant \int_\Omega f_i (v_i - u_i^\varepsilon) dx \quad \forall \underline{v} \in K_{F\varepsilon} \end{cases}$$

Now we study the asymptotic behaviour as $\varepsilon \searrow 0$ by applying the method used in chap. 5, sect. 9 for the expansion of an integral identity. We recall that, if $\sim$ is the usual mean operator on Y, we have :

(6.3) $$\lim_{\varepsilon \searrow 0} \int_\Omega F(x , \frac{x}{\varepsilon}) \, dx = \int_\Omega \tilde{F}(x) \, dx$$

for functions F which are Y-periodic in y.

Now, we postulate the formal expansion (6.4) for $\underline{u}^\varepsilon$ and we take test functions of the form (6.5) :

(6.4) $\underline{u}^\varepsilon(x) = \underline{u}^0(x) + \varepsilon\,\underline{u}^1(x\,,\,y) + \varepsilon^2 \ldots \qquad y = x/\varepsilon$

(6.5) $\underline{v}^\varepsilon(x) = \underline{v}^0(x) + \varepsilon\,\underline{v}^1(x\,,\,y) + \varepsilon^2 \ldots \qquad y = x/\varepsilon$

such that $\underline{u}^i(x\,,\,y)$ are defined on $\Omega \times Y_F$ and are Y-periodic in y. Moreover, we note that because $\underline{u}^0$ and $\underline{v}^0$ do not depend on y (and they are of course smooth, in the general framework of the asymptotic expansions), $\underline{u}^\varepsilon \in K_{F\varepsilon}$ is equivalent in the asymptotic expansion to :

(6.6) $[\underline{u}^1 \,.\, N] \geqslant 0 \qquad$ on F.

Now, we introduce the spaces $V_Y$, $\tilde{V}_Y$ and the convex sets $K_Y$, $\tilde{K}_Y$ :

(6.7) $V_{YF} = \{\underline{v}\ ;\ v_i \in H^1(Y_F) \quad ;\quad \text{Y-periodic}\ \}$

(6.8) $\tilde{V}_{YF} = \{\underline{v}\ ;\ \underline{v} \in V_{YF} \quad ;\quad \tilde{\underline{v}} = 0\ \}$

(6.9) $K_{YF} = \{\underline{v}\ ;\ \underline{v} \in V_{YF} \quad ;\quad [\underline{v}\,.\,\underline{N}]\big|_F \geqslant 0 \quad \text{a.e.}\ \}$

(6.10) $\tilde{K}_{YF} = \{\underline{v}\ ;\ \underline{v} \in K_{YF} \quad ;\quad \tilde{\underline{v}} = 0\ \}$

<u>Remark 6.2</u> - The notation of (6.7) is a little ambiguous ; in fact the functions $\underline{v}_i$ are Y-periodic functions defined on $R^3$ belonging to $H^1_{loc}(R^3)$. ■

According to (6.6), we define the expansions (6.4), (6.5) in a more precise way

(6.11) The expansions((6.4), (6.5) are understood in the sense that $\underline{u}^0$ , $\underline{v}^0 \in (H^1_0(\Omega))^3$ and $\underline{u}^1$, $\underline{v}^1$ are smooth functions of x defined on $\Omega$ with values in $K_{YF}$

Now, we replace (6.4) and (6.5) into (6.2) and by taking the limit value as $\varepsilon \searrow 0$ according to (6.3), we obtain :

$$(6.12)\quad \int_\Omega a_{ijkh} \frac{\partial u^0_k}{\partial x_h} \frac{\partial (v^0_i - u^0_i)}{\partial x_j}\, dx + \int_\Omega a_{ijkh} \frac{\partial u^0_k}{\partial x_h} \left(\frac{\partial (v^1_i - u^1_i)}{\partial y_j}\right)^{\sim} dx +$$

$$+ \int_\Omega a_{ijkh} \left(\frac{\partial u^1_k}{\partial y_h}\right)^{\sim} \frac{\partial (v^0_i - u^0_i)}{\partial x_j}\, dx + \int_\Omega a_{ijkh} \left(\frac{\partial u^1_h}{\partial y_k} \frac{\partial (v^1_i - u^1_i)}{\partial y_j}\right)^{\sim} dy -$$

$$- \int_\Omega f_i (v^0_i - u^0_i)\, dx \geqslant 0$$

for any $\underline{v}^0$, $v\underline{^1}$ satisfying (6.11).

If, in particular, we take $\underline{v}^1 = \underline{u}^1$ in (6.12), we have

(6.13) $$\int_\Omega a_{ijkh}\left(\frac{\partial u_k^0}{\partial x_h} + \frac{\partial u_k^1}{\partial y_h}\right)^\sim \frac{\partial (v_i^0 - u_i^0)}{\partial x_j} dx - \int_\Omega f_i (v_i^0 - u_i^0)\, dx = 0$$

(note that this expression is $\geqslant 0$ ; but, in fact, $\underline{v}^0 - \underline{u}^0$ is arbitrary and then we have = 0. This gives the homogenized equation and boundary condition :

(6.14) $$0 = \frac{\partial \tilde{\sigma}_{ij}^0}{\partial x_j} + f_i \qquad \text{where } \sigma_{ij}^0 = a_{ijkh}\left(\frac{\partial u_k^0}{\partial x_h} + \frac{\partial u_k^1}{\partial y_h}\right)$$

(6.15) $$\underline{u}^0 = 0 \qquad \text{on } \partial\Omega$$

Note that the problem has been "homogenized" and there are no more fissures !

Now, we search for the "homogenized" strain-stress relation between $\tilde{\sigma}^0$ and $u^0$ by studying the local problem in Y.

We replace (6.13) into (6.12) ; moreover, we take $\underline{v}^1(x , y)$ of the form :

(6.16) $$\begin{cases} \underline{v}^1(x , y) = \underline{u}^1(x , y) + \theta(x)\, [\underline{w}(y) - \underline{u}^1(x , y)] \equiv (1 - \theta)\underline{u}^1 + \theta \underline{w} \\ \theta \in \mathcal{D}(\Omega) \quad , \quad 0 \leqslant \theta \leqslant 1 \quad , \quad \underline{w} \in K_{YF}. \end{cases}$$

which is a smooth function of x with values in $K_{YF}$ (see (6.11)). Then, (6.12) becomes :

(6.17) $$\int_\Omega \left[a_{ijkh}\left(\frac{\partial u_k^0}{\partial x_h} + \frac{\partial u_k^1}{\partial y_h}\right) \frac{\partial (w_i - u_i^1)}{\partial y_i}\right]^\sim \theta\, dx \geqslant 0$$

for any $\theta \in \mathcal{D}(\Omega)$ with $0 \leqslant \theta(x) \leqslant 1$. This is equivalent to

(6.18) $$\left[a_{ijkh}\left(\frac{\partial u_k^0}{\partial x_h} + \frac{\partial u_k^1}{\partial y_h}\right) \frac{\partial (w_i - u_i^1)}{\partial y_j}\right]^\sim \geqslant 0 \qquad \forall\, \underline{w} \in K_{YF}$$

which is the local problem. Now, we shall write it under a more explicit form. (6.18) is not modified by adding a constant to the unknown $\underline{u}^1$ or to the test function $\underline{w}$. It is then natural to work in $\tilde{V}_{YF}$ and $\tilde{K}_{YF}$ instead of $V_{YF}$, $K_{YF}$ (in order to obtain coerciveness properties).

Local problem - Let $u^0(x)$ be given. Find $u^1 \in \tilde{K}_{YF}$ such that

(6.19) $$\int_Y a_{ijkh}\left(\frac{\partial u_k^0}{\partial x_h} + \frac{\partial u_k^1}{\partial y_h}\right) \frac{\partial (w_i - u_i^1)}{\partial y_j}\, dy \geqslant 0 \qquad \forall\, \underline{w} \in \tilde{K}_{YF}$$

(problem in y, x playing the role of a parameter)

Theorem 6.1 - The local problem (6.19) has one and only one solution.

Proof - The form

(6.20) $$\int_Y a_{ijkh} \frac{\partial u_k}{\partial y_h} \frac{\partial v_i}{\partial y_j}\, dy$$

is continuous and coercive on $\tilde{V}_{YF}$ (which is equipped with the classical norm of $(H^1(Y_F))^3$) ; the proof of this assertion is analogous to that of lemma 1.2 and will not be given here. Note that (6.20) is not coercive on $V_{YF}$ ; this is the reason why we introduced the space $\tilde{V}_{YF}$ of functions with zero mean value.

On the other hand,

$$\frac{\partial u_k^o}{\partial x_h} \int_Y a_{ijkh} \frac{\partial v_i}{\partial y_j} \, dy$$

is obviously a bounded linear form on $\tilde{V}_{YF}$. The existence and uniqueness follow from theorem 5.1. ■

Proposition 6.1 - If we postulate an expansion of the form (6.4), (6.11) for the solution $\underline{u}^\varepsilon$ of the problem (6.2), its first term $\underline{u}^o(x)$ satisfy the equation (6.14) and the boundary condition (6.15). Moreover, for given $\underline{u}^o(x)$, the function $\underline{u}^1(x , y)$ is the solution of the nonlinear local problem (6.19). Taking into account the second relation (6.14), this defines $\tilde{\sigma}^o_{ij}$ as a function of $\underline{u}^o(x)$ (in fact of $\underline{\mathrm{grad}}_x \, \underline{u}^o(x)$) which is thus a nonlinear elastic strain-stress law.

The strain-stress law is studied in detail in next section.

## 7.- Study of the homogenized strain-stress law and consequences

We study the nonlinear function

$$\underline{\mathrm{grad}}_x \, \underline{u}^o \longmapsto \tilde{\sigma}^o_{ij} \tag{7.1}$$

of the preceeding section and we show that it is a hyperelastic law. Moreover, we establish an existence and uniqueness theorem for the limit problem in $\underline{u}^o$.

First, we change a little the notations and the formulation of the local problem (6.19).

By taking into account the symmetry of the coefficients $a_{ijkh}$, the local problem (6.19) may be written (see (2.8) - (2.10)) :

$$\left\{ \begin{array}{l} \text{find } \underline{u}^1 \in \tilde{K}_{YF} \quad \text{such that} \\ \displaystyle\int_Y a_{ijkh} \left[ e_{khx}(\underline{u}^o) + e_{khy}(\underline{u}^1) \right] e_{ijy}(\underline{w} - \underline{u}^1) \, dy \geqslant 0 \qquad \forall \underline{w} \in \tilde{K}_{YF} \end{array} \right.$$

now we remark that x is a parameter. Then we write

(7.2)
$$\begin{array}{llll} E_{kh} & \text{instead of} & e_{khx}(\underline{u}^0) \\ \underline{u} & " \quad " & \underline{u}^1 \\ e_{kh} & " \quad " & e_{khy} \\ \sigma_{ij} & " \quad " & \sigma^0_{ij} \end{array}$$

Because $E_{kh}$ is a symmetric tensor in $R^3$, it may be considered as an arbitrary element of $R^6$. Then, the local problem becomes

(7.3) Local problem (equivalent to (6.19)) : For given $E_{kh} \in R^6$, find $\underline{u} \in \tilde{K}_{YF}$ such that

$$\int_Y a_{ijkh} [e_{kh}(\underline{u}) + E_{kh}] e_{ij}(\underline{w} - \underline{u}) \, dy \geqslant 0 \qquad \forall \underline{w} \in \tilde{K}_{YF}$$

Moreover, $\sigma_{ij}$ (i.e. $\sigma^0_{ij}$ with the preceeding notation) is defined by

(7.4)
$$\sigma_{ij} = a_{ijkh}(E_{kh} + e_{kh}(\underline{u}))$$

and of course

(7.5)
$$\tilde{\sigma}_{ij} = \frac{1}{|Y|} \int_Y \sigma_{ij}(y) \, dy$$

(7.6) The homogenized stress-train law is the function (which takes $R^6$ into $R^6$) :

$$E_{kh} \rightarrow \tilde{\sigma}_{ij}$$

In order to study the function (7.6), let us define the function (from $R^6$ into R) :

(7.7)
$$W(E_{kh}) = \frac{1}{2|Y|} \int_Y a_{ij\ell m} [e_{ij}(\underline{u}) + E_{ij}][e_{\ell m}(\underline{u}) + E_{\ell m}] \, dy$$

where it is understood that, forgiven E, we calculate $\underline{u}$ by (7.3) and then (7.7).

The principal result of this section is the following :

<u>Theorem 7.1</u> - The function W defined by (7.7) is of class $C^1$, positive, convex. Moreover, $\tilde{\sigma}_{ij}$ is given by

(7.8)
$$\tilde{\sigma}_{ij} = \frac{\partial W}{\partial E_{ij}}$$

In mechanics this means that the homogenized law is <u>hyperelastic</u>. The proof of this theorem is given in two lemmas.

Lemma 7.1 - The function W is of class $C^o$, convex, and $\tilde{\sigma}$ is a subgradient of W :

Proof - The fact that W is continuous is a consequence of the fact that $\underline{u}$ is a continuous function of E (defined on $R^6$) with values in $V_{YF}$ (which is immediately proved by taking two points $E^1_{ij}$, $E^2_{ij}$ and the test function $\underline{w}$ in (7.3) for each one equal to the solution corresponding to the other).

Now, we remark that if a function has a subgradient at any point, it is necessarily convex. Thus, it suffices to verify that $\tilde{\sigma}$ is a subgradient. Let E be a fixed point of $R^6$ ; we consider a variation $\delta E = E^1 - E$ : if we write

$$e^*_{ij} = e_{ij}(\underline{u}) + E_{ij} \tag{7.9}$$

the variation of W is (by virtue of the symmetry of the coefficients and by using (7.4) and (1.4)) :

$$\boxed{\delta W} = \frac{1}{|Y|}\int_Y a_{ijkh}\, e^*_{kh}\, \delta e^*_{ij}\, dy + \frac{1}{2|Y|}\int_Y a_{ijkh}\, \delta e^*_{kh}\, \delta e^*_{ij}\, dy \geqslant \tag{7.10}$$

$$\geqslant \frac{1}{|Y|}\int_Y a_{ijkh}\, e^*_{kh}\, [\delta e_{ij} + \delta E_{ij}]\, dy =$$

$$\boxed{= \frac{1}{|Y|}\int_Y a_{ijkh}\, [e_{kh} + E_{kh}]\delta e_{ij}\, dy + \tilde{\sigma}_{ij}\, \delta E_{ij}}$$

On the other hand by taking $\underline{w} = \underline{u}(E + \delta E) \in \tilde{K}_{YF}$ in (7.3), we have $\underline{w} - \underline{u} = \delta\underline{u}$ and we see that the first term in the right hand side of (7.10) is $\geqslant 0$ ; thus

$$\delta W \geqslant \tilde{\sigma}_{ij}\, \delta E_{ij} \tag{7.11}$$

which shows that $\tilde{\sigma}_{ij}$ is a subgradient. The lemma is proved. ■

The following lemma has a general character and is independent of the special function W considered here.

Lemma 7.2 - Let W(E) be a $C^o$, convex function from $R^N$ into R. If $\tilde{\sigma}$ is a subgradient of W in each point and this subgradient is continuous (from $R^N$ into $R^N$), then W is differentiable and its gradient is $\tilde{\sigma}$ .

Proof - We shall prove this by contradiction. If W is not differentiable, in a point $E^o$, there exist a sequence $E^i \to E^o$ such that

$$|W(E^i) - W(E^o) - \tilde{\sigma}(E^o) \cdot (E^i - E^o)| \geqslant o(|E^i - E^o|)$$

but $\tilde{\sigma}$ is a subgradient, and then, the expression whose modulus is in the left hand side is $\geqslant 0$ and we may write

$$W(E^i) - W(E^o) - \tilde{\sigma}(E^o) \,.\, (E^i - E^o) \geqslant \gamma |E^i - E^o| \tag{7.12}$$

for some positive constant $\gamma$ . On the other hand $\tilde{\sigma}(E^i)$ is a subgradient at the point $E^i$ , and then

$$W(E^i) - W(E^o) - \tilde{\sigma}(E^i) \,.\, (E^i - E^o) \leqslant 0 \tag{7.13}$$

and by hypothesis, $| \tilde{\sigma}(E^i) - \tilde{\sigma}(E^o) | \to 0$ and then (7.13) implies :

$$W(E^i) - W(E^o) - \tilde{\sigma}(E^o).(E^i - E^o) \leqslant o(|E^i - E^o|) \tag{7.14}$$

which is in contradiction to (7.12). ∎

Of course, $\tilde{\sigma}(E)$ is a continuous function from $R^6$ into $R^6$ (because $\underline{u}$ is continuous from $R^6$ into $V_{YF}$). Theorem 7.1 is proved.

Remark 7.1 - It is almost evident that the function W is not of class $C^2$, i.e.; that $\tilde{\sigma}(E)$ is not differentiable. In fact, the local problem (7.3) amounts to search for a Y-periodic strain field satisfying the one-side constraint on F for given $E_{ij}$ (which is the limit mean value of $e_{ij}(\underline{u}^\varepsilon)$. Then, in a case as

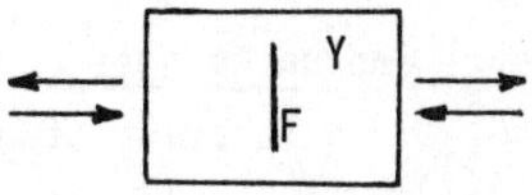


the traction and compression elastic moduli must be different. ∎

Other properties of the function W and of the strain-stress law are given by :

Theorem 7.2 -

a) W(E) is positively homogeneous of degree 2, i.e.

$$W(\lambda E) = \lambda^2 W(E) \qquad \text{for} \quad \lambda \geqslant 0 \tag{7.15}$$

b) W and its gradient $\tilde{\sigma}$ satisfy the relation

$$2\,W(E) = \tilde{\sigma}_{ij} E_{ij} \tag{7.16}$$

c) There exists two positive constants $\beta$ and $\gamma$ such that

$$\beta |E|^2 \leqslant W(E) \leqslant \gamma |E|^2 \tag{7.17}$$

d) There exists a positive constant C such that

$$|\tilde{\sigma}(E^1) - \tilde{\sigma}(E^2)| \leqslant C\, |E^1 - E^2| \qquad \forall\, E^1 , E^2 \in R^6 \tag{7.18}$$

e) W(E) is strictly convex, e.i. :

$$\delta W \geqslant \tilde{\sigma}_{ij}\, \delta E_{ij} \quad , \text{ with } = \Rightarrow \delta E = 0 \tag{7.19}$$

f) $\tilde{\sigma}(E)$ is a strictely monotonous operator (from $R^6$ into $R^6$), i.e.

(7.20) $[\tilde{\sigma}(E^1) - \tilde{\sigma}(E^2)].\ [E^1 - E^2] \geq 0$ , with $= \Rightarrow E^1 - E^2 = 0$

Proof - It is straight forward and we will not give it in detail.
a) is self-evident ; then, b) is a consequence of the Euler's theorem. c),d) is a nonlinear version of chap. 5, (3.10) (see Remark 11.1 later). In order to prove e), we write (see (7.4) and (7.10))

$$\delta W = \frac{1}{|Y|} \int_Y \sigma_{kh}(\delta e_{kh} + \delta E_{kh})\, dy + \frac{1}{2|Y|} \int_Y a_{ijkh}\, \delta e^*_{kh}\ \delta e^*_{ij}\, dy \geq$$

$\geq$ [by taking $\underline{w} = \underline{u}(E + \delta E)$ in (7.3), we see that the term in $\delta e_{kh}$ is

$$\geq 0\ ] \geq \tilde{\sigma}_{kh}\ \delta E_{kh} + \frac{1}{2|Y|} \int_Y a_{ijkh}\, \delta e^*_{kh}\ \delta e^*_{ij}\, dy$$

Then, the sign = implies $\delta e^*_{kh} = 0$ in Y for any kh and consequently (see Remark 11.1 later), we have $\delta E_{kh} = 0$. Finally, f) is a classical consequence of e) (c.f. Ekeland et Temam [1], chap 1, prop. 5.4).■

Now, it is easy to prove the existence and uniqueness of $\underline{u}^0$ (see Proposition 6.1) satisfying (6.14), (6.15), where $\tilde{\sigma}^0_{ij}$ is a function of $\mathrm{grad}_x \underline{u}^0$ defined by (7.8) (see (7.6) and theorems 7.1 and 7.2). For, we write (6.14) under the form :

(7.21) $$A(\underline{u}^0) = \underline{f} \qquad \text{where} \qquad (A(\underline{u}^0))_i = \frac{\partial \tilde{\sigma}^0_{ij}}{\partial x_j}$$

we then have :

Theorem 7.3 - (Existence and uniqueness of $\underline{u}^0$) - For given $\underline{f} \in (H^{-1}(\Omega))^3$, there exists a unique $\underline{u}^0 \in (H^1_0(\Omega))^3$.

Proof - We shall prove that A is an operator from $V = (H^1_0)^3$ into $V' = (H^{-1})^3$ which is bounded, hemicontinuous, strictly monotonous and coercive on $(H^1_0)^3$ (i.e.,

(7.22) $$\frac{\langle A(\underline{v})\ ,\ \underline{v} \rangle_{V'V}}{\|\underline{v}\|_V} \to \infty \qquad \text{as } \|\underline{v}\|_V \to \infty \quad ).$$

the theorem then follows from a classical theorem (c.f. Lions [4], chap. 2, sect. 2.1, 2.2). The proof is straight forward :

(7.23) $$\langle A(\underline{v})\ ,\ \underline{w} \rangle_{V'V} = \int_\Omega \sigma_{ij}(\underline{v}) \frac{\partial w_i}{\partial x_j}\, dx \equiv \int_\Omega \sigma_{ij}(\underline{v})\, e_{ij}(\underline{w})\, dx$$

the boundedness then follows from (7.16) and (7.17). The coerciveness follows from the first relation (7.17). Strict monotonicity is a consequence of (7.20). Hemicontinuity follows from (7.18).■

Remark 7.2 - The convergence $\underline{u}^\varepsilon \to \underline{u}^0$ is an open question. The results of section 6 and 7 are consequences of the formal expansion (6.4). ■

Remark 7.3 - The existence and uniqueness of $\underline{u}^0$ may also be obtained by the method of the hidden variables used in the following section for the study of a problem in viscoelasticity. But the method given here shows the physical properties of the homogenized strain-stress law (hyperelastic law). ■

## 8.- Viscoelastic fissured body. Hidden variables

We consider here the problem of the preceeding sections (5,6,7) but with the strain-stress law (4.5) (viscoelastic law with instantaneous memory) instead of (1.3) (elastic law). In particular, the equation of the movement is (5.5), and consequently the inertia term is neglected (i.e. the problem is studied in the quasi-static approximation). Moreover, $\underline{u}^\varepsilon$ is supposed to be zero at t = 0.

For the sake of definiteness, we repeat the equations and boundary conditions:

(8.1) $$0 = \frac{\partial \sigma^\varepsilon_{ij}}{\partial x_j} + f_i$$

(8.2) $$\sigma^\varepsilon_{ij} = a^0_{ijkh}\, e_{kh}(\underline{u}^\varepsilon) + a^1_{ijkh}\, e_{kh}\left(\frac{\partial \underline{u}^\varepsilon}{\partial t}\right)$$

(8.3) $$\underline{u}^\varepsilon = 0 \qquad \text{on } \partial\Omega$$

(8.4) $$\underline{u}^\varepsilon = 0 \qquad \text{for } t = 0$$

(8.5) one-side conditions (5.7) - (5.9) on $\varepsilon F$.

where the coefficients $a^m_{ijkh}$ are supposed to be constants and satisfy the symmetry and positivity conditions (1.3), (1.4) for m = 0, 1.

It is clear that the present problem will involve memory effects (as in sect. 4) and non linear effects (as in sect. 6 and 7). The variational formulation of (8.1) - (8.5) is :

(8.6) $$\begin{cases} \text{Find } \underline{u}^\varepsilon(t) \text{ such that for each } t,\ \underline{u}^\varepsilon(t) \in K_{F\varepsilon} \\ \displaystyle\int_\Omega \left[ a^1_{ijkh} \frac{\partial}{\partial x_k}\left(\frac{\partial u^\varepsilon_h}{\partial t}\right) + a^0_{ijkh} \frac{\partial u^\varepsilon_h}{\partial x_k} \right] \frac{\partial (v_i - u^\varepsilon_i)}{\partial x_j}\, dx \geq \int_\Omega f_i (v_i - u^\varepsilon_i)\, dx \\ \text{for any } \underline{v} \in K_{F\varepsilon} \text{ : moreover, } \underline{u}^\varepsilon(0) = 0 . \end{cases}$$

This problem is in the framework of theorem 5.2 if we choose $H = V = V_{F\varepsilon}$ .

In order to obtain the asymptotic behaviour as $\varepsilon \searrow 0$, we postulate an expansion of the form

(8.7) $\underline{u}^{\varepsilon}(x\ ,\ t) = \underline{u}^{0}(x\ ,\ t) + \varepsilon \underline{u}^{1}(x,y,t) + \ldots \qquad y = x/\varepsilon$

where $\underline{u}^0 \in (H_0^1(\Omega))^3$ and $\underline{u}^1$ is a smooth function of x defined on $\Omega$ with values in $K_{YF}$.

The asymptotic process of section 6 leads to the equation and initial and boundary conditions for the limit problem :

(8.8) $$0 = \frac{\partial \tilde{\sigma}^0_{ij}}{\partial x_j} + f_i$$

(8.9) where $\sigma^0_{ij} = (a^0_{ijkh} + a^1_{ijkh} \frac{\partial}{\partial t})\left(\frac{\partial u^0_k}{\partial x_h} + \frac{\partial u^1_k}{\partial y_h}\right)$

(8.10) $\underline{u}^0 = 0$ on $\partial\Omega$

(8.11) $\underline{u}^0 = 0$ for $t = 0$

and the relation between $\underline{u}^0$ and $\tilde{\sigma}^0$ must be given by the local problem, which is now an evolution variational inequality.

(8.12) Local problem - Let $\underline{u}^0(x\ ,\ t)$ be given. Find

$\underline{u}^1 \in \tilde{K}_{YF}$ for each t, $\underline{u}^1 = 0$ for $t = 0$ such that

$$\frac{1}{|Y|}\int_Y (a^0_{ijkh} + a^1_{ijkh} \frac{\partial}{\partial t})\left(\frac{\partial u^0_k}{\partial x_h} + \frac{\partial u^1_k}{\partial y_h}\right) \frac{\partial (v_i - u^1_i)}{\partial y_j}\, dy \geqslant 0 \quad \forall \underline{v} \in \tilde{K}_{YF}$$

(problem in y, t, where x plays the role of a parameter)

Theorem 8.1 - The local problem has one and only one solution in the framework of theorem 5.2 (with $H = V = \tilde{V}_{F\varepsilon}$ , $K = \tilde{K}_{F\varepsilon}$)

The proof of this theorem is analogous to that of theorem 6.1.

For fixed x, the function $t \to \text{grad}_x\ u^0$ determines in a unique way the function $t \to \underline{u}^1(y)$ and consequently, by (8.9), the function $t \to \tilde{\sigma}^0$. It is then clear that the homogenized strain-stress law is non linear and with memory.

Consequently, the limit problem (8.8) - (8.11) with the strain-stress law defined by the preceeding process is a nonlinear integro-differential equation. Instead of trying to solve such a problem, we shall study the system formed by (8.8) - (8.11) and the local problem (8.12) with parameter x. It is then possible to obtain the existence and uniqueness of $\underline{u}^0(x\ ,\ t)$ and $\underline{u}^1(x,y,t)$. If $\underline{u}^0(x\ ,\ t)$ is considered to be the very variable, $\underline{u}^1$ appears as an auxiliary variable, that is to say, a hidden variable in the terminology of rheology. The hidden variable satisfies the local problem (8.12) and the (homogenized) strain tensor $\sigma^0_{ij}$ is defined by (8.9) as a function of $\text{grad}_x\ \underline{u}^0$ and of the hidden variable.

In order to obtain an existence and uniqueness theorem for $\underline{u}^0$, $\underline{u}^1$, we introduce the space

(8.13) $$\mathcal{V} = (H^1_o(\Omega))^3 \times L^2(\Omega\ ;\ \tilde{V}_{YF})$$

and we shall search for $\underline{u}^0$, $\underline{u}^1$ belonging to this space for each t. Note that

$$\underline{u}^1 \in L^2(\Omega\ ;\ \tilde{V}_{YF})$$

means that for each $x \in \Omega$, $u^1$ is an element of $\tilde{V}_{YF}$. Moreover, we introduce the closed convex $\mathcal{K}$ :

(8.14) $$\mathcal{K} = \{ (\underline{u}^0\ ,\ \underline{u}^1) \in \mathcal{V}\ ;\ \ \underline{u}^1 \in \tilde{K}_{YF} \quad \text{a.e. in } x \}$$

Now, (8.8), (8.10) may be written under the form :

(8.15) $$\left| \begin{array}{l} \underline{u}^0 \in (H^1_o(\Omega))^3 \\ \displaystyle\int_\Omega \tilde{\sigma}^o_{ij} \frac{\partial(v^o_i - u^o_i)}{\partial x_j}\, dx \geqslant \int_\Omega f_i(v^o_i - u^o_i)\, dx \qquad \forall \underline{v}^0 \in (H^1_o)^3 \end{array} \right.$$

where the equation has been written as an inequation (this is equivalent because the convex set is the whole space). Moreover, by writing in an explicite way the mean operator $\sim$ , (8.15) becomes :

(8.16) $$\int_{\Omega\times Y} (a^o_{ijkh} + a^1_{ijkh}\frac{\partial}{\partial t})\left(\frac{\partial u^o_k}{\partial x_h} + \frac{\partial u^1_k}{\partial y_h}\right) \frac{\partial(v^o_i - u^o_i)}{\partial\, x_j} dxdy \geqslant \int_\Omega f_i(v^o_i - u^o_i)\, dx$$

On the other hand, if in the local problem (8.12) we take test functions depending on the parameter x, we may integrate (8.12) on $\Omega$ and we obtain an <u>equi valent</u> relation (note that we may come back to (8.12) by taking special test functions of the type (6.16)) :

(8.17) $$\int_{\Omega\times Y} (a^o_{ijkh} + a^1_{ijkh}\frac{\partial}{\partial t})\left(\frac{\partial u^o_k}{\partial x_h} + \frac{\partial u^1_k}{\partial y_h}\right) \frac{\partial(v^1_i - u^1_i)}{\partial y_j}\, dx\, dy \geqslant 0$$

Then, by adding (8.16) and (8.18) we see that <u>a variational formulation of the problem in $u^0$, $u^1$ is</u> :

Find $(u^0\ ,\ u^1)$ function of t with values in $\mathcal{V}$, belonging to $\mathcal{K}$ a.e. in t and such that $u^0$ , $u^1$ are zero for t = 0 and :

(8.18) $$\int_{\Omega\times Y} (a^o_{ijkh} + a^1_{ijkh}\frac{\partial}{\partial t})\left(\frac{\partial u^o_k}{\partial x_h} + \frac{\partial u^1_k}{\partial y_h}\right)\left(\frac{\partial(v^o_i - u^o_i)}{\partial x_j} + \frac{\partial(v^o_i - u^1_i)}{\partial y_j}\right) dx\, dy \geqslant$$
$$\geqslant \int_\Omega f_i(v^o_i - u^o_i)\, dx \qquad\qquad \forall\ (v^0\ ,\ v^1) \in \mathcal{K}\ .$$

Under this form, we do not know if a unique solution exists or not, because the coerciveness of the forms is not easy to prove.

On the other hand, existence and uniqueness are immediate if we consider $\mathcal{V}^*$ and $\mathcal{H}^*$ instead of $\mathcal{V}$ and $\mathcal{H}$, where $\mathcal{V}^*$ is the completed of $\mathcal{V}$ for the norm

$$\int_{\Omega\times Y} \sum_{ij} \left| \frac{\partial u_i^0}{\partial x_j} + \frac{\partial u_i^1}{\partial y_j} \right|^2 dx\, dy \tag{8.19}$$

and $\mathcal{H}^*$ is the adherence of $\mathcal{H}$ in $\mathcal{V}^*$. The fact that (8.19) is a norm is easily obtained by using the Remark 11.1 at the end of the chapter. The consideration of $\mathcal{V}^*$ is suitable according to the classical criteria (see Ladyzhenskaya and Uralceva [ 1 ].

9.- <u>The Maxwell's system. Asymptotic expansion.</u>

We consider here the Maxwell system in a homogeneous medium with periodic structure. It appears that the <u>homogenized system is in general integrodifferential</u>. But, if the conductivity is zero (i.e. there is no dissipation) it is differential.

To fix ideas, we consider the Maxwell system in $R^3$, with zero initial conditions and source terms of bounded support (for any t) ; other problems may be handled in an analogous way. We consider, for a given period Y, the following problem.

$$\frac{\partial \underline{D}^\varepsilon}{\partial t} = \underline{\mathrm{rot}}\ \underline{H}^\varepsilon - \underline{J}^\varepsilon + \underline{F} \tag{9.1}$$

$$\frac{\partial \underline{B}^\varepsilon}{\partial t} = - \underline{\mathrm{rot}}\ \underline{E}^\varepsilon + \underline{G} \tag{9.2}$$

$$\underline{E}(x\ ,\ 0) = 0 \quad , \quad \underline{H}(x\ ,\ 0) = 0 \tag{9.3}$$

where

$$D_i^\varepsilon = \eta_{ij}\left(\frac{x}{\varepsilon}\right) E_j^\varepsilon \quad ; \quad B_i^\varepsilon = \mu_{ij}\left(\frac{x}{\varepsilon}\right) H_j^\varepsilon \quad ; \quad J_i^\varepsilon = \sigma_{ij}\left(\frac{x}{\varepsilon}\right) E_j^\varepsilon \tag{9.4}$$

where $\eta$ , $\mu$ , $\sigma$ are smooth Y-periodic functions of $x/\varepsilon$ satisfying

$$\begin{cases} \eta_{ij} = \eta_{ji} \quad ; \quad \eta_{ij}\ \xi_i\ \xi_j \geqslant \gamma\ |\xi|^2 \qquad \forall \xi \in R^3 \\ \text{and analogous relations for } \mu \text{ and } \sigma . \end{cases} \tag{9.5}$$

Remark 9.1 - It is well known that (9.1) - (9.5) is a hyperbolic system with bounded (with respect to $\varepsilon$) speed of propagation. Then, if the source given terms $\underline{F}$ and $\underline{G}$ have their supports in a bounded domain for any t, the supports of $\underline{E}^\varepsilon$ , $\underline{H}^\varepsilon$ are bounded for each t. On the other hand, the smoothness hypothesis for the coefficients may of course be relaxed.■

$\underline{E}$, $\underline{H}$ are for the electric and magnetic fields, $\underline{D}$, $\underline{B}$ are the displacement and induction and $\underline{J}$ is the electric current.

We search for solutions of the form :

$$(9.10)\qquad \begin{cases} \underline{E}^\varepsilon(x , t) = \underline{E}^0(x , y , t) + \varepsilon \underline{E}^1(x , y , t) + \quad \dots \\ \underline{H}^\varepsilon(x , t) = \underline{H}^0(x , y , t) + \varepsilon \underline{H}^1(x , y , t) + \dots \end{cases}$$

with $y = x/\varepsilon$ , Y-periodic in y, and of course analogous expansions for $\underline{D}^\varepsilon$, $\underline{B}^\varepsilon$, $\underline{J}^\varepsilon$ (obtained from the preceeding ones by multiplying them by $\eta(y)$ and so on).

The system (9.1), (9.2), with (9.10) and

$$(9.11)\qquad \frac{d}{dx_i} = \frac{\partial}{\partial x_i} + \frac{1}{\varepsilon}\frac{\partial}{\partial y_i}$$

gives, at the orders $\varepsilon^{-1}$ and $\varepsilon^0$ respectively :

$$(9.12)\qquad \underline{\text{rot}}_y\, \underline{H}^0 = 0 \qquad ; \qquad \underline{\text{rot}}_y\, \underline{E}^0 = 0$$

$$(9.13)\qquad \begin{cases} \dfrac{\partial \underline{D}^0}{\partial t} = \underline{\text{rot}}_x\, \underline{H}^0 + \underline{\text{rot}}_y\, \underline{H}^1 - \underline{J}^0 + \underline{F} \\ \dfrac{\partial \underline{B}^0}{\partial t} = - \underline{\text{rot}}_x\, \underline{E}^0 - \underline{\text{rot}}_y\, \underline{E}^1 + \underline{G} \end{cases}$$

and by taking the mean value $\sim$ on Y in (9.13) we obtain :

$$(9.14)\qquad \begin{cases} \dfrac{\partial \underline{\tilde{D}}^0}{\partial t} = \underline{\text{rot}}_x\, \underline{\tilde{H}}^0 - \underline{\tilde{J}}^0 + \underline{F} \\ \dfrac{\partial \underline{\tilde{B}}^0}{\partial t} = - \underline{\text{rot}}_x\, \underline{\tilde{E}}^0 + \underline{G} \end{cases}$$

because the terms in $\underline{H}^1$, $\underline{E}^1$ of (9.13) have a zero mean value as is easily seen by integration by parts and taking into account the Y-periodicity :

$$\frac{1}{|Y|}\int_Y \underline{\text{rot}}_y\, \underline{H}^1\, dy = \frac{1}{|Y|}\int_Y \underline{n} \wedge \underline{H}^1\, dS = 0$$

<u>(9.14) is the homogenized system,</u> which must be completed with the initial conditions obtained from (9.3) :

$$(9.15)\qquad \underline{\tilde{E}}^0(x , 0) = 0 \qquad ; \qquad \underline{\tilde{H}}^0(x , 0) = 0$$

We see that the homogenized fields $\underline{\tilde{E}}^o$, $\underline{\tilde{H}}^o$, $\underline{\tilde{D}}^o$, $\underline{\tilde{B}}^o$, $\underline{\tilde{J}}^0$ satisfy a Maxwell system. We have to obtain the homogenized constitutive laws analogous to (9.4) for the homogenized system.

Let us consider the divergence of (9.1), (9.2):

$$\operatorname{div}\left(\frac{\partial \underline{D}^\varepsilon}{\partial t} + \underline{J}^\varepsilon\right) = \operatorname{div} \underline{F}$$

$$\operatorname{div}\left(\frac{\partial \underline{B}^\varepsilon}{\partial t}\right) = \operatorname{div} \underline{G}$$

and by using (9.10) and (9.11), we have, at the order $\varepsilon^{-1}$

(9.16) $$\operatorname{div}_y\left(\frac{\partial \underline{D}^0}{\partial t} + \underline{J}^0\right) = 0$$

(9.17) $$\operatorname{div}_y \frac{\partial \underline{B}^0}{\partial t} = 0 \implies \text{with (9.3)} \implies \operatorname{div}_y \underline{B}^0 = 0$$

on the other hand, (9.12) shows that $\underline{H}^0$ and $\underline{E}^0$ are gradients (in y ; x is a parameter). Let us write

(9.18) $$\underline{E}^o - \underline{\tilde{E}}^0 = \underline{\text{grad}}\, \phi \quad ; \quad \underline{H}^o - \underline{\tilde{H}}^0 = \underline{\text{grad}}\, \psi$$

It is clear that the mean values of $\underline{\text{grad}}\, \phi$, and $\underline{\text{grad}}\, \psi$ are zero and this implies that $\phi$, $\psi$ are themselves periodic (to see this it is sufficient to integrate $\partial\phi/\partial y_1$ on Y to see that

$\phi(y_1^0, y_2, y_3) - \phi(0, y_2, y_3) = 0$ ). Then, (9.16) and (9.17) become :

(9.19) $$\frac{\partial}{\partial y_i}\left\{\left[\eta_{ij}(y)\frac{\partial}{\partial t} + \sigma_{ij}(y)\right]\left[\frac{\partial \phi}{\partial y_j} + \tilde{E}_j^0\right]\right\} = 0$$

(9.20) $$\frac{\partial}{\partial y_i}\left\{\mu_{ij}(y)\left[\frac{\partial \psi}{\partial y_j} + \tilde{H}_j^0\right]\right\} = 0$$

Equation (9.20) contains x and t as parameters. It is exactly the same as (2.1) of chap. 5 with $\mu$, $\psi$, $\tilde{H}_j^0$ instead of a, $u^1$, $\partial u^0/\partial x_j$. Then, the mean values of $\underline{H}^0$ and $\underline{B}^0$ are related by

(9.21) $$\tilde{B}_i^0 = \mu_{ij}^h \tilde{H}_j^0$$

where $\mu_{ij}^h$ are the homogenized coefficients associated to $\mu_{ij}(y)$ by using the classical formulae of chap. 5, (3.4) (see also (3.1) of chap. 5).

On the other hand, equation (9.19) is analogous to equation (4.16), (4.21) of the viscoelasticity problem, and will be solved in an analogous way. We introduce the space

(9.22) $$\tilde{V}_Y = \{\theta ; \theta \in H^1_{loc}(R^3) \quad , \text{ Y-periodic}, \quad \tilde{\theta} = 0\}$$

equipped with the scalar product (this point is not essential) :

$$(9.23) \qquad (\phi\ ,\ \theta)_{\tilde{V}_Y} = \int_Y \eta_{ij}(y)\,\frac{\partial\phi}{\partial y_i}\,\frac{\partial\theta}{\partial y_j}\,dy$$

and (9.19) with the Y-periodicity condition for $\phi$ becomes

$$(9.23) \qquad \frac{d}{dt}\int_Y \eta_{ij}\left(\frac{\partial\phi}{\partial y_j} + \tilde{E}^o_j\right)\frac{\partial\theta}{\partial y_i}\,dy + \int_Y \sigma_{ij}\left(\frac{\partial\phi}{\partial y_i} + \tilde{E}^o_j\right)\frac{\partial\theta}{\partial y_i}\,dy = 0$$

We then define the operator A (which is bounded and symmetric from $\tilde{V}_Y$ into itself) and the elements $f^m_j$ $(m = 1\ ,\ 2)$ of $\tilde{V}_Y$ by :

$$\int_Y \sigma_{ij}\,\frac{\partial\phi}{\partial y_j}\,\frac{\partial\theta}{\partial y_i}\,dy = (A\phi\ ,\ \theta)_{\tilde{V}_Y}$$

$$\int_Y \sigma_{ij}\,\frac{\partial\theta}{\partial y_i}\,dy = (f^2_j\ ,\ \theta)_{\tilde{V}_y} \quad ; \quad \int_Y \eta_{ij}\,\frac{\partial\theta}{\partial y_i}\,dy = (f^1_j\ ,\ \theta)_{\tilde{V}_Y} \quad .$$

Equation (9.23) then becomes

$$\frac{d}{dt}\left(\phi + f^1_j\ \tilde{E}^o_j(t)\right) + \left(A\ \phi + f^2_i\ \tilde{E}^o_j(t)\right) = 0$$

which is a differential equation in $\tilde{V}_Y$ . The unknown $\phi$ must of course satisfy the initial condition $\phi(0) = 0$. According to semigroup theory, the solution is :

$$(9.24) \qquad \phi(t) = -\ f^1_j\ \tilde{E}^o_j(t) + \int_0^t e^{-A(t-s)}\ f^3_j\ \tilde{E}^o_j(s)\ ds \quad ; \qquad f^3_j \equiv Af^1_j - f^2_j$$

If $\phi(t)$ is known, $\underline{E}^o$ is calculated from (9.18), and we have

$$D^o_i = \eta_{ij}\ E^o_j \qquad ; \qquad J^o_i = \sigma_{ij}\ E^o_j$$

and the mean values of $D^o_i$ and $J^o_i$ may be calculated ; we finaly obtain :

$$(9.25) \qquad \begin{cases} \tilde{D}^o_k(t) = b^\eta_{kj}\ \tilde{E}^o_j(t) + \displaystyle\int_0^t g^\eta_{kj}\ (t - s)\ \tilde{E}^o_j(s)\ ds \\ \tilde{J}^o_k(t) = b^\sigma_{kj}\ \tilde{E}^o_j(t) + \displaystyle\int_0^t g^\sigma_{kj}(t - s)\ \tilde{E}^o_j(s)\ ds \end{cases}$$

where

$$b^\eta_{kj} = \left[\eta_{kj}(y) - \eta_{ki}(y)\ \frac{\partial f^1_j(y)}{\partial y_i}\right]^\sim$$

$$g^\eta_{kj}(\zeta) = \left[\eta_{ki}(y)\ \frac{\partial}{\partial y_i}\ (e^{-A\zeta}\ f^3_j)(y)\right]^\sim$$

and analogous expressions for $b^\sigma$ , $g^\sigma$ with coefficients $\sigma$ instead of $\eta$ .

(9.21), (9.25) are the homogenized constitutive laws to be joined to the homogenized system (9.14) and to the initial conditions (9.15) to obtain an integro-differential system for the homogenized variables $\tilde{\underline{E}}^o$, $\tilde{\underline{H}}^o$, $\tilde{\underline{D}}^o$, $\tilde{\underline{B}}^o$, $\tilde{\underline{J}}^o$ .

Remark 9.2 - As in sect. 4, if we take the Laplace transform (from f(t) into $\hat{f}(p)$) of (9.19) and (9.25), we see that $\underline{\hat{\tilde{J}}}^o(p) + p\,\underline{\hat{\tilde{D}}}^o(p)$ is given as a function of $\underline{\hat{\tilde{E}}}^o(p)$ by the same formulae that the homogenized coefficients associated with $\sigma_{ij}(y) + p\,\eta_{ij}(y)$ in the elliptic problems of chapter 5. Consequently

$$p\,b^{\eta}_{ij} + p\,\hat{g}^{\eta}_{ij}(p) + b^{\sigma}_{ij} + \hat{g}^{\sigma}_{ij}(p) = (\sigma_{ij} + p\,\eta_{ij})^h \tag{9.26}$$

where $^h$ denotes, as usual, "homogenized". These functions are well defined for $\text{Re}\,p > 0$. ∎

Remark 9.3 - Moreover, as in proposition 4.1, if we write

$$(\sigma_{ij} + p\,\eta_{ij})^h = p\,\alpha_{ij}(p) \quad ,$$

$\alpha_{ij}(p)$ are holomorphic functions of $1/p$ for $p$ in the vicinity of $\infty$ . $\alpha_{ij}(\infty)$ is a symmetric, positive definite matrix. ∎

Theorem 9.1 - The homogenized integro-differential system (9.14), (9.15), (9.21), (9.25) has one and only one solution.

We shall not give here the proof of this theorem, which may be seen in Sanchez-Hubert [1]. We remark that it is obtained by Laplace transform, which gives :

$$\left\{\begin{aligned} p\,\alpha(\infty)\,\underline{\hat{\tilde{E}}}^o &= \underline{\text{rot}}\,\underline{\hat{\tilde{H}}}^o + \hat{F} - p(\alpha(p) - \alpha(\infty))\underline{\hat{\tilde{E}}}^o \\ p\,\mu^h\,\underline{\hat{\tilde{H}}}^o &= -\,\underline{\text{rot}}\,\underline{\hat{\tilde{E}}}^o + \hat{G} \end{aligned}\right. \tag{9.27}$$

The operator (rot, -rot) in the right hand side of (9.27) is a skew self adjoint operator in $(L^2)^6$ (equipped with an appropriate norm ; see Duvaut-Lions [1], chap. 7). The system (9.27) may then be solved for sufficiently large Rep by iteration if we use the estimate given by proposition 4.1, c) of chapter 2 (which holds of course for skew selfadjoint operators).

## 10.- Proof of the convergence

Now we prove the convergence of $\underline{E}^\varepsilon$, $\underline{H}^\varepsilon$ to $\underline{\tilde{E}}^o$, $\underline{\tilde{H}}^o$ in the framework of the preceeding section. The proof is based on the "compensation method" of Tartar.

If $\Omega$ is a bounded open domain of $R^3$ , we introduce the spaces

$$H(\Omega, \text{div}) = \{\underline{u}\ ;\ u_i \in L^2(\Omega)\quad ;\quad \text{div}\,\underline{u} \in L^2(\Omega)\}$$

$$H(\Omega, \underline{\text{rot}}) = \{\underline{u}\ ;\ u_i \in L^2(\Omega)\ ;\ (\underline{\text{rot}}\,\underline{u})_i \in \underline{L}^2(\Omega)\}$$

equipped with the hilbert norms

$$\|\underline{u}\|^2_{H(\Omega,div)} = \|\underline{u}\|^2_{L^2} + \|\text{ div } u\|^2_{L^2}$$

$$\|u\|^2_{H(\Omega,\underline{rot})} = \|\underline{u}\|^2_{L^2} + \|\underline{rot}\ \underline{u}\|^2_{L^2}$$

respectively. We then have :

Theorem 10.1 - Let $\underline{u}^k$, $\underline{v}^k$ be sequences of vectors such that

$$\underline{u}^k \to \underline{u}^* \quad \text{in} \quad H(\Omega, \text{div}) \text{ weakly}$$

$$\underline{v}^k \to \underline{v}^* \quad \text{in} \quad H(\Omega, \underline{\text{rot}}) \text{ weakly}$$

Then,

$$\underline{u}^k \cdot \underline{v}^k \to \underline{u}^* \cdot \underline{v}^* \quad \text{in} \quad \mathcal{D}'(\Omega)$$

The proof of this theorem, which is obtained by Fourier transform, may be seen in Bensoussan-Lions-Papanicolaou [ 2 ] , chap. 1, sect. 11.4.

Theorem 10.2 - In the framework of the preceeding section, if $\underline{F}, \underline{G} \in L^2(0, T; (L^2(R^3))^3)$ have for each t their support in a compact domain G of $R^3$, we have

$$\underline{E}^\varepsilon, \underline{H}^\varepsilon \to \underline{\tilde{E}}^o, \underline{\tilde{H}}^o \quad \text{in} \quad L^\infty(0, T; (L^2(R^3))^3) \text{ weakly } *$$

where $\underline{\tilde{E}}^o$, $\underline{\tilde{H}}^o$ is the solution of the homogenized problem given by theorem 9.1.

In order to prove this theorem, we continue $\underline{F}$ and $\underline{G}$ by zero for $t > T$ ; the solutions are then defined for $t > 0$, $x \in R^3$ and have, for fixed t, compact support. We multiply (9.1) by $\underline{E}^\varepsilon$ and (9.2) by $\underline{H}^\varepsilon$ and we integrate on $R^3$. The terms containing $\underline{rot}$ cancel by integration by parts and we have (the parentheses are for the scalar product in $L^2(R^3)$) :

$$(\eta_{ij} \frac{\partial E_j}{\partial t}, E_i) + (\mu_{ij} \frac{\partial H_j}{\partial t}, H_i) = -(\sigma_{ij} E_j, E_i) + (F_i, E_i) + (G_i, H_i)$$

and by using (9.5) we easily obtain that $\underline{E}^\varepsilon(t)$, $\underline{H}^\varepsilon(t)$ are bounded in $(L^2(R^3))^3$ by constants independent of ε, t. Then, (after extracting subsequences, as usual) we have

Lemma 10.1 -

$$(10.1) \quad \begin{cases} E_i^\varepsilon \to E_i^* & \text{in } L^\infty(0, \infty; L^2(R^3)) \text{ weakly } * \\ H_i^\varepsilon \to H_i^* & \text{in } L^\infty(0, \infty; L^2(R^3)) \text{ weakly } * \end{cases}$$

As a consequence of the uniqueness of the solution of the homogenized problem, theorem 10.2 will be proved if we show that $\underline{E}^*$, $\underline{H}^*$ is a solution of that problem. The Laplace transform of (9.1), (9.2) which is defined for Rep > 0, is :

(10.2) $$\begin{cases} p\,\hat{\underline{D}}^{\varepsilon} + \hat{\underline{J}}^{\varepsilon} = \underline{\text{rot}}\,\hat{\underline{H}}^{\varepsilon} + \hat{\underline{F}} \\ p\,\hat{\underline{B}}^{\varepsilon} = -\,\underline{\text{rot}}\,\hat{\underline{E}}^{\varepsilon} + \hat{\underline{G}} \end{cases}$$

and from (10.1) :

(10.3) $$\begin{cases} \hat{\underline{E}}^{\varepsilon} \to \hat{\underline{E}}^{*} & \text{in } L^2 \text{ weakly (for Rep} > 0) \\ \hat{\underline{H}}^{\varepsilon} \to \hat{\underline{H}}^{*} & \text{in } L^2 \text{ weakly (for Rep} > 0) \end{cases}$$

Moreover, from (9.4) we see that $\underline{D}^{\varepsilon}$, $\underline{B}^{\varepsilon}$, $\underline{J}^{\varepsilon}$ remain bounded, and then, by extracting subsequences:

(10.4) $$D_i^{\varepsilon}\ ,\ B_i^{\varepsilon}\ ,\ J_i^{\varepsilon} \to D_i^{*}\ ,\ B_i^{*}\ ,\ J_i^{*} \quad \text{in } L^{\infty}(0\ ,\infty;\ L^2(R^3)) \text{ weakly } *$$

and by taking the Laplace transform

(10.5) $$\hat{D}_i^{\varepsilon}\ ,\ \hat{B}_i^{\varepsilon}\ ,\ \hat{J}_i^{\varepsilon} \to \hat{D}_i^{*}\ ,\ \hat{B}_i^{*}\ ,\ \hat{J}_i^{*} \quad \text{in } L^2(R^3) \text{ weakly (Rep} > 0)$$

Now, (10.2) shows that $\underline{\text{rot}}\ \hat{\underline{H}}^{\varepsilon}$ and $\underline{\text{rot}}\ \hat{\underline{E}}^{\varepsilon}$ remain bounded ; by taking subsequences, they converge to limits that are the $\underline{\text{rot}}$ of $\hat{H}^{*}$ and $\hat{E}^{*}$ , i.e. :

(10.6) $$\begin{cases} \underline{\text{rot}}\ \hat{\underline{H}}^{\varepsilon} \to \underline{\text{rot}}\ \hat{\underline{H}}^{*} & \text{in } L^2 \text{ weakly} \quad (\text{Rep} > 0) \\ \underline{\text{rot}}\ \hat{\underline{E}}^{\varepsilon} \to \underline{\text{rot}}\ \hat{\underline{E}}^{*} & \text{in } L^2 \text{ weakly} \quad (\text{Rep} > 0) \end{cases}$$

Then, by passing to the limit in (10.2), we have

Lemma 10.2) - $\hat{\underline{E}}^{*}$ , $\hat{\underline{H}}^{*}$, $\hat{\underline{D}}^{*}$, $\hat{\underline{B}}^{*}$, $\hat{\underline{J}}^{*}$, satisfy a system analogous to (10.2), (which is the Laplace transform of the Maxwell's system ).

Consequently, it only remains to prove that $\underline{E}^{*}$, $\underline{H}^{*}$, $\underline{D}^{*}$, $\underline{B}^{*}$, $\underline{J}^{*}$ are related by the homogenized laws (9.21), (9.25). This ammounts to prove (see remark 9.2) that

(10.7) $$p\ \hat{\underline{D}}^{*} + \hat{\underline{J}}^{*} = (p\ \eta + \sigma)^h\ \hat{E}^{*}$$

(10.8) $$\hat{\underline{B}}^{*} = \mu^h \hat{\underline{H}}^{*}$$

We shall prove (10.7)((10.8) is of course proved in an analogous way). Moreover, (10.7) is an identity involving Laplace transforms, which are holomorphic functions for Rep > 0. Then, it is sufficient to prove (10.7) for real $p > 0$.

Let us chose an arbitrary open bounded domain $\Omega$ . From (10.3) and (10.6) we have

(10.9) $$\hat{\underline{E}}^{\varepsilon} \to \hat{\underline{E}}^{*} \quad \text{in } H(\Omega,\ \underline{\text{rot}}) \text{ weakly} \quad (p > 0)$$

On the other hand, with an obvious notation (note that $\eta$, $\sigma$, $\mu$ are matrix)

(10.10) $\quad p\,\underline{\hat{D}}^{\varepsilon} + \underline{\hat{J}}^{\varepsilon} \equiv (p\eta + \sigma)^{\varepsilon}\,\underline{\hat{E}}^{\varepsilon} \rightarrow p\,\underline{\hat{D}}^{*} + \hat{J}^{*}$ in $L^2(\Omega)$ weakly

and by taking the div of the first (10.2) :

$$\operatorname{div}[\,(p\,\eta + \sigma)^{\varepsilon}\,\hat{E}^{\varepsilon}] = \operatorname{div}\hat{F} \Longrightarrow$$

(10.11) relation (10.10) holds in $H(\Omega\,,\operatorname{div})$ weakly $(p > 0)$.

Now for given $g \in H^{-1}(\Omega)$ we construct $v^{\varepsilon} \in H_0^1(\Omega)$ such that

(10.12) $\quad -\operatorname{div}[\,(p\,\eta + \sigma)^{\varepsilon}\operatorname{grad} v_{\varepsilon}] = g$

for fixed $p > 0$, which is a classical elliptic, selfadjoint problem in the framework of chapter 5, sect. 1-4. Consequently as $\varepsilon \searrow 0$, we have :

(10.13) $\quad \underline{\operatorname{grad}}\, v^{\varepsilon} \rightarrow \underline{\operatorname{grad}}\, v$ in $L^2(\Omega)$ weakly

(10.14) $\quad (p\,\eta + \sigma)^{\varepsilon}\,\underline{\operatorname{grad}}\, v^{\varepsilon} \rightarrow (p\,\eta + \sigma)^{h}\,\underline{\operatorname{grad}}\, v$ in $L^2(\Omega)$ weakly

where $v \in H_0^1$ is the solution of

(10.15) $\quad -\operatorname{div}[\,(p\,\eta + \sigma)^{h}\,\underline{\operatorname{grad}}\, v\,] = g$

Moreover, equation (10.12) shows that :

(10.16) relation (10.14) holds in $H(\Omega\,,\operatorname{div})$ weakly, and rot $\underline{\operatorname{grad}} = 0 \Longrightarrow$

(10.17) relation (10.13) holds in $H(\Omega\,,\underline{\operatorname{rot}})$ weakly

Now, we write

(10.18) $\quad (p\,\eta + \sigma)^{\varepsilon}_{ij}\,\hat{E}^{\varepsilon}_{j}\,\dfrac{\partial v^{\varepsilon}}{\partial x_j}$

which is an element of $L^1(\Omega)$. We then calculate its limit as $\varepsilon \to 0$ in two different ways :

First, by coupling $(p\,\eta + \sigma)$ to E, and using (10.11), (10.17) and theorem 10.1, we have

(10.19) $\quad (p\,\underline{\hat{D}}^{*} + \underline{\hat{J}}^{*})_i\,\dfrac{\partial v}{\partial x_i} = \lim_{\varepsilon \to 0}$ (10.18) in $\mathcal{D}'(\Omega)$

Second, by coupling $(p\,\eta + \sigma)$ to $\underline{\operatorname{grad}}\, v$, and using (10.9), (10.16) and theorem 10.1, we have

(10.20) $\quad \hat{E}^{*}_{j}(p\,\eta + \sigma)^{h}_{ij}\,\dfrac{\partial v}{\partial x_i} = \lim_{\varepsilon \to 0}$ (10.18) in $\mathcal{D}'(\Omega)$

and by comparing (10.19) and (10.20) we see that the left hand sides of those expressions are the same. Moreover, this holds for any $v \in H_0^1(\Omega)$ (because g was arbitrarily chosen in $H^{-1}$ and (10.15) is an isomorphism between $H_0^1$ and $H^{-1}$) and consequently we have (10.7) a.e. in $\Omega$ . Theeorem 10.2 is proved.

11.- Comments and bibliographical notes - Homogenization problems in classical elasticity may be seen in Artola et Duvaut [1], Duvaut [1], [2], [3], Metellus et Duvaut [1], Lené [1], Ohayon [1], as well as Bensoussan, Lions and Papanicolaou [2], where a large class of elliptic systems are studied (chap. 1, sect. 9). Eigenvalue and bifurcation problems for homogenized plates is studied in Mignot, Puel and Suquet [1]. For scattering of elastic waves by periodic obstacles, see Codegone [2]. For problems with couple-wise applied forces (sect. 3), see Fleury, Pasa et Polisevshi [1]. Long-memory effects in the homogenization of viscoelastic bodies with instantaneous memory is analogous to the phenomena found in Bensoussan, Lions and Papanicolaou [1] (see also Sanchez-Palencia et Sanchez-Hubert [1]). The study of fissured elastic bodies (sect. 6, 7, 8) is new ; it was proposed to us by F. Sidoroff. The idea of the simultaneous study of $u^0(x)$ and $u^1(x, y)$, which amounts to use hidden variables, (sect. 8) is taken from Lions [6], [7], [8]. The homogenization of the Maxwell system and the method of compensation is owed to Tartar [3], [6]; see also Murat [1], Bensoussan, Lions and Papanicolaou [2] and Tartar [1]. In sections 9 and 10 we study the case of a dissipative system, which gives memory effects (see Sanchez-Palencia and Sanchez-Hubert [1] and Sanchez-Palencia [9]).

Remark 11.1 - The following considerations are useful in the proof of (7.17) and (7.19). If

$$(11.1) \qquad \delta e^*_{kh} \equiv \delta e_{kh}(\underline{u}) + \delta E_{kh} = 0$$

with $\underline{u} = \underline{u}(y)$, we have (see Germain [1], sect. 5.4.4) :

$\delta u_i = - y_m \delta E_{im}$ + solid displacement, and consequently $\delta \underline{u}$ is continuous across F (note that $Y-\overline{F}$ is connected). Then, by integrating (11.1) by parts and using the Y-periodicity, we have

$$|Y| \delta E_{kh} = 0 \quad . \blacksquare$$

# CHAPTER 7

# FLUID FLOW IN POROUS MEDIA

In this chapter we study asymptotic expansions for fluid flow in the (small) canals of a rigid porous solid. The case of a deformable porous solid is handled in chapter 8. A proof of the convergence of the asymptotic expansion for this type of problems is given in the Appendix of L. Tartar at the end of this volume. Sect. 1 contains generalities about the linearized equations of fluid flow. The main expansion is given in sect. 2. Sect. 3 and 4 contain non linear and compressibility effects. Boundary layers and boundary conditions at the surface of the porous medium are studied in sect. 5. Acoustics in porous media, which leads to integro-differential homogenized equations is considered in sect. 6 and 7.

## 1.- Notions about the Stokes equations

Fluid flow in porous media is often very slow, and inertia effets (non linear terms) may be neglected. This is the reason why we study here the Stokes (linear) rather than the Navier-Stokes (non linear) equations. On the other hand, flow in porous media is often coupled with free flow out of the porous body ; we shall study such problems by using a formal matching procedure, and consequently flow in bounded domains is sufficient for our purpose. The results of the present section are classical. Proofs may be found in Temam [ 1 ] , chap. 1 (see also Ladyzhenskaya [ 2 ], chap. 1 and 2, and Tartar [ 5 ]).

Let $\Omega$ be a bounded domain of $R^3$ with smooth boundary. We assume that $\Omega$ is connected, but not necessarily simply-connected. The Stokes equations are

(1.1) $$0 = - \underline{\text{grad}}\, p + \Delta \underline{v} + \underline{f} \qquad \text{in } \Omega$$

(1.2) $$\text{div}\, \underline{v} = 0 \qquad \text{in } \Omega$$

(1.3) $$\underline{v}\,\big|_{\partial\Omega} = 0 \qquad \text{on } \partial\Omega$$

where $\underline{v}$ (the velocity vector) and p (the pressure) are unknown functions defined in $\Omega$ , and $\underline{f}$ is a given function of x (the exterior force by unit volume).

Remark 1.1 - In fact, the viscosity coefficient $\mu$ appears in front of the term $\Delta\underline{v}$. We have taken $\mu = 1$ (or alternatively, $p/\mu$ and $f/\mu$ are the pressure and force).

Equation (1.1) is the momentum equation (in fact, equilibrium equation, because the acceleration term is zero), (1.2) is the incompressibility condition and (1.3) is the no-slip condition at the boundary.

Our aim is to obtain a variational formulation for the problem (1.1)-(1.3). In fact, the variational formulation is the "virtual power" equation of mechanics. Relations (1.2) and (1.3) are in fact "constraints" and it appears that the pressure force in (1.1) gives zero power for the virtual velocity fields which are compatible with (1.2), (1.3). This is the reason why we begin our study by some orthogonality properties in functional spaces associated with the divergence-free condition (1.2).

We denote by $\underline{L}^2(\Omega)$, $\underline{H}^1(\Omega)$, etc. the space of the vectors such that each component belongs to $L^2(\Omega)$, $H^1(\Omega)$ ... Of course, $\underline{L}^2(\Omega) = (L^2(\Omega))^3$ and so on. We also introduce the spaces

(1.4) $\qquad H(\Omega\ , \text{div}) \equiv \{\underline{u}\ ;\ \underline{u} \in \underline{L}^2(\Omega)\quad ,\quad \text{div}\ \underline{u} \in L^2(\Omega)\}$

equipped with the norm associated with the scalar product

(1.5) $\qquad (\underline{u}\ ,\ \underline{v})_{H(\Omega,\text{div})} = (\underline{u}\ ,\ \underline{v})_{\underline{L}^2(\Omega)} + (\text{div}\ \underline{u}\ ,\ \text{div}\ \underline{v})_{L^2(\Omega)}$

Let $\mathcal{V}$ be the vector space (without any norm) formed by the smooth free-divergence vectors with compact support in $\Omega$ , i.e. :

(1.6) $\qquad \mathcal{V} = \{\underline{u}\ ;\ \underline{u} \in \underline{\mathcal{D}}(\Omega)\ ,\ \text{div}\ \underline{u} = 0\ \}$

and let H (resp. V) be the adherence of $\mathcal{V}$ in $\underline{L}(\Omega)$ (resp. $\underline{H}^1_0(\Omega)$). (This amounts to take the completion of $\mathcal{V}$ for the norms of $\underline{L}^2$ or $\underline{H}^1_0$).

Theorem 1.1 - If $\underline{n}$ denotes the unit outer normal to $\Omega$ , the trace operator $\underline{u} \mapsto \underline{u}\ .\ \underline{n}$ is defined on $H(\Omega\ , \text{div})$ and is continuous from $H(\Omega\ , \text{div})$ into $H^{-1/2}(\partial\Omega)$. (In particular, if $\underline{u} \in \underline{L}^2(\Omega)$ and $\text{div}\ \underline{u} = 0$, the trace $\underline{u}\ .\ \underline{n}$ makes sense and is an element of $H^{-1/2}(\partial\Omega)$).

Remark 1.2 - It is clear that $H(\Omega\ , \text{div})$ is larger than $\underline{H}^1(\Omega)$ and the classical trace theorem (chap. 1, theorems 3.1 and 3.2) does not hold. In fact, only the combination $\underline{u} \cdot \underline{n}$ makes sense, and this in $H^{-1/2}$ (the dual of $H^{1/2}$ when $L^2$ is identified with its dual). We shall give an outline of the proof. We recall that if $\phi \in H^{1/2}(\partial\Omega)$, there exists a "lift" of $\phi$ which is continuous from $H^{1/2}(\partial\Omega)$ into $H^1(\Omega)$, i.e. a function $\Phi \in H^1(\Omega)$ such that $\Phi|_{\partial\Omega} = \phi$ and

(1.7) $$\|\Phi\|_{H^1} \leq C \|\phi\|_{H^{1/2}}$$

where C is a constant which depends only on $\Omega$ . Of course $\Phi$ is not unique.

Let $\underline{u}$ be a smooth function defined on $\Omega$ . By integrating by parts, we have

(1.8) $$\int_\Omega u_i \frac{\partial \Phi}{\partial x_i} dx + \int_\Omega \frac{\partial u_i}{\partial x_i} \Phi \, dx = \int_{\partial\Omega} (n_i u_i) \phi \, ds$$

The right hand side is a scalar product in $L^2$, but $\phi$ belongs to $H^{1/2}$ and consequently, it may be considered as a duality product between $H^{-1/2}$ and $H^{1/2}$. Then, we rewrite (1.8) as :

$$< n_i u_i , \phi >_{H^{-1/2}, H^{1/2}} = (\underline{u} , \underline{\text{grad}} \, \Phi)_{\underline{L}^2} + (\text{div} \, \underline{u} , \Phi)_{L^2} \implies$$

(1.9) $$|< n_i u_i , \phi >| \leq 2 \|u\|_{H(\Omega,\text{div})} \|\Phi\|_{H^1} \leq 2C \|u\|_{H(\Omega,\text{div})} \|\phi\|_{H^{1/2}}$$

where (1.7) has been used. It appears from (1.9) that the norm of $u_i n_i$ in $H^{-1/2}$ is bounded by $2C \|u\|_{H(\Omega,\text{div})}$ . Then, by using the fact that the smooth functions form a dense subset of $H(\Omega, \text{div})$ (this fact will not be proved here), we may define $u_i n_i$ by continuity to any function of $u \in H(\Omega, \text{div})$. ■

<u>Theorem 1.2</u> - The following identity holds :

(1.10) $$H \equiv \{\underline{u} \; ; \underline{u} \in \underline{L}^2(\Omega) \; , \; \text{div} \, \underline{u} = 0 \; , \; \underline{u} . \underline{n}|_{\partial\Omega} = 0 \}$$

this amounts to say that $\mathcal{V}$ is dense in the space at the right hand side of (1.10).

It is of course clear (c.f. theorem 1.1) that the right hand side of (1.10) is a closed subspace of $H(\Omega, \text{div})$ and of $\underline{L}^2(\Omega)$.

<u>Theorem 1.3</u> - The space $\underline{L}^2(\Omega)$ admits the orthogonal decomposition

(1.11) $$\underline{L}^2(\Omega) = H \oplus H^\perp \qquad \text{where}$$

(1.12) $$H^\perp = \{\underline{u} \; , \; \underline{u} = \underline{\text{grad}} \, \phi \; , \; \phi \in H^1(\Omega)\}$$

As an exercise, to understand the sense of theorem 1.3, let $\phi \in H^1(\Omega)$, $\underline{u} \in H$ (see (1.10)). By integrating by parts, we formally obtain the orthogonality property :

$$\int_\Omega \underline{\text{grad}}\ \phi \ . \ \underline{u}\ dx = \int_\Omega \frac{\partial}{\partial x_i}(\phi\ u_i)\ dx = \int_{\partial\Omega} \phi\ u_i\ n_i\ dS = 0\ .$$

It is also easily seen that $\underline{g} \in H^1 \Longrightarrow \underline{\text{rot}}\ \underline{g} = 0$ (as a distribution).

Now, let us recall that $\underline{H}^{-1}(\Omega)$ is the dual of $\underline{H}^1_0(\Omega)$ if $\underline{L}^2$ is identified to its dual. We have :

Theorem 1.4 - The following identity holds :

(1.13) $$V = \{\ \underline{u}\ ;\ \underline{u} \in \underline{H}^1_0(\Omega)\ ,\ \text{div}\ \underline{u} = 0\ \}$$

Moreover, if the brakets are for the duality product between $\underline{H}^{-1}$ and $\underline{H}^1_0$ ,

(1.14) $$\begin{cases} <\underline{g}\ ,\ \underline{u}> = 0 \qquad \forall\ \underline{u} \in V & \Longleftrightarrow \\ \underline{g} = \underline{\text{grad}}\ \phi\ ,\ \phi \in L^2(\Omega) & \end{cases}$$

Remark 1.3 - It is clear that the function $\phi$ in (1.12) and (1.14) is defined up to an additive constant. If $\phi$ is chosen in such a way that the mean value

(1.15) $$\tilde{\phi} = \frac{1}{|\Omega|} \int_\Omega \phi\ dx$$

is zero, we have

(1.16) $$\|\phi\|_{H^1} \leq C\ \|\underline{\text{grad}}\ \phi\|_{L^2}\ ;\ \|\phi\|_{L^2} \leq C\ \|\underline{\text{grad}}\ \phi\|_{\underline{H}^{-1}}\ ;\quad \tilde{\phi} = 0$$

which are proved by using Poincaré's type inequalities. If we consider the equivalence class $\phi$ + cost. (instead of $\phi$), the inequalities (1.16) hold (with, at the left hand sides, the norms in the spaces of the equivalence class $H^1$/cost. and $L^2$/cost). ■

Let us recall that any locally integrable function, considered as a distribution, has distributional derivatives of any order. Then, we have

Lemma 1.1 - If $\underline{u} \in \underline{H}^1_0(\Omega)$ and $\phi \in L^2(\Omega)$, the distributions $\Delta\ \underline{u}$ and $\underline{\text{grad}}\ \phi$ belong to $\underline{H}^{-1}(\Omega)$.

Proof - Let us take $\underline{w} \in \underline{H}^1_0(\Omega)$. Then, by definition of distributional derivatives

(1.17) $$-<\Delta\ \underline{v}\ ,\ \underline{w}> = \int_\Omega \frac{\partial v_i}{\partial x_k} \frac{\partial w_i}{\partial x_k}\ dx$$

(1.18) $$-<\frac{\partial p}{\partial x_i}\ ,\ w_i> = \int_\Omega p\ \text{div}\ \underline{w}\ dx$$

which define linear and bounded functionals on $\underline{H}_0^1$ (by the Cauchy's inequality). ∎

Now, we may consider the stokes problem (1.1) - (1.3) in a functional framework. First, if we consider $v \in V$, conditions (1.2) and (1.3) are satisfied (c.f. (1.13)). Then we have

Stokes problem (1.1) - (1.3) in variational formulation :

Find $\underline{v} \in V$ such that (1.1) is satisfied in the distributional sense (in fact, in $\underline{H}^{-1}(\Omega)$).

Theorem 1.5 - If $\underline{f} \in \underline{H}^{-1}(\Omega)$ (in particular if $\underline{f} \in \underline{L}^2$) the Stokes problem (1.1) - (1.3) has a solution $\underline{v} \in V$, $p \in L^2(\Omega)$. This solution is unique up to an additive constant for p.

Proof - Let us define the bilinear and continuous form on $\underline{H}_0^1(\Omega)$ :

$$a(\underline{u}\ ,\ \underline{w}) \equiv \int_\Omega \frac{\partial u_i}{\partial x_k} \frac{\partial w_i}{\partial x_k}\, dx \tag{1.19}$$

(note that it is also coercive by virtue of the Friedrichs inequality (chap. 3, prop. 1.1)).

Let us take any test function $\underline{w} \in V$. We take the duality product of (1.1) with $\underline{w}$. Calculations analogous to (1.17), (1.18) show that

$$a(\underline{v}\ ,\ \underline{w}) = \langle \underline{f}\ ,\ \underline{w} \rangle \qquad \forall\, \underline{w} \in V \tag{1.20}$$

which is in fact the "virtuál power" equation. By the Lax-Milgram theorem, we may find a unique $\underline{v} \in V$ satisfying (1.20). Then, if $\underline{v}$ is obtained, (1.20) writes

$$\langle \Delta\underline{v} + \underline{f}\ ,\ \underline{w} \rangle = 0 \qquad \forall\, \underline{w} \in \underline{H}_0^1$$

and then $p \in L^2$ satisfying (1.1) exists by virtue of (1.14).

Moreover, regularity theory holds for the system (1.1) - (1.3). In particular

Theorem 1.6 - If $\underline{f} \in \underline{L}^2(\Omega)$, the solution $\underline{v}$ , p given by theorem 1.5 is such that

$$\underline{v} \in \underline{H}^2(\Omega) \qquad ; \qquad p \in H^1(\Omega)$$

and consequently every term of (1.1) belongs to $L^2(\Omega)$.

## 2.- Asymptotic expansion for flow in porous media. Darcy's law

We consider the problem of the preceeding section in a fluid domain $\Omega_{\varepsilon f}$ formed by the cavities of a (rigid) porous solid defined in the following way.

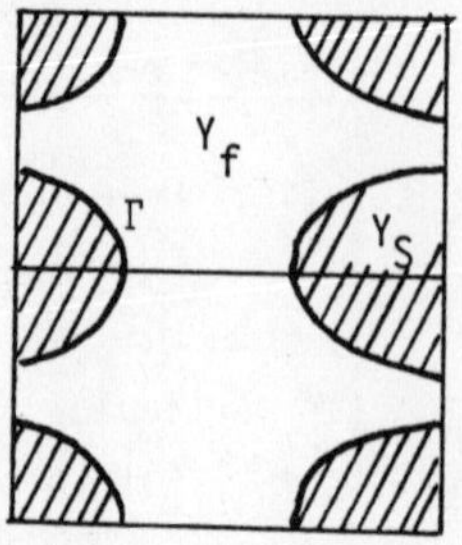


fig. 1

In the classical framework of homogenization problems, we consider the parallelepipedic period Y of the space of variables $y_1$, $y_2$, $y_3$. Each period is made of a fluid and a solid parts, $Y_f$, $Y_s$ with smooth boundary $\Gamma$. Moreover, the unions of all the $Y_f$ parts and of all the $Y_s$ parts are connected. This means that the "solid part is of one piece" ; the same is true for the "fluid part". This situation is impossible in two dimensions but it is possible in three dimensions (see fig. 2, where $Y_f$ is made of three tubes parallel to the axis with a non empty intersection. Nevertheless, we often use two-dimensional drawings as in fig. 1.

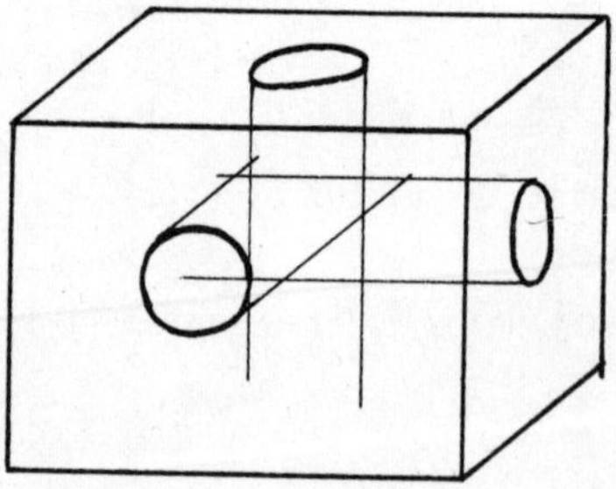

fig. 2

In the sequel, $Y_f$ denotes the fluid part of a period as well as the union of the fluid parts of all the periods.

Now, we consider a small positive parameter $\varepsilon$ ; the fluid domain is defined by

(2.1) $$\Omega_{\varepsilon f} \equiv \Omega \cap \{ x \; ; \quad x \in \varepsilon Y_f \}$$

where $\Omega$ is a given domain. In this section we only consider a formal expansion of the velocity and pressure out of a neighbourhood of $\partial\Omega$ ; consequently, we may consider $\Omega = R^3$

(2.2) $$\Omega_{\varepsilon f} = \{ x \; ; \quad x \in \varepsilon \, Y_f \}$$

Let f(x) be a given function of $L^2(\Omega)$. We consider the problem of finding $\underline{v}^\varepsilon$ , $p^\varepsilon$ satisfying

(2.3) $$0 = - \underline{\text{grad}} \; p^\varepsilon + \Delta \underline{v}^\varepsilon + \underline{f} \qquad \text{in } \Omega_{\varepsilon f}$$

(2.4) $$\text{div } \underline{v}^\varepsilon = 0 \qquad \text{in } \Omega_{\varepsilon f}$$

(2.5) $$\underline{v}^\varepsilon = 0 \qquad \text{on } \partial\Omega_{\varepsilon f}$$

Now, we postulate an asymptotic expansion

(2.6) $$\underline{v}^\varepsilon(x) = \varepsilon^2 \, \underline{v}^0(x \, , \, y) + \varepsilon^3 \, \underline{v}^1(x \, , \, y) + \ldots$$

(2.7) $$p^\varepsilon(x) = p^0(x) + \varepsilon \, p^1(x \, , \, y) + \ldots$$

$$\Big\} \; y = \frac{x}{\varepsilon}$$

with $\underline{v}^i$, $p^i$, Y-periodic in y, for $x \in \Omega$ , $y \in Y_f$. (Note that we postulate that $p^0$ does not depend on y ; this is natural, because the very function in (2.3) is $\underline{\text{grad}}$ $p^\varepsilon$ , which depends on y from the first term because of the classical relation

(2.8) $$\frac{d}{dx_i} = \frac{\partial}{\partial x_i} + \frac{1}{\varepsilon} \frac{\partial}{\partial y_i}$$

The same relation (2.8) shows that the expansion of $\Delta$ is

(2.9) $$\Delta = \frac{1}{\varepsilon^2} \, \Delta_{yy} + \frac{1}{\varepsilon} \ldots$$

where $\Delta_{yy}$ denotes the laplacian with respect to the variables $y_i$ ($x_i$ being paramaters). The form of the relation (2.9) shows that the first significative term in (2.6) must be the $\varepsilon^2$ term ; this explain the form chosen for the postulated asymptotic expansion (2.6), (2.7). As usual in homogenization problems, if we postulate an expansion beginning by $\varepsilon^0$ terms, the two first terms will be find to be zero.

To study the problem (2.3) - (2.5), we replace (2.6), (2.7) into (2.3) - (2.5). Then, by taking the $\varepsilon^0$ term of (2.3), the $\varepsilon$ term of (2.4) and the $\varepsilon^2$ term of (2.5), we have

(2.10) $$0 = - \frac{\partial p^1}{\partial y_i} + \Delta_y \, v_i^0 + \left(f_i - \frac{\partial p^0}{\partial x_i}\right) \qquad \text{in } Y_f$$

(2.11) $$\text{div}_y \, \underline{v}^0 = 0 \qquad \text{in } Y_f$$

(2.12) $$\underline{v}^0|_\Gamma = 0 \qquad \text{on } \Gamma$$

with the supplementary conditions $\underline{v}^0$, $p^1$ are Y-periodic in y. This is the <u>local problem</u>, where x is a parameter and the term in parenthesis of (2.10) plays the role of "given force", $\underline{v}^0$ and $p^1$ are the unknowns. We shall see later that (2.10)-(2.12) leads to the Darcy's law. Before studying this, we consider the $\varepsilon^0$ term of (2.4) :

(2.13) $$\text{div}_x\, \underline{v}^0 + \text{div}_y\, \underline{v}^1 = 0$$

and we apply the classical mean value operator

(2.14) $$\tilde{\cdot} = \frac{1}{|Y|} \int_Y \cdot \, dy$$

Note that $\underline{v}^i$ as functions of y are defined on $Y_f$ : it is natural to extend them to Y with value zero on $Y_s$. (They are zero on $\Gamma$ by virtue of (2.12)). Then, we have

(2.15) $$(\text{div}_y\, \underline{v}^1)^\sim = \frac{1}{|Y|} \int_{Y_f} \frac{\partial v_i^1}{\partial y_i}\, dy = \frac{1}{|Y|} \int_{\partial Y_f} n_i\, v_i^1\, dS = 0$$

To see that the surface integral of (2.15) is zero it suffices to see that $n_i v_i^1$ is zero on $\Gamma$ and that the integral on the parts of $\partial Y_f$ lying on $\partial Y$ (see fig. 1) annihilate by periodicity. On the other hand, the operator $\partial/\partial x_i$ commutes with $\sim$ as usual. Then, by applying $\sim$ to (2.13) we obtain

(2.15) $$\text{div}_x\, \tilde{\underline{v}}^0 = 0$$

which is the <u>macroscopic equation</u>.

<u>Remark 2.1</u> - It is easy to obtain again (2.15) by the "method of the conservation law" of chap. 5, sect.10. It suffices to apply equation (2.11), which is equivalent to the conservation of mass in a macroscopic domain (after observation that volume and surface mean values of $\underline{v}$ equal, as in chap. 5, remark 10.1).(See also sect. 5, "proof of part b)" of the present chapter). <u>(2.15) then appears as the macroscopic form of the conservation of mass</u>.

<u>Now we study the local problem (2.10) - (2.12)</u>. We define an appropriate space of Y-periodic functions:

(2.16) $$V_Y = \{\underline{u}\ ;\ \underline{u} \in H^1(Y_f)\ ;\ \underline{u}|_\Gamma = 0,\ \text{div}_y\, \underline{u} = 0\ ;\ \text{Y-periodic}\}$$

(2.17) $$(\underline{u}\,,\,\underline{w})_{V_Y} = \int_{Y_f} \frac{\partial u_i}{\partial y_k} \frac{\partial w_i}{\partial y_k}\, dy$$

which is a Hilbert space ; the associated norm is equivalent to the $H^1(Y_f)$ norm. (Compare with (1.19)).

Remark 2.2 - If we continuate $\underline{u}$ by zero to $Y_s$, we may write Y instead of $Y_f$ in (2.16). On the other hand, Y may be considered as a period as well as the $R^3$ space; on the first case, Y-periodic means that the traces of $\underline{u}$ on the opposite faces of each period are the same (see chap. 5, remark 2.1) ; in the second case, it is understood that $\underline{u}$ belongs locally to $H^1$. ∎

To obtain a variational formulation of (2.10) - (2.12), we take a test function $\underline{w} \in V_Y$ and we multiply (2.10) by $w_i$ ; by integrating over $Y_f$ we have (note that the integrals over $\partial Y_f$ are zero because $w\big|_\Gamma = 0$ and the periodicity conditions) :

$$(2.18) \qquad -\int_{Y_f} \frac{\partial p^1}{\partial y_i} w_i \, dy = (\text{by } (2.11)) = -\int_{Y_f} \frac{\partial}{\partial y_i} (p^1 w_i) dy = \\ = \int_{\partial Y_f} p^1 w_i n_i \, dS = 0$$

$$(2.19) \qquad \int_{Y_f} \Delta_y v_i^0 w_i \, dy = \int_{Y_f} \left[ \frac{\partial}{\partial y_k}\left(\frac{\partial v_i^0}{\partial y_k} w_i\right) - \frac{\partial v^o}{\partial y_k} \frac{\partial w_i}{\partial y_k} \right] dy = \\ = \int_{\partial Y_f} n_k \frac{\partial v_i^0}{\partial y_k} w_i \, dS - \int_{Y_f} \frac{\partial v_i^0}{\partial y_k} \frac{\partial w_i}{\partial y_k} dy = -\int_{Y_f} \frac{\partial v_i^0}{\partial y_k} \frac{\partial w_i}{\partial y_k} dy$$

Then, by using (2.17), the product of (2.10) by $w_i$ gives :

$$(2.20) \qquad (\underline{v}^0 , \underline{w})_{V_Y} = \left(f_i - \frac{\partial p^0}{\partial x_i}\right) \int_{Y_f} w_i \, dy \qquad \forall \, \underline{w} \in V_Y$$

(note that the function in the parenthesis does not depend on y).

Conversely, if $\underline{v}^0 \in V_Y$ and satisfies (2.20), by integrating by parts it satisfies

$$(2.21) \qquad \int_{Y_f} \left[\Delta_y v_i^0 + \left(f_i - \frac{\partial p^0}{\partial x_i}\right)\right] w_i \, dy = 0 \qquad \forall \, \underline{w} \in V_Y$$

and this shows by theorem 1.4 that there exists $p^1$ of class $L^2$ satisfying (2.10) (note that we may take in (2.21) any $\underline{w} \in H_0^1(Y_f)$ with div $\underline{w} = 0$, and theorem 1.4 holds). Moreover, by theorem 1.6, $\underline{v}^0$ (resp. $p^1$) is of class $H^2$ (resp. $H^1$) (note that the regularity properties are local ; consequently, theorem 1.6 applies without taking into account the periodicity conditions). Now, we show that the function $p^1$ just found is Y-periodic. In fact, grad $p^1$ is periodic because $p^1$ satisfies (2.10). We multiply (2.10) by $w_i$ and we integrate on $Y_f$. By comparing with (2.21) we have

$$0 = -\int_{Y_f} \frac{\partial p^1}{\partial y_i} w_i \, dy$$

and because div $\underline{w} = 0$, we have

(2.22) $$0 = \int_{Y_f} \frac{\partial}{\partial y_i} (p^1 w_i) dy = \int_{\partial Y_f} p^1 w_i n_i \, dS$$

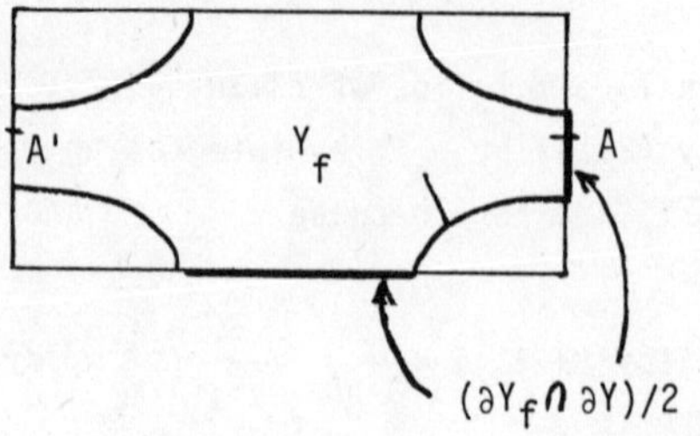


fig. 3

The surface integral in (2.22) is zero on the parts $\Gamma$ because $\underline{w}$ is zero there. Moreover, on the parts contained in $\partial Y$, we consider only half the faces of the period, and we take into account that in two points homologous by periodicity, such that A and A', $\underline{w}$ takes the same value; consequently, (2.22) becomes

(2.23) $$0 = \int_{(\partial Y_f \cap \partial Y)/2} [\, p^1(A') - p^1(A)] \, w_i n_i \, dS$$

where the domain of integration is marked in fig. 3. But on this domain, $w_i n_i$ may be taken arbitrarily (Note that, taking into account div $\underline{w} = 0$, this would not be possible without the hypothesis that $Y_f$ is connected) and consequently, $p^1$ takes the same value in A and A', i.e., $p^1$ is periodic.

Consequently, the local problem (2.10) - (2.12) is equivalent to the following variational problem :

(2.24) Find $v^0 \in V_Y$ satisfying (2.20).

Moreover, by using the standard linearity property, we have

Proposition 2.1 - If we postulate an asymptotic expansion (2.6), (2.7), the first term $\underline{v}^0(x\,,\,y)$ is given by $\underline{f}_i(x)$ and $\partial p^0/\partial x_i(x)$ as

(2.25) $$\underline{v}^0 = (f_i - \frac{\partial p^0}{\partial x_i})\, \underline{v}^i$$

where $\underline{v}^i(y)$ (i = 1, 2, 3) are the solutions of :

(2.26) Find $\underline{v}^i \in V_Y$ such that

$$(\underline{v}^i, \underline{w})_{V_Y} = \int_{Y_f} w_i \, dy \qquad \forall\, \underline{w} \in V_Y$$

The existence and uniqueness of the solutions of (2.24) or (2.26) are immediate consequences of the Lax-Milgram theorem, because the right hand sides of (2.24), (2.26) are linear and bounded functionals on $V_Y$.

Now, if we apply the mean operator $\sim$ (defined by (2.14)) to (2.25), we have

$$\tilde{v}^o_j = K_{ij}(f_i - \frac{\partial p^0}{\partial x_i}) \quad ; \quad K_{ij} = \tilde{v}^i_j \tag{2.27}$$

(as usual, the indexes i denote the components of the vectors).

Relation (2.27) is the Darcy's law. The mean value of the velocity of the fluid is equal to $\underline{f}$ - grad $p^0$ multiplied by a constant tensor with components $K_{ij}$ which only depend on the geometry of the period Y. It is noticeable that (2.27) was obtained from (2.3), i.e., for the viscosity coefficient $\mu$ equal to one. If we consider

$$0 = - \text{grad } p^\varepsilon + \mu \Delta \underline{v}^\varepsilon + \underline{f} \tag{2.28}$$

instead of (2.3), we obtain

$$\tilde{v}^o_j = \frac{K_{ij}}{\mu} (f_i - \frac{\partial p^0}{\partial x_i}) \tag{2.29}$$

instead of (2.27). It is also necessary to introduce the coefficient $\mu^{-1}$ at the right hand side of (2.25).

Proposition 2.2 - The matrix $K_{ij}$, defined by $(2.27)_2$ will be called "permeability tensor". It is a symmetric, positive definite matrix.

Proof - If in (2.26) we take $\underline{w} = \underline{v}^j$ and in the equation analogous to (2.26) for $\underline{v}^j$ we take $\underline{w} = \underline{v}^i$, we obtain, by the symmetry of the scalar product :

$$v^j_i = v^i_j \Leftrightarrow K_{ij} = K_{ji}$$

On the other hand, the matrix $K_{ij}$ is positive because, by virtue of (2.26), we have :

$$K_{ij}\, \xi_i\, \xi_j = \xi_i\, \xi_j\, \tilde{v}^i_j = (\xi_i\, \underline{v}^i\, ,\, \xi_j\, \underline{v}^j)_{V_Y} = \| \xi_i\, \underline{v}^i \|^2_{V_Y} \geq 0$$

Then, to prove that the matrix is positive, definite, it suffices to show that $\xi_i \underline{v}^i = 0$ only if $\xi_i = 0$. To show this, we multiply (2.26) by $\xi_i$ and we take a test function $\underline{w}$ such that $\tilde{w}_i = \xi_i$ (this is possible by the hypothesis that $Y_f$ is connected) and we have

$$(\xi_i\, \underline{v}^i\, ,\, \underline{w})_{V_Y} = \xi_i \int_{Y_f} w_i\, dy = \xi_i\, \xi_i$$

and $\xi_i\, \underline{v}^i = 0 \Rightarrow \xi_i = 0$ , Q.E.D. ∎

Finally, the macroscopic equation (2.15) with (2.27) becomes

$$\frac{\partial}{\partial x_i}\left[K_{ij}\left(f_j - \frac{\partial p^0}{\partial x_j}\right)\right] = 0 \qquad (2.30)$$

Proposition 2.3 - The macroscopic equation (2.15) may be written in the form

$$K_{ij}\frac{\partial^2 p^0}{\partial x_i\,\partial x_j} = K_{ij}\frac{\partial f_j}{\partial x_i} \qquad (2.31)$$

which is an elliptic equation for the unknown $p^0(x)$. If $p^0(x)$ is obtained, the velocity field $\underline{v}^0(x , y)$ is given by (2.25) and the mean value of the velocity satisfies the Darcy's law (2.27).

For the proof of the convergence, see the Appendix by L. Tartar.

Remark 2.3 - In order to obtain $p^0(x)$, we must adjoin boundary conditions to the equation (2.31). This question is studied in sect. 5, as well as the corresponding boundary layers. Nevertheless, in the case of an impervious boundary (as in in (2.5)) the appropriate boundary condition is

$$\tilde{v}_i^0\, n_i = 0 \qquad (2.32)$$

which implies, by (2.27), the Neumann boundary condition for (2.30) (or (2.31)) :

$$K_{ij}\left(f_j - \frac{\partial p^0}{\partial x_j}\right) n_j = 0 \qquad (2.33)$$

on the boundary of the porous medium. ■

## 3.- Effects of compressibility

It is clear that the study of the preceeding section is associated with the problem (2.3) - (2.5) and it is not to be considered as a general theory of flow in porous media. The same kind of techniques may be used for other problems. Moreover, certain parts of the preceeding study are not modified.

As an example, we consider an analogous problem for a compressible barotropic fluid (this means that the density $\rho$ is a given function of the pressure p ; in the general case $\rho$ depends also on the temperature, and the problem involves a thermal equation). As in the preceeding section, we study the steady (i.e. independent of time) motion by neglecting the nonlinear terms.

We give a positive, smooth and increasing function F and

$$\rho^\varepsilon(x) = F(p^\varepsilon(x)) \qquad (3.1)$$

moreover, if $\underline{f}(x)$ is a given function, and $\mu$ and $\eta$ are the viscosity coefficients (for a compressible fluid there are two viscosity coefficients), we consider the equations :

$$(3.2) \qquad 0 = -\frac{\partial p^{\varepsilon}}{\partial x_i} + \mu \Delta v_i^{\varepsilon} + \eta \frac{\partial}{\partial x_i} \operatorname{div} \underline{v}^{\varepsilon} + f_i \qquad \text{in } \Omega_{\varepsilon f}$$

$$(3.3) \qquad \operatorname{div}(\rho^{\varepsilon} \underline{v}^{\varepsilon}) = 0 \qquad \text{in } \Omega_{\varepsilon f}$$

$$(3.4) \qquad \underline{v}^{\varepsilon} = 0 \qquad \text{on } \partial\Omega_{\varepsilon f}$$

in the domain $\Omega_{\varepsilon f}$ described in the preceeding section. We postulate the following asymptotic expansion for the unknowns $\underline{v}^{\varepsilon}$, $p^{\varepsilon}$, $\rho^{\varepsilon}$ :

$$(3.5) \qquad \underline{v}^{\varepsilon}(x) = \varepsilon^2 \underline{v}^0(x , y) + \varepsilon^3 \ldots$$

$$(3.6) \qquad p^{\varepsilon}(x) = p^0(x) + \varepsilon p^1(x , y) + \ldots$$

$$(3.7) \qquad \rho^{\varepsilon}(x) = \rho^0(x) + \varepsilon \rho^1(x , y) + \ldots$$

where from (3.1) we have :

$$(3.8) \qquad \rho^0(x) = F(p^0(x))$$

and other relations for the other terms of $p^{\varepsilon}$ and $\rho^{\varepsilon}$ . We introduce (3.5) - (3.7) into (3.2), (3.3) by using the classical relation (2.8) for the derivatives.

We consider the $\varepsilon^0$ term of (3.2), the $\varepsilon^1$ term of (3.3) and the $\varepsilon^2$ term of (3.4), and we obtain

$$(3.9) \qquad 0 = -\frac{\partial p^0}{\partial x_i} - \frac{\partial p^1}{\partial y_i} + \mu \Delta_y v_i^0 + \eta \frac{\partial}{\partial y_i} \operatorname{div}_y \underline{v}^0 + f_i \qquad \text{in } Y_f$$

$$(3.10) \qquad \rho^0 \operatorname{div}_y \underline{v}^0 = 0 \qquad \text{in } Y_f$$

$$(3.11) \qquad v^0\big|_{\Gamma} = 0 \qquad \text{on } \Gamma$$

which is the system for the local problem in a period. As $\rho^0 > 0$, equation (3.10) is in fact (2.11) : (this means that, at the local level, the flow is incompressible) ; by replacing it into (3.9), this equation becomes (2.10) (i.e., the new term in $\eta$ desappears) and consequently the system (3.9) - (3.11) is the same as (2.10) - (2.12). Consequently, <u>the local behaviour is given by proposition 2.1 (with $\mu^{-1}$ as a factor of the right hand side of (2.25)) and (2.29)</u>.

In order to obtain the macroscopic behaviour, we consider the $\varepsilon^2$ term of the expansion of (3.3) ; this gives :

$$(3.12) \qquad \rho^0(\operatorname{div}_x \underline{v}^0 + \operatorname{div}_y \underline{v}^1) + \rho^1 \operatorname{div}_y \underline{v}^0 + \left(\frac{\partial \rho^0}{\partial x_i} + \frac{\partial \rho^1}{\partial y_i}\right) v_i^0 = 0$$

But $\operatorname{div}_y \underline{v}^0 = 0$ as we have seen. We apply the mean operator $\sim$ to equation (3.12). As usual, by the periodicity conditions, $(\operatorname{div}_y \underline{v}^1)^{\sim} = 0$ ; moreover, by using (3.10) and the periodicity conditions, we have :

$$\int_{Y_f} \frac{\partial \rho^1}{\partial y_i} v_i^o \, dy = \int_{Y_f} \frac{\partial}{\partial y_i} (\rho^1 v_i^o) \, dy = \int_{\partial Y_f} \rho^1 v_i^o \; n_i \, dS = 0$$

Finally, we obtain :

(3.13) $$\mathrm{div}_x(\rho^o \, \underset{\sim}{\tilde{v}}^o) = 0$$

which is the macroscopic equation. By using (2.29) and (3.8), it becomes

(3.14) $$\frac{\partial}{\partial x_i}\left[ \frac{K_{ij}}{\mu} F(p^o)(f_i - \frac{\partial p^o}{\partial x_i} )\right] = 0$$

which is the equation for the unknown $p^o(x)$.

## 4.- Non linear effects

Let us consider again the flow of an incompressible fluid as in sect. 2. It is clear that the viscosity coefficient $\mu$ is fixed (see (2.29)) and in fact it was taken equal to one through out the study. This point of view amounts to consider that the only small parameter is $\varepsilon$ . In practice, when we deal with a porous medium, $\varepsilon$ is small, but it does not "tend" to zero. If $\varepsilon$ has a very small value, it is natural to assume that the flow is "nearly" the asymptotic flow studied in section 2. Nevertheless, in a practical problem, another parameter ($\mu$, for instance) may take also a small value. In this case, it is natural to consider also $\mu$ as a small parameter in the expansion. Other parameters may also be small (for instance, the density $\rho$ of a gas, etc.).

In fact, if $\mu$ is small, the Darcy's law (2.29) shows that the velocity is large and consequently, the appropriate expansion for $\underline{v}^\varepsilon$ may begin by terms of order $\neq \varepsilon^2$ (see (2.6)). Then, if the velocity is not small, the nonlinear terms which were disregarded in (2.3) (and in fact, they were not written) may be important. This is the reason why we study now a problem with small viscosity in a nonlinear framework. We shall see that, in certain cases, the Darcy's law is non linear, moreover, it may desappear as a determinist law.

Let us consider the nonlinear Navier-Stokes system for an incompressible fluid in time-dependent flow :

(4.1) $$\rho(\frac{\partial v_i^\varepsilon}{\partial t} + v_k^\varepsilon \frac{v_i^\varepsilon}{\partial x_k}) = - \frac{\partial p^\varepsilon}{\partial x_i} + \nu \, \varepsilon^\beta \, \Delta \, v_i^\varepsilon + f_i \qquad \text{in } \partial\Omega_{\varepsilon f}$$

(4.2) $$\mathrm{div}\, \underline{v}^\varepsilon = 0 \qquad \text{in } \Omega_{\varepsilon f}$$

(4.3) $$\underline{v}^\varepsilon = 0 \qquad \text{on } \partial\Omega_{\varepsilon f}$$

where $\rho$ is the (constant) density and $\nu\varepsilon^\beta$ is the viscosity coefficient ($\beta > 0$ is undeterminated for the time being).

We consider the following asymptotic expansion (where $\alpha$ is undetermined) :

$$(4.4)\qquad p^{\varepsilon}(x\,,\,t) = p^{0}(x\,,\,t) + \varepsilon\, p^{1}(x\,,\,y\,,\,t) + \ldots$$

$$(4.5)\qquad \underline{v}^{\varepsilon}(x\,,\,t) = \varepsilon^{\alpha}(\underline{v}^{0}(x\,,\,y\,,\,t) + \ldots$$

We consider the classical formula (2.8) for the derivatives and we write the principal terms of the expansion of each term of (4.1). We have :

$$\frac{\partial v_i^{\varepsilon}}{\partial t} = \varepsilon^{\alpha}\frac{\partial v_i^{0}}{\partial t} + \ldots \quad ; \quad v_k^{\varepsilon}\frac{\partial v_i^{\varepsilon}}{\partial x_k} = \varepsilon^{2\alpha-1}\, v_k^{0}\frac{\partial v_i^{0}}{\partial y_k} + \ldots$$

$$\varepsilon^{\beta}\,\Delta\, v_i^{\varepsilon} = \varepsilon^{\beta+\alpha-2}\,\Delta_y\, v_i^{0} + \ldots \quad ; \quad \frac{\partial p^{\varepsilon}}{\partial x_i} = \frac{\partial p^{0}}{\partial x_i} + \frac{\partial p^{1}}{\partial y_i} + \ldots$$

Then, if we choose

$$(4.6)\qquad \alpha = 1/2 \quad ; \quad \beta = 3/2$$

the $\varepsilon^{0}$ term of (4.1), the $\varepsilon^{\alpha-1}$ term of (4.2) and the $\varepsilon^{\alpha}$ term of (4.3) give :

$$(4.7)\qquad \rho\, v_k^{0}\frac{\partial v_i^{0}}{\partial y_k} = -\frac{\partial p^{1}}{\partial y_i} + \nu\Delta_y\, v_i^{0} + \left(f_i - \frac{\partial p^{0}}{\partial x_i}\right) \qquad \text{in } Y_f$$

$$(4.8)\qquad \operatorname{div}_y \underline{v}^{0} = 0 \qquad \text{in } Y_f$$

$$(4.9)\qquad \underline{v}^{0}\big|_{\Gamma} = 0 \qquad \text{on } \Gamma$$

This means that if the viscosity coefficient is of order $\varepsilon^{3/2}$, the appropriate expansion for $v^{\varepsilon}$ is in $\varepsilon^{1/2}$ ; in this case, the local problem (4.7) - (4.9) is non linear (we shall see that the corresponding Darcy's law is nonlinear).

Moreover, if we apply the mean operator $\sim$ to the $\varepsilon^{\alpha}$ term of (4.2) we obtain (note that the terms of the form $(\operatorname{div}_y \underline{v})^{\sim}$ are zero as in (2.15))

$$(4.10)\qquad \operatorname{div}_x \underline{\tilde{v}}^{0} = 0$$

which is the macroscopic equation.

Let us give an outline of the study of the local problem (4.7) - (4.9). The variational formulation of this problem is easily obtained as (2.24) :

$$(4.11)\qquad \left|\begin{array}{l} \text{Find } \underline{v}^{0} \in V_Y \text{ such that} \\[2mm] -\rho\displaystyle\int_{Y_f} v_k^{0}\, v_i^{0}\,\frac{\partial w_i}{\partial y_k}\,dy + \nu(\underline{v}^{0}\,,\,\underline{w})_{V_Y} = \left(f_i - \frac{\partial p^{0}}{\partial x_i}\right)\tilde{w}_i \quad \forall\, \underline{w} \in V_Y \end{array}\right.$$

We have the following estimate for the nonlinear term (note that the existence of the constant C, which depends only on $Y_f$ is a consequence of the Sobolev theorem (chapter 1, theorem 3.1) :

$$(4.12) \qquad \left| \int_{Y_f} v_k^o v_i^o \frac{\partial w_i}{\partial y_k} dy \right| \leqslant \| \underline{v}^o \|^2_{L^4} \| \underline{w} \|_{H^1} \leqslant C^2 \| \underline{v}^o \|^2_{V_Y} \| \underline{w} \|_{V_Y}$$

Consequently, every term in (4.11) is a linear and bounded functional of $\underline{w}$, and by the Riesz theorem (4.11) is equivalent to

$$(4.13) \qquad \rho\, A(\underline{v}^o) + \nu\, v^o = F_i (f_i - \frac{\partial p^o}{\partial x_i})$$

where A(v) is a well determined non linear operator which send $V_Y$ into itself, and $F_i$ are given elements of $V_Y$. This is a steady-state Navier Stokes equation, which may be studied by the classical methods (see Ladyzhenskaya [2], Temam [1], Tartar [5]).

If the values of $f_i - \frac{\partial p^o}{\partial x_i}$ are not very large (for given $\rho$ and $\nu$), eq. (4.13) has one and only one solution (which depends in a nonlinear fashion on $f_i - \frac{\partial p^o}{\partial x}$ ). For large values of $f_i - \frac{\partial p^o}{\partial x_i}$, the uniqueness of the solution disappear (bifurcations may arise) and consequently $v^o(x\ ,\ y)$ is no longer determined by $\underline{f} - \underline{\text{grad}}_x\, p^o$.

<u>Proposition 4.1</u> - For small $\underline{f} - \underline{\text{grad}}_x\, p^o$, $\underline{v}^o(x\ ,\ y)$ is uniquely determined, and consequently, its mean value $\tilde{\underline{v}}^o$ is a well determined (non linear) function of $\underline{f} - \underline{\text{grad}}_x\, p^o$. This is a nonlinear Darcy's law, which gives, with (4.10), a non linear equation for $p^o(x)$ (note that t appears only as a parameter). Moreover, for large values of $\underline{f} - \underline{\text{grad}}_x\, p^o$, bifurcations (turbulence !) may appear and there is no Darcy's law ; in this case, the expansion (4.4), (4.5) is no longer appropriate.

## 5.- Considerations about boundary conditions and boundary layers

We consider again the problem of sect. 2. We have seen (remark 2.2) that the appropriate boundary condition at a solid boundary is that the normal component of the mean velocity $\underline{v}^o$ is zero. Now we study in detail the fluid flow in the vicinity of this region (boundary layer) and we shall see that the preceeding boundary condition is the necessary and sufficient condition for the existence of the boundary layer (this problem recalls that of chap. 5, sect. 7). Next, we shall study other questions about boundary conditions.

We only consider the particular case where the boundary $\partial\Omega$ (which we denote by S) is parallel to the basis of the period. It is clear that the no-slip condition $\underline{v} = 0$ must be imposed on <u>$\Gamma$ and S</u> (see (2.5)) and not only on $\Gamma$. Consequently, the Y-periodicity condition is no longer appropriate. Instead of this, we shall impose a $B_Y$ periodicity (as in chap. 5, sect. 7) where $B_Y$ is the "semi-infinite" domain formed by the superposition of periods in the direction normal to S (i.e. in the $y_2$ direction).

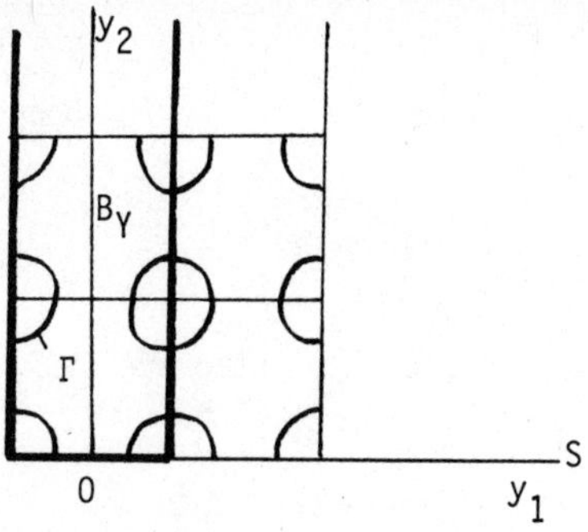


fig. 4

Then, in the vicinity of S, we shall search for an expansion analogous to (2.6), (2.7) :

$$(5.1) \qquad \left\{ \begin{array}{l} v^\varepsilon(x) = \varepsilon^2 \underline{v}^* + \dots \\ p^\varepsilon(x) = p^0(x) + \varepsilon\, p^*(x\ ,\ y) + \dots \end{array} \right\} \quad y = \frac{x}{\varepsilon}$$

$B_y$-periodic in y. The local problem is analogous to (2.10) - (2.12), but in addition to the $B_y$-periodicity condition, the solution must tend to the Y-periodic solution (2.6), (2.7) as $y_2 \to \infty$ :

$$(5.2) \qquad 0 = -\frac{\partial p^*}{\partial y_i} + \Delta_y\, v_i^* + (f_i - \frac{\partial p^0}{\partial x_i})$$

$$(5.3) \qquad \text{div}_y\, \underline{v}^* = 0$$

$$(5.4) \qquad \underline{v}^* \Big|_\Gamma = 0 \quad ; \quad \underline{v}^* \xrightarrow[y_2 \to \infty]{} \underline{v}^0 \quad ; \quad v^*\Big|_S = 0$$

(Note that the second condition (5.4) is a "matching condition" in the sense of the asymptotic expansions in Mechanics ; see Van Dyke [ 1 ]). In order to consider a problem with homogeneous "boundary condition at infinity", we consider the new unknowns

$$(5.5) \qquad \underline{u} = \underline{v}^* - \underline{v}^0 \quad ; \quad q = p^* - p^1$$

instead of $\underline{v}^*$ , $p^*$. From (2.10) - (2.12) and (5.2) - (5.4) we obtain :

$$(5.6) \qquad 0 = -\frac{\partial q}{\partial y_i} + \Delta_y\, u_i$$

$$(5.7) \qquad \text{div}_y\, \underline{u} = 0$$

$$(5.8) \qquad \underline{u}\Big|_\Gamma = 0 \quad ; \quad \underline{u} \xrightarrow[y_2 \to +\infty]{} 0 \quad ; \quad \underline{u}\Big|_S = -\underline{v}^0\Big|_S$$

Proposition 5.1 - a) If the boundary layer exists (i.e., if $\underline{u}$, satisfying (5.6) - (5.8) exist), $\underline{v}^0|_S$ is such that

$$(5.9) \qquad \int_{S \cap B_Y} v_2^0 \, dy_1 \, dy_3 = 0$$

(i.e. the flux of the vector $\underline{v}^0$ through the part of the boundary S corresponding to a $B_Y$ period is zero).

b) Condition (5.9) coïncides with (2.32) or (2.33).

Proof of part a) - It suffices to integrate (5.7) on $B_Y$ and use the $B_Y$-peridicity and (5.8).

Proof of part b) - This is a analogous to chapter 5, prop.10.1 and remark 10.1 : by virtue of the incompressibility condition (2.11) of the Y-periodic flow, the surface and volume mean values of $\underline{v}^0$ are equal.

For, by integrating $\text{div}_y \, \underline{v}^0 = 0$ on the region R of the figure, we have

$$0 = \int_R \text{div}_y \, \underline{v}^0 \, dy = \int_{\partial R} v_i^0 \, n_i \, dS$$

and Y-periodicity

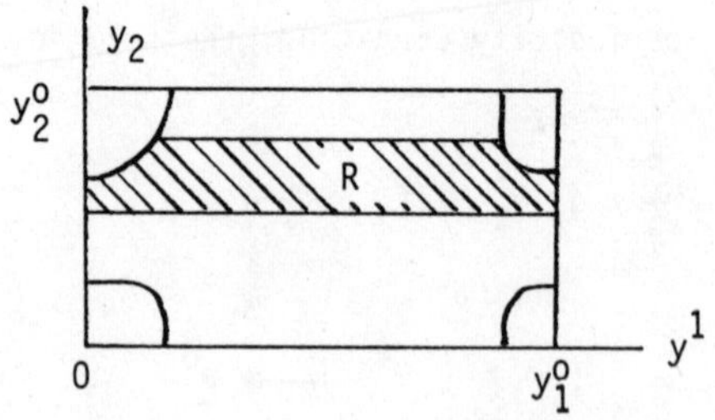


fig. 5

$$(5.10) \qquad I \equiv \int_{Y_f \cap \{y_2 = c\}} v_2^0 \, dy_1 \, dy_3 \quad \text{is independent of } c$$

Then, we have

$$(5.11) \qquad \boxed{\frac{1}{|Y|} \int_{Y_f} v_2^0 \, dy = \frac{y_2^0 \, I}{|Y|} = \frac{1}{y_1^0 \, y_3^0} \int_{Y_f \cap \{y_2 = c\}} v_2^0 \, dy_1 \, dy_3}$$

which shows the desired identity between volume and surface means. Condition (5.9) amounts to say that the right hand side of (5.11) is zero, e.i., the left hand side is zero, which is condition (2.32) or (2.33). ∎

Proposition 5.2 - Condition (5.9) is also a sufficient condition for the existence of the boundary layer.

Outline of the proof - This problem is analogous to that of the intermediate layer in the acoustics in porous media (Levy and Sanchez [2], sect. 3.1). Details may be found there (see also chap. 5, sect. 7).

We note that $(5.8)_2$ with (5.6) implies that q tends to a constant as $y_2 \to \infty$ ; consequently, q (the gradient of which is $B_Y$-periodic) is $B_Y$-periodic. Moreover, we consider an auxiliary function a(y) defined on $B_y$, satisfying the boundary conditions (5.8), div $\underline{a} = 0$ and equal to zero for sufficiently large $y_2$. (Such a function exists by virtue of (5.9) : see Ladyzhenskaya [2] chap. 1, sect.2 sect. 2). We then take as new unknown $\underline{w} = \underline{u} - \underline{a}$. We also construct a Hilbert space V formed by $B_Y$-periodic vectors with zero divergence which tend to zero as $y_2 \to \infty$ and are zero on $\partial B_Y$ and $\Gamma$. the scalar product in V is

$$(\underline{w}\ ,\ \underline{\omega})_V = \int_{B_Y} \frac{\partial w_i}{\partial y_j} \frac{\partial \omega_i}{\partial y_j}\, dy$$

Then, problem (5.6) - (5.8) amounts to find $\underline{w} \in V$ such that,

$$(\underline{w}\ ,\ \underline{\omega})_V = \int_{B_Y} \frac{\partial a_i}{\partial y_j} \frac{\partial \omega_i}{\partial y_j}\, dy \qquad \forall\, \underline{\omega} \in V$$

which has a unique solution by the Lax-Milgram theorem. ■

Now we consider the problem of the interface conditions to be imposed at the boundary between a fluid in a porous medium and the fluid in a free domain.

First, we consider a porous body B (as in sect. 2) which is contained in an outer domain D filled by the fluid.

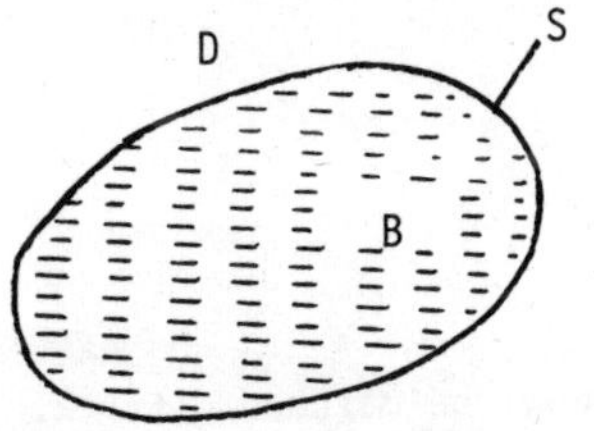


fig. 6

The formal expansion (2.6), (2.7) is appropriate for the region B, but not for the region D. In the region D, we of course have an expansion

$$(5.12)\qquad \begin{aligned} v^{\varepsilon}(x) &= v^{D}(x) + \dots \\ p^{\varepsilon}(x) &= p^{D}(x) + \dots \end{aligned}$$

which is independent of $\varepsilon$ (at least for the first terms). The expansions (5.12) must match on S with expansions (2.6), (7.7). The matching of the velocity vector implies (note that $\varepsilon \to 0$ !) that

(5.13) $$\underline{v}^D\Big|_S = 0$$

which is the boundary condition to be imposed on $\underline{S}$ if B was a solid (impervious) body. This is natural because the velocity vector inside B is very small in front of $\underline{v}^D$. Then, the flow in the region D may be determined (at least its first terms $\underline{v}^D$, $p^D$ !) with (5.13). Next, the matching of $p^\varepsilon$ shows that the appropriate boundary condition for $p^\varepsilon$ is the Dirichlet) condition

(5.14) $$p^\varepsilon\Big|_S = p^D\Big|_S$$

where the right hand side is known. This determines (with equation (2.30)) the first term $p^0$ of the pressure in D (and the first term $\underline{v}^0$ of the velocity in D is then given by the Darcy's law (2.29)).

But there is another class of problems where an expansion of the type (2.12) is impossible. Such are problems where the motion in the region D is a consequence of the motion in the porous body B (for instance D is a cavity in the porous body B). Then the appropriate expansion in D is

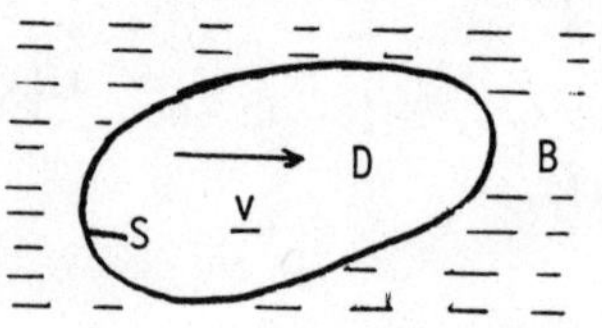


fig.7

(5.15) $$\underline{v}^\varepsilon(x) = \varepsilon^2\, \underline{v}^D(x) + \ldots$$
$$p^\varepsilon(x) = p^D(x) + \ldots$$

instead of (5.12). This expansion must be matched with (2.6), (2.7). In fact, it is clear that $v^D$ and $p^D$ in (5.15) does not depend on the variable y, and consequently, expansions (5.15) and (2.6), (2.7) are of a very different nature.

If we introduce (5.15) into equation (2.3), we obtain at the first order in $\varepsilon$ :

(5.16) $$0 = -\,\underline{\text{grad}}\; p^D + \underline{f} \qquad ;$$

which shows that $p^D$ is in fact the hydrostatic pressure associated with velocity zero.

A deeper study of these questions, in particular a study of the structure of the layers in a neighbourhood of S, considerations about dimensionless numbers and examples may be found in Levy and Sanchez [1] and Ene and Sanchez [1], (sections 6 - 9).

## 6.- Acoustics in porous media

Acoustics deals with small perturbations of a compressible fluid at rest. The perturbations are assumed to be sufficiently small to be studied in a linearized framework. In this section we study the case of a barotropic fluid, that is to say, a fluid such that its density is a function of the pressure only (general fluids, where the density is a function of pressure and temperature are studied in next section).

The viscosity of air is usually negligibly small, and consequently, acoustics is usually studied for an inviscid fluid. As in sect. 4, a small viscosity coefficient may be important at the local level (for the problem in y). We shall take $\nu\varepsilon^2$ and $\eta\varepsilon^2$ ($\nu,\eta$ constants) as viscosity coefficients, ($\varepsilon$ is, as usual, the characteristic length of the period). This choice of the viscosity coefficients leads to integrodifferential equations at the macroscopic level.

If p, $\rho$, are the (constant) values of pressure and density at the rest state, the equations of acoustics are

$$(6.1) \qquad \rho \frac{\partial \underline{v}^\varepsilon}{\partial t} = - \underline{\text{grad}}\, p^\varepsilon + \nu\, \varepsilon^2 \Delta \underline{v}^\varepsilon + \eta\, \varepsilon^2\, \underline{\text{grad}}\, \text{div}\, \underline{v}^\varepsilon + \underline{f}$$

$$(6.2) \qquad \frac{\partial \rho^\varepsilon}{\partial t} + \rho\, \text{div}\, \underline{v}^\varepsilon = 0$$

$$(6.3) \qquad p^\varepsilon = c^2\, \rho^\varepsilon$$

which are obtained by linearization of the Navier-Stokes equations for a compressible fluid. Here, $\rho^\varepsilon$, $p^\varepsilon$, $\underline{v}^\varepsilon$ are the perturbations of density, pressure and velocity ($\underline{v}^\varepsilon$ is also the velocity itself) and the positive constant c is the "sound velocity" such that $c^2$ is equal to $dp/d\rho$ for the rest state. f is the given force.

By eliminating $\rho^\varepsilon$, (6.2) becomes

$$(6.4) \qquad \frac{\partial p^\varepsilon}{\partial t} + \rho\, c^2\, \text{div}\, \underline{v}^\varepsilon = 0$$

and we have the system (6.1), (6.4) for $\underline{v}^\varepsilon$, $p^\varepsilon$.

Now, we consider a porous domain $\Omega_{\varepsilon f}$ as in sect. 2 and we postulate the expansion :

$$(6.5)\quad \left\{\begin{array}{l} \underline{v}^{\varepsilon}(x, t) = \underline{v}^{0}(x, y, t) + \varepsilon v^{1} + \ldots \\ p^{\varepsilon}(x, t) = p^{0}(x, t) + \varepsilon p^{1}(x, y, t) + \ldots \end{array}\right\} y = \frac{x}{\varepsilon}$$

Y-periodic in y. We of course impose the no-slip boundary condition :

$$(6.6)\quad \underline{v}^{\varepsilon} = 0 \qquad \text{on } \partial\Omega_{\varepsilon f}$$

and an initial condition, for instance $v^{\varepsilon}\big|_{t=0} = 0$.

Now, we introduce (6.5) into (6.1), (6.4), (6.6) by using the classical relation (2.8) between the derivatives. Equation (6.4) gives at orders $\varepsilon^{-1}$ and $\varepsilon^{0}$:

$$(6.7)\quad \text{div}_y\, \underline{v}^{0} = 0 \qquad \text{in } Y_f$$

$$(6.8)\quad \frac{\partial p^{0}}{\partial t} + \rho\, c^{2}(\text{div}_x\, \underline{v}^{0} + \text{div}_y\, \underline{v}^{1}) = 0 \quad \text{in } Y_f$$

Relation (6.7) shows that the flow at the microscopic level is incompressible. Moreover, we apply the operator

$$(6.9)\quad \tilde{\cdot} = \frac{1}{|Y|} \int_{Y_f} \cdot \; dy$$

which coincides with the mean value operator if we continuate $\underline{v}^{0}$, $\underline{v}^{1}$, $p^{0}$ to $Y_s$ with value zero. We define the porosity $\pi = |Y_f| / |Y|$. As in (2.15), the meanvalue of the term $\text{div}_y$ of (6.8) is zero, and we obtain : (note that $p^{0} = \tilde{p}^{0}$ because $p^{0}$ is independent of y ) :

$$(6.10)\quad \pi \frac{\partial p^{0}}{\partial t} + \rho\, c^{2}\, \text{div}_x\, \underline{\tilde{v}}^{0} = 0$$

which will be the macroscopic equation of acoustic (note that it is analogous to (6.4) but with the mean value of $\tilde{v}^{0}$).

On the other hand, equation (6.1) at order $\varepsilon^{0}$ gives

$$(6.11)\quad \rho \frac{\partial \underline{v}^{0}}{\partial t} = - \underline{\text{grad}}_y\, p^{1} + \nu\, \Delta_y\, \underline{v}^{0} + (\underline{f} - \underline{\text{grad}}_x\, p^{0})$$

which gives, with (6.7) and

$$(6.12)\quad \underline{v}^{0}\big|_{\Gamma} = 0 \quad ; \quad v^{0}\big|_{t=0} = 0$$

the <u>local problem : (6.7), (6.11), (6.12)</u>.

In order to obtain a variational formulation of this problem, we define a new space $H_Y$ (in addition to $V_Y$ defined in (2.16)), which is the completion of $V_Y$ for the norm associated with the scalar product :

$$(6.13)\quad (\underline{u}, \underline{w})_{H_Y} = \int_{Y_f} u_i\, w_i\, dy$$

(this space is the Y-periodic counterpart of the space H defined in (1.10)).

The variational formulation of (6.7), (6.11), (6.12) is easily obtained (in the same way that (2.20) by multiplying by a test function $\underline{w} \in V_Y$ ; it is:

Find $\underline{v}^o$ (function of t with values in $V_Y$) such that

$$(6.14)\quad \begin{cases} \rho\left(\frac{\partial \underline{v}^o}{\partial t} , \underline{w}\right)_{H_Y} + \nu (\underline{v}^o , \underline{w})_{V_Y} = (f_i - \frac{\partial p^o}{\partial x_i}) \int_{Y_f} w_i \, dy \qquad \forall \underline{w} \in V_Y \\ \underline{v}^o(o) = 0 \end{cases}$$

In order to write $v^o$ as a function of f - grad $p^o$, it is useful to introduce the three elements $\phi^i$ (i = 1 , 2 , 3) of $H_Y$ defined by

$$(6.15)\qquad \int_{Y_f} w_i \, dy = (\underline{\phi}^i , \underline{w})_{H_Y} \qquad \forall \underline{w} \in H_Y$$

and the selfadjoint (unbounded) operator A of $H_Y$ associated by the second representation theorem with the form $(\underline{v} , \underline{w})_{V_Y}$. Then, (6.14) becomes :

$$(6.16)\quad \begin{cases} \frac{\partial v^o}{\partial t} + \rho^{-1} \nu A \underline{v}^o = \rho^{-1}(f_i - \frac{\partial p^o}{\partial x_i})(t) \, \underline{\phi}^i \\ \underline{v}^o(o) = 0 \end{cases}$$

which is solved by the classical formula (c.f. chapt 4, (3.2)) of semigroup theory :

$$(6.17)\qquad \underline{v}^o(t) = \rho^{-1} \int_o^t e^{-\rho^{-1}\nu A(t-s)} \, \underline{\phi}^i (f_i - \frac{\partial p^o}{\partial x_i})(s) \, ds$$

which gives $\underline{v}^o(t)$ as a functional of ($\underline{f}$ - grad $p^o$) for the time preceeding t. This is the analogous of (2.25) for the present problem.

Next, we consider the mean value of (6.17). In fact, we apply ~ to the left hand side ; we note that (6.15) shows that the action of the operator ~ amounts to multiply by $\underline{\phi}^i$ in $H_Y$. We then have :

$$(6.18)\qquad \tilde{v}^o_k(t) = \int_0^t g_{ki}(t - s)(f_i - \frac{\partial p^o}{\partial x_i})(s) \, ds$$

where

$$(6.19)\qquad g_{ki}(\xi) \equiv \rho^{-1}(e^{-\rho^{-1}\nu A\xi}\underline{\phi}^i , \underline{\phi}^k)_{H_Y}$$

Proposition 6.1 - If we postulate the asymptotic expansion (6.5), $\underline{v}^o(y , t)$ (x plays the role of a parameter) is the function of t with values in $V_Y$ defined by (6.17). In particular, the mean value $\tilde{\underline{v}}^o(t)$ is given by (6.18), where $g_{ki}(\xi)$ are well-defined functions of $\xi$, which decrease exponentially as $\xi \to + \infty$ ; moreover, $g_{ki} = g_{ik}$.

<u>Proof</u> - We only have to prove the part concerning the properties of the functions $g_{ki}(\xi)$. The function (with values in $H_Y$)$\exp(-\rho^{-1}\nu\, A\xi)\underline{\phi}^i$ is the solution of

$$\begin{cases} \left(\frac{dv}{dt}, \underline{w}\right)_{H_Y} + \rho^{-1}\nu(\underline{v}, \underline{w})_{V_Y} = 0 & \forall\, \underline{w} \in V_Y \\ \underline{v}(o) = \underline{\phi}^i \end{cases}$$

which tends to zero as $t \to +\infty$ in the $H_Y$-norm. For, we take $\underline{w} = \underline{v}$ and by the embedding of $V_Y$ in $H_Y$ we see that there exists $\gamma > 0$ such that

$$\frac{1}{2}\frac{d}{dt}\|v\|^2_{H_Y} + \gamma\|v\|^2_{H_Y} \leqslant 0$$

and the conclusion follows.

As for the symmetry in k, i, it suffices to note that $\exp(-\rho^{-1}\nu A\xi)$ is a bounded selfadjoint operator, it is then equal to the square of its square root and by selfadjointness, (6.19) may be written

$$g_{ki}(\xi) = \rho^{-1}\left(e^{-1/2\,\rho^{-1}\nu A\xi}\,\underline{\phi}^i\,,\; e^{-1/2\,\rho^{-1}\nu A\xi}\,\underline{\phi}^k\right)_{H_Y}$$

which shows the symmetry property. ∎

Finally, <u>the equation for the homogenized pressure is</u> (6.10) which becomes, with (6.18)

$$\pi\frac{\partial p^o}{\partial t} + \rho c^2 \frac{\partial}{\partial x_k}\int_0^t g_{ki}(t-s)\left(f_i - \frac{\partial p^o}{\partial x_i}\right)(s)\, dS = 0 \tag{6.20}$$

which is the integro-differential equation of acoustical waves in the porous medium.

As in the preceeding section, we may consider an interface between a porous body B and a free fluid. In the region D, the equations of motion are (6.1) - (6.3) and the appropriate expansion is of the type (6.5), but the functions do not depend on the variable y :

$$\begin{cases} \underline{v}^\varepsilon(x) = \underline{v}^e(x, t) + \varepsilon \ldots \\ p^\varepsilon(x) = p^e(x, t) + \varepsilon \ldots \end{cases} \tag{6.21}$$

and this expansion must match with (6.5) in the region B. It is not difficult to see that the suitable matching conditions on S are

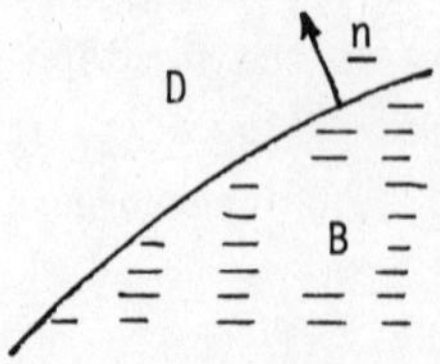


<u>fig</u>. 8

(6.22) $\quad p^e\Big|_S = p^o\Big|_S \quad ; \quad \underline{v}^e \cdot \underline{n}\Big|_S = \underline{\tilde{v}}^o \cdot \underline{n}\Big|_S$

i.e., the continuity of the pressure and of the normal component of the mean velocity. It is clear that $(6.22)_1$ appears as the matching condition for $(6.5)_2$ and $(6.21)_2$. Condition (6.22) is in fact the interface form of the conservation of mass. It may be obtained by the method of the conservation laws (chapter 5, sect.10) applied to conservation of mass. It is also possible to obtain it by considering the flow in a narrow layer near the interface S. As in (6.7), it is easy to see that the flow is asymptotically incompressible, and the matching of the layer with the flow on the two sides of S gives $(6.22)_2$. On the other hand, it is possible to study the layer near S ; it has a B-periodic structure analogous to that of sect.5 ; conditions (6.22) appears as the necessary and sufficient conditions for such a layer to exist (see Levy and Sanchez [ 2 ] ).

It is worthwhile to obtain the form of the functions $g_{ki}(\xi)$ (6.19) in the "macroscopically isotropic" case where the periods are cubes and the domain $Y_f$ is symmetric with respect to the planes of coordinates. It is easy to see that in this case

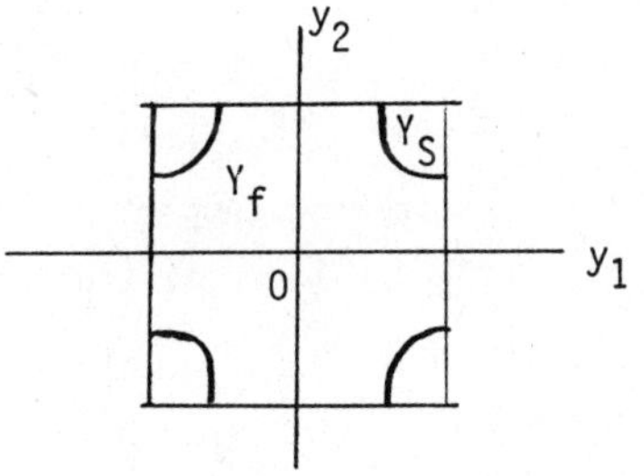


fig. 9

(6.33) $\quad g_{ki}(\xi) = \delta_{ij}\, g(\xi)$

where

(6.24) $\quad g(\xi) = g_{11}(\xi) = \rho^{-1}(e^{-\rho^{-1}\nu A\xi}\underline{\phi}^1 \, , \underline{\phi}^1)_{H_Y}$

Let $\lambda^i$ and $\underline{e}^i$ be the eigenvalues and eigenvectors of the operator $\rho^{-1}\nu A$ ; we consider them arranged in increasing order

$$0 < \lambda^1 \leqslant \lambda^2 \leqslant \dots \lambda^i \leqslant \dots \to +\infty$$

and $\underline{e}^i$ are orthonormed in $H_Y$. We consider the expansion of the vector $\underline{\phi}^1$ in the basis $\underline{e}^i$ :

$$\underline{\phi}^1 = a_i\, \underline{e}^i \quad ; \quad \| \phi^1 \|^2_{H_Y} = \sum_i a_i^2 < +\infty$$

and (6.24) becomes (see (4.1) of chapter 4 ) :

(6.25) $$\boxed{g(\xi) = \rho^{-1}(\sum_i e^{-\lambda^i\xi} a_i \underline{e}^i, \sum_j a_j \underline{e}^j)_{H_Y} = \sum_i e^{-\lambda^i\xi} a_i^2}$$

which has the form indicated in the figure. It is positive, exponentially decreasing as $\xi \to \infty$ and its slope at $\xi = +0$ is $-\infty$.

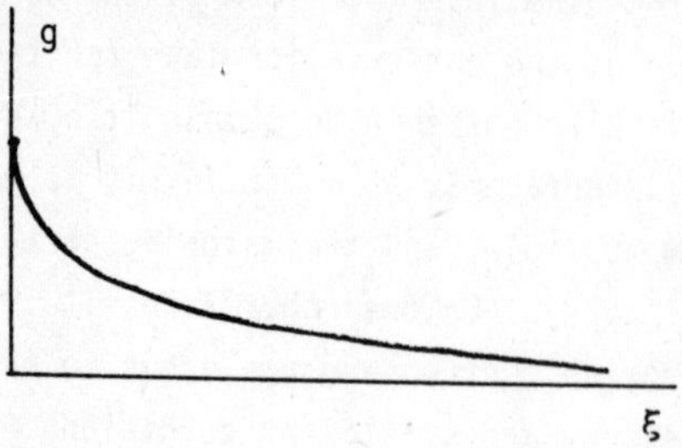


fig. 10

7.- Thermal effects

We pointed out at the beginning of the preceeding section that air is not a barotropic gas, and its density is a function of pressure and temperature, given by the "state equation"

$$\frac{p^*}{\rho^*} = R\,\theta^*$$

where $p^*$, $\rho^*$, $\theta^*$ are the pressure, density and temperature, respectively, and R is the constant of the gas, (by unit mass). By linearization near the rest state we obtain

(7.1) $$\frac{p^\varepsilon}{\rho} - p\,\frac{\rho^\varepsilon}{\rho^2} = R\,\theta^\varepsilon$$

where p, $\rho$ are, as in the preceeding section, the pressure and density at the rest state, and $p^\varepsilon$, $\rho^\varepsilon$, $\theta^\varepsilon$ are the perturbations.

Equation (7.1) replaces (6.3) in the present case ; but it introduces the new function $\theta$. We must join the linearized form of the energy equation (see for instance Liepmann and Roshko [1] , sect. 13.13)

(7.2) $$C_p\,\frac{\partial\theta^\varepsilon}{\partial t} - \frac{1}{\rho}\,\frac{\partial p^\varepsilon}{\partial t} = \frac{k}{\rho}\,\varepsilon^2\,\Delta\,\theta^\varepsilon$$

where $k\varepsilon^2$ is coefficient of thermal conductivity (of order $\varepsilon^2$, as the viscosity coefficients in (6.1) and $C_p$ is the specific heat at constant pressure)

As a result, the equations of a acoustics are (6.1), (6.2), (7.1) and (7.2), for the unknowns $\underline{v}^\varepsilon$, $p^\varepsilon$, $\rho^\varepsilon$, $\theta^\varepsilon$. Moreover, we must consider boundary condition for $\theta^\varepsilon$. In fact $\underline{v}^\varepsilon$, $p^\varepsilon$, $\rho^\varepsilon$ are defined in $\Omega_{f\varepsilon}$ only, but $\theta^\varepsilon$ is also defined in $\Omega_{S\varepsilon}$. Nevertheless, in the problem under consideration here, if the solid has, as usual, a large density with respect to the gas density, the heat transfert at the surface $\Gamma$ between them cannot modify the temperature of the solid part, which may be taken equal to its value at the rest state. Consequently, $\theta^{\varepsilon}$ is searched only in $\Omega_{f\varepsilon}$, and we have the boundary condition (in addition to (6.6)) :

(7.3) $$\theta^{\varepsilon} = 0 \qquad \text{on } \partial\Omega_{\varepsilon f}$$

Now, to study the asymptotic behaviour of (6.1), (6.2); (7.1), (7.2), (6.3), (7.3), we consider the expansions (6.5) and

(7.4) $$\begin{cases} \rho^{\varepsilon}(x , t) = \rho^{0}(x , y , t) + \varepsilon\rho^{1} + \dots \\ \theta^{\varepsilon}(x , t) = \theta^{0}(x , y , t) + \varepsilon\theta^{1} + \dots \end{cases}$$

with the usual Y-periodicity conditions. It is clear that the boundary condition (7.3) must be imposed on the boundary $\Gamma$ of $Y_f$ in each period; consequently, $\theta^0$ must depend on y (but $p^0$ does not !) and (7.1) shows that $\rho^0$ must also depend on y.

In order to obtain equations for $\theta^0$ and $\rho^0$, we introduce (7.4) and (2.8) into (7.1), (7.2), and (7.3). Then (7.2) at order $\varepsilon^0$ and (7.3) at order $\varepsilon^{-1}$ give :

(7.5) $$\begin{cases} C_p \dfrac{\partial\theta^0}{\partial t} - \dfrac{k}{\rho}\, \Delta_y\, \theta^0 = \dfrac{1}{\rho}\dfrac{\partial p^0}{\partial t} \\ \theta^0 = 0 \qquad \text{on } \Gamma \text{ and } t = 0 \end{cases}$$

which is in fact an equation for $\theta^0(y , t)$, with the parameter x (note that $p^0$ does not depend on y). The system (7.5) is solved as (6.11), (6.12) but in an easier framework. If $L^2_Y$ (resp. $H^1_Y$) denotes the space of Y-periodic functions defined on $Y_f$ (resp. defined on $Y_f$ and which are zero on $\Gamma$) with the scalar products:

$$(\theta , \zeta)_{L^2_Y} = \int_{Y_f} \theta\,\zeta\, dy \quad ; \quad (\theta , \zeta)_{H^1_Y} = \int_{Y_f} \frac{\partial\theta}{\partial y_i}\frac{\partial\zeta}{\partial y_i}\, dy \quad ,$$

the variational formulation of (7.5) is

(7.6) $$\begin{cases} \text{Find } \theta^0(t) \text{ with values in } H^1_Y \text{ such that} \\ C_p\left(\dfrac{\partial\theta^0}{\partial t} , \zeta\right)_{L^2_Y} + \dfrac{k}{\rho}\,(\theta^0 , \zeta)_{H^1_Y} = \dfrac{1}{\rho}\,\dfrac{\partial p^0}{\partial t}\,(\psi , \zeta)_{L^2_Y} \qquad \forall \zeta \in H^1_Y \end{cases}$$

where $\psi$ is the function of $L^2_Y$ equal to 1. We have introduced an initial condition $\theta^0(0) = 0$, as in (6.12). To solve this problem, let B be the selfadjoint operator of $L^2_Y$ associated to the form $( \quad , \quad )_{H^1_Y}$ according to the second representation theorem. (7.6) becomes

$$C_p \frac{\partial \theta^o}{\partial t} + \frac{k}{\rho} B \, \theta^o = \frac{1}{\rho} \frac{\partial p^o}{\partial t} \psi \quad ; \quad \theta^o(0) = 0 \quad \Longrightarrow$$

(7.7) $$\theta^o(t) = \frac{1}{\rho C_p} \int_0^t e^{-\frac{k}{\rho C_p} B(t-S)} \psi \frac{\partial p^o}{\partial t}(S) \, dS$$

The function $\theta^o(y)$ is continuated to $Y_S$ with value zero ; moreover, we apply the mean operator (2.14) to (7.7) ; this amounts to multiply by $\psi$ in the $L^2_Y$-space ; we obtain :

(7.8) $$\tilde{\theta}^o(t) = \int_0^t f(t - s) \frac{\partial p^o}{\partial t}(S) \, dS$$

where

(7.9) $$f(\xi) \equiv \frac{1}{\rho \, C_p} \left( e^{-\frac{k}{\rho C_p} B \, \xi} \psi, \psi \right)_{L^2_Y}$$

is a function having a structure analogous to that of g of (6.25)

On the other hand, equation (6.2) at the order $\varepsilon^0$ gives

(7.10) $$\frac{\partial \rho^o}{\partial t} + \rho(\mathrm{div}_x \, \underline{v}^o + \mathrm{div}_y \, \underline{v}^1) = 0$$

and we apply to it the mean operator. In fact, $\rho^o(y)$ is only defined on $Y_f$ and not on $Y_S$. The appropriate definition of the mean operator is

(7.11) $$\tilde{\cdot} = \frac{1}{|\,Y\,|} \int_{Y_f} \cdot \; dy$$

which coincides with (2.14) for functions which are zero on $Y_S$, (as $\theta^o$ and $\underline{v}^o$). The term in $\mathrm{div}_y \, \underline{v}^1$ vanishes as in (6.10) and (7.10) gives

(7.12) $$\frac{\partial \tilde{\rho}^o}{\partial t} + \rho \, \mathrm{div}_x \, \tilde{\underline{v}}^o = 0$$

which is <u>the macroscopic equation</u>.

On the other hand, from (7.1), at order $\varepsilon^0$ :

$$\rho^{-1} \, p^o - \frac{p}{\rho^2} \rho^o = \theta^o$$

and by applying the mean operator defined by (7.11) we have

(7.13) $$\rho^{-1} \, \pi \, p^o - \frac{p}{\rho^2} \tilde{\rho}^o = R \tilde{\theta}^o \quad ; \quad \pi \equiv \frac{|Y_f|}{|\,Y\,|} = \text{porosity}$$

From (7.8) and (7.13) we have

$$(7.14) \qquad \tilde{\rho}^{o}(t) = \int_0^t F(t-S)\,\frac{\partial p^o}{\partial t}(S)\,dS \quad ; \quad F(\xi) \equiv \frac{\rho\pi}{p} - \frac{\rho^2 R}{p} f(\xi)$$

We then replace (7.14) and (6.18) into (7.12) to obtain the macroscopic equation as a relation for $p^o(x , t)$ :

$$(7.15) \qquad \frac{\partial}{\partial t}\int_0^t F(t-S)\,\frac{\partial p^o}{\partial t}(S)\,dS + \rho\,\frac{\partial}{\partial x_k}\int_0^t g_{ki}(t-S)\left(f_i - \frac{\partial p^o}{\partial x_i}\right)(S)\,dS$$

which is the equation of acoustics in porous media when thermal conductivity is taken into account. Interface conditions at a boundary S between a porous medium and a free fluid are again (6.22).

The structure of the layer near S, as well as some examples of solutions of (7.15) for functions depending on time in the form $e^{i\omega t}$ may be seen in Levy-Sanchez [ 2 ] . The existence and uniqueness of solutions and some properties of propagation at finite velocity for equation (7.15) where studied in Sanchez Hubert [ 1 ].

8.- Comments and bibliographical notes - An extensive mathematical study of fluid flow may be seen in Ladyzhenskaya [ 2 ], Tartar [ 5 ] and Temam [ 1 ]. Sect. 2 is based on Ene and Sanchez-Palencia [ 1 ]: for a proof of the convergence, see the Appendix by L. Tartar at the end of this volume. For non linear effects see also Ene and Sanchez-Palencia [ 1 ] . Boundary conditions at the interface between a porous body and a liquid is a very controversed question ; our study shows that several situations may appear following the orders of the velocity in the different regions. Roughly speaking, the problem is in some sense a "stiff problem" (see chapter 13) as a consequence of the fact that viscosity effects are very important in the (narrow) canals of the porous body. For a study of this problem, see Levy and Sanchez-Palencia [ 1 ] , and also Ene and Sanchez-Palencia [ 1 ] .

In unsteady motions, as in acoustics, the Darcy's law is not instantaneous but with memory and the corresponding homogenized equations are integro-differential ; equivalently, if only a dependence in $e^{i\omega t}$ in time is considered, the homogenized coefficients depend on $\omega$ (see Levy and Sanchez-Palencia [ 2 ] and Sanchez-Palencia [ 3 ] ). For acoustics in porous media, see Levy [ 1 ] and Levy and Sanchez-Palencia [ 2 ] , where boundary conditions are discussed (see also Vogel [ 1 ] for other boundary conditions proposed in such problems). Some mathematical properties of the integro-differential equations of acoustics in porous media, such as existence, uniqueness and finite speed of propagation are considered in Sanchez-Hubert [ 1 ] and Sanchez-Palencia and Sanchez-Hubert [ 1 ] . Other homogenization problems in fluid mechanics are studied in Fleury [ 1 ] an Tartar [ 2 ] .

# CHAPTER 8

# VIBRATION OF MIXTURES OF SOLIDS AND FLUIDS

This chapter is devoted to the study of several problems about motion of mixtures of elastic solids and viscous fluids in the framework of small motions linearized with respect to a rest state where the geometric distribution of the solid and fluid parts is periodic (with a small period, of course).

A great variety of different problems arises according to the orders of the viscosity coefficients (see in this connection the considerations at the beginning of chapter 7, sect. 4) and the topological properties of the mixture : the solid part may be connected (formed by a single body hollowed by canals) or not (formed by a suspension of solid particles). Very different expansions then appear ; we have chosen some model problems showing different macroscopic behaviours, and bibliographical notes on other problems are given in the last section.

It is to be noticed that thermal effects are not considered. The fluid is considered to be barotropic (i.e., the density is a given function of the pressure). It is clear that this assumption is not appropriate for certain physical problems.

Formal asymptotic expansions are obtained by the method of expansion of an integral identity (chapter 5, sect. 9). In physical problems, the method of the conservation laws (chapter 5, sect. 10) shows the physical meaning of the macroscopic equations : this method is often used in the bibliography listed in the last section.

Rigorous proofs of the convergence of some problems are given in sect. 1 and 3. The former is based on the ideas of chap. 6, the latter on the proof of the Appendix by Tartar. The question of the convergence in the problems of sect. 4 and 5 is open.

## 1.- Mixture of an elastic solid and a viscous fluid. Case of large viscosity -

This section is devoted to the study of the vibration (in the theory of small, linearized perturbations) of a mixture of an elastic body and a viscous barotropic fluid. We consider a reference state at rest, and we study the displacement vector

$\underline{u}$ with respect to that state. In the solid part of the mixture, the equations are :

$$\rho^s \frac{\partial^2 u_i}{\partial t^2} = \frac{\partial \sigma^s_{ij}}{\partial x_j} + f_i \tag{1.1}$$

$$\sigma^s_{ij} = a^s_{ijkh}\, e_{kh}(\underline{u}) \quad ; \quad e_{kh}(\underline{u}) \equiv \frac{1}{2}\left( \frac{\partial u_k}{\partial x_n} + \frac{\partial u_h}{\partial x_k} \right) \tag{1.2}$$

where $\underline{f}$ is the given force, the coefficients $a^s_{ijkh}$ (the superscript "s" is for "solid") are constant (this assumption is not essential) and satisfy the usual properties of symmetry and positivity :

$$a^s_{ijkh} = a^s_{jikh} = a^s_{jihk} = a^s_{khij} \tag{1.3}$$

$$a_{ijkh}\, e_{ij}\, e_{kh} \geqslant \alpha\, e_{ij}\, e_{ij} \quad ; \quad \alpha > 0 \qquad e_{ij} \text{ (symmetric)} \tag{1.4}$$

In the fluid part, the equations are :

$$\rho^f \frac{\partial^2 u_i}{\partial t^2} = \frac{\partial \sigma^f_{ij}}{\partial x_j} + f_i \tag{1.5}$$

$$\sigma^f_{ij} = -\,\delta_{ij}\, p + (\eta\ \delta_{ij}\, \delta_{kh} + 2\,\mu\, \delta_{ik}\, \delta_{jh})\, e_{kh}\left(\frac{\partial \underline{u}}{\partial t}\right) \tag{1.6}$$

where $\eta$ and $\mu$ are the viscosity coefficients, which satisfy the condition

$$\mu > 0 \quad ; \quad \frac{\eta}{\mu} > -\frac{2}{3}\alpha \quad ; \quad 0 < \alpha < 1 \tag{1.7}$$

Note that the equations for the fluid are usually written for the velocity vector

$$\underline{v} = \frac{\partial \underline{u}}{\partial t} \tag{1.8}$$

but here we use a description with $\underline{u}$ which is consistent with (1.1). $\rho^s$ and $\rho^f$ are the densities of the fluid and solid at the reference state ; they are supposed to be constant but this is not essential.

In equation (1.6), p is the perturbation of pressure with respect to the reference state. It is related to the perturbation of density $\rho$ by the relation

$$p = c^2 \rho \tag{1.9}$$

where $c > 0$ is the "velocity of sound" in the reference state. Moreover, $\rho$ and $\underline{v}$ are related by the continuity equation (conservation of mass) which writes, in the linearized framework :

$$\frac{\partial \rho}{\partial t} + \rho^f \operatorname{div} \underline{v} = 0$$

and by integration (note that $\rho = 0$ for $\underline{u} = 0$) :

(1.10) $$\rho + \rho^f \operatorname{div} \underline{u} = 0$$

and (1.9) becomes

(1.11) $$p = - c^2 \ \rho^f \operatorname{div} \underline{u}$$

which may be replaced in (1.6) ; $\sigma^f_{ij}$ then appears as a function of $\underline{u}$ and $\partial \underline{u}/\partial t$.

Moreover, at the interface between the solid and the fluid, we must have the continuity of displacement and stress :

(1.12) $$[\underline{u}] = 0 \quad , \quad [\sigma_{ij} \ n_j] = 0$$

where the brakets mean, as usual, "discontinuity of ".

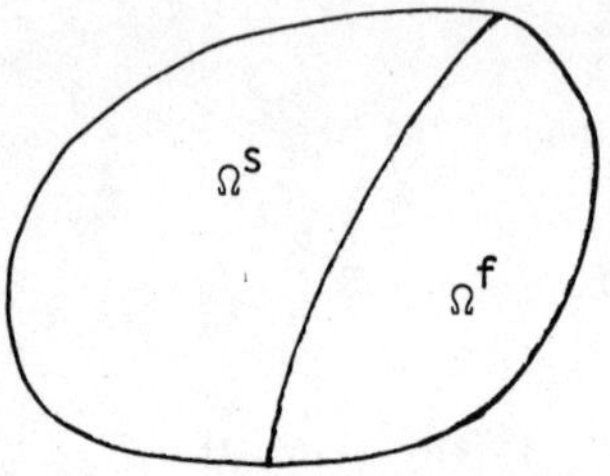


In order to obtain a well posed initial - boundary value problem, we must adjoin initial and boundary conditions. We shall take them homogeneous, but inhomogeneous conditions may be considered as well.:

(1.13) $$\underline{u} = 0 \qquad \text{on } \partial\Omega$$

(1.14) $$\underline{u} = \frac{\partial u}{\partial t} = 0 \qquad \text{for } t = 0$$

where $\Omega$ is the total domain which will be taken bounded and is formed by $\Omega^s$ and $\Omega^f$. The variational formulation of the problem (1.1), (1.5), (1.12), (1.13), (1.14) is easily obtained as in chapter 6, sect. 1 :

Find $\underline{u}$, function of t with values in $\underline{H}^1_0(\Omega)$ such that

(1.15) $$\int_\Omega \rho \frac{\partial^2 u_i}{\partial t^2} w_i \, dx + \int_\Omega \sigma_{ij} \frac{\partial w_i}{\partial x_j} dx = \int_\Omega f_i \, w_i \, dx \qquad \forall \underline{w} \in \underline{H}^1_0$$

(1.16) $$\underline{u}(0) = \underline{u}'(0) = 0$$

where $\rho$ is a function of x which takes the values $\rho^s$ and $\rho^f$ in $\Omega^s$ and $\Omega^f$ resp. In the same way, $\sigma_{ij}$ is defined by (1.2) and (1.6) in $\Omega^s$ and $\Omega^f$. Moreover, as a consequence of the symmetry properties in i j ,

We may write

(1.17) $$e_{ij}(\underline{w}) \quad \text{instead of} \quad \partial w_i/\partial x_j \quad \text{in (1.15)}$$

If we replace the expression (1.11) of p into (1.6) we see that $\sigma_{ij}$ is uniquely determined by $\underline{u}$ and its derivatives, and we have ($\underline{v} \equiv \partial\underline{u}/\partial t$ is the velocity) :

(1.18) $$\int_\Omega \sigma_{ij} \frac{\partial w_i}{\partial x_j}\, dx \equiv a(\underline{u}\,,\,\underline{w}) + b(\frac{\partial \underline{u}}{\partial t}\,,\,w) \qquad \text{where}$$

(1.19) $$\begin{cases} a(\underline{u}\,,\,\underline{w}) \equiv \int_\Omega a_{ijkh}(x)\, e_{kh}(\underline{u})\, e_{ij}(\underline{w})\, dx \\ b(\underline{v}\,,\,\underline{w}) \equiv \int_\Omega b_{ijkh}(x)\, e_{kh}(\underline{v})\, e_{ij}(\underline{w})\, dx \end{cases}$$

(1.20) $$\begin{cases} a_{ijkh}(x) = \begin{cases} a^s_{ijkh} & \text{if } x \in \Omega_s \\ a^f_{ijkh} \equiv \rho^f c^2 \delta_{ij}\, \delta_{kh} & \text{if } x \in \Omega_f \end{cases} \\ b_{ijkh}(x) = \begin{cases} b^s_{ijkh} \equiv 0 & \text{if } x \in \Omega_s \\ b^f_{ijkh} \equiv 2\, \mu\delta_{ik}\, \delta_{jh} + \eta\, \delta_{ij}\, \delta_{kh} & \text{if } x \in \Omega_f \end{cases} \end{cases}$$

Remark 1.1 - It is clear that the forms a and b are not coercive on $\underline{H}^1_0$ ; nevertheless, the sum of these forms is coercive, i.e., there exists $\gamma > 0$ such that

(1.21) $$a(\underline{u}\,,\,\underline{u}) + b\,(\underline{u}\,,\,\underline{u}) \geq \gamma\, \|u\|^2_{\underline{H}^1_0}$$

This is a consequence of (1.7) ; with (1.7) and (1.4) it is easily established that

$$a(\underline{u}\,,\,\underline{u}) + b(\underline{u}\,,\,\underline{u}) \geq \delta \int_\Omega e_{ij}(\underline{u})\quad e_{ij}(\underline{u})\, dx$$

and (1.21) is then proved as lemma 1.2 of chapter 6. ■

Remark 1.2 - Under the hypothesis (1.3), (1.4), (1.7), the initial-boundary value problem (1.15) has a solution and only one, which satisfies the a priori estimates which are easily obtained by taking $\underline{w} = \underline{u}'$ in (1.15).

This result may be obtained by semigroup theory (see Sanchez-Hubert [2] for details). The existence and uniqueness is also easily obtained by Laplace transform. The Laplace transform ($t \Rightarrow \lambda$) of (1.15) is : Find $\hat{\underline{u}} \in \underline{H}^1_0$ such that

$$(1.22) \qquad \lambda^2 \int_\Omega \rho\, \hat{\underline{u}} \cdot \underline{w}\, dx + a(\hat{\underline{u}}\,,\, \underline{w}) + \lambda\, b(\hat{\underline{u}}\,,\, \underline{w}) = \int_\Omega \hat{\underline{f}} \cdot \underline{w}\, dx \quad \forall \underline{w} \in \underline{H}_o^1$$

For fixed $\lambda$, (with $\mathrm{Re}\,\lambda > 1$, for instance) equation (1.22) is associated with an operator with compact inverse (see (1.21)) and uniqueness implies existence (see Proposition 1.4 of Chap. 11). If we take $\underline{w} = \hat{\underline{u}}$ in (1.22), we have

$$\lambda \int_\Omega \rho\, \hat{\underline{u}} \cdot \hat{\underline{u}}\, dx + \frac{1}{\lambda}\, a(\hat{\underline{u}}\,,\, \hat{\underline{u}}) + b(\hat{\underline{u}}\,,\, \hat{u}) = \frac{1}{\lambda} \int_\Omega \hat{\underline{f}} \cdot \hat{\underline{u}}\, dx$$

and by taking the real part, we have

$$(1.23) \qquad (\mathrm{Re}\,\lambda) \int_\Omega \rho\, \hat{\underline{u}} \cdot \hat{\underline{u}}\, dx \leq \frac{1}{|\lambda|}\, \| \hat{f} \|_{\underline{L}^2} \| \hat{u} \|_{L^2}$$

which implies uniqueness. Consequently, $\hat{\underline{u}}(\lambda)$ exists for $\mathrm{Re}\,\lambda > 1$ . Moreover, because $\hat{f}$ is a Laplace transform, $\| \hat{f} \|$ is bounded by a polynomial in $|\lambda|$ and (1.23) implies that the same holds for $\| \hat{\underline{u}} \|_{L^2}$. We replace this in (1.22) with $\underline{w} = \hat{\underline{u}}$ (the term in $\lambda^2$ is written in the right hand side) and by taking the real part and using (1.21) we have :

$$\| \hat{u} \|_{\underline{H}_o^1} \leq \text{Polynomial } (\,|\lambda|\,)$$

Moreover, $\hat{\underline{u}}(\lambda)$ is a holomorphic function of $\lambda$ with values in $\underline{H}_o^1(\Omega)$ and then $\underline{u}(t)$ is obtained by inverse Laplace transform. ■

Now, we consider a homogenization problem in the preceeding framework. We define in the standard way the basic period Y which is formed by a solid part $Y_s$ and a fluid part $Y_f$. We define the coefficients :

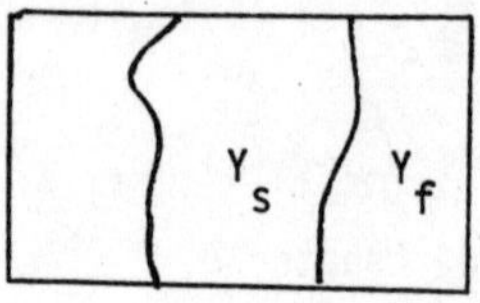


$$(1.24) \qquad \rho(y) = \begin{cases} \rho^s & \text{if } y \quad Y_s \\ \rho^f & \text{if } y \quad Y_f \end{cases}$$

$$(1.25) \qquad \begin{cases} a_{ijk\ell}(y) = \begin{cases} a^s_{ijk\ell} & \text{if } y \in Y_s \\ a^f_{ijk\ell} & \text{if } y \in Y_f \end{cases} \\ b_{ijk\ell}(y) = \text{an analogous expression.} \end{cases}$$

Moreover, we define the coefficients :

(1.26) $$\begin{cases} \rho^{\varepsilon}(x) = \rho(\frac{x}{\varepsilon}) \quad \text{and analogous expressions} \\ \text{for} \quad a^{\varepsilon}_{ijk\ell}(x) \text{ and } b^{\varepsilon}_{ijk\ell}(x) \end{cases}$$

and the forms :

(1.27) $$\begin{cases} c^{\varepsilon}(\underline{u}\,,\,\underline{w}) \equiv \int_{\Omega} \rho^{\varepsilon}\, u_i\, w_i \, dx \\ a^{\varepsilon}(\underline{u}\,,\,\underline{w}) \equiv \int_{\Omega} a^{\varepsilon}_{ijk\ell}\, e_{k\ell}(\underline{u})\, e_{ij}(\underline{w})\, dx \\ b^{\varepsilon}(\underline{u}\,,\,\underline{w}) \equiv \quad \text{an analogous expression} \end{cases}$$

We then consider the problem

(1.28) $$\begin{cases} \text{Find } \underline{u}^{\varepsilon}, \text{ function of t with values in } \underline{H}^1_0(\Omega) \text{ such that} \\ c^{\varepsilon}\left(\frac{\partial^2 \underline{u}^{\varepsilon}}{\partial t^2}\,,\,\underline{w}\right) + a^{\varepsilon}(\underline{u}^{\varepsilon}\,,\,\underline{w}) + b^{\varepsilon}\left(\frac{\partial \underline{u}^{\varepsilon}}{\partial t}\,,\,\underline{w}\right) = \int_{\Omega} \underline{f}\,.\,\underline{w}\, dx \quad \forall \underline{w} \in \underline{H}^1_0(\Omega) \\ \underline{u}^{\varepsilon}(0) = \frac{\partial \underline{u}^{\varepsilon}}{\partial t}(0) = 0 \end{cases}$$

Theorem 1.1 - Let $\underline{f} \in L^2(0\,,\,T\,;\,\underline{L}^2(\Omega))$ be a fixed function, and $\underline{u}^{\varepsilon}$ the corresponding solution of (1.28). Then,

(1.29) $$\underline{u}^{\varepsilon} \to \underline{u} \quad \text{in} \quad L^{\infty}(0\,,\,T\,;\,\underline{H}^1_0(\Omega)) \text{ weakly } *$$

where $\underline{u}$ is the unique solution of the problem :

(1.30) $$\begin{cases} \text{Find } \underline{u}(t) \text{ with values in } \underline{H}^1_0, \text{ zero for negative t and such that} \\ \int_{\Omega} \tilde{\rho}\, \underline{u}\, \underline{w}\, dx + \int_{\Omega} (\beta_{ijk\ell} * e_{k\ell}(\underline{u}))\, e_{ij}(\underline{w})\, dx = \int_{\Omega} \underline{f}\, \underline{w}\, dx \\ \qquad \forall\, \underline{w} \in \underline{H}^1_0 \end{cases}$$

where $\sim$ is the classical mean value on the period Y :

$$\tilde{\cdot} = \frac{1}{|Y|} \int_Y \cdot \; dy$$

the star $*$ denotes the convolution product in time and $\beta_{ijk\ell}(t)$ are distributions with support in $t \geqslant 0$, such that the Laplace transforms of $\beta_{ijk\ell}(t)$ are

(1.31) $$\hat{\beta}_{ijk\ell}(\lambda) = (a_{ijk\ell}(y) + \lambda\, b_{ijk\ell}(y))^h$$

i.e. the homogenized coefficients associated with $a + \lambda\, b$ in the framework of homogenization in elasticity (chapter 6, sect. 2).

Remark 1.3 - It is clear that (1.30) is a viscoelasticity problem with strain-stress relation given by

(1.32) $$\sigma_{ij} = \beta_{ijk\ell} * e_{k\ell}(\underline{u})$$

and $\sigma(t)$ is a functional of the history (but not of the future !, see chapter 4, sect. 6) of $e_{k\ell}(\underline{u})$. ■

Outline of the proof of theorem 1.1 - We only give some indications about the formal asymptotic expansion and the proof of the convergence. A more explicit treatment may be seen in Sanchez - Hubert [ 2 ] .

As in chapter 6, we consider an asymptotic expansion

(1.33) $$\underline{u}^{\varepsilon}(x , t) = \underline{u}^{0}(x , t) + \varepsilon u^{1}(x , y , t) + \dots \qquad y = x/\varepsilon$$

Y-periodic in y. Moreover, $\underline{u}^{1}(x , y , t)$ is a smooth function of x and t with values in $V_Y$ (or $\tilde{V}_Y$) (see chapter 6, (2.16) or (4.22)).

Moreover, in order to expand the integrals (1.30), we take, in the framework of chap. 5, sect. 9, the test function $\underline{w}$ in the form :

(1.34) $$\underline{w} = \underline{w}^{0}(x) + \varepsilon\, \theta(x)\, \underline{w}^{1}(y) \qquad (\underline{w}^{1} \in \tilde{V}_Y , \quad \theta \in \mathcal{D}(\Omega))$$

and (1.30) at order $\varepsilon^{0}$ gives :

(1.35) $$\int_{\Omega} \rho^{\varepsilon} \frac{\partial^2 \underline{u}^{0}}{\partial t^2} \underline{w}^{0} \, dx = \int_{\Omega} \sigma^{o}_{ij}\left(\frac{\partial w^{o}_{i}}{\partial x_j} + \theta \frac{\partial w^{1}_{i}}{\partial y_j}\right) dx + \int_{\Omega} f_i \, w^{o}_{i} \, dx$$

(1.36) where $$\sigma^{o}_{ij} \equiv \left(a^{\varepsilon}_{ijk\ell} + b^{\varepsilon}_{ijk\ell} \frac{\partial}{\partial t}\right)\left(\frac{\partial u^{o}_{k}}{\partial x_\ell} + \frac{\partial u^{1}_{k}}{\partial y_\ell}\right)$$

By taking $\theta(x) = 0$ in (1.35) and applying the limit process $\varepsilon \to 0$, we obtain (according to the method of chapter 5, sect. 9, i.e. taking the mean value of the integrand) :

$$\int_{\Omega} \tilde{\rho} \frac{\partial^2 \underline{u}^{0}}{\partial t^2} \underline{w}^{0} \, dx = \int_{\Omega} \tilde{\sigma}^{o}_{ij} \frac{\partial w^{o}_{i}}{\partial x_j} dx + \int_{\Omega} f_i \, w^{o}_{i} \, dx$$

which gives the homogenized equation

(1.37) $$\tilde{\rho} \frac{\partial^2 \underline{u}_i}{\partial t^2} = \frac{\partial \tilde{\sigma}^{o}_{ij}}{\partial x_j} + f_i$$

Moreover, by taking $\underline{w}^{0} = 0$ in (1.35) and $\varepsilon \to 0$, we obtain, in the same way :

$$\int_{\Omega} \widetilde{\left[\sigma^{o}_{ij} \frac{\partial w^{1}}{\partial y_j}\right]} \theta(x) \, dx = 0 \qquad \forall \theta \in \mathcal{D}(\Omega)$$

and consequently the integrand must be zero, i.e., with (1.36) :

$$(1.38) \qquad \int_Y (a_{ijk\ell}(y) + b_{ijk\ell}(y) \frac{\partial}{\partial t}) \left( \frac{\partial u_k^0}{\partial x_\ell} + \frac{\partial u_k^1}{\partial y_\ell} \right) \frac{\partial w_i^1}{\partial y_j} \, dy = 0 \qquad \forall \underline{w}^1 \in \tilde{V}_Y$$

which is the local problem. It is an evolution problem in t, which is formally analogous to the local problem in viscoelasticity (chapter 6, (4.24)) but a little different because in the present case the form associated with the coefficients b is not coercive on $\tilde{V}_Y$. In fact, (see remark 1.1), a + b is coercive on $\tilde{V}_Y$, and (4.24) is easily solved by Laplace transform (as in remark 1.2). We then obtain the homogenized stress - strain relation, the Laplace transform of which is (1.31). The proof of the convergence is easily obtained as in chapter 6, sect. 4.

2.- Mixture of two compressible, slightly viscous fluids - The only small parameter in the problem of the preceeding section is $\varepsilon$ , associated with the length of the period. But it is also possible to consider problems with other small coefficients, for instance, the viscosity coefficients of the fluid (see, in this context, the beginning of sect. 4 of chapter 7).

In fact, if we consider the mixture of two fluids with viscosity coefficients of order $\varepsilon^2$, we obtain an asymptotic expansion very different of that of the preceeding section. In fact, this problem recalls that of the "acoustics in porous media" of chap. 7, sect. 6, but we have two fluids instead of a fluid and a rigid solid. As in acoustics in porous media, we consider compressible fluids with sound velocity ($c$ in the notation of chapter 7, sect. 6) of order 1, i.e., independent of $\varepsilon$. As in the preceeding section, it is possible to eliminate the pressure (see (1.28)), but pressure plays an important role in the present problem (in fact, there is a sort of "Darcy's law" for the velocity) and it is better to keep the pressure in the formulation of the problem.

To fix ideas, we consider an open, bounded domain $\Omega$ of $R^3$ and a parallelepipedic period Y as usual, composed of two regions $Y_1$ and $Y_2$ corresponding to the two fluids. The density $\rho(y)$, the sound velocity $c(y)$ and the viscosity coefficients $\varepsilon^2\,\mu(y)$, $\varepsilon^2\,\eta(y)$ take constant values on $Y_1$ and $Y_2$ :

$$(2.1) \qquad \left\{ \begin{array}{l} \rho(y) = \left\{ \begin{array}{ll} \rho^1 & \text{if } y \in Y_1 \\ \rho^2 & \text{if } y \in Y_2 \end{array} \right\} \quad \rho^\varepsilon(x) \equiv \rho(\frac{x}{\varepsilon}) \\ \text{analogous formulae for } c^2(y), \mu(y), \eta(y) \end{array} \right.$$

The pressure $p^\varepsilon(x , t)$ and the displacement $\underline{u}^\varepsilon(x , t)$ are related by (compare with (1.11)) :

$$(2.2) \qquad p^\varepsilon(x , t) \equiv -\gamma^\varepsilon(x) \text{ div } \underline{u}^\varepsilon , \quad \text{where } \gamma(y) = c^2(y)\,\rho(y) \; ; \; \gamma^\varepsilon(x)=\gamma(x/\varepsilon)$$

Moreover, the tensor $\sigma^\varepsilon$ is defined by (compare with (1.6) :

$$(2.3)\quad \begin{cases} \sigma^\varepsilon_{ij} \equiv -\delta_{ij}\, p^\varepsilon + \varepsilon^2 [\, \eta^\varepsilon(x)\, \delta_{ij}\, \delta_{k\ell} + 2\, \mu^\varepsilon(x) \delta_{ik}\, \delta_{j\ell}] e_{k\ell}(\underline{v}^\varepsilon) \\ \underline{v}^\varepsilon \equiv \text{ velocity } \equiv \partial \underline{u}^\varepsilon / \partial t \end{cases}$$

Moreover, we consider the boundary and initial conditions

$$(2.4)\quad \underline{u}^\varepsilon \Big|_{\partial\Omega} = 0 \qquad \text{on } \partial\Omega$$

$$(2.5)\quad \underline{u}^\varepsilon = \underline{v}^\varepsilon = 0 \qquad \text{for } t = 0$$

Then, $\underline{u}^\varepsilon$ is defined by (compare with (1.15) and (1.18)) :

$$(2.6)\quad \left|\; \begin{array}{l} \text{Find } \underline{u}^\varepsilon(t) \text{ with values in } \underline{H}^1_o(\Omega) \text{ such that} \\ \qquad \underline{u}^\varepsilon(0) = \dfrac{\partial \underline{u}^\varepsilon}{\partial t}(0) = 0 \quad \text{and} \\ \displaystyle\int_\Omega \rho^\varepsilon \frac{\partial^2 \underline{u}^\varepsilon}{\partial t^2} \underline{w}\, dx + \int_\Omega \sigma^\varepsilon_{ij} \frac{\partial w_i}{\partial x_j} dx = \int_\Omega f_i\, w_i\, dx \quad \forall\, \underline{w} \in \underline{H}^1_o \end{array} \right.$$

where

$$(2.7)\quad \int_\Omega \sigma^\varepsilon_{ij} \frac{\partial w_i}{\partial x_j} dx \equiv - \int_\Omega p^\varepsilon \operatorname{div} \underline{w}\, dx + \varepsilon^2\, b^\varepsilon(\underline{v}^\varepsilon, \underline{w})$$

$$(2.8)\quad \begin{cases} b^\varepsilon(\underline{v}\, , \underline{w}) \equiv \displaystyle\int_\Omega b^\varepsilon_{ijk\ell}(x)\, e_{k\ell}(\underline{v})\, e_{ij}(\underline{w})\, dx \\ b_{ijk\ell}(y) \equiv 2\mu(y)\, \delta_{ik}\, \delta_{jh} + \eta(y)\, \delta_{ij}\, \delta_{k\ell} \end{cases}$$

Remark 2.1 - We of course assume that the viscosity coefficients $\mu(y)$, $\eta(y)$ satisfy relations of the type (1.7). The form $b^\varepsilon(\underline{v}\, , \underline{w})$ is then coercive on $\underline{H}^1_o$ and the existence and uniqueness of $\underline{u}^\varepsilon$ (for fixed $\varepsilon > 0$) then follows as in remarks 1.1 and 1.2, and even as in chapter 6, sect. 4. ■

In order to obtain the asymptotic behaviour of $\underline{u}^\varepsilon$ , we introduce the expansions

$$(2.9)\quad \begin{cases} \underline{u}^\varepsilon(x\, , t) = \underline{u}^0(x\, , y\, , t) + \varepsilon\, \underline{u}^1(x\, , y\, , t) + \varepsilon^2 \ldots \\ p^\varepsilon(x\, , t) = p^0(x\, , y\, , t) + \varepsilon\, p^1(x\, , y\, , t) + \varepsilon^2 \ldots \end{cases}$$

for $y = x/\varepsilon$ , Y-periodic in y.

We remark that (2.6) implies the classical equation

$$(2.10)\quad \rho^\varepsilon \frac{\partial^2 u^\varepsilon_i}{\partial t^2} = \frac{\partial \sigma^\varepsilon_{ij}}{\partial x_j} + f_i$$

and taking into account (2.3) and (2.9) we see that the leading term in (2.10) is

(2.11) $$\varepsilon^{-1} \frac{\partial p^0}{\partial y_i} = 0$$ and we then have :

(2.12) $$p^0 = p^0(x\ ,\ t)$$

Moreover, by expanding (2.2) we have :

$$p^0(x) + \varepsilon p^1 + \ldots = -\gamma \frac{1}{\varepsilon} \mathrm{div}_y \underline{u}^0 - \gamma(\mathrm{div}_x \underline{u}^0 + \mathrm{div}_y \underline{u}^1) - \varepsilon \ldots$$

and consequently

(2.13) $$\mathrm{div}_y\, \underline{u}^0 = 0$$

(2.14) $$p^0(x\ ,\ t) = -\gamma(y)\ (\mathrm{div}_x \underline{u}^0 + \mathrm{div}_y \underline{u}^1) \qquad \text{(indep. of y !)}$$

<u>Remark 2.2</u> - We have seen that $p^0$ does not depend on y, as in acoustics in porous media. This implies (2.13), i.e. the flow is incompressible at the local level. This important features are consequences of the fact that the viscosity coefficients are of order $\varepsilon^2$ (see (2.3), (2.10), (2.13)) and are very different of the behaviour in the preceeding section, where the coefficients were of order $\varepsilon^0$. This is the reason why we keeped the pressure in the formulation of the problem. ■

Now, in order to study the local problem, we define two Hilbert spaces $V_Y$ and $H_Y$ of Y-periodic vectors :

(2.15) $$\begin{cases} V_Y = \{\ \underline{v}\ ;\ \underline{v} \in \underline{H}^1_{loc}(R^3),\ \mathrm{div}\ \underline{v} = 0\ ,\ \text{Y-per.}\ \} \\ \|\ \underline{v}\ \|_{V_Y} = \|\underline{v}\ \|_{\underline{H}^1(Y)} \end{cases}$$

(2.16) $H_Y$ is the completion of $V_Y$ for the norm (equivalent to the $L^2(Y)$ norm) associated with the scalar product
$$(\underline{v}\ ,\ \underline{w})_{H_Y} = \int_Y \rho(y)\ v_i\ w_i\ dy$$

Now, <u>we consider $\underline{u}^0(x\ ,\ y\ ,\ t)$ as a smooth function of x and t for $x \in \Omega$, with values in $V_Y$</u> (Note that this implies that (2.13) and the periodicity condition are satisfied). Then, we consider (2.6) with test functions of the form : (compare with chap. 5, sect. 9)

(2.17) $$\begin{cases} \underline{w}\ (x) = \theta(x)\ \underline{\omega}\ (x/\varepsilon) \\ \theta \in \mathcal{D}(\Omega)\ ;\ \underline{\omega} \in V_Y \end{cases}$$

We replace this into (2.6), with (2.7) and (2.8) ; we note that

$$\mathrm{div}_y\ \underline{\omega} = 0 \implies \mathrm{div}\ \underline{w} = \underline{\mathrm{grad}}_x\ \theta \cdot \underline{\omega}$$

and we obtain, at order $\varepsilon^0$ :

$$(2.18)\qquad \int_\Omega \rho^\varepsilon \frac{\partial^2 u_i^0}{\partial t^2}\,\omega_i\,\theta\,dx - \int_\Omega p^0 \frac{\partial\theta}{\partial x_i}\,\omega_i\,dx + \int_\Omega b_{ijkh}^\varepsilon\, e_{khy}(\underline{v}^0)\, e_{ijy}(\omega)\,\theta\,dx = \int_\Omega f_i\,\omega_i\,\theta\,dx$$

Before taking $\varepsilon \to 0$, we integrate by parts the term in $p^0$ ; because $\operatorname{div}\underline{\omega} = 0 \Rightarrow -\int_\Omega p^0 \frac{\partial\theta}{\partial x_i}\,\omega_i\,dx = \int_\Omega \frac{\partial p^0}{\partial x_i}\,\omega_i\,\theta\,dx$

that we replace into (2.18) ; then, we take the limit as $\varepsilon \to 0$ (that is to say, we take the mean value of the integrand in Y, according to chapter 5, sect. 9) and we obtain :

$$(2.19)\qquad \int_\Omega \Big[\rho(y) \frac{\partial^2 u_i^0}{\partial t^2}\,\omega_i\Big]^\sim \theta\,dx + \int_\Omega \frac{\partial p^0}{\partial x_i}\,\tilde{\omega}_i\,\theta\,dx + \int_\Omega \big[\, b_{ijkh}(y)\, e_{khy}(\underline{v}^0)\, e_{ijy}(\underline{\omega})\big]^\sim \theta\,dx = \int_\Omega f_i\,\tilde{\omega}_i\,\theta\,dx$$

where $\theta \in \mathcal{D}(\Omega)$ is arbitrary ; (2.19) then implies that

$$\Big[\rho(y) \frac{\partial^2 u_i^0}{\partial t^2}\,\omega_i\Big]^\sim + \frac{\partial p^0}{\partial x_i}\,\tilde{\omega}_i + \big[b_{ijkh}\, e_{khy}(\underline{v}^0)\, e_{ijy}(\underline{\omega})\big]^\sim = f_i\,\tilde{\omega}_i$$

We rewrite this relation in terms of $\underline{v}$ instead of $\underline{u}$ ; moreover, we transform the term in $b_{ijkh}$ by using the fact that the divergences of $\underline{v}^0$ and $\underline{\omega}$ are zero and the Y-periodicity condition ; we obtain

$$(2.20)\qquad \int_Y \rho(y) \frac{\partial v_i^0}{\partial t}\,\omega_i\,dy + \frac{\partial p^0}{\partial x_i}\int_Y \omega_i\,dy + \int_Y \mu(y)\,\frac{\partial v_i^0}{\partial y_j}\frac{\partial \omega_i}{\partial y_j}\,dy = f_i \int_Y \omega_i\,dy$$

$$\forall\, \underline{\omega} \in V_Y$$

which is the local equation. It recalls eq. (6.14) of chapiter 7, and it is solved in an analogous way. We note that

$$(2.21)\qquad a(\underline{v}^0 , \underline{\omega}) \equiv \int_Y \mu(y)\,\frac{\partial v_i^0}{\partial y_j}\frac{\partial \omega_i}{\partial y_j}\,dy$$

is a bilinear symmetric form on $V_Y$ and that

$$a(\underline{v}^0 , \underline{\omega}) + (\underline{v}^0 , \underline{\omega})_{H_Y}$$

is coercive on $V_Y$. Let A be the selfadjoint operator of $H_Y$ associated with the form a according to the first representation theorem. Moreover let $\underline{\phi}^i \in H_Y$ be defined by

$$(2.22)\qquad \int_Y v_i\,dy = (\underline{\phi}^i , \underline{v})_{H_Y} \quad \forall\, \underline{v} \in H_Y$$

Then, (2.20), with the initial condition issued from (2.5), becomes

$$\frac{\partial v^0}{\partial t} + A\, \underline{v}^0 = (f_i - \frac{\partial p^0}{\partial x_i})\, \phi^i \quad ; \quad v^o(0) = 0 \tag{2.23}$$

and from semigroup theory we obtain the following expression for $\underline{v}^0(y\, ,\, t)$ (which also depends of course on the "parameter" x) :

$$\underline{v}^0(t) = \int_0^t e^{-A(t-s)}\, \underline{\phi}^i (f_i - \frac{\partial p^0}{\partial x_i})(s)\, ds \tag{2.24}$$

Moreover, we consider the mean value in Y ; we remark that (as in chapter 7, sect. 6) to take the mean value of the i-component amounts to take the scalar product by $\underline{\phi}^i$ in $H_Y$. (up to the factor $|Y\,|$). This gives

$$\begin{cases} \tilde{v}^o_k(t) = \int_0^t g_{ki}(t-s)(f_i - \frac{\partial p^0}{\partial x_i})(s)\, ds \\ |Y| g_{ki}(\xi) = (e^{-A\xi}\underline{\phi}^i\, ,\, \underline{\phi}^k)_{H_Y} \end{cases} \tag{2.25}$$

which is a "Darcy's law" for our problem : the mean value of the velocity is a functional of the "history" of $\underline{\text{grad}}_x p^0$, i.e., of the values of $\underline{\text{grad}}_x p^0$ for the preceeding time.

Remark 2.3 - As in cnapter 7, proposition 6.1, the functions $g_{ik}(\xi)$ are symmetric in ik. Nevertheless, there is an important difference between the present case and that of chapter 7. Here, $g_{ik}(\xi)$ do not tend to zero as $\xi \to \infty$ as a consequence of the fact that the form a is not coercive on $V_Y$. This fact is natural from a physical point of view. Indeed, let us consider the case $\underline{f} = 0$ for a homogeneous fluid : the functions do not depend on y and (2.20) is then equivalent to

$$\rho \frac{\partial v^o_i}{\partial t} = - \frac{\partial p^0}{\partial x_i} \Rightarrow \underline{v}^0(t) = -\rho^{-1} \int_0^t \underline{\text{grad}}_x p^0(s)\, ds$$

as for an inviscid fluid. We then see that the memory does not vanish.■

Remark 2.4 - It is useful to obtain an explicit form of the Darcy's law (2.25) in the case of isotropy (see chapter 7, at the end of sect. 6). We then have the formula (6.25) of chapter 7, but the first eigenvalue of A is zero. Moreover, the associated eigenfunctions are $\underline{e}_i$ (i.e., the unitary vectors in the directions of the coordinates) and the multiplicity of that eigenvalue is 3 (or 2, in the twodimensional case). It is easy to see (by the symmetry and positivity properties), that

$$(\underline{\phi}^1\, ,\, \underline{e}_j) = \begin{cases} = 0 \text{ if } j = 2\, ,\, 3 \\ > 0 \text{ if } j = 1 \end{cases}$$

and we obtain :

$$(2.26) \qquad g_{ik}(\xi) = \delta_{ik}\, g(\xi) \quad ; \quad g(\xi) = a_1^2 + \sum_{i=4}^{\infty} e^{-\lambda^i \xi}\, a_i^2$$

where $a_1 > 0$ and $\lambda^i > 0$ are the eigenvalues of the problem. In particular, we see that the memory is not vanishing (see remark 2.3).■

Now, we seach for a macroscopic equation in $p^o(x\ ,\ t)$. Usualy, the macroscopic equation is obtained by eliminating a higher order term in an appropriate equation by using periodicity and integration on Y. This is easily obtained from (2.14). We multiply by $\gamma^{-1}(y)$ and we integrate on Y. The periodicity condition for $\underline{u}^1$ gives :

$$(2.27) \qquad \int_Y \mathrm{div}_y\, \underline{u}^1 = \int_{\partial Y} n_i\, u_i^1\, ds = 0$$

and using the definition of $\gamma$, given by (2.2), we obtain (the mean value operator $\sim$ commutes with differentiation in x) :

$$(2.28) \qquad \left[\frac{1}{\rho(y)\, c^2(y)}\right]^{\sim} p^o + \mathrm{div}_x\, \underline{\tilde{u}}^o = 0$$

which, by taking into account (2.25) and differenciating with respect to t is the integro-differential equation

$$(2.29) \qquad \left(\frac{1}{\rho c^2}\right)^{\sim} \frac{\partial p^o}{\partial t}(x\ ,\ t) + \frac{\partial}{\partial x_i} \int_0^t g_{ik}(t - s)\left(f_k - \frac{\partial p^o}{\partial x_k}\right)(s)\, ds = 0$$

which is a unique equation for $p^o(x\ ,\ t)$. It is the equation of "acoustics" in the present problem.

On the other hand, it is easy to see that the appropriate boundary condition at a solid boundary is, as in chap. 7, sect. 5

$$(2.30) \qquad \underline{\tilde{v}}^o \cdot \underline{n} = 0 \qquad \text{on } \partial\Omega$$

Similarly, for problems with a surface S limiting a mixture of fluids and a homogeneous fluid, we have (as in chapter 7, (6.22)) :

$$(2.31) \qquad [\, p^o\, ] = 0 \quad ; \quad [\, \underline{\tilde{v}}^o \cdot \underline{n}\, ] = 0$$

where [ ] denotes discontinuity across S.

Remark 2.5 - It is not difficult to see that (2.27) is in fact a homogenized form of the conservation of mass (see (1.10)). In fact, the perturbation of density has an expansion

$$\rho^o(x\ ,\ y\ ,\ t) + \varepsilon\ \rho^1(x\ ,\ y\ ,\ t) + \varepsilon^2 \ldots \qquad \text{with}$$

(2.32) $$\rho^{o} = \frac{1}{c^{2}(y)}\, p^{o}$$

and (2.14) may be written

$$\frac{\rho^{o}(x\,,\,y\,,\,t)}{\rho(y)} + \mathrm{div}_x\, \underline{u}^{o} + \mathrm{div}_y\, \underline{u}^{1} = 0$$

and by taking the mean value and bearing in mind (2.27) we have

(2.33) $$\left(\frac{\rho^{o}}{\rho}\right)^{\sim} + \mathrm{div}_y\, \tilde{\underline{u}}^{o} = 0$$

which is another form of (2.28). In fact, this relation was formerly obtained in Levy [ 4 ] by using a macroscopic form of the conservation of mass. ∎

Moreover, <u>it is easy to obtain an existence and uniqueness theorem for $p^{o}(x\,,\,t)$</u>. For the sake of simplicity, we only consider the isotropic case where the Darcy's law (2.25) is given by (2.26) and we give an outline of the proof. By taking the Laplace transform from t into $\lambda$ of (2.29) we have

(2.34) $$\lambda\left(\frac{1}{\gamma}\right)^{\sim} \hat{p}^{o} - \hat{g}\,\Delta\, \hat{p}^{o} = -\hat{g}\ \mathrm{div}\, \hat{\underline{f}} \ \text{ in } \Omega \ ; \quad \frac{\partial \hat{p}^{o}}{\partial n} - n_i\, \hat{f}_i = 0 \text{ on } \partial\Omega$$

where the Neumann boundary condition is a consequence of (2.30). The variational formulation of (2.34) is

(2.35) $$\begin{cases} \hat{p}^{o} \in H^{1}(\Omega) \quad \text{and} \\ \lambda\left(\frac{1}{\gamma}\right)^{\sim} (\hat{p}^{o}\,,\,q)_{L^2(\Omega)} + \hat{g}(\underline{\mathrm{grad}}\ \hat{p}^{o}\,,\,\underline{\mathrm{grad}}\ q)_{L^2(\Omega)} = -\hat{g}(\mathrm{div}\ \underline{f}\,,\,q)_{L^2(\Omega)} \\ \forall q \in H^{1}(\Omega) \end{cases}$$

As usual ( $\Omega$ is bounded), existence for (2.35) follows from uniqueness ; by taking $q = \hat{p}^{o}$ we obtain

(2.36) $$\mathrm{Re}\ \{\lambda\left(\frac{1}{\gamma}\right)^{\sim} \|\hat{p}^{o}\|^{2}_{L^2} + \hat{g}(\lambda)\ \|\underline{\mathrm{grad}}\ \hat{p}^{o}\|^{2}_{L^2}\} \leq |\hat{g}|\ \|\mathrm{div}\ \hat{\underline{f}}\|_{L^2}\ \|\hat{p}^{o}\|_{L^2}$$

with (from (2.26)) :

(2.37) $$\hat{g}(\lambda) = \frac{a_1^2}{\lambda} + \sum_{i=4}^{\infty} \frac{a_i^2}{\lambda + \lambda^{i}}$$

then, if we take $\left(\frac{1}{\gamma}\right)^{\sim}$ Re $\lambda > 1$ (and also > abcissa of summability of $\underline{f}$) we have Re $\hat{g} > 0$ and then

(2.38) $$\|\hat{p}^{o}\|_{L^2} \leq |\hat{g}|\ \|\mathrm{div}\ \hat{\underline{f}}\|_{L^2}$$

consequently, $\hat{p}^{o}(\lambda)$ exists and is unique (and holomorphic) for Re $\lambda > 1$ ; moreover, its norm is bounded by a polynomial in $|\lambda|$ , and $p^{o}$ exists and is unique.

We summarize the results of this section :

Proposition 2.1 - We consider the problem (2.6) ($p^\varepsilon$ is given by (2.2)). If we postulate an asymptotic expansion of the form (2.9), $p^0$ is a function of x and t which satisfies eq. (2.29) and $\underline{\tilde{v}}^0$ is given by (2.25). Moreover, the boundary condition (2.30) is satisfied. In the isotropic case, the existence and uniqueness of $p^0(x, t)$ (and then of $\underline{v}^0(x, y, t)$ is proved.

3.- Proof of the convergence - We consider again the problem of the preceeding section and we prove the convergence of $\underline{u}^\varepsilon(x, t)$ to the limit $\underline{u}^0(x, t)$ in an appropriate topology. In order to simplify the problem, we consider the isotropic case where the Darcy's law takes the form (2.26) (the existence and uniqueness of the solution of the limit problem was proved in this case). We have

Theorem 3.1 - Let us consider the problem of sect. 2 in the isotropic case (see remark 2.4). Let us give $\underline{f} \in C^0([0, T]; \underline{C}^\infty(\Omega))$. If $\underline{u}^\varepsilon$ is the corresponding solution of (2.6) and $\underline{u}^0$ the solution of (2.29) with the boundary condition (2.30), which takes the form

$$\frac{\partial p}{\partial n} = 0 \qquad \text{on } \partial\Omega \tag{3.1}$$

Then, we have (as $\varepsilon \to 0$) :

$$\left\{\begin{array}{lll} \underline{u}^\varepsilon \to \underline{u}^0 & \text{in } L^\infty(0, T; \underline{L}^2(\Omega)) & \text{weakly} * \\ \operatorname{div} \underline{u}^\varepsilon \to \operatorname{div} \underline{u}^0 & \text{in } L^\infty(0, T; \underline{L}^2(\Omega)) & \text{weakly} * \\ \underline{u}^{\varepsilon\prime} \to \underline{u}^{0\prime} & \text{in } L^\infty(0, T; \underline{L}^2(\Omega)) & \text{weakly} * \end{array}\right. \tag{3.2}$$

In order to prove this theorem, we obtain a priori estimates for $\underline{u}^\varepsilon$ and we take the Laplace transform ($t \Rightarrow \lambda$). As usual, the proof is reduced to the proof of a corresponding theorem for the Laplace transforms with real $\lambda > 0$.

First, we continue f with value 0 for $t > T$. The solution $\underline{u}^\varepsilon$ is then defined for all $t > 0$. By taking $\underline{w} = \underline{u}^{\varepsilon\prime}$ in (2.6) we have :

$$\frac{1}{2}\frac{d}{dt}\left[\int_\Omega \rho^\varepsilon |\underline{u}^{\varepsilon\prime}|^2 dx + \int_\Omega \gamma^\varepsilon |\operatorname{div} \underline{u}^\varepsilon|^2 dx\right] + \varepsilon^2 b^\varepsilon(\underline{u}^{\varepsilon\prime}, \underline{u}^{\varepsilon\prime}) = \int_\Omega f_i u_i^{\varepsilon\prime} dx \tag{3.3}$$

The apriori estimates are obtained in two steps, for $t \in [0, T]$ and for $t > T$. First, for $t \in [0, T]$, by neglecting the positive term b in (3.3) we obtain (C denotes several constants)

$$\frac{1}{2}\frac{d}{dt}[\quad] \leq C[\quad]^{1/2} \Rightarrow \frac{d}{dt}[\quad]^{1/2} \leq C$$

and bearing in mind that $\underline{u}^\varepsilon(0) = \underline{u}^{\varepsilon\prime}(0)$, the function in the brakets of (3.3) remains bounded (independently of $\varepsilon$) for $t \in [0, T]$. For $t > T$, f is zero and

(3.3) shows that that function remains bounded for any $t > 0$. We then have

$$(3.4) \quad \begin{cases} \|\underline{u}^{\varepsilon\prime}\|_{\underline{L}^2(\Omega)} \leq C & \text{(indep. of } t \text{ and } \varepsilon) \\ \|\text{div } \underline{u}^{\varepsilon}\|_{L^2(\Omega)} \leq C & \text{(indep. of } t \text{ and } \varepsilon) \end{cases}$$

and by integrating in t with $\underline{u}^{\varepsilon}(0) = 0$ :

$$(3.5) \quad \|\underline{u}^{\varepsilon}\|_{\underline{L}^2(\Omega)} \leq C \qquad \text{(indep. of } \varepsilon,\ t \in [0,T\,])$$

We then see that we can extract a subsequence (in fact the whole sequence, because the limit will be proved to be unique !) converging in the topologies of (3.2). Moreover, from (3.4) we see that the Laplace transform of $\underline{u}^{\varepsilon}$ is defined for Re $\lambda > 0$. From (3.2), the Laplace transforms of $\underline{u}^{\varepsilon\prime}$, div $\underline{u}^{\varepsilon}$ converge for Re $\lambda > 0$ in the weak topology of $L^2(\Omega)$. On the other hand, the Laplace transform of $\underline{u}^0$ was proved to be well defined for sufficiently large Re $\lambda$ (see the end of sect. 2) ; consequently,

(3.6) in order to prove (3.2) it suffices to prove the corresponding convergence (in $L^2(\Omega)$ weakly) of the Laplace transforms for real $\lambda > 0$ (the property then holds for any $\lambda$ by analytic continuation, as usual, see chapter 6, sect. 4).

Then, for fixed $\lambda > 0$, we consider the Laplace transform of (2.6) (we use the notation $\underline{u}^{\varepsilon}$ for the Laplace transform as well as for the function because no ambiguity is possible), obtained from (2.6) by writting $\lambda$ instead of $\partial/\partial t$ :

For fixed $\lambda > 0$, let $\underline{u}^{\varepsilon} \in \underline{H}_0^1(\Omega)$ and $p^{\varepsilon}$ be the solution of :

$$(3.7)_1 \qquad p^{\varepsilon} = -\gamma^{\varepsilon} \text{ div } u^{\varepsilon}$$

$$(3.7)_2 \qquad \lambda^2 \int_{\Omega} \rho^{\varepsilon}\, u_i^{\varepsilon}\, w_i \, dx + \int_{\Omega} \gamma^{\varepsilon} \text{ div } \underline{u}^{\varepsilon} \text{ div } \underline{w} \, dx + \varepsilon^2 \lambda\, b^{\varepsilon}(\underline{u}^{\varepsilon},\underline{w}) = \int_{\Omega} f_i \, w_i \, dx \text{ for any } \underline{w} \in \underline{H}_0^1(\Omega)$$

Lemma 3.1 - If $\underline{u}^{\varepsilon}$, $p^{\varepsilon}$ are defined by (3.7), after extracting a subsequence, we have for $\varepsilon \to 0$ :

$$\underline{u}^{\varepsilon} \to \underline{u}^{*} \qquad \text{in } \underline{L}^2(\Omega) \text{ weakly}$$

$$\text{div } \underline{u}^{\varepsilon} \to \text{div } \underline{u}^{*} \qquad \text{in } L^2(\Omega) \text{ weakly}$$

$$p^{\varepsilon} \to p^{*} \qquad \text{in } L^2(\Omega) \text{ strongly}$$

$$(3.11) \qquad \underline{\text{grad}}\ p^{\varepsilon} \to \underline{\text{grad}}\ p^{*} \qquad \text{in } \underline{H}^{-1}(\Omega) \text{ strongly}$$

and $\underline{u}^{*}$ satisfies the boundary condition

$$(3.12) \qquad \underline{u}^{*} \cdot \underline{n} = 0 \quad \text{on } \partial\Omega$$

Proof - We first remark that, if (3.8) and (3.9) are proved, $\underline{u}^{\varepsilon}$ converges in the weak topology of the space $H(\Omega , \text{div})$ (see chapter 7, (1.4)) and by theorem 1.1 of chapter 7, $\underline{u}^{\varepsilon} . \underline{n}$ converges in the weak topology of $H^{-1/2}(\partial\Omega)$ ; moreover $\underline{u}^{\varepsilon}$ is zero on $\partial\Omega$ and we have (3.12) in the sense of $H^{-1/2}(\partial\Omega)$, which is then a consequence of (3.8), (3.9).

We take $\underline{w} = \underline{u}^{\varepsilon}$ in (3.7) and we immediately obtain :

$$(3.13) \qquad \|\underline{u}^{\varepsilon}\|^2_{L^2} + \|\text{div}\, \underline{u}^{\varepsilon}\|^2_{L^2} + \varepsilon^2 \|\underline{u}^{\varepsilon}\|^2_{\underline{H}^1_0} \leqslant C \|\underline{u}^{\varepsilon}\|_{L^2}$$

and we see that $\underline{u}^{\varepsilon}$ remains bounded in $\underline{L}^2$ ; consequently, the right hand side of (3.13) (and then each term of the left hand side) is bounded. We then obtain (3.8) (3.9) and

$$(3.14) \qquad \|\underline{u}^{\varepsilon}\|_{\underline{H}^1_0} \leqslant \frac{C}{\varepsilon}$$

Moreover, from the definition of $p^{\varepsilon}$ in (3.7) and (3.9) we see that $p^{\varepsilon}$ remains bounded in the norm of $L^2$ and consequently

$$(3.15) \qquad p^{\varepsilon} \to p^{*} \quad \text{in } L^2(\Omega) \text{ weakly}$$

To prove the strong convergence in (3.10) and (3.11) is a little more difficult. We recall that $H^{-1}(\Omega)$ is the dual space of $H^1_0(\Omega)$ when $L^2(\Omega)$ is identified with its dual, i.e. for the duality product associated with the distributions on $\Omega$. Moreover, if $q \in L^2$, its gradient (which is a distribution) belongs to $\underline{H}^{-1}$. This means that the distribution $\underline{\text{grad}}\, q \in \mathcal{D}'$ may act not only on any $\underline{w} \in \mathcal{D}$ but also on any $\underline{w} \in \underline{H}^1_0$ ; consequently according to the definition of derivatives of distributions, we have (the braket denotes the duality product between $\underline{H}^{-1}$ and $\underline{H}^1_0$):

$$(3.16) \quad \left< -\frac{\partial q}{\partial x_i} , w_i \right> = \left< q , \frac{\partial w_i}{\partial x_i} \right> = \int_{\Omega} q \,\text{div}\, \underline{w}\, dx \qquad \forall \underline{w} \in \underline{H}^1_0$$

which may be described by saying that "the product of $\underline{\text{grad}}$ q by $\underline{w}$ may be integrated by parts by using $\underline{w}|_{\partial\Omega} = 0$".

Now, we write $(3.7)_2$ by using $(3.7)_1$ :

$$\int_{\Omega} p^{\varepsilon} \,\text{div}\, \underline{w}\, dx = \lambda^2 \int_{\Omega} \rho^{\varepsilon}\, \underline{u}^{\varepsilon} . \underline{w}\, dx + \varepsilon^2 \lambda\, b^{\varepsilon}(\underline{u}^{\varepsilon} , \underline{w}) - \int_{\Omega} \underline{f} . \underline{w}\, dx$$

and by using (3.8) and (3.14)

$$(3.17) \qquad \left| \int_{\Omega} p^{\varepsilon} \,\text{div}\, \underline{w}\, dx \right| \leqslant C \|\underline{w}\|_{L^2} + \varepsilon^2 \frac{C}{\varepsilon} \|w\|_{\underline{H}^1_0}$$

and by applying (3.16) (because $p^{\varepsilon} \in L^2$)

$$(3.18) \qquad |\langle \underline{\text{grad}}\, p^{\varepsilon} , \underline{w} \rangle| \leq C(\| \underline{w} \|_{\underline{L}^2} + \varepsilon \|\underline{w}\|_{\underline{H}_0^1}) \quad \forall \underline{w} \in \underline{H}_0^1$$

Moreover, if we take 1 instead of $\varepsilon$ in the right hand side of (3.18) we see that the norm of $\underline{\text{grad}}\, p^{\varepsilon}$ in the dual of $\underline{H}_0^1$ (i.e. in $\underline{H}^{-1}$) is bounded, and by extracting a subsequence :

$$(3.19) \qquad \underline{\text{grad}}\, p^{\varepsilon} \to \underline{\text{grad}}\, p^{*} \quad \text{in } \underline{H}^{-1}(\Omega) \text{ weakly.}$$

Next, we take a sequence of test functions depending on $\varepsilon$ and such that

$$(3.20) \qquad \underline{w}^{\varepsilon} \to \underline{w}^{*} \qquad \text{in } \underline{H}_0^1 \text{ weakly}$$

Consequently,

$$(3.21) \qquad \underline{w}^{\varepsilon} \to \underline{w}^{*} \qquad \text{in } \underline{L}^2 \text{ strongly}$$

by the Rellich's theorem. We write

$$(3.22) \qquad |\langle \underline{\text{grad}}\, p^{\varepsilon} , \underline{w}^{\varepsilon} \rangle - \langle \underline{\text{grad}}\, p^{*} , \underline{w}^{*} \rangle| \leq$$

$$\leq |\langle \underline{\text{grad}}\, p^{\varepsilon} , \underline{w}^{\varepsilon} - \underline{w}^{*} \rangle| + |\langle \underline{\text{grad}}\, p^{\varepsilon} - \underline{\text{grad}}\, p^{*} , \underline{w}^{*} \rangle|$$

but the second term in the right hand side tends to zero by (3.19). As for the first, by using (3.18), it is bounded by

$$C( \| \underline{w}^{\varepsilon} - \underline{w}^{*} \|_{\underline{L}^2} + \varepsilon \| \underline{w} - \underline{w}^{*}\|_{\underline{H}_0^1} )$$

which tends to zero as $\varepsilon \to 0$ by virtue of (3.20) and (3.21). Consequently, (3.22) gives :

$$(3.23) \qquad \langle \underline{\text{grad}}\, p^{\varepsilon} , \underline{w}^{\varepsilon} \rangle \to \langle \underline{\text{grad}}\, p^{*} , w^{*} \rangle$$

and this implies that the convergence in (3.19) is strong, that is to say, <u>we have (3.11)</u> . In fact, this is a classical result in Hilbert space theory. If V and V' are a Hilbert space and its dual, and

$$(3.24) \qquad y^{i} \to y^{*} \qquad \text{in V' weakly and}$$

$$(3.25) \qquad \langle y^{i} , x^{i} \rangle \to \langle y^{*} , x^{*} \rangle$$

for any sequence $x^{i}$ such that

$$x^{i} \to x^{*} \qquad \text{in V weakly}$$

Then, (3.24) holds in V' strongly. To prove this, we identify V with its dual and we take $x^{i} = y^{i} - y^{*}$ ; $x^{*} = 0$ ; (3.25) gives

$$\|y^{i} - y^{*}\|_V^2 = \langle y^{i} - y^{*} , y^{i} - y^{*} \rangle = \langle y^{i} , y^{i} - y^{*} \rangle + \text{(term which tends to zero)}$$

which tends to zero by (3.25). If V is not identified with its dual, the same result may be obtained by using the duality function.

It only remains to prove (3.10). We use the inequality (see Temam [1], Lemma 6.1 of chapter 1, or Tartar [5]) :

$$(3.26)\qquad \|q\|_{L^2(\Omega)} \leqslant C(\Omega)\left( \left|\int_\Omega q\,dx\right| + \|\underline{\mathrm{grad}}\, q\|_{H^{-1}(\Omega)} \right)$$

with $q = p^\varepsilon - p^*$. We note that the integral in (3.26) is then the scalar product in $L^2(\Omega)$ of $p^\varepsilon - p^*$ by the function equal to one, which tends to zero by (3.15). The norm of $\underline{\mathrm{grad}}\,(p^\varepsilon - p^*)$ in $\underline{H}^{-1}$ also tends to zero by 3.11, and we obtain (3.10). Lemma 3.1 is proved. ∎

Lemma 3.2 - Under the hypothesis of Lemma 3.1, we have the following relation between $p^*$ and $\underline{u}^*$ :

$$(3.27)\qquad \left[\frac{1}{\gamma(y)}\right]^{\sim} p^* + \mathrm{div}\,\underline{u}^* = 0$$

(Note that this is (2.28), or, moreover, the Laplace transform of (2.28)).

Proof - We write $(3.7)_1$ under the form :

$$(3.28)\qquad \frac{1}{\gamma(\frac{x}{\varepsilon})}\,p^\varepsilon(x) + \mathrm{div}\,\underline{u}^\varepsilon = 0$$

we multiply by $\theta \in \mathcal{D}(\Omega)$ and by integrating on $\Omega$ we have

$$(3.29)\qquad \int_\Omega \theta\,\mathrm{div}\,\underline{u}^\varepsilon\,dx = -\int_\Omega \frac{1}{\gamma(\frac{x}{\varepsilon})}\,p^\varepsilon\,\theta\,dx$$

From (3.10) and lemma 4.1 of chapter 5

$$p^\varepsilon\theta \to p^*\theta \text{ in } L^2(\Omega) \text{ strongly}$$

$$\frac{1}{\gamma(\frac{x}{\varepsilon})} \to \left[\frac{1}{\gamma(y)}\right]^{\sim} \quad \text{in } L^2(\Omega) \text{ weakly}$$

we then pass to the limit in (3.29) by using (3.9) and we see that (3.28) is satisfied in the sense of distributions. ∎

Now, we come back to (3.6). We remark that by (3.12) and (3.27) we have proved that the limit function $\underline{u}^*$ satisfies the (Laplace transforms of the) boundary condition (2.30) and the equation (2.28) ; it only remains to prove that $\underline{u}^*$ satisfies the Laplace transform of the Darcy's law (2.25). To this end, we use the standard technique in homogenization proofs. We consider the Laplace transform of the local equation (2.20) where we denote $\underline{\alpha} = \underline{\mathrm{grad}}_x p^0 - \underline{f}$ the given term. Moreover, as in the appendix, p. 369 we write it under the form :

$$(3.30)\qquad \begin{cases} \lambda^2 \rho(y) v_i + \alpha_i + \dfrac{\partial q}{\partial y_i} = \lambda \dfrac{\partial}{\partial y_j}\left[ b_{ijkh}(y)\, e_{khy}(\underline{v}) \right] \\ \mathrm{div}_y\,\underline{v} = 0 \qquad ; \qquad \underline{v}\,,\ q \text{ are Y-periodic} \end{cases}$$

(Note that the term in $\partial q/\partial y_i$ is associated with the fact that (2.20) is only satisfied for test functions with zero divergence ; this introduces a term $\underline{\text{grad}}_y\, q$ as in chapter 7, (1.14) ; moreover, q is Y-periodic as in chapter 7, (2.23)). Then, the Laplace transform of the Darcy's law is given by

$$\tilde{\underline{v}} = - K \, \underline{\alpha} \tag{3.31}$$

(Note that we only consider the isotropic case).

Next, we write (3.30) in terms of $x = \varepsilon y$ :

$$\underline{v}^{\varepsilon}(x) = \underline{v}\left(\frac{x}{\varepsilon}\right) \quad ; \quad q^{\varepsilon}(x) = q\left(\frac{x}{\varepsilon}\right) \quad : \tag{3.22}$$

$$\lambda^2 \rho^{\varepsilon} v_i^{\varepsilon} + \alpha_i + \varepsilon \frac{\partial q^{\varepsilon}}{\partial x_i} = \varepsilon^2 \lambda \frac{\partial}{\partial x_j} [ b_{ijkh}^{\varepsilon}\, e_{khx}(\underline{v}^{\varepsilon}) ] \tag{3.33}$$

$$\text{div}_x\, \underline{v}^{\varepsilon} = 0 \tag{3.34}$$

where $\underline{v}^{\varepsilon}$ , $q^{\varepsilon}$ are $\varepsilon$Y-periodic, defined on $R^3$. Because $\underline{v}(y)$ is a fixed function, it is easily seen that the following estimates hold (where $L^2$ is for $L^2(D)$, D any bounded domain fixed in the x-variables, for instance $\Omega$ ) :

$$\| \underline{v}^{\varepsilon} \|_{L^2} \leqslant C \quad ; \quad \| q^{\varepsilon} \|_{L^2} \leqslant C \tag{3.35}$$

$$\| \varepsilon\, \underline{\text{grad}}_x\, \underline{v}^{\varepsilon} \|_{\underline{L}^2} \leqslant C \quad ; \quad \| \varepsilon\, \underline{\text{grad}}_x\, q^{\varepsilon} \|_{L^2} \leqslant C \tag{3.36}$$

Now, we consider $\phi \in \mathcal{D}(\Omega)$ and we take the product of (3.33) by $\phi\, u_i^{\varepsilon}$ and we integrate on $\Omega$ ($\underline{u}^{\varepsilon}$ is of course the solution of (3.7)). (We note that $\phi \underline{u}^{\varepsilon}$ has not zero divergence and it cannot be taken as test function in the Laplace transform of (2.20) ; this is the reason why we wrote it under the form (2.30)). We obtain after integrating by parts (in fact the duality product between $\underline{H}^{-1}(\Omega)$ and $H_0^1(\Omega)$) :

$$\lambda^2 \int_{\Omega} \rho^{\varepsilon}\, \underline{v}^{\varepsilon}\, \phi\, \underline{u}^{\varepsilon}\, dx + \alpha_i \int_{\Omega} \phi\, u_i^{\varepsilon}\, dx - \varepsilon \int_{\Omega} q^{\varepsilon} \frac{\partial}{\partial x_i}(\phi\, u_i^{\varepsilon})\, dx + \\ + \varepsilon^2 \lambda \int_{\Omega} b_{ijkh}^{\varepsilon} \frac{\partial v_k^{\varepsilon}}{\partial x_h} \frac{\partial (\phi u_i^{\varepsilon})}{\partial x_j}\, dx = 0 \tag{3.37}$$

On the other hand, we consider (3.7) with $\phi \underline{v}^{\varepsilon}$ as test function $\underline{w}$ :

$$\lambda^2 \int_{\Omega} \rho^{\varepsilon}\, \underline{u}^{\varepsilon}\, \phi\, \underline{v}^{\varepsilon}\, dx - \int_{\Omega} p^{\varepsilon}\, \text{div}(\phi\, \underline{v}^{\varepsilon})\, dx + \\ + \varepsilon^2 \lambda \int_{\Omega} b_{ijkh}^{\varepsilon} \frac{\partial u_k^{\varepsilon}}{\partial x_h} \frac{\partial (\phi v_i^{\varepsilon})}{\partial x_j}\, dx = \int_{\Omega} f_i \phi\, v_i^{\varepsilon} dx \tag{3.38}$$

and we compare (3.37) with (3.38). First we remark that, by $(3.35)_2$, (3.8), (3.9) :

$$\left| \varepsilon \int_{\Omega} q^{\varepsilon} \frac{\partial}{\partial x_i} (\phi\, u_i^{\varepsilon}) dx \right| = \left| \varepsilon \int_{\Omega} q^{\varepsilon} (\underline{\text{grad}}\, \phi \,.\, \underline{u}^{\varepsilon} + \phi\, \text{div}\, \underline{u}^{\varepsilon}) dx \right| < C\, \varepsilon$$

On the other hand, by (3.14) and (3.35)$_1$

$$\varepsilon^2 \lambda \left| \int_\Omega b^\varepsilon_{ijkh} \frac{\partial u^\varepsilon_k}{\partial x_h} \frac{\partial(\phi v^\varepsilon_i)}{\partial x_j} dx - \int_\Omega b^\varepsilon_{ijkh} \frac{\partial u^\varepsilon_k}{\partial x_h} \phi \frac{\partial v^\varepsilon_i}{\partial x_j} dx \right| =$$
$$= \varepsilon^2 \lambda \left| \int_\Omega b^\varepsilon_{ijkh} \frac{\partial u^\varepsilon_k}{\partial x_h} v^\varepsilon_i \frac{\partial \phi}{\partial x_j} dx \right| \leq C \varepsilon$$

An analogous calculation with $\underline{v}^\varepsilon$ (resp. $\underline{u}^\varepsilon$) at the place of $\underline{u}^\varepsilon$ (resp. $\underline{v}^\varepsilon$) holds by using (3.8) and (3.36)$_1$. As a consequence, the difference between the terms in b of (3.37) and (3.38) tends to zero as $\varepsilon \to 0$. The comparaison of (3.37) and (3.38) then gives :

$$(3.39) \quad \alpha_i \int_\Omega \phi\, u^\varepsilon_i \, dx + \int_\Omega p^\varepsilon \operatorname{div}(\phi\, \underline{v}^\varepsilon) dx + \int_\Omega f_i \phi\, v^\varepsilon_i \, dx \to 0 \quad \text{as } \varepsilon \to 0.$$

We consider the limit of the term containing $p^\varepsilon$ ; by recalling that $\operatorname{div} \underline{v}^\varepsilon = 0$ we have

$$p^\varepsilon \operatorname{div}(\phi\, \underline{v}^\varepsilon) = p^\varepsilon v^\varepsilon_i \frac{\partial \phi}{\partial x_i}$$

moreover, $\underline{v}^\varepsilon$ is an $\varepsilon Y$-periodic function and by lemma 4.1 of chapter 5, it tends in $\underline{L}^2(\Omega)$ weakly to $\tilde{\underline{v}}$ , that is to say, to $-K\underline{\alpha}$ by (3.31). On the other hand, by (3.10), $p^\varepsilon$ $\underline{\text{grad}}\ \phi$ converges to $p^*$ $\underline{\text{grad}}\ \phi$ in $\underline{L}^2(\Omega)$ strongly (note that $\phi$ is a smooth fixed function) and the limit of the term containing $p^\varepsilon$ in (3.39) is

$$- \int_\Omega p^* \frac{\partial \phi}{\partial x_i} \alpha_i \, K \, dx$$

then, by passing to the limit in (3.39) we have (by (3.8)) :

$$\alpha_i \int_\Omega u^*_i \phi \, dx - \alpha_i K \int_\Omega p^* \frac{\partial \phi}{\partial x_i} dx - \alpha_i K \int_\Omega f_i \phi \, dx = 0$$

and, as $\phi$ is any function of $\mathcal{D}(\Omega)$, we have in the sense of distributions :

$$(3.40) \qquad \alpha_i \left[ u^*_i + K\left(\frac{\partial p^*}{\partial x_i} - f_i\right) \right] = 0$$

and because $\underline{\alpha} \in R^3$ is arbitrary, the functions in the brackets of (3.40) are zero, that is to say, $\underline{u}^*$ and $p^*$ satisfy the Laplace transform of the Darcy's law with the constant K of (3.31). <u>The proof of theorem 3.1 is finished</u>.

## 4. - <u>Suspension of rigid particles in a slightly viscous compressible fluid</u>.

We consider a compressible, slightly viscous fluid as in sect. 2 with solid, rigid particles in suspension. The small vibrations of such a mixture may be studied by a method analogous to that of sect. 2. We shall not give details.; they may be seen in Fleury [ 3 ] , where a different, but equivalent treatment is used.

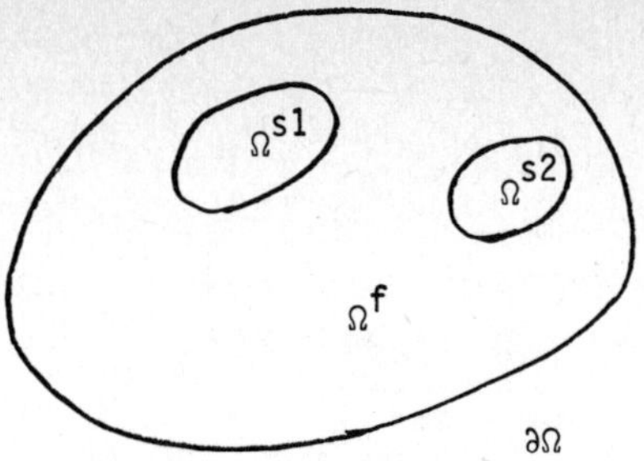


The bounded domain $\Omega$ is the union of the "solid domain $\Omega^s$", which is formed by one or several connected components $\Omega^{si}$ and the fluid domain $\Omega^f$ (see figure). We define the spaces V and H in the following way.: V is the space of the functions $\underline{u}$ belonging to $\underline{H}_0^1(\Omega)$ such that, on each solid portion $\Omega^{si}$ are equal to a solid displacement, i.e. are of the form

(4.1) $$\underline{u} = \underline{\gamma}^i + \underline{\zeta}^i \wedge \underline{x} \qquad \text{on} \quad \Omega^{si}$$

where $\underline{\gamma}^i$ and $\underline{\zeta}^i$ are elements of $R^3$ (one for each $\Omega^{si}$) which depend on the function $\underline{u}$. The scalar product in V is that of $\underline{H}_0^1(\Omega)$. The space H is defined as the completion of V for the norm of $\underline{L}^2(\Omega)$.

It is clear that V (and H) is the space of the "displacements compatible with the constraints", in the language of analytic mechanics.

We define the function $\rho$ equal to $\rho^s$ and $\rho^f$ in $\Omega^s$ and $\Omega^f$ respectively ($\rho^s$ and $\rho^f$ are the densities of the solid and of the fluid in the unperturbed state).

If the stress tensor in the fluid is defined, as in sect. 2, by

(4.2) $$\sigma_{ij} = -\delta_{ij}\, p + \varepsilon^2 [\eta\, \delta_{ij}\, \delta_{k\ell} + 2\, \mu\, \delta_{ik}\, \delta_{j\ell}]\, e_{k\ell}(\underline{v})$$

it is easy to see that the motion of the system is given by the "virtual power identity" (we suppose that the system is at rest for $t < 0$ and that $\underline{u} = 0$ on the boundary) :

(4.3) $$\begin{cases} \text{Find } \underline{u}(t) \text{ with values in V such that} \\ \displaystyle\int_\Omega \rho \frac{\partial^2 u_i}{\partial t^2} w_i \, dx + \int_{\Omega^f} \sigma_{ij}\, e_{ij}(\underline{w}) dx = \int_\Omega f_i\, w_i \, dx \qquad \forall \underline{w} \in V \end{cases}$$

and the pressure in the fluid is given by

(4.4) $$p = - c^2 \rho^f \operatorname{div} \underline{u}$$

Remark 4.1 - It is classical that a rigid solid is a mechanical system with ideal constraints. Consequently, the forces associated with these constraints give a zero power for the virtual velocities which satisfy the constraints. This is the reason why the term in $\sigma_{ij}$ of (4.3) is only written on $\Omega^f$.

On the other hand, it is easy to see that (4.3) is equivalent to

(4.5) $$\rho^f \frac{\partial^2 u_i}{\partial t^2} = \frac{\partial \sigma_{ij}}{\partial x_j} + f_i \qquad \text{in } \Omega^f$$

and the classical relations of equilibrium of forces and moments : (with the inertia forces) :

$$\int_{\Omega^{sk}} \left( \rho^s \frac{\partial^2 u_i}{\partial t^2} - f_i \right) dx - \int_{\partial\Omega^{sk}} \sigma_{ij} n_j \, dS = 0$$

$$\int_{\Omega^{sk}} \underline{x} \wedge \left( \rho^s \frac{\partial^2 \underline{u}}{\partial t^2} - \underline{f} \right) dx - \int_{\partial\Omega^{sk}} \underline{x} \wedge (\sigma \bullet n) \, dS = 0$$

on each $\Omega^{sk}$ .■

The existence and uniqueness of $\underline{u}$ satisfying (4.3) (when the initial conditions $\underline{u}(0)$ and $\underline{u}'(0)$ are given) is easily obtained by standard techniques.

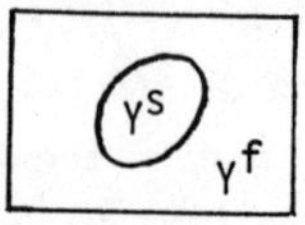


Now, we study a homogenization problem in the preceeding framework. We define in the standard way the period Y, formed by a part $Y^s$ (the "solid"part") which is strictely contained in Y (i.e., the boundary of $Y^s$ does not intersect the boundary of the period) and the fluid part $Y^f$. As usual, we consider the space $R^3$ of the variable y filled by the periods Y. Moreover, the region $\Omega$ of the space $R^3$ of the variable x is decomposed into $\Omega^{s\varepsilon}$ (the solid part) and $\Omega^{f\varepsilon}$ (the fluid part), where $\Omega^{s\varepsilon}$ is the intersection of $\Omega$ and the set of the points x such that $x/\varepsilon \in Y^s$. (For the sake of definiteness, it is useful to consider that the solid particles $\Omega^{sf}$ do not intersect $\partial\Omega$ , but this is irrelevant for the formal study that follows).

We define as usual :

$$\rho^\varepsilon(x) = \rho\left(\frac{x}{\varepsilon}\right) = \begin{cases} \rho^s & \text{if } x/\varepsilon \in Y^s \\ \rho^f & \text{if } x/\varepsilon \in Y^f \end{cases}$$

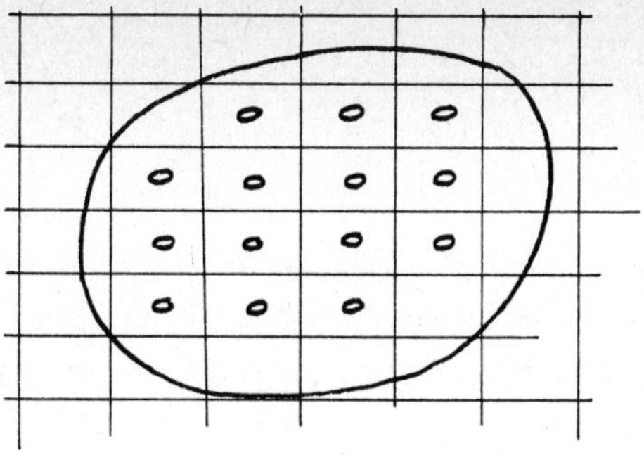

and the form

(4.6) $$b^{\varepsilon}(\underline{v}\,,\,\underline{w}) = \int_{\Omega^{f\varepsilon}} b_{ijkh}\, e_{kh}(\underline{v})\, e_{ij}(\underline{w})\, dx \qquad \text{where}$$

$$b_{ijkh} = 2\,\mu\,\delta_{ik}\,\delta_{jh} + \eta\,\delta_{ij}\,\delta_{kh}$$

We also define the space $V^{\varepsilon}$ which is the subspace of $\underline{H}_0^1(\Omega)$ formed by the vectors which are equal to a solid displacement in each connected component of $\Omega^{s\varepsilon}$.

We now consider the $\varepsilon$-dependent form of the problem 4.3 with zero initial conditions :

Find $\underline{u}^{\varepsilon}(t)$ with values in $V^{\varepsilon}$ such that

(4.7) $$\int_{\Omega} \rho^{\varepsilon} \frac{\partial^2 u_i^{\varepsilon}}{\partial t^2}\, w_i\, dx - \int_{\Omega^{f\varepsilon}} p^{\varepsilon} \,\mathrm{div}\, \underline{w}\, dx + \varepsilon^2\, b^{\varepsilon}(\underline{v}^{\varepsilon}, \underline{w}) = \int_{\Omega} f_i\, w_i\, dx \quad \forall \underline{w} \in V^{\varepsilon}$$

with $\underline{u}^{\varepsilon}(0) = \underline{u}^{\varepsilon\prime}(0) = 0$ , where $\underline{v}^{\varepsilon} = u^{\varepsilon\prime}$ and

(4.8) $$p^{\varepsilon} \equiv -c^2\,\rho^f\,\mathrm{div}\,\underline{u}^{\varepsilon} \qquad \text{on } \Omega^{f\varepsilon}.$$

As in sect. 2 we consider the expansion

(4.9) $$\underline{u}^{\varepsilon}(x\,,\,t) = \underline{u}^0(x\,,\,y\,,\,t) + \varepsilon\underline{u}^1(x\,,\,y\,,\,t) + \ldots \;;\quad y = \frac{x}{\varepsilon}$$

(4.10) $$p^{\varepsilon}(x\,,\,t) = p^0(x\,,\,t) + \varepsilon\, p^1(x\,,\,y\,,\,t) + \ldots \;;\quad y = \frac{x}{\varepsilon}$$

Y-periodic in y. It is clear that $p^{\varepsilon}$ is defined only on $\Omega^{f\varepsilon}$ ; in the expansion (4.10), the different terms are defined only for $y \in Y^f$. In fact, as in (2.11), it is seen that $p^0$ does not depend on y (it was written so in (2.10) !). Then, this term is in fact defined on $\Omega$ , but (4.8) only holds in $\Omega^{f\varepsilon}$.

Moreover, the expansion of (4.8) gives

(4.11) $$\mathrm{div}_y\, \underline{u}^0 = 0 \qquad \text{on } Y^f$$

(4.12) $$p^0(x\,,\,t) = -\,c^2\,\rho^f(\mathrm{div}_x\,\underline{u}^0 + \mathrm{div}_y\,\underline{u}^1) \qquad \text{on } Y^f$$

As in sect. 2, the macroscopic equation (which is associated with the conservation of mass) is easily obtained from (4.12). This equation is valid for $y \in Y^f$ ; in order to use a property of the type (2.27), we must write an analogous relation in $Y^s$ ; this is easily obtained by noting that in the $\Omega^{s\varepsilon}$ region, the divergence of $\underline{u}^\varepsilon$ is zero ; then, from (4.9) we have :

(4.13) $$0 = \mathrm{div}_x\,\underline{u}^0 + \mathrm{div}_y\,\underline{u}^1 \qquad \text{on } Y^s$$

We then divise (4.12) by $-\,c^2\,\rho^f$ and we consider the sum of the integral of (4.12) on $Y^f$ and (4.13) on $Y^s$ ; by (2.27) this gives

$$\frac{|Y^f|}{-c^2\,\rho^f}\,p^0 = \mathrm{div}_x \int_Y \underline{u}^0\,dx \quad \Longrightarrow$$

(4.14) $$0 = \frac{|Y^f|}{|Y|}p^0 + c^2\,\rho^f\,\mathrm{div}_x\,\tilde{\underline{u}}^0$$

where, as usual :

(4.15) $$\tilde{\cdot} = \frac{1}{|Y|}\int_Y \cdot\;dy$$

The obtainement of the local equation, which leads to a "Darcy's law with memory" is a little more complicated.

Equation (4.11) shows that the local problem is "incompressible". Then, it is natural to define the space $V_Y$ as the space of the Y-periodic, divergence-free vectors of $\underline{H}^1_{loc}(R^3)$ such that they are equal to a solid displacement on $Y^s$ (see 4.1). $V_Y$ is equipped with the standard norm of $\underline{H}^1(Y)$. Moreover, $H_Y$ is defined as the completion of $V_Y$ for the norm of $\underline{L}^2(Y)$.

In order to obtain the asymptotic expansion of (4.7), we first continue $p^\varepsilon$ to $\Omega$ in such a way that expansion (4.10) with the periodicity conditions hold; this amounts to continuate $p^0$ (this was yet made !), $p^1$ on $Y^s$, etc. Moreover, div $\underline{w}$ is zero on $\Omega^{sf}$, and then, we may write

$$\int_{\Omega^{f\varepsilon}} p^\varepsilon\,\mathrm{div}\,\underline{w}\,dx = \int_\Omega p^\varepsilon\,\mathrm{div}\,\underline{w}$$

and integrating by parts ($\underline{w}$ is zero on $\partial\Omega$) :

(4.16) $$\int_{\Omega^{f\varepsilon}} p^\varepsilon\,\mathrm{div}\,\underline{w}\,dx = -\int_\Omega \underline{\mathrm{grad}}\,p^\varepsilon\,.\,\underline{w}\,dx$$

Now, in order to expand (4.7), as in sect. 2, we take

$$(4.17)\quad \begin{cases} \underline{w} = \underline{w}^{\varepsilon} = \theta(x)\, \underline{\omega}\, (\frac{x}{\varepsilon}) + \varepsilon\, \underline{w}^1(x\, ,\, \frac{x}{\varepsilon}) + \dots \\ \text{with} \quad \theta \in \mathcal{D}(\Omega) \qquad : \ \underline{\omega} \in V_Y \end{cases}$$

The expansion of (4.7) (with (4.16)) at order $\varepsilon^0$ gives :

$$\int_{\Omega} \rho^{\varepsilon} \frac{\partial^2 u_i^0}{\partial t^2} \omega_i\, \theta\, dx + \int_{\Omega} (\underline{grad}_x\, p^0 + \underline{grad}_y\, p^1) \cdot \underline{\omega}\, \theta\, dx +$$
$$+ \int_{\Omega^{f\varepsilon}} b_{ijkh}\, e_{khy}(\underline{v}^0)\, e_{ijy}(\underline{\omega})\, \theta\, dx = \int_{\Omega} f_i\, \omega_i\, \theta\, dx$$

and we take the limit value as $\varepsilon \to 0$ by taking, as usual (chap. 5, sect. 9) the mean value on Y (note that the $b_{ijkh}$ must be considered to be zero on $Y^s$). Because $\theta \in \mathcal{D}(\Omega)$ is arbitrary, we obtain (after multiplying by $|Y|$ ) :

$$(4.18)\quad \int_Y \rho \frac{\partial^2 u_i^0}{\partial t^2} \omega_i\, dy + \int_Y (\frac{\partial p^0}{\partial x_i} + \frac{\partial p^1}{\partial y_i})\, \omega_i\, dy +$$
$$+ \int_{Y^f} b_{ijkh}\, e_{khy}(\underline{v}^0)\, e_{ijy}(\underline{\omega})\, dy = \int_Y f_i\, \omega_i\, dy$$

Moreover, $\underline{\omega} \in V_Y \Rightarrow \mathrm{div}_y\, \underline{\omega} = 0$ and by the periodicity :

$$\int_Y \frac{\partial p^1}{\partial y_i} \omega_i\, dy = \int_Y \frac{\partial}{\partial y_i}(p^1\, \omega_i)\, dy = \int_{\partial Y} p^1\, \omega_i\, n_i\, dS = 0$$

and (4.18) becomes (in terms of $\underline{v}$ instead of $\underline{u}$) :

$$(4.19)\quad \int_Y \rho \frac{\partial v_i^0}{\partial t} \omega_i\, dy + \frac{\partial p^0}{\partial x_i} \int_Y \omega_i\, dy + \int_{Y_f} b_{ijkh}\, e_{khy}(\underline{v}^0)\, e_{ijy}(\underline{\omega}) dy = \int_Y f_i\, \omega_i\, dy$$

which must be satisfied for any $\underline{\omega} \in V_Y$ ; moreover, from (4.11) it is natural to search $\underline{u}^0$ with values in $V_Y$. We then have the local problem (4.19), which has a structure analogous to that of (2.20). We immediately obtain a Darcy's law of the type (2.25) (Remark 2.3 also holds).

By replacing this into the macroscopic equation (4.14) we obtain the <u>equation of vibrations of the suspension</u> :

$$(4.20)\quad \frac{1}{c^2 \rho^f} \frac{|Y^f|}{|Y|} \frac{\partial p^0(x,t)}{\partial t} + \frac{\partial}{\partial x_i} \int_0^t g_{ki}(t - s)(f_k - \frac{\partial p^0(x\, ,\, t)}{\partial x_k})(s)\, ds = 0$$

with the boundary condition (Neumann condition)

$$\tilde{\underline{v}}^0 \cdot \underline{n} = 0$$

where $\tilde{\underline{v}}^0$ is given by (2.25).

<u>Remark 4.2</u> - We have seen that the asymptotic expansion gives (4.11) and consequently it is natural to search for $\underline{u}^0(x\, ,\, y\, ,\, t)$ as a function of x (and t) with

values in $V_Y$. Then, in order to obtain consistent relations, the test function is searched under the form (4.17). It is noticeable that $\theta(x)\ \underline{\omega}\ (x/\varepsilon)$ with $\theta \in \mathcal{D}(\Omega)$, $\underline{\omega} \in V_Y$ does not belong to $V^\varepsilon$ : in fact, it is not equal to a solid displacement on each connected component of $\Omega^{S\varepsilon}$ ; this is the reason why we introduced the term $\underline{w}^1(x\ ,\ x/\varepsilon)$ in (4.17) (compare with (2.17)), which was irrelevant in the sequel.

In fact, to belong to $V_\varepsilon$ means that the components $e_{ij}$ of the deformation tensor are zero for $x \in \Omega$ , $x/\varepsilon \in Y^S$ . It is possible to prove that, if $\theta \in \mathcal{D}$ and $\underline{\omega} \in V_Y$ are given, $\underline{w}^1$ may be found in such a way that (4.17) belongs to $V^\varepsilon$, at least at the first order. Bearing in mind the definition of $V_Y$ we have in (4.17)

$$e_{ij}(\underline{w}) = e_{ijx}(\theta\ \underline{\omega}) + e_{ijy}(\underline{w}^1) + \varepsilon\ \ldots.$$

we then have :

$$e_{ijy}(\underline{w}^1) = -\ e_{ijx}(\theta\ \underline{\omega}) = -\ \frac{1}{2}\left(\omega_i \frac{\partial\theta}{\partial x_j} + \omega_j \frac{\partial\theta}{\partial x_i}\right) \tag{4.21}$$

for $y \in Y^S$. But, x being a parameter, the right hand sides of (4.21) are linear functions of y in $Y^S$ and then the compatibility conditions for the construction of $\underline{w}^1$ are satisfied (see, for instance, Germain [ 1 ] sect. V.4.4).■

## 5.- Connected elastic solid with canals filled with a slightly viscous fluid.

In this section we study a problem analogous to that of sect. 1 but with slightly viscous compressible fluid (with viscosity coefficients of order $\varepsilon^2$) under the hypothesis that $\Omega^S$ and $\Omega^f$ are both connected, i.e., the solid part of the total domain $\Omega$ is a one-piece body ; and the fluid part also. As it was noted in chapter 7, sect. 2, this is possible in the three-dimentional case.

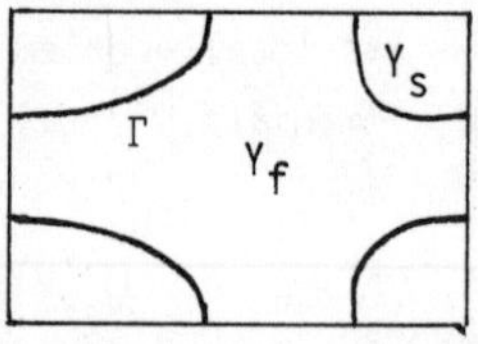


With standard notations (as in the preceeding sections), the problem for fixed ε is (we consider homogeneous Dirichlet boundary conditions) :

Find $\underline{u}^\varepsilon(x\ ,\ t)$ with values in $\underline{H}^1_0(\Omega)$ such that

$$\int_\Omega \rho^\varepsilon(x) \frac{\partial^2 u_i^\varepsilon}{\partial t^2} w_i\ dx + a^\varepsilon(\underline{u}^\varepsilon, \underline{w}) + \varepsilon^2 b\left(\frac{\partial \underline{u}^\varepsilon}{\partial t}, \underline{w}\right) = \int_\Omega f_i\ w_i\ dx \quad \forall \underline{w} \in \underline{H}^1_0(\Omega) \tag{5.1}$$

$$p^\varepsilon(x\ ,\ t) \equiv -\ \gamma\ \mathrm{div}\ \underline{u}^\varepsilon \qquad (\text{in } \Omega^{f\varepsilon}) \tag{5.2}$$

Where

$$a^{\varepsilon}(\underline{u}\ ,\ \underline{w}) = \int_{\Omega^{s\varepsilon}} a^{s}_{ijkh} \frac{\partial u_k}{\partial x_h} \frac{\partial w_i}{\partial x_j}\, dx + \int_{\Omega^{f\varepsilon}} \gamma \operatorname{div} \underline{u}^{\varepsilon} \operatorname{div} \underline{w}\, dx$$

$$b^{\varepsilon}(\underline{v}\ ,\ \underline{w}) = \int_{\Omega^{f\varepsilon}} [\ \eta \operatorname{div} \underline{v} \operatorname{div} \underline{w} + \mu\, e_{ij}(\underline{v})\, e_{ij}(\underline{w})\ ]\, dx$$

the coefficients $a^s$ are of course the elastic coefficients of the material, $\varepsilon^2\mu$ and $\varepsilon^2\eta$ the viscosity coefficients of the fluid and $\gamma$ is given by

$$\gamma = c^2\, \rho^f \tag{5.3}$$

where c, $\rho^f$ are the velocity of sound and the density of the fluid in the unperturbed state.

It is natural to assume that the appropriate asymptotic expansion in the solid part is analogous to that of elastic mixtures (chapter 6, sect. 2) ; on the other hand, the fluid will have a vibration analogous to that of the acoustics in a porous medium (chapt. 7, sect. 6) (as usual other types of expansion do not work !). We then try :

$$\left\{\begin{array}{l} \Omega^{s\varepsilon} : \underline{u}^{\varepsilon}(x\ ,\ t) = \underline{u}^{0}(x\ ,\ t) + \varepsilon\, \underline{u}^{1}(x\ ,\ y\ ,\ t) + \varepsilon^2 \ldots \\ \Omega^{f\varepsilon}\ ;\ u^{\varepsilon}(x\ ,t) = \underline{u}^{0}(x\ ,\ y\ ,\ t) + \varepsilon\, \underline{u}^{1}(x\ ,\ y\ ,\ t) + \ldots \end{array}\right\}\ y = \frac{x}{\varepsilon} \text{ in } \Omega \tag{5.4}$$

where $\underline{u}^0$ do not depend on y in the solid region but it does in the fluid region. If we consider a period Y (with x and t fixed) the displacement in $Y^s$ is a constant vector, whereas in the fluid it is not ; it is then useful to consider the relative displacement of the fluid with respect to the solid :

$$\underline{u}^{0}(x\ ,\ y\ ,\ t) = \underline{u}^{0}(x\ ,\ t) + \underline{u}^{rel}(x\ ,\ y\ ,\ t)$$

where $\underline{u}^{rel}$ is zero on $\Gamma$ ; moreover, it is useful to continue $\underline{u}^{rel}$ in $Y^s$ with value 0. We may write (5.4) under the form (which is suitable in $\Omega^{f\varepsilon}$ as well as in $\Omega^{s\varepsilon}$) :

(5.5) $\underline{u}^{\varepsilon}(x\ ,\ t) = \underline{u}^{0}(x\ ,\ t) + \underline{u}^{rel}(x\ ,\ y\ ,\ t) + \varepsilon\, \underline{u}^{1}(x\ ,\ y\ ,\ t) + \ldots$
where $y = x/\varepsilon$ ; all functions are Y-periodic in y. The vector $\underline{u}^{rel}$ for fixed x, t takes values in $H^1(Y)$ and is zero on $Y^s$.

$$p^{\varepsilon}(x\ ,\ t) = p^{0}(x\ ,\ t) + \varepsilon\, p^{1}(x\ ,\ y\ ,\ t) + \ldots \tag{5.6}$$

We replace (5.5), (5.6) into (5.2) and we have :

$$\operatorname{div}_y \underline{u}^{rel} = 0 \tag{5.7}$$

$$\frac{1}{\gamma}\, p^{0}(x\ ,\ t) + \operatorname{div}_x(\underline{u}^{0} + \underline{u}^{rel}) + \operatorname{div}_y \underline{u}^{1} = 0 \quad \text{in } \Omega^{f\varepsilon} \tag{5.8}$$

Now, we obtain from this a macroscopic equation which is in fact the macroscopic form of the conservation of mass. The standard way to do this is to integrate on Y by using

$$\int_Y \mathrm{div}_y \, \underline{u}^1 \, dy = \int_{\partial Y} \underline{u}^1 \, . \, \underline{n} \, dS = 0$$

which is zero by periodicity. In fact, (5.8) only hold in $Y^f$. We write the obvious identity in $Y^s$ :

(5.9) $$\mathrm{div}_y \, \underline{u}^1 = \mathrm{div}_y \, \underline{u}^1 \quad \text{in } Y^s$$

and we take the sum of the integrals of (5.8) over $Y^f$ and (5.9) over $Y^s$ ; by dividing by $|Y|$ we have

(5.10) $$\frac{\Pi}{\gamma} p^0 + \Pi \, \mathrm{div}_x \, \underline{u}^0 + \frac{1}{|Y|} \int_Y \mathrm{div}_x \, \underline{u}^{rel} \, dy = \frac{1}{|Y|} \int_{Y_s} \mathrm{div}_y \, \underline{u}^1 dy \quad \text{where}$$

(5.11) $$\Pi \equiv |Y^f| / |Y| \equiv \text{porosity}$$

Equation (5.10) will be written under another form later.

Now we consider, as usual, test functions depending on $\varepsilon$ in a form analogous to (5.5) :

(5.12) $$\underline{w}(x) = \underline{w}^0(x) + \underline{w}^{rel}(x \, , \, y) + \varepsilon \, \underline{w}^1(x \, , \, y) \; ; \quad \mathrm{div}_y \underline{w}^{rel} = 0$$

Then, (5.1), (5.2) at order $\varepsilon^0$ gives

(5.13) $$\begin{aligned} &\int_\Omega \rho^\varepsilon \frac{\partial^2 (u_i^0 + u_i^{rel})}{\partial t^2} (w_i^0 + w_i^{rel}) \, dx + \\ &+ \int_\Omega a^s_{ijkh} \left(\frac{\partial u_k^0}{\partial x_h} + \frac{\partial u_k^1}{\partial y_h}\right)\left(\frac{\partial w_i^0}{\partial x_j} + \frac{\partial w_i^1}{\partial y_j}\right) dx \; - \\ &- \int_{\Omega^{f\varepsilon}} p^0 \, [\, \mathrm{div}_x \underline{w}^0 + \mathrm{div}_x \underline{w}^{rel} + \mathrm{div}_y \underline{w}^1 \,] dx \; + \\ &+ \int_{\Omega^{f\varepsilon}} \mu \frac{\partial^2 \, u_i^{rel}}{\partial t \; \partial y_j} \frac{\partial w_i^{rel}}{\partial y_j} \, dy = \int_\Omega f_i (w_i^0 + w_i^{rel}) \, dx \end{aligned}$$

Now, we study the relative motion of the fluid. To this end, we take in (5.12) :

(5.14) $$\begin{cases} \underline{w}^0 = \underline{w}^1 = 0 \\ \underline{w}^{rel} = \theta(x) \, \underline{\omega}^{rel}\left(\frac{x}{\varepsilon}\right) \end{cases}$$

(5.15) | where $\theta \in \mathcal{D}(\Omega)$ and $\underline{\omega}^{rel}(y)$ is any divergence-free vector of $H^1(Y)$, Y-periodic and zero on $Y^s$.

In fact, it is useful to modify the corresponding p-term in (5.13) as in sect. 4. From (5.6), (5.14) we see that it is the $\varepsilon^0$ term of

$$-\int_{\Omega^{f\varepsilon}} p^\varepsilon \operatorname{div} w^{rel} \, dx = \int_\Omega \underline{\operatorname{grad}}\, p^\varepsilon \, . \, \underline{w}^{rel} \, dx$$

where we integrated by parts bearing in mind that $\underline{w}^{rel}$ is zero on $\Omega^{s\varepsilon}$. The corresponding $\varepsilon^0$-term is

$$\int_\Omega (\underline{\operatorname{grad}}_x\, p^0 + \underline{\operatorname{grad}}_y\, p^1) \, . \, \underline{w}^{rel} \, dx$$

Then, (5.13) with (5.14) gives

$$\int_{\Omega^{f\varepsilon}} \rho^f \frac{\partial^2 (u_i^0 + u_i^{rel})}{\partial t^2} \, \omega_i^{rel} \, \theta \, dx + \int_{\Omega^{f\varepsilon}} (\underline{\operatorname{grad}}\, p^0 + \underline{\operatorname{grad}}_y\, p^1) . \, \underline{\omega}^{rel} \, \theta \, dx +$$

$$+\int_{\Omega^{f\varepsilon}} \mu \frac{\partial^2 u_i^{rel}}{\partial t \, \partial y_j} \frac{\partial \omega_i^{rel}}{\partial y_j} \, \theta \, dx = \int_\Omega f_i \, \omega_i^{rel} \, \theta \, dx$$

and the classical $\varepsilon \to 0$ process gives (note that the term in $p^1$ is zero by periodicity and $\operatorname{div}_y \underline{\omega} = 0$)

$$(5.16) \quad \int_{Y^f} \rho^f \left(\frac{\partial^2 u_i^0}{\partial t^2} + \frac{\partial^2 u_i^{rel}}{\partial t^2}\right) \omega_i^{rel} \, dy + \int_{Y^f} \frac{\partial p^0}{\partial x_i} \, \omega_i^{rel} \, dx +$$

$$+ \mu \int_{Y^f} \frac{\partial^2 u_i^{rel}}{\partial t \, \partial y_j} \frac{\partial \omega_i^{rel}}{\partial y_j} \, dy = \int_{Y^f} f_i \, \omega_i^{rel} \, dy$$

which is an equation analogous to the Darcy's law in acoustics (chap. 7, (6.14)) with the supplementary term in $\partial^2 u^0 / \partial t^2$. This is natural because we deal with a relative motion and the inertia force is to be taken into account. As in chapter 7, (6.18) we have :

$$(5.17) \qquad \frac{\partial \tilde{u}_k^{rel}}{\partial t} = \int_0^t g_{ki}(t - S)\left(f_i - \frac{\partial p^0}{\partial x_i} - \rho^f \frac{\partial^2 u_i^0}{\partial t^2}\right)(S) \, dS$$

which is the Darcy's law.

Next, we study the local state in the solid. To this end, we take in (5.12) $\underline{w}^0 = \underline{w}^{rel} = 0$ ; $\underline{w}^1 = \theta(x) \, \underline{\omega}^1(x , y)$ were $\theta \in \mathcal{D}(\Omega)$ and $\underline{w}^1(y)$ is Y-periodic of class $H^1$. The asymptotic process $\varepsilon \to 0$ in (5.13) gives

$$(5.18) \quad \int_{Y^s} a_{ijkh}^s \left(\frac{\partial u_k^0}{\partial x_h} + \frac{\partial u_k^1}{\partial y_h}\right) \frac{\partial \omega_i^1}{\partial y_j} \, dy - p^o \int_{Y^f} \operatorname{div}_y \underline{\omega}^1 \, dy = 0$$

but by the Y-periodicity of $\underline{\omega}^1$

$$\int_Y \operatorname{div}_y \underline{\omega}^1 \, dy = 0$$

and consequently (5.18) becomes

$$(5.19) \int_{Y^s} a^s_{ijkh} \left( \frac{\partial u^0_k}{\partial x_h} + \frac{\partial u^1_k}{\partial y_h} \right) \frac{\partial \omega^1_i}{\partial y_j} \, dy + p^0 \int_{Y^s} \mathrm{div}_y \, \underline{\omega}^1 \, dy = 0$$

which is an equation on $Y^s$. In fact, if $\frac{\partial u^0_k}{\partial x_h}$ and $p^0$ (which are functions of x and t only) are considered as given constants, (5.19) is a well posed problem for $\underline{u}^1$ (which is a Y-periodic function of $H^1(Y^s)$). This problem recalls the local problem of homogenization in elasticity (chapter 6, sect. 2), but we have a supplementary given constant $p^0$, and the functions are only defined on $Y^s$ (not on Y). In fact this problem is studied in the same way (see for details Auriault - Sanchez [1]). $\underline{u}^1$ is determined (up to an additive constant vector) by $e_{ijx}(u^0)$ and $p^0$ (note that the compatibility conditions are satisfied because (5.19) take a value 0 if $\underline{\omega}^1$ is taken constant). In fact, we have :

Lemma 5.1 - If x and t are considered as parameters, $\underline{u}^1(y)$ is determined up to an additive constant vector by $e_{ijx}(\underline{u}^0)$ and $p^0$. If we define

$$(5.20) \quad \tilde{\sigma}^{so}_{ij} = \frac{1}{|Y|} \int_Y a^s_{ijkh} \, [ e_{khx}(\underline{u}^0) + e_{khy}(\underline{u}^1) ] \, dy$$

there exists the coefficients $a^h_{ijkh}$ ; $\beta_{ij}$ , $\beta$ such that :

$$(5.21) \qquad \tilde{\sigma}^{so}_{ij} = a^h_{ijkh} \, e_{khx}(\underline{u}^0) + \beta_{ij} \, p^0$$

$$(5.22) \qquad \frac{1}{|Y|} \int_{Y^s} \mathrm{div}_y \, \underline{u}^1 \, dy = \beta_{ij} \, e_{ijx}(\underline{u}^0) - \beta \, p^0$$

(Note that the coefficients $\beta_{ij}$ in (5.21) and (5.22) are the same).

Moreover, the coefficients $a^h_{ijkh}$ satisfy the classical symmetry and positivity conditions of elasticity, (1.3), (1.4).

Proof - We only give an outline of the proof, (in particular of the fact that the coefficients $\beta_{ij}$ are the same in (5.21) and (5.22) ; the properties of $a^h_{ijkh}$ are proved as in the case of elasticity).

In this proof, we change a little the notations

| | | |
|---|---|---|
| $e_{ijx}(\underline{u}^0)$ | will be denoted | $E_{ij}$ |
| $\underline{u}^1$ | " " " | $\underline{u}$ |
| $e_{ijy}(u^1)$ | " " " | $e_{ij}$ |
| $e_{ijx}(\underline{u}^0) + e_{ijy}(\underline{u}^1)$ | " " | $\hat{e}_{ij}$ |
| $p^0$ | " " " | P |

Then, by analogy with chapter 6, (2.25), we consider the following function W of $E_{ij}$ and P:

(5.23) $$W(E_{ij}\ ,\ P) = \frac{1}{2\,|Y|}\int_{Y^s} a^s_{ijkh}\ \hat{e}_{kh}\ \hat{e}_{ij}\ dy + \frac{P}{|Y|}\int_{Y^s} \operatorname{div} \underline{u}\ dy$$

where, as usual, it is understood that $\underline{u}$ is the corresponding solution of (5.19) and W is then a bilinear function of $E_{ij}$, P. By differenciating and using the symmetry properties of $a^s_{ijkh}$ we have :

(5.24) $$\delta W = \frac{1}{|Y|}\int_{Y^s} a^s_{ijkh}\ \hat{e}_{kh}\,\delta\,\hat{e}_{ij}\,dy + \frac{\delta P}{|Y|}\int_{Y^s} \operatorname{div} \underline{u}\ dy + \frac{P}{|Y|}\int_{Y^s} \operatorname{div} \delta\,\underline{u}\ dy$$

where

$$\delta\,\hat{e}_{ij} = \delta\,E_{ij} + \delta\,e_{ij}$$

but from (5.19) with $\underline{\omega}^1 = \delta\underline{u}$ we see that the term containing $\delta e_{ij}$ and $\delta\underline{u}$ cancel and (5.24) becomes :

$$\delta W = \tilde{\sigma}^{so}_{ij}\,\delta\,E_{ij} + \left(\frac{1}{|Y|}\int_{Y^s} \operatorname{div} \underline{u}\ dy\right)\delta\,P \Longrightarrow$$

(5.25) $$\tilde{\sigma}^{so}_{ij} = \frac{\partial W}{\partial E_{ij}} \quad ; \quad \frac{1}{|Y|}\int_{Y^s} \operatorname{div} \underline{u}\ dy = \frac{\partial W}{\partial P}$$

and we see that the coefficients $a^h$, $\beta_{ij}$, $\beta$ in (5.21) and (5.22) are :

$$a^h_{ijkh} = \frac{\partial^2 W}{\partial E_{ij}\,\partial E_{kh}} \quad ; \quad \beta_{ij} = \frac{\partial^2 W}{\partial P\,\partial E_{ij}} \quad ; \quad -\,\beta = \frac{\partial\,W}{\partial P^2}$$

in particular, the coefficients $\beta_{ij}$ in (5.21) and (5.22) are the same as crossed derivatives of W. Lemma 5.1 is proved.■

Moreover, we define $\tilde{\sigma}^T_{ij}$ as the mean value of the "total" stress tensor (equal to $\sigma^{so}_{ij}$ in $Y^s$ and to $-p^o\delta_{ij}$ in $Y^f$), i.e. :

(5.26) $$\tilde{\sigma}^T_{ij} = \frac{1}{|Y|}\int_{Y^s} a^s_{ijkh}\,[\,e_{khx}(\underline{u}^0) + e_{khy}(\underline{u}^1)\,]\,dy - \frac{p^o}{|Y|}\int_{Y^f} dy\ \delta_{ij}$$

we have :

(5.27) $$\boxed{\tilde{\sigma}^T_{ij} = a^h_{ijkh}\ e_{khx}(\underline{u}^0) - \alpha_{ij}\ p^o \quad ; \quad \alpha_{ij} \equiv \Pi\ \delta_{ij} - \beta_{ij}}$$

We are now able to rewrite (5.10) under a new form. By using (5.22), (5.27) and the fact that $\partial/\partial x$ commutes with $\sim$, it becomes

(5.28) $$\boxed{\left(\frac{\Pi}{\gamma} + \beta\right)p^o + \operatorname{div}_x \tilde{\underline{u}}^{rel} + \alpha_{ij}\ e_{ijx}(\underline{u}^0) = 0}$$

which is equivalent to the conservation of mass.

Finally, we obtain a new equation (in fact the conservation of momentum) by writing (5.13) with $\underline{w}^{rel} = \underline{w}^1 = 0$. The asymptotic process $\varepsilon \to 0$ gives ($\tilde{\sigma}^{so}_{ij}$ is defined by (5.20)) :

$$(5.29)\qquad \int_\Omega \tilde{\rho}^\varepsilon \left(\frac{\partial^2 u_i^o}{\partial t^2} + \frac{\partial^2 u_i^{rel}}{\partial t^2}\right) w_i^o \, dy + \int_\Omega \tilde{\sigma}_{ij}^{so} \frac{\partial w_i^o}{\partial x_j} dx - \int_\Omega \Pi \, p^o \, \mathrm{div}_x \, \underline{w}^o \, dx = \int_\Omega f_i \, w_i^o \, dx \qquad \forall w_i^o \in \underline{H}_o^1(\Omega)$$

By integrating by parts the terms containing derivatives of $\underline{w}^o$, we have

$$\int_\Omega \tilde{\sigma}_{ij}^{so} \frac{\partial w_i^o}{\partial x_j} dx - \int_\Omega \Pi \, p^o \, \mathrm{div} \, \underline{w}^o \, dx = - \int_\Omega \frac{\partial}{\partial x_j}(\tilde{\sigma}_{ij}^{so} - \Pi \, \delta_{ij} \, p^o) w_i^o \, dx$$

$$= [\text{by } (5.26)\,] = - \int_\Omega \frac{\partial \sigma_{ij}^T}{\partial x_j} w_i^o \, dx$$

and (5.29) gives ($\underline{w}^o$ is arbitrary) :

$$(5.30)\qquad \boxed{\tilde{\rho} \, \frac{\partial^2 u_i^o}{\partial t^2} + \rho^f \, \frac{\partial^2 \tilde{u}_i^{rel}}{\partial t^2} - \frac{\partial \sigma_{ij}^T}{\partial x_j} = f_i}$$

In conclusion, the homogenized motion of the mixture may be described by the displacement in the solid $\underline{u}^o(x\ ,\ t)$, the pressure in the fluid $p^o(x\ ,\ t)$ and the mean value of the relative displacement of the fluid with respect to the solid, $\tilde{\underline{u}}^{rel}$, which satisfy equations (5.17) (Darcy's law). (5.28) (conservation of mass) and (5.30) (conservation of momentum). The coefficients $a_{ijkh}^h$, $\beta_{ij}$, $\beta$ depend on the local structure (see lemma 5.1) and the "total" stress $\sigma_{ij}^T$ is defined by (5.27).

6.- Comments and bibliographical notes - As we said at the beginning of the chapter, there is a great variety of homogenization problems in this domain. This field is not yet sufficiently studied, and a great number of questions are open, in particular convergence of the homogenization processes, non linear motions and in certain cases, role of the connectness of the phases and of the orders of the different constitutive constants (such as density, viscosity, ...). For a more explicit study of the problem of sect. 1, see Sanchez-Hubert [ 2 ] . The problem of sect. 2 is studied in Levy [ 4 ] ; the proof of the convergence (sect. 3) follows the ideas of the Appendix of L. Tartar at the end of this volume. The suspension of rigid particles (sect. 4) is studied (including the case of elastic particles) in Fleury [ 2 ] , [ 3 ] . The problem of section 5 is taken from Levy [ 3 ] ; it deals with a mixture of an elastic solid and a slightly viscous fluid, in the case where both phases are connected (if the elastic phase is not connected, it cannot support a global stress and we have the very different study of Fleury [ 3 ] ). Other expansions hold if the process is slowly varying in time (see Auriault and Sanchez-Palencia [ 1 ] and if the ratio of the densities of the fluid and the solid is very small (see Levy [ 2 ] ).

# PART III

# MISCELLANEOUS PERTURBATION PROBLEMS

## CHAPTER 9

## EXAMPLES OF PERTURBATIONS FOR ELLIPTIC PROBLEMS

In this chapter we introduce certain kinds of perturbation problems and physical examples. We give the first properties of the solutions. A deeper study of these problems and in particular their spectral properties will be given in subsequent chapters.

1.- A class of singular perturbations - Let V and W be two (real or complex) Hilbert spaces satisfying

(1.1) $$V \subset W$$

with continuous and dense injection. Let us consider the (linear or sesquilinear) forms a(u , v) and b(u , v) which are continuous on V and W respectively, with

(1.2) $$\begin{cases} |a(u\,,\,v)| \leqslant M \|u\|_V \|v\|_V & \forall\, u,v \in V \\ |b(u\,,\,v)| \leqslant M \|u\|_W \|v\|_W & \forall\, u,v \in W \end{cases}$$

(1.3) $$\begin{cases} \mathrm{Re}\ a(v\,,\,v) \geqslant 0 \\ \mathrm{Re}\ a(v\,,\,v) + \|v\|_W^2 \geqslant \alpha \|v\|_V^2 \quad , \quad \alpha > 0 \quad ; \ \forall\, v \in V \end{cases}$$

(1.4) $$\mathrm{Re}\ b(v\,,\,v) \geqslant \beta \|v\|_W^2 \ ; \quad \beta > 0 \quad ; \quad \forall\, v \in W$$

Let $f^\varepsilon$, $f^0$ be (linear or antilinear) continuous forms on W (and hence also on V) such that

(1.5) $$f^\varepsilon \to f^0 \text{ in } W' \text{ weakly (resp. } W' \text{ strongly)}$$

If $\varepsilon$ is a real positive parameter, we consider the problems (where [ ] denotes the duality product between W' and W) :

<u>Problem ($\varepsilon$)</u> - Find $u^\varepsilon \in V$ such that

(1.6) $$\varepsilon\, a(u^\varepsilon , v) + b(u^\varepsilon , v) = [f^\varepsilon , v] \qquad \forall v \in V$$

<u>Problem (0)</u>.- Find $u^0 \in W$ such that

(1.7) $$b(u^0 , v) = [f^0 , v] \qquad \forall v \in W$$

It is clear that by virtue of the Lax-Milgram theorem, $u^\varepsilon$ and $u^0$ exists and are unique. Then, we have :

<u>Theorem 1.1</u>. - Under the hypothesis (1.1) - (1.5), as $\varepsilon \to 0$, we have

(1.8) $$u^\varepsilon \to u^0 \text{ in } W \text{ weakly (resp. } W \text{ strongly)}$$

<u>Proof</u> - By taking $v = u^\varepsilon$ in (1.6), we have for the real part :

$$\text{Re } [\varepsilon\, a(u^\varepsilon , u^\varepsilon) + b(u^\varepsilon , u^\varepsilon)] \leq |[f^\varepsilon , u^\varepsilon]|$$

and by using (1.3) and (1.4) :

(1.9) $$\varepsilon\, \alpha\, \|u^\varepsilon\|_V^2 + (\beta - \varepsilon)\, \|u^\varepsilon\|_W^2 \leq \|f^\varepsilon\|_{W'}\, \|u^\varepsilon\|_W$$

From (1.5) we see that $\|f^\varepsilon\|_{W'}$ is bounded ; then, by neglecting the term $\alpha$ in (1.9) we have :

(1.10) $$\|u^\varepsilon\|_W \leq C \qquad \text{(independent of } \varepsilon)$$

and then, from (1.9), we have

(1.11) $$\varepsilon \|u^\varepsilon\|_V^2 \leq C \qquad \text{(independent of } \varepsilon)$$

we then see that $u^\varepsilon$ belongs to a bounded set of W (i.e., a precompact set for the weak topology of W). We shall prove that any accumulation point of the set $u^\varepsilon$ is $u^0$. Let $u^*$ be such an accumulation point :

(1.12) $$u^\varepsilon \to u^* \text{ in } W \text{ weakly}$$

(in fact, we have (1.11) for a subsequence as $\varepsilon \to 0$). By fixing $v \in V$ in (1.6), we have by virtue of (1.12), (1.11) and (1.5), (1.3) :

$$|\varepsilon \, a(u^\varepsilon , v) | \leq \varepsilon \, C \, \|u^\varepsilon\|_V \leq C \, \varepsilon^{1/2} \to 0$$

$$b(u^\varepsilon , v) \to b(u^* , v)$$

$$[ f^\varepsilon , v ] \to [ f^0 , v ]$$

(the second is immediate because $b(u , v) \equiv (u , B v)_W$ for a $B \in \mathcal{L}(W , W)$) and then we have (1.7) with $u^*$ instead of $u^0$ for any $v \in V$. Then, because of the density of V in W, we may take a sequence belonging to V such that

$$v^i \to v \in W \quad \text{in W strongly}$$

and by passing to the limit we see that, $u^*$ satisfies Pb(0). Because of the uniqueness of the solution of this problem, we have $u^* = u^0$ and (1.12) shows that the theorem is proved in the case where (1.5) holds in the weak topology.

Moreover, if we have (1.5) in W strongly, taking into account (1.6) and (1.7) with $v = u^\varepsilon$ , we write the identity

$$\varepsilon \, a(u^\varepsilon , u^\varepsilon) + b(u^\varepsilon - u^0 , u^\varepsilon - u^0) = [ f^\varepsilon - f^0 , u^\varepsilon ] + b(u^\varepsilon - u^0 , -u^0)$$

which tends to zero by virtue of (1.5) and (1.12). Then, by bounding from below the real part of the left hand side we have :

$$\beta \, \|u^\varepsilon - u^0 \|_W^2 \to 0 \qquad \text{Q.E.D.} \blacksquare$$

It is clear that in general the solution $u^0$ of (1.6) does not belong to V, and thus the convergence in (1.8) cannot hold in V. In fact, as we shall see in the example, this impossibility is associated with the fact that in general $u^0$ does not satisfy the boundary conditions required for $u^\varepsilon$. In concrete problems, it is possible to modify $u^\varepsilon$ by <u>boundary layers</u> or <u>correctors</u> in order to obtain a convergence in V. We shall not study here such problems which can be seen in Lions [ 3 ].

<u>Instead of this, let us prove an easier result</u>. Let us consider only the real case, with a and b symmetric. Moreover, $f^\varepsilon = f^0$ and the last inequality (1.3) is in the form :

$$a(u , u) > \alpha \, \|v \|_V^2$$

Then, <u>if $u^0 \in V$, we have</u> :

(1.13) $$u^\varepsilon \to u^0 \quad \text{in V weakly}$$

It is clear that to say $u^0 \in V$ is equivalent to assuming that $f^0$ satisfies certain conditions. For certain $f^0$, the solution of Pb(0) will be in V.

Proof of (1.13) - We shall write (as a consequence of (1.6) and (1.7)) :

(1.14) $$\varepsilon\, a(u^\varepsilon , v) + b(u^\varepsilon , v) = b(u^0 , v) \qquad \forall\, v \in V$$

then, we identify W to its dual, and let A be the operator associated to the form a in the framework of the first representation theorem ; (1.14) writes

$$\varepsilon A\, u^\varepsilon + u^\varepsilon = u^0$$

which is an equation between elements of W, and $u^\varepsilon \in D(A)$. Multiplying it by $A\,u^\varepsilon$ in W and by taking into account that $u^0 \in V = D(A^{1/2})$, by virtue of the second representation theorem, we have

$$\varepsilon \, \|A\, u^\varepsilon\|_W^2 + a(u^\varepsilon , u^\varepsilon) = (u^0 , A\, u^\varepsilon)_W = (A^{1/2}\, u^0, A^{1/2}\, u^\varepsilon)_W = $$
$$= a(u^0 , u^\varepsilon) \Longrightarrow$$

$$\alpha \, \| u^\varepsilon\|_V^2 \leqslant | a(u^0 , u^\varepsilon)| \leqslant M \, \|u^0\|_V \, \|u^\varepsilon \|_V$$

$u^\varepsilon$ therefore remains in a bounded set of V (i.e., a precompact for the weak topology of V) and from (1.8) we deduce (1.13). ■

2.- Example.- Plate of small rigidity - Let $\Omega$ be a bounded domain of $R^2$. We consider the following boundary value problem :

(2.1) $$\varepsilon\, \Delta^2 u^\varepsilon - \Delta u^\varepsilon = F \quad \text{in } \Omega$$

(2.2) $$u^\varepsilon = 0 \quad \text{on } \partial\Omega$$

(2.3) $$\frac{\partial u^\varepsilon}{\partial n} = 0 \quad \text{on } \partial\Omega$$

where F is a given function defined on $\Omega$ (for example, $F \in L^2(\Omega)$). $\varepsilon$ is a small parameter that measures the rigidity of a plate in $\Omega$. The plate has a tension unity in its plane (the second term in (2.1)) ; moreover, it is clamped at its boundary (conditions (2.2) and (2.3)).

The limit problem, is as we shall see,

(2.4) $$- \Delta u^0 = F \quad \text{in } \Omega$$

(2.5) $$u^0 = 0 \quad \text{on } \partial\Omega$$

which is the analogous problem for a membrane.

This problem is in the framework of theorem 1.1 by taking :

$$W = H_0^1(\Omega) = \{ v \in H^1(\Omega) \ ; \ u|_{\partial\Omega} = 0 \}$$

$$V = H_0^2(\Omega) = \{ v \in H^2(\Omega) \ ; \ u|_{\partial\Omega} = \frac{\partial u}{\partial n}\Big|_{\partial\Omega} = 0 \}$$

equipped with the classical norms of $H^1$ and $H^2$.

$$b(u\ ,\ v) = \int_\Omega \frac{\partial u}{\partial x_i}\frac{\partial v}{\partial x_i}\,dx$$

$$a(u\ ,\ v) = \int_\Omega \left(\frac{\partial^2 u}{\partial x_1^2}\frac{\partial^2 v}{\partial x_1^2} + 2\,\frac{\partial^2 u}{\partial x_1\,\partial x_2}\,\frac{\partial^2 v}{\partial x_1\,\partial x_2} + \frac{\partial^2 u}{\partial x_2^2}\frac{\partial^2 v}{\partial x_2^2}\right) dx$$

The coerciveness condition (1.4) is classical (c.f. chap. 3, sect. 1) and (1.3) is then evident. Moreover, the functionals $f^\varepsilon = f^0$ are defined by

$$[f^\varepsilon\ ,\ v\,]_{W',\,W} = \int_\Omega F\,v\,dx$$

The theorem 1.1 then proves that

$$u^\varepsilon \to u^0 \text{ in } H^1(\Omega) \text{ strongly} \tag{2.6}$$

Moreover, if F is such that the solution $u^0$ of (2.4), (2.5) is in $H_0^2$ (i.e., if $\partial u^0/\partial n = 0$ on $\partial\Omega$ ), we have, as in (2.13)

$$u^\varepsilon \to u^0 \text{ in } H^2(\Omega) \qquad \text{weakly}$$

3.- <u>Convergence of the inverses in the norm</u> - We here prove a property which will be useful for the study of the spectral properties (cf. chap. 11, sect. 3 )

In the framework of theorem 1.1, let us introduce the Hilbert space H, identified to its dual and such that <u>$W \subset H$ densely and with compact injection</u>. We then have

$$V \subset W \subset H \equiv H' \subset W' \subset V' \tag{3.1}$$

Let $A_\varepsilon$ (resp. $A_0$) be the operators associated with the forms $\varepsilon a + b$ (resp. b) considered as operators in H, which are unbounded, maximal accretive (c.f. first representation theorem, chap. 2, theorem 5.2)

<u>Theorem 3.1.</u> - Under the hypothesis of theorem 1.1 and (3.1), if the injection of W in H is compact, we have

$$\| A_\varepsilon^{-1} - A_0^{-1} \|_{\mathcal{L}(H,W)} \to 0 \quad \text{and hence} \tag{3.2}$$

$$\| A_\varepsilon^{-1} - A_0^{-1} \|_{\mathcal{L}(H,H)} \to 0 \tag{3.3}$$

<u>Proof</u> - In the proof of theorem 1.1 with $f^\varepsilon = f^0 = f \in W'$, from (1.8) we have

$$\| u^\varepsilon \|_W \leq C \, \| f \|_{W'} \qquad \text{i.e.}$$

$$\| A_\varepsilon^{-1} \|_{\mathcal{L}(W',W)} \leq C \quad \text{and of course} \tag{3.4}$$

(3.5) $$\| A_o^{-1} \|_{\mathcal{L}(W',W)} \leqslant C$$

(3.2) amounts to prove that for any given $\delta$, there exists an $\varepsilon_o$ such that for $\varepsilon \leqslant \varepsilon_o$,

(3.6) $$\| (A_\varepsilon^{-1} - A_o^{-1}) f \|_W < \delta \quad \text{for} \quad \| f \|_H = 1$$

We shall prove this by contradiction. If it is not true, there exists a $\delta_1$ and a sequence

(3.7) $\varepsilon_i \to 0$ and an associated sequence $f_i$ with $\| f_i \|_H = 1$ such that

$$\| (A_{\varepsilon_i}^{-1} - A_o^{-1}) f_i \|_W > \delta_1 \qquad i = 1,2,\ldots$$

Then, we can extract a subsequence such that $f_i \to f^*$ in H weakly. The injection of W in H is compact and then that of H in W' is too (see chap. 2, (6.12)) and so

(3.8) $$f_i \to f^* \quad \text{in W' strongly}$$

By using (3.4) and (3.5), we have

$$\| (A_{\varepsilon_i}^{-1} - A_o^{-1}) f_i \|_W = \| A_{\varepsilon_i}^{-1}(f_i - f^*) + (A_\varepsilon^{-1} - A_o^{-1}) f^* + A_o^{-1}(f^* - f_i) \|_W \leqslant$$

$$\leqslant C \| f_i - f^* \|_{W'} + \| (A_\varepsilon^{-1} - A_o^{-1}) f^* \|_W$$

which tends to 0 by virtue of (3.8) and theorem 1.1. We then have a contradiction with (3.7). ■

Remark 3.1 - The example of sect. 2 is in the framework of theorem 3.1 if we choose $H = L^2(\Omega)$. ■

4.- The case where the limit problem is not coercive - We now consider a case where the limit problem is not coercive ; in the particular case W = V, it contains a type of "stiff problems" (c.f. Lions ) but the results for this "singular stiff perturbation" are much more restricted than those of ordinary stiff problems.

Let V and W be two real Hilbert spaces with

(4.1) $$V \subset W$$

with continuous and dense injection. Let a (resp. b) be a symmetric continuous form on V (resp. W) with :

$$(4.2) \quad \begin{cases} |a(u, v)| \leq M \|u\|_V \|v\|_V \\ |b(u, v)| \leq M \|u\|_W \|v\|_W \end{cases}$$

$$(4.3) \quad \begin{cases} a(v, v) \geq \alpha \|v\|_V^2 & ; \quad \alpha > 0, \quad \forall v \in V \\ b(v, v) \geq 0 & \forall v \in W \end{cases}$$

Let us define

$$(4.4) \qquad W^o = \{ v \in W \; ; \; b(v, v) = 0 \}$$

Because of (4.2), this is a closed subspace of W. Let $W^1$ be its orthogonal, i.e.

$$(4.5) \qquad W = W^o \oplus W^1 \qquad \text{and let}$$

$$(4.6) \qquad v = v_o + v_1$$

be the associated decomposition of the elements of W. Moreover, we suppose that

$$(4.7) \qquad b(v, v) + \| v_o \|_W^2 \geq \beta \|v\|_W^2 \quad ; \beta > 0 \quad , \forall v \in W$$

<u>Remark 4.1</u> - From the symmetry and bilinearity of b, we have

$$(4.8) \qquad b(u, v) = b(u_o + u_1, v_o + v_1) = b(u_1, v_1)$$

the other terms are zero because of (4.4) and

$$|b(u, v)| \leq b(u, u)^{1/2} \, b(v, v)^{1/2}$$

then, (4.7) may be interpreted as a "<u>coerciveness on $W^1$</u>" $b(v, v) \geq \beta \| v_1 \|_W^2$. ■

<u>Let $\tilde{W}$ be the space of the equivalence classes obtained from W by identifying all the elements of $W^o$</u>. This is a Hilbert space for the norm (c.f. Kato [2], p. 140)

$$(4.9) \qquad \|\tilde{v}\|_{\tilde{W}}^2 = \inf_{v \in \tilde{v}} \|v\|_W^2 = \inf_{v \in \tilde{v}} ( \|v_o\|_W^2 + \|v_1\|_W^2) = \|v_1\|_W^2 \text{ for any } v \in \tilde{v}$$

Moreover, let us define $f^\varepsilon$ , $f^o \in W'$ with

$$(4.10) \qquad [f^o, v] = [f^\varepsilon, v] = 0 \quad \text{for} \quad v \in W^o$$

This means that $f^\varepsilon$ and $f^o$ take the same values on any $v \in \tilde{v}$ and they are in fact linear functionals <u>on $\tilde{W}$</u> .

We now define the problems Pb.($\varepsilon$) and the limit problem Pb.(0) :

Problem (ε) : Find $u^\varepsilon \in V$ such that

(4.11) $$\varepsilon a(u^\varepsilon , v) + b(u^\varepsilon , v) = [f^\varepsilon , v]_{W'W} \qquad \forall v \in V$$

Problem (0) : Find $\tilde{u}^0 \in \tilde{W}$ such that

(4.12) $$b(\tilde{u}^0 , v) = [f^0 , v] \qquad \forall v \in W$$

Remark 4.2 - The expression $b(\tilde{u} , v)$ means $b(u , v)$ for any $u \in \tilde{u}$ ; note that, by virtue of (4.8), all these values are the same. In the same way, taking into account (4.10), we see that we may write $\tilde{v}$ (resp. $\tilde{W}$) instead of $v$ (resp. $W$) in (4.12). Moreover, from (4.2) and (4.7) we see that b is bilinear on $\tilde{W}$, and $f^0$ is continuous and linear on $\tilde{W}$. From (4.7) - (4.9) we have

$$b(\tilde{v} , \tilde{v}) = b(v_1 , v_1) \geqslant \beta \|\tilde{v}\|^2_{\tilde{W}}$$

and therefore the form b is coercive on $\tilde{W}$. ∎

By the Lax-Milgram theorem, $u^\varepsilon$ and $\tilde{u}^0$ are well determined elements of V and $\tilde{W}$.

We then have

Theorem 4.1 - Under the preceeding hypothesis, for each $u^\varepsilon \in V \subset W$ we consider the equivalence class $\tilde{u}^\varepsilon \in \tilde{W}$.

(4.13) If $f^\varepsilon \to f^0$ in W' weakly (resp. in W' strongly)

Then

(4.14) $\tilde{u}^\varepsilon \to \tilde{u}^0$ in $\tilde{W}$ weakly (resp. in $\tilde{W}$ strongly)

and

$$\varepsilon a(u^\varepsilon , u^\varepsilon) \leqslant C \qquad (\text{resp. } \to 0)$$

Proof - It is analogous to that of theorem 1.1. As in (1.9) we have

$$\varepsilon \alpha \|u^\varepsilon\|^2_V + \beta \|u_1^\varepsilon\|^2_W \leqslant C \|u_1^\varepsilon\|_W$$

and from (4.9)

(4.15) $$\begin{cases} \varepsilon \alpha \|u^\varepsilon\|^2 \leqslant C \\ \|\tilde{u}^\varepsilon\|_{\tilde{W}} \leqslant C \end{cases}$$

(4.16) $$\tilde{u}^\varepsilon \to \tilde{u}^* \quad \text{in } \tilde{W} \text{ weakly}$$

we then fix $v \in V$ in (4.11). We have

$$|\,\varepsilon\, a(u^\varepsilon , v)\,| \leqslant \varepsilon\, a(u^\varepsilon , u^\varepsilon)^{1/2}\, a(v , v)^{1/2} \to 0$$

$$b(u^\varepsilon , v) = b(u_1^\varepsilon , v_1) = b(\tilde{u}^\varepsilon , v) \to b(\tilde{u}^* , v)$$

$$[f^\varepsilon , v] \to [f^0 , v]$$

and we have (4.12) with $v \in V$, and we can pass to the limit for $v \in W$. Then, we see that the limit $\tilde{u}^*$ is $\tilde{u}^0$. The theorem is proved for (4.13) in the weak topology. Moreover, in the case of the strong convergence, by taking into account that

$$b(u , v) = b(u_1 , v_1) = b(\tilde{u} , \tilde{v}) \text{ we have}$$

$$\varepsilon a(u^\varepsilon , u^\varepsilon) + b(\tilde{u}^\varepsilon - \tilde{u} , \tilde{u}^\varepsilon - \tilde{u}) = [f^\varepsilon , u^\varepsilon] - [f^0 , \tilde{u}^\varepsilon] - b(\tilde{u}^\varepsilon - \tilde{u}, \tilde{u}) \to 0$$

and the left hand side is bounded from below by $\beta \| \tilde{u}^\varepsilon - \tilde{u} \|^2_{\tilde{W}}$ . The theorem is proved. ■

## 5.- Examples of the preceeding section -

Example 1. - Consider the bounded domains of $R^2$ as in the figure, and

$$\Omega = \Omega_1 \cup \Omega_0 \cup \Sigma \qquad ; \qquad \Gamma = \partial\Omega$$

We take

$$V = H_0^1(\Omega) \quad ; \quad W = L^2(\Omega)$$

$$a(u , v) = \int_\Omega \frac{\partial u}{\partial x_i} \frac{\partial v}{\partial x_i}\, dx \qquad ; \qquad b(u , v) = \int_{\Omega_1} u\, v\, dx$$

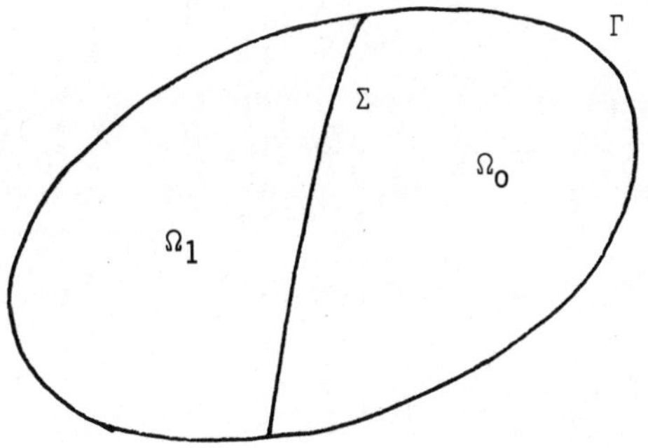

Then

$$W^0 = \{ v \in L^2(\Omega) \quad ; \quad v|_{\Omega_1} = 0 \}$$

$$W^1 = \{ v \in L^2(\Omega) \quad ; \quad v|_{\Omega_0} = 0 \}$$

Moreover, let $F^\varepsilon$ , $F^0$ be functions of $L^2(\Omega)$, zero on $\Omega_0$, such that

(5.1) $$F^\varepsilon \to F^0 \qquad L^2(\Omega_1) \text{ weakly (resp. strongly)}$$

the functionals $f^\varepsilon$ , $f^0$ are defined by

(5.2) $$[f , v] = \int_\Omega F v \, dx$$

The problems $Pb^\varepsilon$ are in terms of boundary value problem :

Problem ($\varepsilon$) $$\begin{cases} -\varepsilon \Delta u^\varepsilon + \chi_1 u^\varepsilon = F \\ u^\varepsilon|_{\partial\Omega} = 0 \end{cases} \quad \text{where } \chi_1(x) = \begin{matrix} 1 \text{ if } x \in \Omega_1 \\ 0 \text{ if } x \in \Omega_o \end{matrix}$$

The problem Problem(0) is equivalent to

$$u_1^0 = F \qquad \text{in } \Omega_1$$

$$u_0^0 \text{ arbitrary in } \Omega_0.$$

since theorem 4.1 holds and we have

$$u^\varepsilon|_{\Omega_1} \to F \quad \text{in } L^2(\Omega_1) \text{ weakly (resp. strongly)}$$

Example 2. - In the same domains as example 1, we consider a problem analogous to that of sect. 2 (plate of small rigidity) but with only the part $\Omega_1$ acting as a membrane in tension (in particular, we must apply on $\Sigma$ a force in the direction of the normal, of intensity 1 by unit length).

Then, we take :

$$V = H_0^2(\Omega)$$

$$W = \{ v \in L^2(\Omega) \quad ; \quad u|_{\Omega_1} \in H^1(\Omega_1) \quad ; \quad u|_{\Gamma_1} = 0 \quad ; \quad u|_{\Omega_0} \in L^2(\Omega_0) \}$$

where $\Gamma_1 = \Gamma \cap \overline{\Omega}_1$ (i.e. the boundary of $\Omega_1$ contained in $\Gamma$). W is a Hilbert space for the norm

$$\|u\|_W^2 = \int_{\Omega_1} \sum_i \left|\frac{\partial u}{\partial x_i}\right|^2 dx + \int_{\Omega_0} |u|^2 \, dx$$

Let

$$W^0 = \{ v \in W \quad ; \quad v|_{\Omega_1} = 0 \}$$

$$W^1 = \{ v \in W \quad ; \quad v|_{\Omega_o} = 0 \}$$

In fact, $W^0$ (resp. $W^1$) is the space of the functions of $L^2(\Omega_0)$ (resp. $H^1(\Omega_1)$ which are zero on $\Gamma_1$) extended by zero to $\Omega_1$ (resp. $\Omega_0$). We now take

$$a(u\ ,\ v) = \int_\Omega \Delta u\, \Delta v\ dx$$

$$b(u\ ,\ v) = \int_{\Omega_1} \frac{\partial u}{\partial x_i}\frac{\partial v}{\partial x_i}\ dx$$

and we define $F^\varepsilon$ , $F^0$ and $f^\varepsilon$ , $f^0$ as in (5.1), (5.2). Theorem 4.1 holds and we have

(5.3) $$u^\varepsilon|_{\Omega_1} \to u^0|_{\Omega_1} \text{ in } H^1(\Omega_1) \text{ weakly (resp. strongly)}$$

It is worthwile to express $u^\varepsilon$ and $u^0$ as solution of boundary value problems.

The limit problem amounts to finding $u_1^0 \in W^1$ such that

$$\int_{\Omega_1} \frac{\partial u}{\partial x_i}\frac{\partial v}{\partial x_i}\ dx = \int_{\Omega_1} F\ v\ dx \qquad \forall\, v \in W^1$$

or

$$\begin{cases} \Delta\, u_1^0 = F \\ u_1^0|_{\Gamma_1} = 0 \\ \dfrac{\partial u_1^0}{\partial n}\Big|_\Sigma = 0 \end{cases}$$

And the Problem ($\varepsilon$) amounts to finding $u^\varepsilon$ defined on $\Omega$ such that

$$\varepsilon\Delta^2 u^\varepsilon - \Delta\, u^\varepsilon = F \qquad \text{on } \Omega_1$$

$$\varepsilon\Delta^2 u^\varepsilon = 0 \qquad \text{on } \Omega_0$$

with the boundary conditions

$$u^\varepsilon|_{\partial\Omega} = \frac{\partial u^\varepsilon}{\partial n}|_{\partial\Omega} = 0 \qquad \text{on } \Gamma$$

and the transmission conditions :

$$u^\varepsilon|_1 = u^\varepsilon|_2 \quad ; \quad \frac{\partial u^\varepsilon}{\partial n}\Big|_1 = \frac{\partial u^\varepsilon}{\partial n}\Big|_2 \qquad \text{on } \Sigma$$

$$\Delta\, u^\varepsilon|_1 = \Delta\, u^\varepsilon|_2 \qquad \text{on } \Sigma$$

$$-\,\varepsilon\frac{\partial\Delta u^\varepsilon}{\partial n}\Big|_1 + \varepsilon\frac{\partial\Delta u^\varepsilon}{\partial n}\Big|_2 + \frac{\partial u}{\partial n} = 0 \qquad \text{on } \Sigma\ .$$

6.- A problem of perturbation of the boundary conditions - We first study an abstract situation and later shall give an example on boundary value problems.

Let V and H be two Hilbert spaces, let H be identified to its dual

(6.1) $$V \subset H \subset V'$$

with continuous and dense imbedding. Let $V_\varepsilon$ be a sequence ($\varepsilon \to 0$) of closed subspaces of V, equipped with the topology induced by V. Let us suppose that $V_\varepsilon$ tends to V in the sense that

(6.2) $$\forall\, w \in V \Rightarrow \inf_{v \in V_\varepsilon} \| v - w \|_V = \eta(\varepsilon, w) \xrightarrow[\varepsilon \to 0]{} 0$$

Of course the $V_\varepsilon$ are not dense in V (otherwise, $V_\varepsilon = V$) but we shall suppose that $V_\varepsilon$ is dense in H (with the topology of H !) and we then have

(6.3) $$V_\varepsilon \subset H \subset V'_\varepsilon$$

Moreover, let $a(u , v)$ be a sesquilinear continuous form on V (and hence also on $V_\varepsilon$) coercive in the sense

(6.4) $$\begin{cases} |a(u , v)| \leqslant M \|u\|_V \|v\|_V \\ \operatorname{Re} a(v , v) \geqslant \alpha \|v\|_V^2 \quad ; \quad \alpha > 0 \ , \ v \in V \end{cases}$$

Then, for $f \in H$, let $u^\varepsilon$ (resp. $u^0$) be the solution (which exists and is unique by the Lax-Milgram theorem) of

Problem ($\varepsilon$) (resp. Problem (0)) : Find $u^\varepsilon \in V^\varepsilon$ (resp. $u^0 \in V$) such that for any $v \in V^\varepsilon$ (resp. $\in V$) we have :

(6.5) $$a(u^\varepsilon , v) = (f , v)_H \quad (\text{resp. } a(u^0 , v) = \ldots)$$

Moreover, let $A_\varepsilon \in \mathcal{L}(V_\varepsilon , V'_\varepsilon)$ (resp. $A_0 \in \mathcal{L}(V , V')$) the operators associated with the form $a(u , v)$ defined on $V^\varepsilon$ (resp. V). As usual, they are considered as unbounded maximal accretive operators on H. We then have :

Theorem 6.1 - Under the hypothesis (6.1) - (6.5),

(6.6) $$u^\varepsilon \to u^0 \quad \text{in } V \text{ strongly}$$

Moreover, if the embedding $V \subset H$ is compact,

(6.7) $$\| A_\varepsilon^{-1} - A_0^{-1} \|_{\mathcal{L}(H,H)} \longrightarrow 0$$

<u>Proof</u> - We write the classical a priori estimate by taking $v = u^\varepsilon$ in (6.5).

$$u^\varepsilon \in V^\varepsilon \subset V$$

$$\alpha \| u^\varepsilon\|_V^2 \leqslant |\operatorname{Re} a(u^\varepsilon , u^\varepsilon) | \leqslant C \|f\|_H \|u^\varepsilon\|_V$$

then, $u^\varepsilon$ remains in a bounded set of V and (after extraction of a subsequence, which is the original sequence because the limit is unique, as we shall show) :

(6.8) $$u^\varepsilon \to u^* \quad \text{in } V \text{ weakly}$$

Moreover, for any $v \in V$, by virtue of (6.2), we can take a sequence

(6.9) $$V_\varepsilon \ni v_\varepsilon \to v \qquad \text{in } V \text{ strongly}$$

and by passing to the limit in (6.5) with $v = v_\varepsilon$, we see that

$$a(u^* , v) = (f , v) \qquad \forall v \in V$$

(6.10) and so $u^* = u$ and we have proved (6.6) in V weakly. Moreover, we construct (by virtue of (6.2)) a sequence

(6.11) $$V^\varepsilon \ni v_\varepsilon \to u \quad \text{in } V \text{ strongly}$$

We then write :

$$a(u^\varepsilon , u^\varepsilon) = (f , u^\varepsilon)_H$$

$$a(-u , u^\varepsilon) = - (f , u^\varepsilon)_H$$

$$a(-u , -u) = (f , u)_H$$

$$a(u^\varepsilon , -u) = -a(u^\varepsilon , v^\varepsilon) + a(u^\varepsilon , v^\varepsilon - u) = - (f , v^\varepsilon)_H + a(u^\varepsilon , v^\varepsilon - u)$$

and thus

$$a(u^\varepsilon - u , u^\varepsilon - u) = (f , u - v^\varepsilon)_H + a(u^\varepsilon , v^\varepsilon - u)$$

which tends to zero by virtue of (6.8) and (6.11) ; Then, the coerciveness condition (6.4) gives (6.6).

The proof of (6.7) is analogous to that of th. 3.1. If (6.7) is not true,

(6.12) $$\left| \begin{array}{l} \exists\, \delta_1 , \varepsilon_i \to 0, f_i \text{ with } \| f_i \|_H = 1 \text{ such that} \\ \\ \| (A_{\varepsilon_i}^{-1} - A_0^{-1}) f_i \|_H > \delta_1 \end{array} \right.$$

Then, we can select a susequence such that $f_i$ converges in H weakly and hence in V' strongly (see chap. 3, (6.12)).

(6.13) $$f_i \to f^* \quad \text{in } V' \text{ strongly}$$

Now, let us recall that a bounded functional on V is a bounded functional on $V_\varepsilon$ , i.e. :

$$V' \subset V'_\varepsilon$$

(the embedding is not dense). Moreover, from (6.4) and (6.5) with $v = u^\varepsilon$ we have

$$\alpha \, \|u^\varepsilon\|_V^2 \leq \|f\|_{V'} \, \|u^\varepsilon\|_V$$

and $A_\varepsilon^{-1}$ remains bounded in the norm of $\mathcal{L}(V', V)$ and then

(6.14) $$\|A_\varepsilon^{-1}\|_{\mathcal{L}(V',H)} \leq C$$

as well as $\| A_o^{-1} \|$ . We then have :

$$\|(A_{\varepsilon_i}^{-1} - A_o^{-1}) f_i\|_H = \|A_{\varepsilon_i}^{-1}(f_i - f^*) + (A_{\varepsilon_i}^{-1} - A_o^{-1}) f^* + A_o^{-1}(f^* - f_i)\|_H \leq$$

$$\leq C \, \|f_i - f^*\|_{V'} + \|(A_{\varepsilon_i}^{-1} - A_o^{-1}) f^*\|_H$$

The first term tends to zero by (6.13). The second term also, because it is clear from the first part of the proof that (6.8) and (6.10) hold for $f^* \in V'$ and the weak convergence in V implies the strong convergence in H.

We then have a contradiction with (6.12) and the theorem is proved. ∎

<u>Example</u> .- Let $\Omega$ be a bounded domain of $R^2$ with smooth boundary and $\Gamma_\varepsilon$ (resp. $\Gamma$) be connected arcs of the boundary $\partial\Omega$ such that

(6.15) $$\Gamma_\varepsilon \subset \Gamma$$

and $\Gamma_\varepsilon$ tends to $\Gamma$ in the sense of the convergence of the extremities.

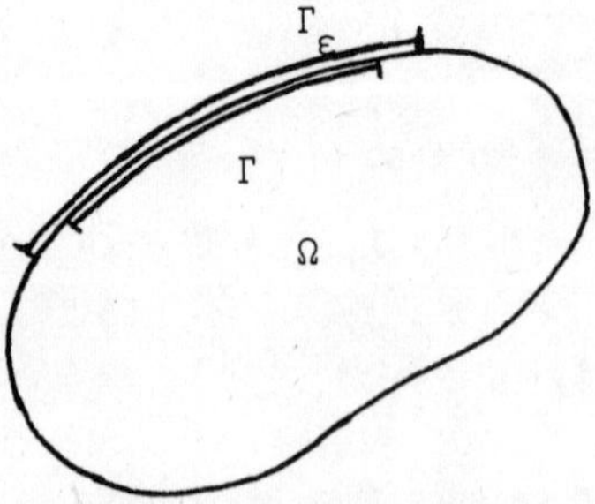

Now we consider the problems

$$(6.16)\qquad \begin{cases} -\Delta u^{\varepsilon} = f \\ u^{\varepsilon}\big|_{\Gamma_{\varepsilon}} = 0 \\ \dfrac{\partial u^{\varepsilon}}{\partial n}\Big|_{\partial\Omega-\Gamma_{\varepsilon}} = 0 \end{cases}$$

and the analogous one with $\Gamma$.

These problems are particular cases of that of chap. 3, sect. 3. We define $H = L^2(\Omega)$

$$V^{\varepsilon} = \{v \in H^1(\Omega) \;;\; v|_{\Gamma_{\varepsilon}} = 0\} \;;$$

which are subspaces of V defined as $V^{\varepsilon}$ with $\Gamma$ instead of $\Gamma_{\varepsilon}$. The hypothesis (6.2) is a consequence of the fact (well known in the theory of Sobolev spaces) that $V^{\varepsilon}$ (and V) is the adherence in $H^1(\Omega)$ of the set of the functions which are zero in a neighbourhood of $\Gamma_{\varepsilon}$.

The problem (6.16) takes the form (6.8) by defining

$$a(u, v) = \int_{\Omega} \frac{\partial u}{\partial x_i} \frac{\partial v}{\partial x_i}\, dx$$

7.- Bibliographical notes - In this chapter, we only introduced some examples of perturbation problems in the framework of functional analysis. Sect. 1 and 2 are based on Huet [1]. A deeper study of singular perturbation, including boundary layers may be seen in Lions [3], Visik and Lyusternik [1], Trenogin [1]. See also Huet [2] and Stummel [1], [2], as well as Van Dyke [1] for a formal treatment of singular perturbations in fluid dynamics). The spectral properties of these problems will be studied in chapter 11, and are based on theorem 3.1 of the present chapter. Sect. 4 and 5 deal with a class of problems intermediate between the singular perturbations and the stiff problems (see chapter 13). Theorem 4.1 seems new. Sect. 6 deals with a kind of perturbation of the boundary conditions, and is based on Nečas [1], chap. 3, sect. 6. where the reader may find many other problems.

Problems analogous to that of this chapter in the parabolic (resp. hyperbolic) case are studied in chap. 10 sect. 2 (resp. sect. 5, remark 5.1). Spectral properties are studied in chapter 11, sect. 3.

# CHAPTER 10

## THE TROTTER-KATO THEOREM AND RELATED TOPICS

This chapter deals with the convergence (with respect to the parameter $\varepsilon$) of the solutions of the "stationary" equations

$$(\lambda - A^{\varepsilon})\, u^{\varepsilon} = f \quad ; \qquad \lambda > 0$$

and the solutions of the "evolution" equations

$$\left.\begin{aligned} \frac{\partial v^{\varepsilon}}{\partial t} &= A^{\varepsilon}\, v^{\varepsilon} \\ v^{\varepsilon}(0) &= w \end{aligned}\right\} \Longleftrightarrow \qquad v^{\varepsilon} = e^{tA^{\varepsilon}}\, w \quad ; \quad t > 0$$

and with their relationships.

1.- The Trotter-Kato theorem - This theorem was stated in chap. 4, sect. 3, and was used in several problems of homogenization. We shall prove it here.

Theorem 1.1.- Let $A_n$, $A$ be generators of contraction semigroups in the Banach space $B$. Then

a) If

(1.1) $$(\lambda - A_n)^{-1} v \xrightarrow[n\to\infty]{} (\lambda - A)^{-1} v \quad \text{in } B \text{ strongly} \quad \forall v \in B$$

for some $\lambda$ with $\operatorname{Re} \lambda > 0$, then

(1.2) $$e^{tA_n}\, v \xrightarrow[n\to\infty]{} e^{tA} v \qquad \text{in } B \text{ strongly} \quad \forall v \in B$$

for any $t > 0$.

b) If (1.2) holds for all $t \geq 0$, then (1.1) holds for all $\lambda$ with $\operatorname{Re} \lambda > 0$.

Proof of a) - For any $v \in B$, $(\lambda - A)^{-1} v \in D(A)$ and by taking it as initial value, the corresponding solution of the initial value problem, $e^{sA}(\lambda - A)^{-1} v$ is differentiable. Here $s$ is for time. Moreover,

$$\frac{d}{ds}\, e^{sA}\, (\lambda - A)^{-1} v = e^{sA}\, A(\lambda - A)^{-1}\, v =$$

$$\equiv e^{sA}\, [-I + \lambda(\lambda - A)^{-1}]\, v$$

We also have an analogous formula with $A_n$, $t-s$ instead of $A$, $s$ .

Then, we have

$$\frac{d}{ds}[e^{(t-s)A_n}(\lambda - A_n)^{-1} e^{sA}(\lambda - A)^{-1}v] =$$

$$= e^{(t-s)A_n} e^{sA}(\lambda - A)^{-1} v - e^{(t-s)A_n}(\lambda - A_n)^{-1} e^{sA}v =$$

$$= e^{(t-s)A_n}[(\lambda - A)^{-1} - (\lambda - A_n)^{-1}] e^{sA} v$$

because some terms cancel. We then integrate with respect to s from 0 to t and we bear in mind that the semigroup commutes with its generator :

(1.3) $$(\lambda - A_n)^{-1}(e^{tA} - e^{tA_n})(\lambda - A)^{-1}v =$$

$$= \int_0^t e^{(t-s)A_n}[(\lambda - A)^{-1} - (\lambda - A_n)^{-1}] e^{sA} v\, ds$$

We wish to pass to the limit $n \to \infty$ in this relation. The semigroups dealt with are of contraction and then the integrand is bounded above in norm by

$$\|(\lambda - A)^{-1}v\| + \|(\lambda - A_n)^{-1} v\|$$

which is bounded above in turn by $2(\mathrm{Re}\lambda)^{-1}\|v\|$ as is easily seen from proposition 6.3 of chapter 4 about Laplace transform. Moreover, for fixed s, the integrand is bounded above by

$$\|[(\lambda - A)^{-1} - (\lambda - A_n)^{-1}]e^{sA} v\|$$

which tends to zero by (1.1). Then, we pass to the limit by dominated convergence and we see that the left hand side of (1.3) tends to zero in B-strongly for any t and $v \in B$. But for $w \in D(A)$, there exists $v \in B$ such that

$$w = (\lambda - A)^{-1} v$$

and then we have

(1.4) $(\lambda - A_n)^{-1}(e^{tA} - e^{tA_n})w \to 0$ in B strongly, $t \geq 0$ for any $w \in D(A)$

Moreover, the operator in (1.4) is bounded above in norm by $2(\mathrm{Re}\,\lambda)^{-1}$, and then

(1.5) (1.4) is true for any $v \in B$

On the other hand, by the contraction property and (1.1) we have

(1.6) $$e^{tA_n}(\lambda - A)^{-1} w - (\lambda - A_n)^{-1} e^{tA_n} w =$$

$$= e^{tA_n}[(\lambda - A)^{-1} - (\lambda - A_n)^{-1}] w \to 0$$

in B strongly $t \geq 0$, $w \in B$.

In the same way

(1.7) $$(\lambda - A_n)^{-1} e^{tA} w - e^{tA}(\lambda - A)^{-1} w =$$

$$= [(\lambda - A_n)^{-1} - (\lambda - A)^{-1}] e^{tA} w \to 0$$

in B strongly, $t \geq 0$, $w \in B$.

We then add (1.4), (1.6), (1.7) to get

$(e^{tA_n} - e^{tA})(\lambda - A)^{-1} w \to 0$ in B strongly, $w \in B$, $t \geq 0$., but, as before, for any $u \in D(A)$, we can take w such that $(\lambda - A)^{-1} w = u$. Thus,

(1.8) $$(e^{tA_n} - e^{tA}) u \to 0 \quad \text{in B strongly, } u \in D(A),\ t \geq 0.$$

But the norm of the operator in (1.8) is $\leq 2$ and (1.8) holds for $u \in B$ Q.E.D. ■

Proof of b) - By prop. 6.3 of chap. 4 about Laplace transform, we have

(1.9) $$(\lambda - A_n)^{-1} v = \int_0^\infty e^{A_n t} e^{-\lambda t} v \, dt \qquad \text{Re}\,\lambda > 0 \ , \ v \in B$$

and an analogous relation for A instead of $A_n$. We can pass to the limit in the right hand side of (1.9) because for fixed t the integrand converges by the hypothesis ; moreover, the integrand is bounded above by

$$e^{-(\text{Re}\lambda)} \|v\|$$

which is independent of n. Then, by dominated convergence the right hand side (and then the left hand side of (1.9) converges and we have (1.1).■

Then, the properties of convergence of the resolvent of the generator for $\text{Re}\,\lambda > 0$ lead to the convergence of the semigroup. It is then very easy to obtain properties of the convergence of the solutions of initial value problems when the free term as well as the initial values changes at the same time that the operator does. We then have

Theorem 1.2.- Let $A_n$, A be generators of contraction semigroups in the Banach space B such that (1.1) holds for a certain $\lambda$ with $\text{Re}\lambda > 0$. Moreover, let $f_n^{(t)}$, $f^{(t)}$ be continuous functions of t with values in B. If

(1.10) $$f_n(t) \to f(t) \qquad \text{in B strongly}$$

uniformly on compact intervals of time and

(1.11) $$v_n \to v \qquad \text{in B strongly}$$

then, the solution $u_n(t)$ of

$$\frac{du_n(t)}{dt} = A_n\, u_n(t) + f_n(t) \quad ; \quad u_n(0) = v_n$$

converges to the solution u(t) of the analogous problem without index n, in B strongly for any $t \geq 0$.

<u>Proof</u> - By chap. 4, (3.2), we have

$$u_n(t) = e^{tA_n}\, v_n + \int_0^t e^{(t-s)A_n}\, f_n(s)\, ds \tag{1.12}$$

and an analogous relation without n. We shall see that for any $t > 0$ the two terms on the right hand side of (1.12) converge in B strongly

$$\| e^{tA_n}\, v_n - e^{tA}\, v \| \leq \| e^{tA_n}\, (v_n - v) \| + \| (e^{tA_n} - e^{tA}) v \|$$

The first term on the right hand side converges because of (1.11) and of the contraction hypothesis. The second term converges as a consequence of theorem 1.

On the other hand

$$\| \int_0^t [ e^{(t-s)A_n}\, f_n(s) - e^{(t-s)A}\, f(s) ]\, ds \| \leq$$

$$\leq \int_0^t [ \| e^{(t-s)A_n}\, (f_n(s) - f(s)) \| + \| (e^{(t-s)A_n} - e^{(t-s)A})\, f(s) \| ]\, dS$$

The first term on the right hand side is bounded above (by virtue of the contraction hypothesis) by

$$\int_0^t \| f_n(S) - f(S) \|\, dS$$

and it converges by virtue of (1.10). As for the second term, it converges to zero for fixed s by theorem 1.1 ; moreover, the integrand is bounded above by $2 \| f(S) \|$ which is integrable from 0 to t, and we pass to the limit by dominated convergence.∎

2.- <u>Examples : singular perturbations for nonstationary problems. Application to acoustics</u> - The preceeding theorems 1.1 and 1.2 apply to the singular perturbations of the class studied in chap. 9 sect. 1, 2, 3. Moreover, with the notations of sect. 3 with

$$W \subset H \tag{2.1}$$

with continuous (not necessarily compact) embedding, we have

$A_\varepsilon^{-1}\, f \to A_0^{-1} f$ in H strongly for any $f \in H$. It is clear from the proof that, if we take

$$b(u , v) + \lambda(u , v)_H$$

for positive $\lambda$ instead of $b(u , v)$, the results hold, and we then have

$(\lambda + A_\varepsilon)^{-1}f \to (\lambda + A_0)^{-1}f$ in H strongly for any $f \in H$. On the other hand, $-A_\varepsilon$, $-A_0$ are generators of contraction semigroups in H (see chap. 4, sect. 4); we may apply the Trotter-Kato theorem and we have :

$$e^{-tA_\varepsilon} v \to e^{-tA_0} v \quad \text{in H strongly for any } v \in H,\ t \geq 0.$$

In the same way we have results for nonhomogeneous equations from theorem 1.2.

The same thing is obtained for the problem of perturbation of boundary conditions of chap. 9 section 6.

For the spectral properties of these perturbations, see chap. 11, sect. 3.

Let us now consider the problem of the acoustic vibration of slightly viscous air in a rigid vessel. We shall only give some indications. For a deeper treatment, see Geymonat - Sanchez [ 1 ] .

Let $\Omega$ be a bounded domain of $R^3$ with smooth boundary. After a suitable process the problem may be stated in non dimensional form. If $\underline{u} = (u_1, u_2, u_3)$ and p are for the velocity components and the pressure, the problem is

$$(2.2) \quad \begin{cases} \dfrac{\partial u_i}{\partial t} + \dfrac{\partial p}{\partial x_i} - \varepsilon \dfrac{\partial \overset{V}{\sigma}_{ij}(\underline{u})}{\partial x_j} = 0 \\ \dfrac{\partial p}{\partial t} + \operatorname{div} \underline{u} = 0 \end{cases}$$

with the boundary conditions

$$(2.3) \qquad u_{|\partial\Omega} = 0 \qquad \text{if } \varepsilon > 0$$

$$(2.4) \qquad \underline{u} \cdot \underline{n}_{|\partial\Omega} = 0 \qquad \text{if } \varepsilon = 0$$

and no boundary conditions for p. Here

$$(2.5) \qquad \overset{V}{\sigma}_{ij}(\underline{u}) \equiv \lambda\, \delta_{ij} \operatorname{div} \underline{u} + 2\,\mu\, e_{ij}(\underline{u})$$

$$(2.6) \qquad e_{ij}(\underline{u}) \equiv \frac{1}{2}\left(\frac{\partial u_i}{\partial x_j} + \frac{\partial u_j}{\partial x_i}\right)$$

where $\varepsilon\lambda$, $\varepsilon\mu$ are the (small) viscosity coefficients, such that there exists $C > 0$ with

$$(2.7) \qquad \int_\Omega \overset{V}{\sigma}_{ij}(u) \frac{\partial \bar{u}_i}{\partial x_j}\, dx \geq C \sum_{ij} \int_\Omega |e_{ij}(\underline{u})|^2\, dx$$

We then introduce as unknown the vector

$$u \equiv (\underline{u} , p) \equiv (u_1 , u_2 , u_3 , p) ; \qquad u_4 = p$$

and we write (2.2) under the form

$$\frac{\partial u}{\partial T} + \mathcal{A}^\varepsilon u = 0 \quad . \tag{2.8}$$

We then have ($\mathcal{A}^0$ is the analogous operator for $\varepsilon = 0$)

Theorem 2.1.- With an appropriate definition see Geymonat-Sanchez [1 ] ) of the boundary contions (2.3), (2.4), the operators $-\mathcal{A}^\varepsilon$ and $-\mathcal{A}^0$ are generators of contraction semigroups in $(L^2(\Omega))^4$. Moreover, the Trotter-Kato theorem holds and we have

$$e^{-\mathcal{A}^\varepsilon t} f \to e^{-\mathcal{A}^0 t} f \qquad (L^2)^4$$

for any $f \in (L^2)^4$, $t \geqslant 0$.

The proof is made via the Lumer-Phillips theorem and the Trotter-Kato theorem. The kay of the proof is the study of solvability of the equation

$$\zeta u^\varepsilon + \mathcal{A}^\varepsilon u^\varepsilon = f \tag{2.9}$$

for $f \in (L^2)^4$ , $\zeta$ real $> 0$ and the study of the convergence $\varepsilon \to 0$. By eliminating the pressure, (2.9) is equivalent to the equation

$$\zeta^2 u_i^\varepsilon - \frac{\partial}{\partial x_i} \operatorname{div} \underline{u}_\varepsilon - \varepsilon\zeta \frac{\partial \overset{V}{\sigma}_{ij}}{\partial x_j} (\underline{u}^\varepsilon) = \zeta f_i - \frac{\partial f_4}{\partial x_i}$$

which is associated to the variational formulation:

$$\zeta^2(\underline{u}^\varepsilon , \underline{v}) + (1 + \varepsilon \zeta \lambda) a^0(\underline{u}^\varepsilon , \underline{v}) + \varepsilon \zeta 2\mu a^1 (\underline{u}^\varepsilon, \underline{v}) = \tag{2.10}$$

$$= \zeta(\underline{f} , \underline{v}) + (f_4 , \operatorname{div} \underline{v} )$$

where

$$(\underline{u} , \underline{v}) = \sum_i^3 (u_i , v_i)_{L^2}$$

$$a^0(\underline{u} , \underline{v}) = (\operatorname{div} \underline{u} , \operatorname{div} \underline{v})_{L^2}$$

$$a^1(\underline{u} , \underline{v}) = \int_\Omega \overset{V}{\sigma}_{ij}(\underline{u}) \frac{\partial \bar{v}_i}{\partial x_j} dx$$

The existence is then obtained by the Lax-Milgram theorem : note that for $\varepsilon > 0$, the form $a^1$ is coercive on $(H^1(\Omega))^3$, but for $\varepsilon = 0$, the form $a^0$ is only coercive on the space of functions with divergence in $L^2(\Omega)$. This is the reason why the boundary conditions (2.3) and (2.4) depend on $\varepsilon$. (Physically, an inviscid fluid can slip on the wall, but a viscous one cannot). A slight modification of the proof of the singular perturbation of chap. 9 th. 1.1 shows that

$$u^{\varepsilon} \underset{\varepsilon \to 0}{\longrightarrow} u^{0}$$

in (2.10), and the Trotter-Kato theorem holds.

3.- Another theorem on convergence of semigroups - In applications, we often deal with sequences of operators whose corresponding sequence of resolvents converges in a weak sense, or only when it is applied to certain elements of the space. It is then useful to obtain results about the convergence of the semigroup.

In this direction we have the following result :

Theorem 3.1.- Let $A_n$, $A$ be generators of contraction semigroups in the Hilbert space $H$. If $v \in H$ is fixed, the necessary and sufficient condition for

(3.1) $$e^{tA_n} v \to e^{tA} v \text{ in } L^{\infty}(0,\infty;H) \text{ weakly } *$$

is

(3.2) $$(\lambda - A_n)^{-1} v \to (\lambda - A)^{-1} v \quad \text{in } H \text{ weakly, for all real positive } \lambda$$

(moreover, (3.2) holds for any $\lambda$ with $\mathrm{Re}\,\lambda > 0$ if it holds for a certain value $\lambda_1$)

Proof - As in part b of the proof of th. 1.1, we have

(3.3) $$(\lambda - A_n)^{-1} v = \int_0^{\infty} e^{-\lambda t} \, e^{-tA_n} v \, dt \qquad \mathrm{Re}\,\lambda > 0$$

and an analogous formula without index n. Then, if (3.1) is satisfied, by taking $e^{-\lambda t} w$ as test function with $\mathrm{Re}\,\lambda > 0$, $w \in H$, the right hand side of (3.3) and then the left hand side converge to the analogous integral without index n, and we have (3.2).

Conversely, let us suppose that (3.2) holds. Because of the contraction hypothesis,

$$\| e^{tA_n} v \| \leq \| v \| \qquad \forall \; n,t$$

and then, after extraction of a subsequence, we have

(3.4) $$e^{tA_n} v \rightharpoonup \chi(t) \qquad \text{in } L^{\infty}(0,\infty;H) \text{ weakly } *$$

and by taking $e^{-\zeta t} w$ as test function for any $\zeta$ with $\mathrm{Re}\,\zeta > 0$, we have

(3.5) $$\int_0^{\infty} e^{-\zeta t} e^{tA_n} v \, dt \to \int_0^{\infty} e^{-\zeta t} \chi(t) \, dt \quad \text{in } H \text{ weakly}$$

but the left hand side of (3.5) is $(\zeta - A_n)^{-1} v$ and by (3.2) we have

$$(\lambda - A)^{-1} v = \int_0^{\infty} e^{-\lambda t} \chi(t) \, dt \qquad \lambda > 0$$

By comparing it with the relation analogous to (3.3) for A, we see that this means that the Laplace transforms of $e^{-tA} v$ and $\chi(t)$ (which are well defined) coincide for real $\lambda$ . But they are holomorphic functions in the half plane $\mathrm{Re}\lambda > 0$, then they coincide in the half plane. Moreover, the inverse Laplace transforms coincide, and we have

(3.6) $$e^{-tA}\, v = \chi(t)$$

and (3.4) then becomes the desired relation (3.1) (note that (3.6) holds for any extracted subsequence).■

Remark 3.1 - In the framework of theorem 3.1 it is easy to consider the initial value problem with variable initial values. But the problem with variable free term $f_n$ seems more difficult. As a result, if we have (3.1) for $v \in H$ fixed and

$$v_n \to v \qquad \text{in } H \text{ strongly}$$

$u_n$, $u$ are the solutions of

$$\frac{du_n}{dt} = A_n u_n \quad ; \quad u_n(0) = v_n$$

and the analogous problem without index n, we have

(3.7) $$u_n \to u \quad \text{in } L^\infty(0,T;H) \text{ weakly } * \text{ for any } T > 0.$$

For,

$$u_n(t) = e^{tA}\, v_n \quad \text{and}$$

$$u_n(t) - u(t) = e^{tA_n}(v_n - v) + (e^{tA_n} - e^{tA})\, v$$

the second term of the right hand side converges by theorem 3.1. As for the first, it is bounded below in norm by $\| v_n - v \|_H$ for any t and we have (3.7).■

4.- A case where the configuration space depends on ε - Let us consider two Hilbert spaces

(4.1) $$E^1 \subset E^0$$

algebraically and topologically, the embedding being dense. We shall note $( \, , \, )_0$, $( \, , \, )_1$ the corresponding scalar products. Moreover, for $\varepsilon \in ]\, 0, \varepsilon_0 ]$, we shall denote $E^\varepsilon$ the space $E^1$ equipped with a new scalar product that depends on $\varepsilon$ and such that the associated norms are equivalent ($\varepsilon$ fixed) to the $E^1$ norm. Moreover

(4.2) $$\| \;\|_{\varepsilon_1} \geq \| \;\|_{\varepsilon_2} \qquad \text{if } \varepsilon_1 \geq \varepsilon_2$$

(4.3) $$\| \;\|_{\varepsilon} \geq \| \;\|_0 \qquad \varepsilon > 0 \quad .$$

Let $\mathcal{A}^\varepsilon$ $\mathcal{A}^0$ be generators of contraction semigroups on $E^\varepsilon$, E (Note that $\mathcal{A}^\varepsilon$ is generator of a semigroup on $E^1$, but it is not necessarily of contraction). Finaly let $v \in E^1$ be a fixed element of $E_1$. We then have

Theorem 4.1 - Under the preceeding hypothesis, we have

a) If

(4.4) $$((\lambda - \mathcal{A}^{\varepsilon})^{-1} v, w)_0 \to ((\lambda - \mathcal{A}^{0})^{-1} v, w)_0$$

$$\forall w \in E^0 , \quad \text{real } \lambda > 0$$

Then

(4.5) $$e^{t\mathcal{A}^{\varepsilon}} v \to e^{t\mathcal{A}^{0}} v \text{ in } L^{\infty}(0,\infty;E^0) \text{ weakly}*.$$

b) Conversely (4.4) follows from (4.5). In addition, (4.4) holds for any complex $\lambda$ with $\mathrm{Re}\,\lambda > 0$.

Note that $(\lambda - \mathcal{A}^{\varepsilon})^{-1} v$ is an element of $E^{\varepsilon}$ and then of $E^0$ ; in the same way, $e^{t\mathcal{A}^{\varepsilon}} v$ is an element of $L^{\infty}(0,\infty;E^{\varepsilon})$ and then of $L^{\infty}(0,\infty;E^0)$, and (4.4), (4.5) make sense.

Proof of a) - From the contraction hypothesis and (4.2), (4.3)

(4.6) $$\| e^{t\mathcal{A}^{\varepsilon}} v \|_0 \leqslant \| e^{t\mathcal{A}^{\varepsilon}} v \|_{\varepsilon} \leqslant \| v \|_{\varepsilon} \leqslant \| v \|_{\varepsilon_0} = C$$

independent on $\varepsilon$ and $t$ : by extraction of a subsequence, we have :

(4.7) $$e^{t\mathcal{A}^{\varepsilon}} v \to \chi(t) \text{ in } L^{\infty}(0,\infty;E^0) \text{ weakly} *$$

Then, by taking $e^{-\zeta t} w$ with $\mathrm{Re}\,\zeta > 0$, $w \in E^0$ as test function, we see that

(4.8) $$\int_0^{\infty} e^{t\mathcal{A}^{\varepsilon}} v\, e^{-\zeta t}\, dt \to \int_0^{\infty} \chi(t)\, e^{-\zeta t}\, dt \text{ in } E^0 \text{ weakly, for } \mathrm{Re}\,\zeta > 0$$

but the left hand side is $(\zeta - \mathcal{A}^{\varepsilon})^{-1} v$ and by (4.4), for real $\zeta$ it converges to $(\zeta - \mathcal{A}^{0})^{-1} v$ which is the Laplace transform of $e^{t\mathcal{A}^{0}} v$. Then, the Laplace transforms of $e^{t\mathcal{A}^{0}} v$ and $\chi(t)$ coincide for $\zeta$ real and thus everywhere (by analytic continuation). Then, the inverse Laplace transforms coincide :

(4.8) $$e^{t\mathcal{A}^{0}} = \chi(t)$$

and (4.7) becomes (4.5). ■

Proof of b) - It is immediate. From (4.5), by taking as test function $e^{-\lambda t} w$, $\mathrm{Re}\,\lambda > 0$, $w \in E^0$ we have :

$$\int_0^{\infty} e^{t\mathcal{A}^{\varepsilon}} v\, e^{-\lambda t}\, dt \to \int_0^{\infty} e^{t\mathcal{A}^{0}} v\, e^{-\lambda t}\, dt \quad \text{in } E^0 \text{ weakly}$$

which is (4.4) with $\mathrm{Re}\,\lambda > 0$ by virtue of the formula for the Laplace transform of the semigroups. ■

5.- Application to a problem of singular perturbation in viscoelasticity -
Let $\Omega$ be a bounded open set of $R^2$ with smooth boundary. We consider the equation

(5.1) $$\frac{\partial^2 u}{\partial t^2} = c_1 \Delta u - \varepsilon c_2 \Delta^2 u - \varepsilon c_3 \Delta^2 \frac{\partial u}{\partial t} : \quad c_i > 0$$

with the boundary conditions

(5.2) $$u_{|\partial\Omega} = 0 \qquad \text{if } \varepsilon = 0$$

(5.3) $$u_{|\partial\Omega} = \frac{\partial u}{\partial n}\Big|_{\partial\Omega} = 0 \qquad \text{if } \varepsilon > 0$$

and the initial values of $u(0)$ and $\frac{\partial u}{\partial t}(0)$ given. Here $\Delta$ denotes the Laplacian in the space variables, and $t$ is time.

This problem for $\varepsilon = 0$ is the classical vibrating membrane problem. For $\varepsilon > 0$, the term $c_2$ is a rigidity term ; we then have a plate of small rigidity in tension, as in chapter 9 sect. 2. Moreover, the term in $c_3$ is a small dissipative term at the level of rigidity. This may happen if the material is viscoelastic.

For $\varepsilon = 0$ we have a particular case of the problem studied in chapter 4, sect. 5, remark 5.1. Let us study it for $\varepsilon > 0$. We shall see that it is associated to a semigroup in an appropriate space $E^\varepsilon$ and we shall apply the theorem of the preceeding section to prove the convergence of the semigroup as $\varepsilon \to 0$.

Let us define the spaces

(5.4) $$H = L^2(\Omega) \quad ; \quad V_0 = H_0^1(\Omega)) \quad ; \quad V_1 = H_0^2(\Omega)$$

and the forms

$$a_0(u , v) = \int_\Omega \frac{\partial u}{\partial x_i} \frac{\partial \bar{v}}{\partial x_i} dx$$

$$a_1(u , v) = \int_\Omega \Delta u \, \Delta \bar{v} \, dx$$

$$a_\varepsilon = c_1 a_0 + \varepsilon c_2 a_1$$

The space $V_0$ is equipped with the scalar product $C_1 a_0$. Moreover, we shall denote $V_\varepsilon$ the space $V_1$ equipped with the scalar product $a_\varepsilon$ . Now, for fixed $\varepsilon > 0$,

$$V_\varepsilon \subset V_0 \subset H \equiv H' \subset V_0' \subset V_\varepsilon'$$

with dense and continuous embeddings. Let

$$A_0 \in \mathcal{L}(V_0 , V_0'), \qquad A_1 \in \mathcal{L}(V_1 , V_1')$$
$$A_\varepsilon \in \mathcal{L}(V_\varepsilon , V_\varepsilon')$$

be the operators associated to the forms $c_1 a_0$, $a_1$, $a_\varepsilon$ respectively. They shall also be considered as selfadjoint unbounded operators in H. We of course have

$$A_0 = - C_1\Delta \quad ; \quad A_1 = \Delta^2 \quad ; \quad A_\varepsilon = - C_1\Delta + \varepsilon C_2 \Delta^2$$

Let us consider the configuration space

(5.5) $$E_\varepsilon = V_\varepsilon \times H$$

If we denote, as usually in second order equations in time :

(5.6) $$u_1 = u \quad ; \quad u_2 = \frac{\partial u}{\partial t} ,$$

equation (5.1) is equivalent to the equation

(5.7) $$\frac{d\underline{u}}{dt} = \mathcal{A}^\varepsilon \underline{u}$$

(5.8) with $$\underline{u} = \begin{pmatrix} u_1 \\ u_2 \end{pmatrix}$$

(5.9) $$\mathcal{A}^\varepsilon = \begin{pmatrix} 0 & I \\ -A_\varepsilon & - \varepsilon C_3 A_1 \end{pmatrix}$$

To define in an exact form $\mathcal{A}^\varepsilon$ as an unbounded operator in $E_\varepsilon$, we define it on the domain :

(5.10) $$D(\mathcal{A}^\varepsilon) = \{(u_1,u_2) \in V_\varepsilon \times H \ ; \ u_2 \in V_\varepsilon \ , - A_\varepsilon u_1 - \varepsilon C_3 A_1 u_2 \in H \}$$

Theorem 5.1 - The operator $\mathcal{A}^\varepsilon$ is generator of a contraction semigroup in $E_\varepsilon$.

Proof - We shall apply the Lumer-Phillips theorem. We must prove that $\mathcal{A}^\varepsilon$ is a densely defined dissipative operator and that

(5.11) $$R(\lambda - \mathcal{A}^\varepsilon) = E_\varepsilon \quad \text{for real} \quad \lambda > 0 .$$

Firstly, $D(\mathcal{A})$ contains $\mathcal{D}(\Omega) \times \mathcal{D}(\Omega)$ and then, it is dense in $V_\varepsilon \times H$.

As for the dissipativity, we have

$$\mathrm{Re}\,(\mathcal{A}^\varepsilon \underline{u}, \underline{u})_{E_\varepsilon} = \mathrm{Re}\,[a_\varepsilon(u_2 , u_1) + (-A_\varepsilon u_1 - \varepsilon C_3 A_1 u_2 , u_2)_H ] =$$

$$= \mathrm{Re}\,[a_\varepsilon(u_2 , u_1) - a_\varepsilon(u_1 , u_2) - \varepsilon C_3 a_1(u_2 , u_2) ] = - \varepsilon C_3 a_1(u_2 , u_2) \leq 0$$

and the operator is dissipative.

To prove (5.11) we must prove that, for any given $\underline{f} \in E_\varepsilon$, there exists $\underline{u} \in D(\mathcal{A}^\varepsilon)$ such that

$$\lambda \underline{u} - \mathcal{A}^{\varepsilon}\, \underline{u} = \underline{f}$$

for real $\lambda > 0$, that is to say

$$(5.12)\qquad \begin{cases} \lambda\, u_1 - u_2 = f_1 \in V_\varepsilon \\ \lambda\, u_2 + A_\varepsilon\, u_1 + \varepsilon\, C_3\, A_1\, u_2 = f_2 \in H \end{cases}$$

and by eliminating $u_2$, we have

$$(5.13)\qquad (A_\varepsilon + \lambda^2 + \varepsilon\, \lambda\, c_3\, A_1)u_1 = f_2 + \lambda\, f_1 + \varepsilon\, C_3\, A_1\, f_1$$

the right hand side is an element of $V'_\varepsilon$ and the left hand side is an operator associated to a form which is hermitian and coercive on $V_\varepsilon$. By the Lax-Milgram theorem, there exists $u_1 \in V_\varepsilon$ solution of (5.13) and then $u_2 \in H$ exists such that (5.12) is satisfied. Moreover, $(u_1\, ,\, u_2) \in D(\mathcal{A}^\varepsilon)$

For $\varepsilon = 0$, as in chap. 4, remark 5.1, the operator

$$(5.14)\qquad \mathcal{A}^0 = \begin{pmatrix} 0 & I \\ -A_o & 0 \end{pmatrix}$$

is the generator of a contraction semigroup (in fact a unitary group) on $E_o = V_o \times H$.

<u>Theorem 5.2</u> - For any $\underline{f} \in E_\varepsilon$ we have

$$e^{t\mathcal{A}^\varepsilon}\, \underline{f} \to e^{t\mathcal{A}^{0}}\, \underline{f} \quad \text{in } L^\infty(0,\infty; E^0) \quad \text{weakly} *$$

<u>Proof</u> - We are in the situation of th.4.1; and it suffices to prove that

$$(5.15)\qquad (\lambda - \mathcal{A}^\varepsilon)^{-1}\, \underline{f} \longrightarrow (\lambda - \mathcal{A}^0)^{-1}\, \underline{f} \quad \text{in } E^0 \text{ weakly}$$

for real $\lambda > 0$. As in (5.12), this means that, for given $f_1 \in V_1$, $f_2 \in H$, the solution $u_1^\varepsilon$, $u_2^\varepsilon$ of

$$\begin{cases} \lambda\, u_1^\varepsilon - u_2^\varepsilon = f_1 \\ \lambda\, u_2^\varepsilon + A_\varepsilon\, u_1^\varepsilon + \varepsilon\, C_3\, A_1\, u_2^\varepsilon = f_2 \end{cases}$$

converges in $V_o \times H$ weakly to the solution $u_1^o$ , $u_2^o$ of

$$\begin{cases} \lambda\, u_1^o - u_2^o = f_1 \\ \lambda\, u_2^o + A_o\, u_1^o = f_2 \end{cases}$$

By eliminating $u_2^\varepsilon$ and $u_2^o$, this amounts to prove that $u_1^\varepsilon \to u_1^o$ in $V_o$ weakly

(5.16) $(\lambda^2 + A_0)u_1^\varepsilon + \varepsilon(C_2 + \lambda C_3)A_1 u_1^\varepsilon = f_2 + \lambda f_1 + \varepsilon C_3 A_1 f_1$

(5.17) $(\lambda^2 + A_0)u_1^0 = f_2 + \lambda f_1$

This proof is merely an adaptation of that of chap. 9 th. 1.1. The associated variational formulations are obtained by multiplying (5.16) (resp. (5.17)) by $v \in V_1$ (resp. $V_0$). By multiplying (5.16) by $u_1^\varepsilon$ we obtain

$$\| u_1^\varepsilon \|_H^2 + \varepsilon \| u_1^\varepsilon \|_{V_1}^2 \leqslant C( \| u_1^\varepsilon \|_H + \varepsilon \| u_1^\varepsilon \|_{V_1})$$

and then

$$\| u_1^\varepsilon \|_H \leqslant C'$$

$$\sqrt{\varepsilon} \; \| u_1^\varepsilon \|_{V_1} \leqslant C'$$

and we finish as in chap. 9 th. 1.1.

Some spectral properties of this perturbation problem will be studied in chapter 11 sect. 5.

Remark 5.1 - All the results of this section hold for the particular case $C_3 = 0$ . We then have results for the hyperbolic problems associated with the singular perturbations of chap. 9 sect. 1, 2, 3. ∎

6.- Conclusions and comments - The Trotter-Kato theorem allows to obtain convergence properties for the solutions of evolution equations from results for the associated steady-state equations. In particular, the results of chapiter 9 are generalized to the corresponding parabolic (resp. hyperbolic) problems in sect. 2 (resp. remark 5.1). A good reference for these questions is Kato [2] .

The application to acoustics is taken from Geymonat and Sanchez-Palencia [1], where the essential spectrum of the problem is also studied. See also in this connection Krein [1] .

Theorem 3.1 is in the framework of C. Bardos, D. Brézis and H. Brézis [1] . An analogous theorem for semigroups in configuration spaces depending on a parameter is given in sect. 4 ; this theorem is then applied to singular perturbations of hyperbolic problems (see remark 5.1) and to problems which are parabolic for $\varepsilon > 0$ and hyperbolic for $\varepsilon = 0$ (theorem 5.2). Spectral properties of this perturbations are given in chapter 11, sect. 5. See also Kato [3] for the convergence of semigroups.

# CHAPTER 11

## SPECTRAL PERTURBATION. CASE OF ISOLATED EIGENVALUES

The aim of spectral perturbation theory is the study of the variations of the spectral properties of an operator when it depends on a parameter. Its physical motivation is the study of variation of eigenvalues and eigenvectors of vibrating systems under perturbation.

In this chapter we give a resumé of the classical results of Rellich and Kato. Proofs are often omited but the reader may find them in Kato's book [ 2 ]. Applications to homogenization and singular perturbations are given in sect. 3. Sect. 5 is devoted to "implicite eigenvalue problems" which are introduced through an example in viscoelasticity. Sect. 6 deals with spectral properties of the homogenization of a boundary.

1.- Resolvent, spectrum and separation - We consider here closed operators A in a Banach space B. The resolvent domain $\rho(A)$ is the set (possibly empty) of complex numbers z such that A - z has a bounded inverse (defined on the whole space B) $(A - z)^{-1}$. The resolvent of A is

$$R(z) = (A - z)^{-1} \tag{1.1}$$

considered as a function of z defined on $\rho(A)$ with values in $\mathcal{L}(B , B)$. Of course, the domain of R(z) is B and the image is D(A) (the domain of A). An operator commutes with its resolvent.

The resolvent satisfies for any $z_1, z_2 \in \rho$ the "resolvent equation"

$$\boxed{R(z_1) - R(z_2)} = R(z_1)(A - z_2) R(z_2) - R(z_1)(A - z_1) R(z_2) = \boxed{= (z_1 - z_2) R(z_1) R(z_2)} \tag{1.2}$$

This shows that if R(z) is compact in a point $z = z_2$, it is compact for any $z_1 \in \rho$. This is the case for many operators in mathematical physics as we shall see in the examples.

Moreover, the resolvent is a holomorphic uniform function of z on $\rho$, with values in $\mathcal{L}(B , B)$. In fact, it is not hard to prove that R(z) posseses the expansion

$$(1.3) \qquad R(z) = [ I - (z - z_0) R(z_0) ]^{-1} R(z_0) = \sum_0^{\infty} (z - z_0) R(z_0)^{n+1}$$

which is absolutely convergent for

$$|z - z_0| < \| R(z_0) \|^{-1}$$

This proves, in particular, that $\rho(A)$ is an open set. In the general case, $\rho(A)$ may have several connected components and R(z) is "piecewise holomorphic". Nevertheless for most of the operators appearing in applications $\rho(A)$ is connected.

The complementary set $\sigma(A)$ of $\rho(A)$ in the complex plane is called the spectrum of A. If z is a point of the spectrum, A - z is either not invertible or is invertible but its range is not the whole space B. In the finite-dimensional case, A is a matrix, and $\sigma(A)$ is the set of the points such that A - z is not invertible ; these points are the roots of a polynomial, and $\sigma$ is a finite set of points. We shall see that certain operators (for instance the operators with compact inverse) in Banach spaces have a spectrum formed only by an infinite sequence of points (discrete spectrum) and the structure is almost the same as that of a matrix. This chapter deals mostly with the dependence of such spectra on a parameter.

It is well known that any matrix may be represented in a special basis in the canonical form of Jordan. In particular, for each eigenvalue, there exists a subspace such that the operator transforms it into itself, and the operator restricted to this subspace has a unique eigenvalue. An analogous situation often appears in the infinite-dimensional case.

To explain this, let us recall that a linear operator P in B is called a projection if

$$PP = P$$

and B has the decomposition

$$B = M \oplus N \qquad ; \qquad M = PB \quad , \quad N = (I - P)B$$

and any $x \in B$ may be expressed in a unique way as $x = x' + x''$ ; $x' \in M$, $x'' \in N$. It is then said that P is the projection on M along N. In a Hilbert space, the projection is said to be orthogonal if M and N are orthogonal subspaces. This amounts to saying that P is selfadjoint.

If P and Q are projections in a Banach space, and $\| P - Q \| < 1$, then PB and QB are isomorphic (of the same dimension, if they are of finite dimension). As an example, if P(t) is a projector which depends continuously in the norm on a parameter t defined in a connected region, the subspaces P(t)B for different t are isomorphic, and the dimension (if finite) is constant. We can now study the isolated singularities of the spectrum.

Let $z_0$ be an isolated point of $\sigma(A)$. The resolvent R(z) is then a holomorphic uniform function in a neighbourhood of $z_0$ and it has a Laurent expansion

(1.4) $$R(z) = \sum_{-\infty}^{+\infty} \zeta^n A_n \quad ; \qquad \zeta = z - z_0$$

where the coefficients are (operators of $\mathcal{L}(B , B)$)

(1.5) $$A_n = \frac{1}{2\pi i} \int_\Gamma \zeta^{-n-1} R(\zeta)\, d\zeta$$

with $\Gamma$ a positively-oriented simple curve enclosing $z_0$. In order to obtain relations between the different coefficients, we remark that the curve $\Gamma$ can be modified in a slightly larger curve $\Gamma'$, and we have, by the resolvent equation (1.2) :

$$A_n A_m = \left(\frac{1}{2\pi i}\right)^2 \int_{\Gamma'} \int_\Gamma \zeta^{-n-1} \zeta'^{-m-1} R(\zeta) \; R(\zeta')\, d\zeta\, d\zeta' =$$
$$= \left(\frac{1}{2\pi i}\right)^2 \int_{\Gamma'} \int_\Gamma \zeta^{-n-1} \; \zeta'^{-m-1} (\zeta' - \zeta)^{-1} [R(\zeta') - R(\zeta)]\, d\zeta\, d\zeta'$$

Then, an easy calculation gives

(1.6) $$A_n A_m = (\eta_n + \eta_m - 1)\, A_{n+m-1}$$

where

$$\eta_n = 1 \;\text{ for }\; n \geqslant 0 \quad , \quad \eta_n = 0 \;\text{ for }\; n < 0$$

Then we have $A_{-1} A_{-1} = - A_{-1}$ and $-A_1 = P$ is a projector. Moreover, by writing $S = A_0$ (resp. $D = -A_{-2}$) we have $A_1 = S^2$, $A_2 = S^3$... (resp. $A_{-3} = - D^2$, $A_{-4} = - D^3$ ...) and the Laurent series becomes :

(1.7) $$R(z) = [-(z - z_0)^{-1} P - \sum_1^\infty (z - z_0)^{-n-1} D^n] + [\sum_0^\infty (z - z_0)^n S^{n+1}]$$

Let us consider the decomposition of the space $B = M \oplus N$

where $M = PB$ , $N = (I - P)B$ ; by taking

$n = - 1$ , $m = - 2$ (resp. $n = -1$, $m = 0$) we obtain

$$PD = DP = D \quad (\text{resp. } PS = SP = 0) \; .$$

It is then evident that D (resp. S) transforms any element of the subspace PB (resp. (1 - P)B) in an element of PB (resp. (I - P)B).

The two terms on the right of (1.7) are the singular and regular parts of $R(z)$ in the vicinity of $z_0$. It is clear from the above that P commutes with $R(z)$ (and hence with A). It is then natural to consider the decomposition

$$B = PB \oplus (I - P)B \tag{1.8}$$

and the singular (resp. regular) parts of $R(z)$ operates from PB to PB (resp. from $(I-P)B$ to $(I-P)B$. Moreover, from the classical properties of the Laurent series, we see that the singular part is convergent for any $z \neq z_0$ and hence, $z_0$ is the only singularity of the resolvent of the operator restricted to PB. If PB is of finite dimension, P is the projection on the Jordan bloc corresponding to the eigenvalue $z_0$. It is also easily seen that if PB is of finite dimension, $z_0$ is an eigenvalue and the principal part of (1.7) is finite, i.e. the singularity is a pole. The dimension of PB is called the algebraic multiplicity of the eigenvalue $z_0$. The geometric multiplicity is the dimension of the subspace spanned by the eigenvectors corresponding to $z_0$ ; it is $\leqslant$ algebraic multiplicity.

In fact, a separation of the spectrum is possible in more general situations, if there exists a finite, closed curve $\Gamma$ enclosing a part of the spectrum (not necessarily an isolated point). Then, if the projection P is defined (as before) by

$$P = \frac{-1}{2\pi i} \int_{\Gamma} (A - z)^{-1}\, dz$$

P and I - P commute with A and we obtain a decomposition of the space B as before.

Some properties of the resolvent useful in applications are the following :

Proposition 1.1. - If $A^*$ is the adjoint of A, $\sigma(A^*)$ and $\rho(A^*)$ are the mirror images of $\sigma(A)$ and $\rho(A)$ with respect to the real axis, and

$$R(z\,,\,A^*) = R(\bar{z}\,,\,A)^* \qquad , \qquad \bar{z} \in \rho(A).$$

Proposition 1.2. - If A is a compact operator, $\sigma(A)$ is a countable set with no accumulation point different from zero. Each non-zero $z \in \sigma(A)$ is an eigenvalue of A with finite multiplicity.

Proposition 1.3. - If the complex plane is extended to the sphere by adjoining to it the point of infinity, we shall define the extended spectrum $\tilde{\sigma}(A)$ as $\sigma(A)$ if A is bounded and as $\sigma(A)$ plus the point of infinity if A is unbounded. Then, if $R(z_0) = (A - z_0)^{-1}$ is considered as an operator, its spectrum is the bounded set obtained from $\tilde{\sigma}(A)$ by the transformation

$$z \to \zeta \qquad ; \qquad \zeta = (z - z_0)^{-1}.$$

Proposition 1.4. - We already said that if $R(z)$ is compact in a point $z_0$, it is compact for any $z \in \rho$. Moreover, if A is of compact resolvent, by proposition 1.3, its spectrum is enterely formed by eigenvalues (with finite multiplicities) with no accumulation point different from infinity. In particular uniqueness implies existence for $(A - z)x = f$ (Fredholm alternative).

2.- Convergence in the gap and convergence of the resolvents in the norm - In applications, it is useful to relate the convergence of a sequence of operators to the convergence of its inverses (if they exist). This is easily made if we work with the graphs of the operators, which are closed subspaces (if the operators are closed). Let us then define a concept of convergence for closed subspaces of a Banach space.

Definition 2.1. - Let M, N be closed subspaces of the Banach space Z. $S_M$, $S_N$ are the unit spheres of M and N (i.e. $S_M = \{ x \; ; \; x \in M \; , \; \|x\| = 1 \}$). The gap $\hat{\delta}$ between M and N is defined as follows :

$$\hat{\delta}(M , N) = \max \{\delta(M , N) , \delta(N , M)\} \tag{2.1}$$

$$\text{where} \quad \delta(M , N) = \sup_{x \in S_M} \text{distance}(x , N).$$

It is easy to see that $0 \leqslant \hat{\delta} \leqslant 1$, and that $\hat{\delta} = 0 \Leftrightarrow M = N$. The gap is not a distance, because it does not satisfy the triangle inequality, but it is possible to defined a distance

$\hat{d}(M , N)$ such that

$$\hat{\delta}(M , N) \leqslant \hat{d}(M , N) \leqslant 2 \hat{\delta}(M , N) \tag{2.2}$$

Thus, $\hat{\delta} \to 0$ amounts to the convergence in a metric space.

Let us now consider two Banach spaces X, Y and two closed operators A, B from X to Y. The graph is defined as we know by

$$G(A) = \{(x , y) \; ; \; x \in D(A) , y = Ax \}$$

which is a closed subspace of $Z = X \times Y$ (equipped with the norm $\| \|_Z^2 = \| \|_X^2 + \| \|_Y^2$).

Definition 2.2. - The gap $\hat{\delta}(A , B)$ between the operators A and B is defined as the gap between the subspaces G(A) and G(B) of Z.

With the above definitions, tedious but not difficult calculations lead to the following theorems (see Kato [2] , chap. 4, sect. 2.4, 2.5) :

Theorem 2.1 - Let A and B be closed operators from X to Y. If $C \in \mathscr{L}(X,Y)$, then,

$$\hat{\delta}(A + C , B + C) \leqslant 2(1 + \|C\|^2) \hat{\delta}(A , B).$$

Theorem 2.2 - Let A and B be closed operators from X to Y. If they are invertible, we have

$$\hat{\delta}(A^{-1} , B^{-1}) = \hat{\delta}(A , B).$$

Theorem 2.3 - Let A and B be closed operators from X to Y. If $A^{-1}$ exists and belongs to $\mathscr{L}(X,Y)$, and

$$\hat{\delta}(A , B) < (1 + \|A^{-1}\|^2)^{1/2}$$

Then $B^{-1}$ exists and belongs to $\mathscr{L}(X , Y)$.

Theorem 2.4 - Let A, $A_n$, (n = 1,2,...) be closed operators from X to Y. Then :

a) If A is bounded, $\hat{\delta}(A , A_n) \to 0$ if and only if $A_n$ is bounded for sufficiently large n and $\|A - A_n\| \to 0$.

b) If $A^{-1} \in \mathscr{L}(X , Y)$ then $\hat{\delta}(A , A_n) \to 0$ if and only if $A_n^{-1} \in \mathscr{L}(X , Y)$, (for sufficiently large n) and

$$\|A^{-1} - A_n^{-1}\| \to 0$$

c) If $\hat{\delta}(A , A_n) \to 0$ and B is bounded, then $\hat{\delta}(A + B , A_n + B) \to 0$.

By using the preceeding theorems, it is not difficult to obtain the convergence of the resolvents in a point if the same convergence in another point holds. In fact, we have :

Theorem 2.5 - Let A, $A_n$ (n = 1,2,...) be closed operators in a Banach space B such that

(2.3) $$\|A_n^{-1} - A^{-1}\| \to 0 \qquad (n \to \infty)$$

and let $\lambda$ be a point of the resolvent set of A and $\lambda_n$ (n = 1,2 ...) a sequence such that

(2.4) $$\lambda_n \to \lambda \qquad n \to \infty$$

Then, for sufficiently large n, $\lambda_n$ belongs to the resolvent set of $A_n$ and

(2.5) $$\| (A_n - \lambda_n)^{-1} - (A - \lambda)^{-1} \| \to 0 \qquad (n \to \infty).$$

Proof - From (2.3), theorem 2.4 part a applied to the inverses and theorem 2.2,

$$\hat{\delta}(A_n^{-1} , A^{-1}) \to 0 \Rightarrow \hat{\delta}(A_n, A) \to 0$$

and by theorem 2.1,

(2.6) $$\hat{\delta}(A_n - \lambda_n , A - \lambda_n) \to 0$$

On the other hand, the resolvent set of A is open, and the resolvent is holomorphic in $\lambda$ ; thus, for sufficiently large n, $\lambda_n \in \rho(A)$ and

$$\| (A - \lambda_n)^{-1} - (A - \lambda)^{-1} \| \to 0$$

and by using again theorem 2.4 part a and theorem 2.2

(2.7) $$\hat{\delta}(A - \lambda_n , A - \lambda) \to 0$$

Then, from (2.6) and (2.7) we obtain (see (2.2)) :

$$\hat{\delta}(A_n - \lambda_n , A - \lambda) \to 0$$

and by theorem 2.3, for sufficiently large n, $\lambda_n \in \rho(A_n)$ and by theorem 2.2 :

(2.8) $$\hat{\delta}((A_n - \lambda_n)^{-1} , (A - \lambda)^{-1}) \to 0$$

moreover, the operators in (2.8) are bounded and by theorem 2.4 part a, we obtain (2.5) .■

If in particular we take $\lambda_n = \lambda$ , we have

Corollary 2.1 - If $A_n$, A are closed operators in a Banach space B and the resolvents $(A_n - \lambda)^{-1}$ converge in the norm to the resolvent $(A - \lambda)^{-1}$ for $\lambda = \lambda_0$, they also converge for any $\lambda \in \rho(A)$. ■

## 3.- Spectral perturbation of operators whose resolvents converge in the norm.- Applications to homogenization and singular perturbations.-

Let us consider a Banach space B and the closed operators A, $A_n$ (n = 1,2 ...) such that $\lambda_0$ belongs to the resolvent set of all of them and

(3.1) $$(A_n - \lambda_0)^{-1} \to (A - \lambda_0)^{-1}$$

in the norm (i.e. in $\mathcal{L}(B , B)$).

Remark 3.1 - This situation appears in several perturbations for elliptic problems in bounded domains, such as :

a) Homogenization (see chap. 5 , sect. 5)

b) Singular perturbations (chap. 9 , sect. 3)

c) Perturbation of boundary conditions (chap. 9 , sect 6)

d) Special types of stiff problems (chap. 13, sect. 3 and 4).■

In the situation (3.1), it is possible to obtain properties of the convergence ($n \to \infty$) of the eigenvalues and eigenvectors of $A_n$ by using the Dunford integral of sect. 1, without any hypothesis of selfadjointness. We have

Theorem 3.1. - Under the hypothesis (3.1), let $\Gamma$ be a simple closed curve contained in the resolvent set of A. Then, for sufficiently large n, $\Gamma$ belongs to the resolvent set of $A_n$. Moreover, the projection

$$P(A_n , \Gamma) = \frac{-1}{2\pi i} \int_\Gamma (A_n - Z)^{-1} dZ \tag{3.2}$$

converges in the norm of $\mathcal{L}(B , B)$ to the corresponding projection for A.

Proof - From (3.1) and the theorem 2.2 and 2.1

$$\alpha_n \equiv \hat{\delta}(A_n , A) \to 0 \qquad n \to \infty$$

Moreover, $\Gamma$ is a bounded curve and then, by theorem 2.2,

$$\hat{\delta}(A_n - z , A - z) \leqslant C \alpha_n \qquad \forall z \in \Gamma \tag{3.3}$$

where C is a constant which depends only on $\Gamma$. On the other hand, the resolvent of A is holomorphic for $z \in \Gamma$ which is compact, hence

$$\| (A - z)^{-1} \| \leqslant C' \qquad \forall z \in \Gamma \tag{3.4}$$

from (3.3), (3.4) and theorem 2.3, we see that for sufficiently large n, $(A_n - z)^{-1}$ for $z \in \Gamma$ exists and is bounded, i.e., $\Gamma$ belongs to $\rho(A_n)$.

Now we see that the projection (3.2) is well defined, and we can pass to the limit in the integral of its right hand side. For, the integrand is pointwise convergent by corollary 2.1 ; moreover,

$$\| (A_n - z)^{-1} \| \qquad z \in \Gamma$$

is bounded by a constant independent of n ; if this was not true, by taking into account that $\Gamma$ is compact, there should exist a subsequence such that

$$\| (A_n - z_n)^{-1} \| \to \infty \qquad z_n \to z_0 \in \Gamma$$

which is not possible by theorem 2.5 (applied to $A_n - \lambda_0$). Then, by dominated convergence, we can pass to the limit in (3.2) and the theorem is proved. ∎

<u>Comments</u>.- In the framework of the preceeding theorem, let $z_0$ be an isolated singularity of the resolvent of A, which is an eigenvalue of finite multiplicity (for instance, any singularity, if A is of compact resolvent). We may take as $\Gamma$ a small circle centered at $z_0$. Then, the projection $P(A , \Gamma)$ given by (3.2) is different from zero (in particular it transforms the eigenvector $v_0$ corresponding to $z_0$ into itself. Then, by theorem 3.1, for sufficiently large n, $\Gamma$ separates the spectrum of $A_n$ and $P(A_n , \Gamma)$ tends to $P(A , \Gamma)$ in norm. From sect. 1, $P(A_n , \Gamma)B$ and $P(A , \Gamma)B$ are isomorphic for n large, and thus $P(A_n , \Gamma)$ is not null : moreover, this means that $\Gamma$ contains in its interior one (or several !) singularities of the resolvent of $A_n$ and the total geometric multiplicity of it (or them !) is the same as for the eigenvalue $z_0$ of A. In particular, if $z_0$ is a simple eigenvalue of A (i.e., of geometric multiplicity one), for n sufficiently large, $\Gamma$ encloses one and only one eigenvalue of $A_n$, which is also simple. In addition,

$$v_n = P(\Gamma , A_n)v_0$$

is an eigenvector of $A_n$ which tends to $v_0$ as $n \to \infty$.

On the other hand, if $z_0$ is a point of the resolvent set of A, the preceeding considerations hold, but the projections are null. This means that $\Gamma$ and its interior belong to $\rho(A_n)$ for sufficiently large n.

It is noticeable that the algebraic multiplicity of the eigenvalues does not necessarily pass to the limit as $n \to \infty$ . This is the reason why <u>the continuity of projections is the appropriate generalization of continuity of eigenvalues and eigenvectors for the case where the eigenvalues are not simple</u>. ∎

4.- <u>Holomorphic families of operators</u> - In applications one often deals with operators which depend holomorphically on a parameter $\mu$ . We now introduce the definitions and principal theorems associated with such a concept.

<u>Definition 4.1</u>. - A function $u(\mu)$ defined on a domain $\Delta$ of the complex plane, with values in a Banach space B is said to be holomorphic if it is differentiable (in the sense of the norm) at each point $\mu \in \Delta$.

It is easy to see that holomorphic functions have the classical Taylor and Laurent expansions.

In fact, a much weaker condition than differenciability in the sense of the norm is sufficient to obtain holomorphy :

Proposition 4.1. - A function $u(\mu)$ with values in B is holomorphic if for any $f \in B'$ (the dual of B), the complex valued function $(u(\mu) , f)$ (where the parenthesis is for the duality product) is differentiable (i.e. holomorphic).

We also have

Proposition 4.2. - $u(\mu)$ is holomorphic in $\Delta$ if and only if each $\mu \in \Delta$ has a neighbourhood in which $u(\mu)$ is bounded in norm and $(u(\mu),f)$ is holomorphic for any f belonging to a set which is dense in B'.

Analogous properties hold for holomorphic functions with values in a space of bounded operators :

Definition 4.2 - Let X and Y be two Banach spaces. A function $A(\mu)$ defined on a domain $\Delta$ of the complex plane with values in $\mathcal{L}(X , Y)$ is said to be bounded holomorphic (or only holomorphic) if it is differentiable in the norm of $\mathcal{L}(X , Y)$ of each $\mu \in \Delta$.

Proposition 4.3. - A function $A(\mu)$ with values in $\mathcal{L}(X , Y)$ is bounded holomorphic if $(A(\mu)\phi , f)$ is a complex-valued holomorphic function for any $\phi \in X$, $f \in Y'$.

Proposition 4.4. - A function $A(\mu)$ with values in $\mathcal{L}(X , Y)$ is bounded holomorphic if each $\mu \in \Delta$ has a neighbourhood in which $A(\mu)$ is bounded in norm and $(A(\mu) \phi, f)$ is holomorphic for each $\phi$ (resp. f) in a set dense in X (resp. in Y).

The following property is useful for the inversion of operators.

Proposition 4.5. - If a function $A(\mu)$ with values in $\mathcal{L}(X , Y)$ is bounded holomorphic in $\Delta$ and $A^{-1}(\mu_0)$ exists and belongs to $\mathcal{L}(Y , X)$ for a $\mu_0 \in \Delta$ , then, $A^{-1}(\mu)$ exists and is bounded holomorphic for sufficiently small $|\mu - \mu_0|$.

For unbounded operators there exists a more complicated concept of holomorphy, which is associated to the convergence in the gap:

Definition 4.3. - Let X and Y be two Banach spaces. A family of closed operators $A(\mu)$ from X into Y depending on the parameter $\mu$ (defined for $\mu$ in a neighbourhood of $\mu_0$ in the complex plane) is said to be holomorphic (in the gap) at $\mu = \mu_0$ if there exists a third Banach space Z and two families of operators $U(\mu) \in \mathcal{L}(Z, X)$, $V(\mu) \in \mathcal{L}(Z, Y)$ which are bounded holomorphic at $\mu = \mu_0$ and such that U maps Z one to one onto $D(A(\mu))$ and

$$V(\mu) = A(\mu)\, U(\mu). \tag{4.1}$$

Moreover, $A(\mu)$ is said to be holomorphic in a domain $\Delta$ if it is holomorphic for each $\mu_0 \in \Delta$.

Proposition 4.6. - If $A(\mu)$ is holomorphic in the gap, $\hat{\delta}(A(\mu) - A(\mu_0)) \to 0$ as $\mu \to \mu_0$.

Remark 4.1. - It is obvious that bounded holomorphic families of operators admit the classical analytic continuation because their concept of analyticity is associated with a Taylor series expansion. For unbounded holomorphic families this is not evident. In fact, in the case where X = Y and there exists a $z_0$ belonging to the resolvent sets of $A(\mu)$ $\forall \mu \in \Delta$, the classical analytic continuation properties hold for $A(\mu)$. ∎

Remark 4.2. - It is noticeable that if X = Y = H is a Hilbert space, an important case is that of $A(\mu)$ defined for $\mu \in \Delta$ and $A^*(\mu) = A(\bar{\mu})$. In such a case, $A(\mu)$ is selfadjoint for real $\mu$ and it is said that the family is selfadjoint. Moreover, if the family is of type (A) of Kato (see below, proposition 4.6), the eigenvectors and eigenvalues are analytic functions of $\mu$ for real $\mu$ (c.f. Riesz - Nagy [ 1 ], sect. 136). ∎

As the definition of holomorphic families in the gap is complicated, it is useful to know simple criteria for the holomorphy of certain classes of families. The following two propositions are useful in this connection.

Proposition 4.6.- Let $A(\mu)$ be a family of closed operators from X to Y defined for $\mu \in \Delta$ ($\Delta$ is a domain of the complex plane) such that $D(A(\mu)$ is independent of $\mu$ and $A(\mu)x$ for any $x \in D(A)$ is a holomorphic function with values in Y. Then, $A(\mu)$ is a holomorphic family in the gap and we shall say that it is of type (A) of Kato.

For sesquilinear forms and their associated operators we have the following definition and criterion of holomorphy :

Definition 4.4. - Let V be a Hilbert space and let $a(\mu ; u , v)$ be a family of bounded sesquilinear forms on V defined for $\mu \in \Delta$ (domain of the complex plane).

We shall say that $a(\mu)$ is bounded holomorphic if $a(\mu ; u , u)$ is holomorphic for each $u \in V$.

Proposition 4.7. - Let V and H be two Hilbert spaces, $V \subset H$ densely, and let $a(\mu , u , v)$ be a bounded holomorphic family of sesquilinear forms on V for $\mu \in \Delta$ (domain of the complex plane). Moreover, let $a(\mu)$ be coercive in the sense that there exists constants M, $\alpha$ and $\beta$ ($\alpha > 0$) such that

$$|a(\mu, u , v)| \leqslant M \|u\|_V \|v\|_V \qquad u , v \in V \ , \quad \mu \in \Delta$$

$$\mathrm{Re}\, a(\mu, u , u) + \beta \|u\|_H^2 \geqslant \alpha \|u\|_V^2 \qquad u \in V \ , \quad \mu \in \Delta$$

Then, if $A(\mu)$ is the family of maximal accretive operators on H associated with the form a by the first representation theorem, $A(\mu)$ is a holomorphic family in the gap and we shall say that it is of type (B) of Kato.

Remark 4.3 - In the situation of prop. 4.7, the domain of the form $a(\mu)$ is V independent of $\mu$ ; but the domain of the operator, $D(A(\mu))$ in general does depend on $\mu$. ■

In order to obtain spectral properties of the holomorphic families of operators, the main property is the following theorem, which gives holomorphic properties of the resolvents :

Theorem 4.1 - Let $A(\mu)$ be a family of closed operators in a Banach space B, defined in a complex neighbourhood of $\mu = \mu_0$ and let $z_0$ be a point of $\rho(A(\mu_0))$, the resolvent set of $A(\mu_0)$. Then, $A(\mu)$ is holomorphic at $\mu = \mu_0$ if and only if $z_0 \in \rho(A(\mu))$ and the resolvent $(A(\mu) - z_0)^{-1}$ is bounded holomorphic for sufficiently small $|\mu - \mu_0|$. Moreover, if $A(z)$ is holomorphic, $(A(\mu) - z)^{-1}$ is bounded holomorphic in the two variables $\mu, z$ on the set of all $\mu, z$ such that $z \in \rho(A(\mu_0))$ and $|\mu - \mu_0|$ is sufficiently small (depending on z).

As an exercise, we shall prove the first part of this theorem. If $A(z)$ is holomorphic, the operators $U(\mu)$ and $V(\mu)$ of def. 4.3 exist (with $X = Y = B$) and $(A(\mu) - z_0)\, U(\mu) = V(\mu) - z_0\, U(\mu)$ is bounded holomorphic from Z to B. But $(A(\mu_0) - z_0)\, U(\mu_0)$ maps Z one to one onto B. Its inverse $[(A(\mu_0) - z_0)\, U(z_0)]^{-1} \equiv [V(\mu) - z_0\, U(\mu))]^{-1}$ is bounded from B to Z and by proposition 4.5, $[(A(\mu) - z_0)\, U(z_0)]^{-1}$ is bounded holomorphic for small $|\mu - \mu_0|$. Hence

$$(A(\mu) - z_0)^{-1} = U(\mu)\, [V(\mu) - z_0\, U(\mu)]^{-1}$$

is bounded holomorphic, Q.E.D.

Conversely if $(A(\mu) - z_0)^{-1}$ is bounded holomorphic for small $|\mu - \mu_0|$, by writing

$$A(\mu)[A(\mu) - z_0]^{-1} = I + z_0 [A(\mu) - z_0]^{-1}$$

we see that the definition of holomorphic in the gap is satisfied if we take Z = B, V the right hand side and U the factor of A in the left hand side. ∎

The following theorem, which is the analogous of theorem 3.1 for operators depending holomorphically on a parameter is a consequence of theorem 4.1 :

Theorem 4.2. - Let $A(\mu)$ be a family of operators from B to B holomorphic in the gap for small $|\mu - \mu_0|$. If $\Gamma$ is a simple closed curve contained in $\rho(A(\mu_0))$, then, for $|\mu - \mu_0|$ sufficiently small, $\Gamma$ is contained in $\rho(A(\mu))$ and the projection

$$P(\mu) = \frac{-1}{2\pi i} \int_\Gamma [A(\mu) - z]^{-1} dz$$

is bounded holomorphic in $\mu$. Moreover, if $\{M(\mu), N(\mu)\}$ where

$$M(\mu) = P(\mu) B \quad ; \quad N(\mu) = (I - P(\mu))B$$

is the associated decomposition of the space in the framework of sect. 1, there exists an operator $U(\mu)$ which is bounded holomorphic together with its inverse $U^{-1}(\mu)$ which transforms $P(\mu_0)$ into $P(\mu)$ :

$$P(\mu) = U(\mu) P(\mu_0) U^{-1}(\mu)$$

the operator

$$\tilde{A}(\mu) = U^{-1}(\mu) A(\mu) U(\mu)$$

commutes with $P(\mu_0)$ and the pair $M(\mu_0)$, $N(\mu_0)$ decomposes $\tilde{A}(\mu)$. The eigenvalue problem for $A(\mu)$ in $M(\mu)$ is equivalent to that of $\tilde{A}(\mu)$ in $M(0)$.

The consequences of this theorem are obvious, as in the comments at the end of section 3, but with holomorphy instead of convergence.

Remark 4.4 - In the framework of the preceeding theorem, if $\Gamma$ encloses only a point $\lambda_0$ of $\sigma(A(\mu_0))$ and P is of finite dimension m, the transformation $U(\mu)$ shows that the structure of the resolvent in the vicinity of $\mu_0$ is analogous to that of the inverse of a holomorphic matrix in a finite dimensional space. In particular, there exists two integers $k \leq m$ and n such that the eigenvalues are (in the vicinity of $\lambda_0$),

$$\lambda_1(\mu) , \ldots . \lambda_k(\mu)$$

and each $\lambda_i(\mu)$ is a holomorphic function of $(\mu - \mu_0)^{1/n}$ taking the value

$\lambda_i(\mu) = \lambda_0$. Moreover, the corresponding projections have Laurent expansions of $(\mu - \mu_0)^{1/n}$ with at most a finite principal part.

This is a consequence of the Weierstrass preparation theorem on the bifurcation of the zeroes of a holomorphic function which depends holomorphically on a parameter. ∎

As a first elementary example of this theory, we consider the vibrations of a membrane with a damping term proportional to the velocity. If $\Omega$ is a bounded open set in $R^2$, we consider

(4.2) $$\frac{\partial^2 u}{\partial t^2} = \Delta u - \varepsilon\, b(x) \frac{\partial u}{\partial t} \qquad \text{in } \Omega$$

(4.3) $$u = 0 \qquad \text{on } \partial\Omega.$$

and of course the appropriate initial conditions. Here $\Delta$ is the laplacian, and b is a bounded positive function.

For $\varepsilon = 0$, this problem is the same as in chapter 4, remark 5.1, if we choose

$$H = L_2(\Omega) \quad ; \quad V = H_0^1(\Omega)$$

and the form a is

$$a(u\ ,\ v) = \int_\Omega \frac{\partial u}{\partial x_i} \frac{\partial v}{\partial x_i}\, dx \quad .$$

Moreover, if B is the operator of $\mathcal{L}(H\ ,\ H)$ defined by

$$Bu(x) = b(x)\ u(x)$$

we can write this problem in matrix form :

(4.4) $$\vec{u} = \begin{pmatrix} u \\ \frac{du}{dt} \end{pmatrix} \quad ; \quad \frac{d\vec{u}}{dt} + A^\varepsilon \vec{u} = 0$$

(4.5) $$A^\varepsilon = \begin{pmatrix} 0 & -I \\ -\Delta & \varepsilon B \end{pmatrix} \qquad \text{with domain :}$$

(4.6) $$D(A^\varepsilon) = \{(u,\ v) \in V \times H \quad ; \quad -\Delta\ u \in H \quad ; \quad v \in V\}$$

which does not depend on $\varepsilon$ . As in chapter 4, sect. 5 we see that $A^\varepsilon$ is accretive and maximal (for $\varepsilon > 0$). The abstract equation (4.4) is then an appropriate realization of (4.2), (4.3). It is clear from (4.5), (4.6) that $A^\varepsilon$ is a holomorphic family of type (A) of Kato. For $\varepsilon = 0$, the eigenvalues are $\pm\, i\, \omega_k$ (where $-\omega_k^2$ are the eigenvalues of the laplacian). By virtue of theorem 4.2, the coresponding projections depend holomorphically on $\varepsilon$ .

More complicated cases appear if the damping term in (4.2) is an unbounded operator. We shall study an example of this case in the following section.

5.- Implicit eigenvalue problems. Application to a singular perturbation in viscoelastic vibrations.- The preceeding methods deal with operators depending on a parameter but acting in a space which does not depend on the parameter. In applications, the space often depends on the parameter and the preceeding methods must be adapted to new cases. The results are then less precise. We shall explain this through an example that has been dealt with in semigroup theory.

Let us consider, as in chap.10, sect. 5 the problem (slightly viscoelastic plate of small rigidity) :

(5.1) $$\frac{\partial^2 u}{\partial t^2} = C_1 \Delta u - \varepsilon C_2 \Delta^2 u - \varepsilon C_3 \Delta^2 \frac{\partial u}{\partial t} \quad ; \quad C_i > 0$$

with the boundary conditions

(5.2) $$u|_{\partial\Omega} = 0 \quad \text{if} \quad \varepsilon = 0$$

(5.3) $$u|_{\partial\Omega} = \frac{\partial u}{\partial n}\Big|_{\partial\Omega} = 0 \quad \text{if} \quad \varepsilon > 0$$

in the bounded domain $\Omega$ of $R^2$. Alternatively, we have the matrix equivalent form (5.7) - (5.9) of chapter 10 . The problem of getting the eigenvalues is equivalent to that of getting the values $\zeta$ such that

(5.4) $$u(x\ t) = e^{\zeta t} v(x)$$

is a non zero solution of (5.1), i.e.

(5.5) $$[A_0 + \varepsilon(C_2 + C_3 \zeta) A_1] v + \zeta^2 = 0$$

where $A_0 = - C_1\Delta$ with the boundary condition (5.2) and $A_1 = \Delta^2$ with the boundary condition (5.3). Such values of $\zeta$ will be called in the sequel the eigenvalues of (5.5). The following result is evident :

Lemma 5.1 - (5.6) $\begin{cases} \text{If } \varepsilon = 0, \text{ the eigenvalues } \zeta \text{ are } \zeta_k = \pm i\, \omega_k \quad k = 1,2... \\ \text{where } \omega_k^2 \text{ are the eigenvalues of } A_0. \end{cases}$

Our aim is to study the eigenvalues for small $\varepsilon$ in the region

(5.7) $$\mathrm{Re}\, \zeta > - \frac{C_2}{C_3}$$

which contains in particular the eigenvalues of the problem for $\varepsilon = 0$. Let us consider the nonhomogeneous equation associated with (5.5)

(5.8) $$[A_0 + \varepsilon (C_2 + C_3 \zeta) A_1] \quad v + \zeta^2 z = f$$

in $L_2(\Omega)$. For fixed $\zeta$ in the region (5.6), the coefficient of $\varepsilon A_1$ in (5.7) has a positive real part, and the singular perturbation theory of chapter 9 , sections 1 and 3 holds because $A_0$ is associated with the space $H_0^1$ which has a compact embedding in $L_2$. Moreover, the considerations of sect. 3 of the present chapter apply and we have :

<u>Theorem 5.1</u> - If $\zeta$ is a fixed point of the region (5.7), different from the eigenvalues of the limit problem (5.6), then, for sufficiently small $\varepsilon$, $\zeta$ is not an eigenvalue of (5.5), and the solution v(f) of (5.8) is well defined.

On the other hand, in the vicinity of each eigenvalue of the limit problem (5.6), for sufficiently small $\varepsilon$ , there exists an eigenvalue of (5.5). More exactly :

<u>Theorem 5.2</u> - Let us consider a fixed eigenvalue of the limit problem (5.6), denoted by $\zeta_0$. Let $\mathcal{V}$ be a neighbourhood of $\zeta_0$ contained in the region (5.7) and such that the origin and all the other eigenvalues of the limit problem are outside $\mathcal{V}$. Then, for sufficiently small $\varepsilon$, v contains at least an eigenvalue of (5.5).

We shall prove this theorem by contradiction. If it is not true, there exists a sequence $\varepsilon_j \to 0$ such that $\mathcal{V}$ does not contain eigenvalues of (5.5) corresponding to the $\varepsilon_j$.

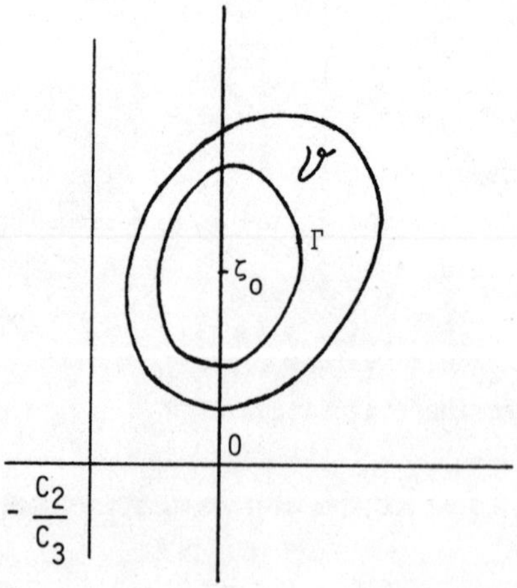

Let $\Gamma$ be a simple curve enclosing $\zeta_0$ and contained in $\mathcal{V}$. Let us consider the function with values in $\mathcal{L}(L^2, L^2)$

$$(A_0 + \zeta^2)^{-1}$$

it is a uniform holomorphic function of $\zeta$ in $\mathcal{V}$ with the unique singularity $\zeta = \zeta_0$. It then has a Laurent series with non zero singular part. Hence, for a certain integer $m \geqslant 0$, the analytic function

$$(5.9) \qquad (\zeta - \zeta_0)^m (A_0 + \zeta^2)^{-1}$$

has a Laurent series with non zero coefficient of $(\zeta - \zeta_0)^{-1}$. As a consequence, the integral of (5.9) on $\Gamma$ is not zero. (It is of course an operator of $\mathcal{L}(L^2, L^2)$). We then have

Lemma 5.2 - There exists an integer $m \geqslant 0$ and two elements $f, g$ of $L^2$ such that

$$(5.10) \qquad \left( \int_\Gamma (\zeta - \zeta_0)^m (A_0 + \zeta^2)^{-1} d\zeta \; f, g \right)_{L^2} \neq 0 \quad .$$

On the other hand, for each fixed $\varepsilon = \varepsilon_i > 0$, the family $A_0 + \varepsilon(C_2 + C_3\zeta)A_1$ is a holomorphic family of class (A) of Kato of the variable $\zeta$ (in the region (5.7)). Moreover, for $\zeta \in \mathcal{V}$, $\zeta^2$ is not an eigenvalue of this operator (and thus it belongs to its resolvent set because the operator is of compact resolvent) and hence, by theorem 4.1

$$[[A_0 + \varepsilon(C_2 + C_3\zeta) A_1] - Z]^{-1}$$

is an holomorphic function of $\zeta$ and $Z$ for $\zeta \in \mathcal{V}$ and $Z$ in the region where $Z^{1/2} \in \mathcal{V}$. By taking $Z = \zeta^2$, we have an holomorphic function of $\zeta$ in $\mathcal{V}$. We then have

Lemma 5.3 - For the sequence $\varepsilon_j \to 0$, and for any integer $m$ and any elements $f, g$ of $L_2$ and (in particular those defined by Lemma 5.2),

$$(5.11) \quad \left( \int_\Gamma (\zeta - \zeta_0)^m [[A_0 + \varepsilon(C_2 + C_3\zeta) A_1] - \zeta^2]^{-1} d\zeta \; f, g \right)_{L^2} = 0$$

Now, to obtain a contradiction between (5.10) and (5.11) it suffices to take the limit of (5.11) as $\varepsilon_j \to 0$ in the integral of (5.11). This is in fact possible. For fixed $\zeta \in \Gamma$, we have the convergence of the resolvents in the norm (as was done in the proof of theorem 5.1). Moreover, we may pass to the limit by dominated convergence. For, if $[[A_0 + \varepsilon_j(C_2 + C_3\zeta) A_1] - \zeta^2]^{-1}$ is not bounded in the norm for $\varepsilon_j \to 0$, $\zeta \in \Gamma$, we can extract a subsequence $\varepsilon_k \to 0$ and $\zeta_k \to \zeta^* \in \Gamma$

such that the corresponding norms tend to infinity, and this is impossible by theorem 2.5, because $\zeta_k^2 \to \zeta^{*2}$ and

$$[A_o + \varepsilon_k(C_2 + C_3\, \zeta_k)A_1]^{-1} \to A_o^{-1}$$

in the norm. This achieves the proof of theorem 5.2.

6. Homogenization of a boundary. Spectral properties - We study here the spectral properties of the problem of homogenization of a boundary studied in chap. 5, sect.7 and 8.

We consider the domains $\Omega_\varepsilon$ and $\Omega_o$ as described in chap. 5, sect. 7. We shall use the following notations :

$(. , .)_o$ , $|.|_o$ (resp. $(. , .)_\varepsilon$ , $|.|_\varepsilon$ ) are the scalar product and norm in $L^2(\Omega_o)$ (resp. $L^2(\Omega_\varepsilon)$). $((. , .))_o$, $\|.\|_o$ (resp. $((. , .))_\varepsilon$ , $\|.\|_\varepsilon$ ) are the scarlar product and norm in $H^1(\Omega_o)$ (resp. $H^1(\Omega_\varepsilon)$).

Moreover, $A_\varepsilon$ and $A_o$ are the operators $-\Delta$ in $\Omega_\varepsilon$ and $\Omega_o$ with the boundary conditions :

$$\frac{\partial u^\varepsilon}{\partial n} + \lambda\, u^\varepsilon = 0 \quad ; \quad \frac{\partial u^o}{\partial n} + \lambda\, \Gamma\, u^o = 0 \tag{6.1}$$

The corresponding forms on $H^1(\Omega_\varepsilon)$ and $H^1(\Omega_o)$ are :

$$a_\varepsilon(u , v) = \int_{\Omega_\varepsilon} \frac{\partial u}{\partial x_i} \frac{\partial v}{\partial x_i}\, dx + \lambda \int_{\partial\Omega_\varepsilon} uv\, dS \tag{6.2}$$

$$a_o(u , v) = \int_{\Omega_o} \frac{\partial u}{\partial x_i} \frac{\partial v}{\partial x_i}\, dx + \lambda\, \Gamma \int_{\partial\Omega_o} u\, v\, dS \tag{6.3}$$

We also consider the operators

$$B_\varepsilon = A_\varepsilon + I \quad ; \quad B_o = A_o + I \tag{6.4}$$

and the associated forms

$$b_\varepsilon(u , v) = a_\varepsilon(u , v) + (u , v)_\varepsilon \quad ; \quad b_o(u , v) = a_o(u , v) + (u , v)_o \tag{6.5}$$

Our problem is to study the spectral properties of operators $A_\varepsilon$ and $A_o$ as $\varepsilon \searrow 0$. This amounts to study the spectral properties of $B_\varepsilon$ and $B_o$ because the respective spectra are translated by a real unity.

The following study may be applied, with the appropriate modifications, to other perturbation problems. In fact, it is analogous to the "discrete convergence" of Stummel [1], [3], and applies to non-selfadjoint operators. In this connection it is noticeable that the self-adjointness is not used to prove theorems 6.1 and 6.2 ; it has only be used in the proof of theorem 6.3.

Lemma 6.1 - If

(6.6) $$B_\varepsilon v^\varepsilon = f^\varepsilon \quad \text{and}$$

(6.7) $$|f^\varepsilon|_\varepsilon \leq C \quad ; \quad f^\varepsilon|_{\Omega_0} \to f^0 \quad \text{in } L^2(\Omega_0) \text{ weakly}$$

then, there exists a function $\delta(\varepsilon)$ ($\delta \searrow 0$ as $\varepsilon \searrow 0$) such that :

(6.8) $$\|v^\varepsilon\|_{L^2(\Omega_\varepsilon - \Omega_0)} \leq \delta(\varepsilon) \quad ; \quad v^\varepsilon|_{\Omega_0} \to v^0 \quad \text{in } H^1(\Omega_0) \text{ weakly}$$

(6.9) $$\text{and} \quad B_0 v^0 = f^0$$

Proof - This lemma is proved exactly as theorem 8.1 of chapter 5, with the only modification that $f^\varepsilon$ is variable. (6.8) is proved as (8.16) of chap. 5.∎

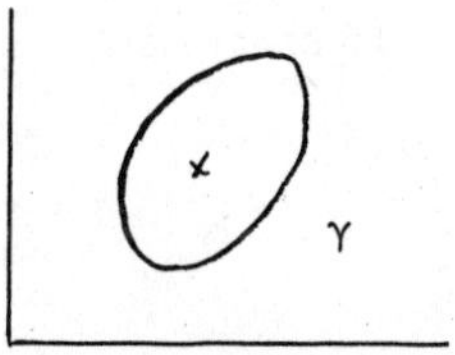


Now, we can study the singularities of the resolvents by a method analogous to that of sect. 3.

Lemma 6.2 - Let $\gamma$ be a simple, closed curve contained in the resolvent set of $B_0$ : $\gamma \subset \rho(B_0)$. Then, for sufficiently small $\varepsilon$, $\gamma$ is also contained in $\rho(B_\varepsilon)$.

Proof - Because the embedding of $H^1(\Omega_\varepsilon)$ into $L^2(\Omega_\varepsilon)$ is compact, $B_\varepsilon^{-1}$ (resp. $B_0^{-1}$) are compact operators in $L^2(\Omega_\varepsilon)$ (resp. $L^2(\Omega_0)$) and the spectrum $\sigma(B_\varepsilon)$ (resp. $\sigma(B_0)$) is formed by isolated points with infinity as only accumulation points. Then, if the lemma is not true, there exists, for a sequence $\varepsilon \searrow 0$, an eigenvalue $z_\varepsilon \in \gamma$. Let $v^\varepsilon$ be a corresponding eigenvector :

(6.10) $$B_\varepsilon v^\varepsilon = z_\varepsilon v^\varepsilon \quad ; \quad |v^\varepsilon|_\varepsilon = 1$$

Because $\gamma$ is a compact set, we may suppose that :

(6.11) $$z_\varepsilon \to z_0 \in \gamma$$

and also that

(6.12) $$v^{\varepsilon}\big|_{\Omega_0} \to v^0 \qquad \text{in } L^2(\Omega_0) \text{ weakly}$$

then, by writting $f^{\varepsilon} = z_{\varepsilon}\, v^{\varepsilon}$ the hypothesis of lemma 6.1 are satisfied. Consequently, we have the analogous of (6.8), (6.9), i.e. :

(6.13) $$v^{\varepsilon}|_{\Omega_0} \to u^0 \quad \text{in } H^1(\Omega_0) \text{ weakly} \;\; ; \;\; \|v^{\varepsilon}\|_{L^2(\Omega_{\varepsilon}-\Omega_0)} \leqslant \delta(\varepsilon)$$

(6.14) $$B_0\, v^0 = z^0\, v^0$$

Moreover, from (6.13) we see that the convergence also holds in $L^2(\Omega_0)$ strongly, and by taking into account the second (6.10) and the second (6.13), we have $\|v^0\|_0 = 1 \quad v^0 \neq 0$ which shows, with (6.14), that $z^0 \in \gamma$ is an eigenvalue of $B_0$, and this is impossible by the hypothesis. ∎

As a consequence of lemma 6.2, we can integrate the resolvents of $B_{\varepsilon}$ (for small $\varepsilon$) and $B_0$ on $\gamma$. Moreover :

Theorem 6.1 - If $\gamma$ is a simple, closed curve contained in the resolvent set of $B_0$, for any $f \in L^2(\Omega_0)$ (we shall continuate f with zero values out of $\Omega_0$, and thus $f \in L^2(\Omega_{\varepsilon})$), we have

(6.15) $$(P_{\varepsilon}\, f)|_{\Omega_0} \to P_0\, f \qquad \text{in } L^2(\Omega_0) \text{ strongly as } \varepsilon \to 0$$

where $P_{\varepsilon}$ is the projection

(6.16) $$P_{\varepsilon} = \frac{-1}{2\pi i} \int_{\gamma} (B_{\varepsilon} - z)^{-1}\, dz$$

and $P_0$ is the analogous projector for $B_0$.

Proof - According to lemma 6.2, $P_{\varepsilon} f$ and $P_0\, f$ make sense, and we have :

(6.17) $$(P_{\varepsilon}\, f)\big|_{\Omega_0} = \frac{-1}{2\pi i} \int_{\gamma} [(B_{\varepsilon} - z)^{-1}\, f]\big|_{\Omega_0}\, dz$$

We shall prove that it is possible to pass to the limit in (6.17). First, we see that, for $z \in \gamma$

(6.18) $$\begin{cases} \text{The integrand of (6.17) is} \\ u^{\varepsilon}\big|_{\Omega_0} \;, \text{ where } \; B_{\varepsilon}\, u^{\varepsilon} = f + z\, u^{\varepsilon} \end{cases}$$

We are proving that, for a certain constant C independent of $\varepsilon$ and any $z \in \gamma$,

(6.19) $$|u^{\varepsilon}|_{\varepsilon} \leqslant C$$

We prove (6.19) by contradiction. If not, there exists a subsequence, such that

$$|u^\varepsilon|_\varepsilon = m(\varepsilon) \nearrow \infty \quad , \quad z_\varepsilon \to z^* \in \gamma$$

where $B_\varepsilon u^\varepsilon = f + z_\varepsilon u^\varepsilon$ . We then define :

(6.20) $$v^\varepsilon = u^\varepsilon/m(\varepsilon) \implies |u^\varepsilon|_\varepsilon = 1 \quad ; \quad z_\varepsilon \to z^* \in \gamma$$

(6.21) $$B_\varepsilon v_\varepsilon = \frac{f}{m(\varepsilon)} + z_\varepsilon v^\varepsilon$$

and after extracting another subsequence,

(6.22) $$v^\varepsilon|_{\Omega_0} \to v^* \quad \text{in } L^2(\Omega_0) \text{ weakly.}$$

We then apply lemma 6.1 to (6.21) and we obtain :

$$|v^*|_0 = 1 \quad ; \quad B_0 v^* = z^* v^*$$

which is impossible because $z^* \in \gamma \subset \rho(B_0)$. Inequality (6.19) is proved.

On the other hand, <u>for a fixed $z \in \gamma$ , we have the convergence of the integrands in (6.17) (see also (6.18))</u>. For, by virtue of (6.19), and extracting a subsequence, we have

(6.23) $$u^\varepsilon|_{\Omega_0} \to u^* \quad \text{in } L^2(\Omega_0) \text{ weakly}$$

and by applying lemma 6.1 to equation (6.18) we have

$$|u^\varepsilon|_\varepsilon \leq C \quad ; \quad u^\varepsilon|_{\Omega_0} \to u^* \quad \text{in } L^2(\Omega_0) \text{ strongly.}$$

(6.24) $$B_0 u^* = f + z\, u^*$$

which shows that $u^*$ is the unique solution of (6.24) ; consequently, in (6.23) we have in fact the whole sequence, and the limit $u^*$ is $P_0 f$.

From here and (6.19) we see that we can pass to the limit in (6.17) by dominated convergence, and the theorem is proved.■

<u>Corollary 6.1</u> - If $z^0$ is an eigenvalue of $B_0$ and $\gamma$ is a simple closed curve enclosing $z^0$ and no other eigenvalue of $B_0$, $\gamma$ encloses at least an eigenvalue of $B_\varepsilon$ for sufficiently small $\varepsilon$ .

<u>Proof</u> - It suffices to take f equal to an eigenvector of $B_0$ corresponding to $z^0$. Then $P_0 f$ is not zero and by virtue of (6.15), $P_\varepsilon f$ is not zero for small $\varepsilon$.

<u>Theorem 6.2</u> - If $z^\varepsilon$ is a sequence of eigenvalues of $B_\varepsilon$ with $\varepsilon \searrow 0$, such that $z^\varepsilon \to z^0$, then $z^0$ is an eigenvalue of $B_0$.

Proof - The proof has been given as a part of the proof of lemma 6.2.

Theorem 6.3 - In the conditions of theorem 6.1, let the curve $\gamma$ enclose only one eigenvalue $z^0$ of $B_0$, the multiplicity of which is m, and let $w_1^0 \dots w_m^0$ be the corresponding eigenvectors. Then, for sufficiently small $\varepsilon$, $\gamma$ encloses eigenvalues of $B_\varepsilon$ with total multiplicity m, and the eigenvalues may be chosen in such a way that

(6.25) $$w_i^\varepsilon \big|_{\Omega_0} \to w_i^0 \quad \text{in } L^2(\Omega_0) \text{ strongly.}$$

Proof - First, we prove that

(6.26) $$m(\varepsilon) \equiv \underline{\lim} \dim P^\varepsilon L^2(\Omega_\varepsilon) \leq \dim P_0 L^2(\Omega_0) = m$$

we take a sequence $\varepsilon \to 0$ such that $\dim P^\varepsilon L^2(\Omega_\varepsilon) = m(\varepsilon)$ and let $w_1^\varepsilon \dots w_{m(\varepsilon)}^\varepsilon$ be a basis formed by eigenvectors, (orthonormalized in $L^2(\Omega_\varepsilon)$). After extracting a subsequence, we may assume that

(6.27) $$w_i^\varepsilon \big|_{\Omega_0} \to w_i^* \quad \text{in } L^2(\Omega_0) \text{ weakly}$$

Then, the reasoning of the proof of lemma (6.2) (with the only modification that the $z^\varepsilon$ are at the interior of $\gamma$ instead of on $\gamma$) shows that the convergence in (6.27) is strong, that the $w_i^*$ are eigenvectors of $B_0$ associated to the eigenvalue $z^0$ and that $\|w_i^\varepsilon\|_{L^2(\Omega_\varepsilon - \Omega_o)} \longrightarrow 0$ and we then see that the $w_i^*$ are linearly independent. (6.26) is proved.

Moreover, we consider again the subsequences (6.27). Let us suppose that $m(\varepsilon) < m$. Then, there exists an eigenvector $\hat{w}^0$ of $B_0$ (for the eigenvalue $z_0$) such that

(6.28) $$(\hat{w}^0 , w_i^*)_0 = 0 \quad ; \quad i = 1 , \dots m(\varepsilon)$$

We then write : (note that the operators are selfadjoint and thus the projections are orthogonal)($\hat{w}^0$ is continuated with zero values to $\Omega_\varepsilon$) :

$$P_\varepsilon \hat{w}^0 = \sum_i^{m(\varepsilon)} (\hat{w}^0 , w_i^\varepsilon)_\varepsilon w_i^\varepsilon$$

we take the restriction to $\Omega_0$ and we multiply by $\hat{w}^0$ :

$$((P_\varepsilon \hat{w}^0)|_{\Omega_0} , \hat{w}_0)_0 = \sum_i^{m(\varepsilon)} (\hat{w}^0 , w_i^\varepsilon) \quad (w_i^\varepsilon|_{\Omega_0} , \hat{w}^0)_0$$

now, we take the limit as $\varepsilon \to 0$. The left hand side tends to 1 by virtue of theorem 6.1. The right hand side is a finite sum of numbers less or equal than 1 multiplied by products which tend to 0 by virtue of (6.27) and (6.28). We then have a contratiction, and consequently, $m(\varepsilon) = m$. The property about the choice of the eigenvectors has been proved in the preceeding reasoning. Theorem 6.3 is proved.■

7.- Notes - The general reference for sect. 1 to 4 is Kato's book [2] . The principal feature of the method used here is that selfadjointness is not used (except in theorem 6.3). Most of the familiar examples in physics are selfadjoint, but general results are also established for non selfadjoint operators. This is for instance the case in homogenization (sect. 3) ; for the selfadjoint case, see Boccardo and Marcellini [1] and Kesavan [1] . Eigenvalue properties of singular perturbations has been extensively studied : see Greenlee [1] , Groen [1] , [2] , Harris [1] , Huet [1] , Stummel [2] ; and, for operators with not necessarily compact resolvent, Kato [1] and the following chapter 12, sect. 2. Implicit eigenvalue problems often appears in physical problems where a small dissipation destroys the skew-selfadjointness of a vibration problem, as in sect. 5. On the other hand, they are analogous to certain perturbation problems for the scattering frequencies (see chapters 15 to 17) ; the method of study is near those of Beale [1] and Sanchez-Palencia [6] . Finally, the problem of sect. 6 is a model problem of eigenvalue perturbations in spaces depending on the parameter ; see in this connection Stummel [1], [3].

# CHAPTER 12

## PERTURBATION OF SPECTRAL FAMILIES

## AND APPLICATIONS TO SELFADJOINT EIGENVALUE PROBLEMS

We study some properties of the spectral families $E_\varepsilon(\lambda)$ of selfadjoint operators $A_\varepsilon$ depending on a parameter $\varepsilon$. Some results are valid for operators with either continuous or discrete spectrum. Sect. 1 is devoted to the classical theorem of Rellich (see Rellich [1], or Riesz - Nagy [1], sect.135), which is applied in sect. 2 to singular perturbations of boundary value problems following the ideas of Kato [1]. Sections 3 and 4 are based on Lobo and Sanchez-Palencia [1]. A problem of homogenization of a boundary in a more singular situation than that of chap. 5, sect. 7 and 8, and chap. 11, sect.6, is studied in sect. 4. Proposition 4.2 is taken from Courant - Hilbert [1], I.

1.- <u>The Rellich's theorem</u> - Let us consider a Hilbert space H and a family of selfadjoint operators $A_\varepsilon \in \mathcal{L}(H, H)$ such that

(1.1) $$\| A_\varepsilon \| \leqslant M$$

and let A be their strong limit, i.e. :

(1.2) $$A_\varepsilon x \to A x \qquad \text{in } H \text{ strongly}, \quad \forall x \in H$$

Let $E_\varepsilon(\lambda)$ and $E(\lambda)$ be the corresponding spectral families. We have :

<u>Theorem 1.1 (Rellich)</u> - Under the hypothesis (1.1), (1.2), if $\mu$ is not an eigenvalue of A, $E_\varepsilon(\mu)$ converges strongly to $E(\mu)$ as $\varepsilon \searrow 0$, e.i. :

(1.3) $$E_\varepsilon(\mu)x \to E(\mu)x \quad \text{in } H \text{ strongly}, \quad \forall x \in H.$$

<u>Proof</u> - By considering the operators $A_\varepsilon - \mu$ we always may consider only the case <u>$\mu = 0$</u> (after a modification of the constant M).

From (1.2), for any polynomial P :

$$P(A_\varepsilon)x \to P(A)x \quad \text{in } H \text{ strongly}, \quad \forall x \in H .$$

Moreover, by (1.1), the spectral families are equal to 0 (resp. I) for $\lambda < M$ (resp. $\lambda > M$). By the Weierstrass theorem, any continuous function $f(\lambda)$ defined on $[-M, +M]$ may be uniformly approximated by polynomials, and hence

(1.4) $$f(A_\varepsilon)x \to f(A)x \quad \text{in } H \text{ strongly.}$$

On the other hand, by the properties of the spectral families it is easily seen that

$$E(0)\,Ax = \int_{-M}^{0} \lambda\, dE(\lambda)x = \int_{-M}^{+M} f(\lambda)\, dE(\lambda)x = f(A)x \quad \text{for}$$

(1.5) $$f(\lambda) = \begin{cases} \lambda & \text{if } \lambda \leqslant 0 \\ 0 & \text{if } \lambda > 0 \end{cases}$$

and analogous formulae for $A_\varepsilon$. But f in (1.5) is continuous and by (1.4)

(1.6) $$E_\varepsilon(0)A_\varepsilon x \to E(0)Ax \quad \text{in } H \text{ strongly}$$

But on the other hand, by (1.2) we have :

$$\| E_\varepsilon(0)A_\varepsilon x - E_\varepsilon(0)Ax \| \leqslant \| A_\varepsilon x - Ax \| \to 0 \qquad \text{as } \varepsilon \to 0$$

and from (1.6) we obtain

(1.7) $$E_\varepsilon(0)Ax \to E(0)Ax \quad \text{in } H \text{ strongly}$$

The operator A is bounded and symmetric ; by hypothesis, zero is not an eigenvalue and hence

(1.8) $$Ay = 0 \Rightarrow y = 0$$

Moreover, its range is dense in H ; for, if y is orthogonal to the range, because A is bounded and symmetric, we have, from (1.8) :

$$\forall x \in H, \quad (Ax\ ,\ y) = 0 = (x\ ,\ Ay) \Rightarrow y = 0 \quad .$$

Then, from (1.7)

(1.9) $$E_\varepsilon(0)x \to E(0)x$$

for any x belonging to a dense set in H ; moreover, $E_\varepsilon(0)$ and $E(0)$ are bounded in norm (by 1 !) and (1.9) holds for any $x \in H$. Q.E.D. ■

Now, we are going to prove a result about the convergence of spectral families of operators whose inverse converge. We first prove a lemma.

Lemma 1.1 - Let A be a positive defined selfadjoint operator. Then the spectral family of the bounded selfadjoint operator $A^{-1}$ is

(1.10) $$E(A^{-1}, \lambda) = I - E(A, \lambda^{-1})^*$$
where the star * means that the spectral family is taken to be strongly continuous on the left (instead of the classical convention : strongly continuous on the right).

Proof - For any spectral family, because of $(E(\lambda)x, x)$ is non decreasing, the left and right limits exist in any discontinuity, and $E(\lambda)^*$ is well defined.

Then, it is easily seen that the right hand side of (1.10) is a right-continuous spectral family. Moreover, by taking $\lambda = \mu^{-1}$

$$A = \int_\alpha^\infty \lambda \, dE(A, \lambda)$$

$$A^{-1} = \int_\alpha^\infty \lambda^{-1} \, dE(A, \lambda) = \int_0^{\alpha^{-1}} \mu \, d[-E(A, \mu^{-1})]^*$$

and by adding the constant I, by the uniqueness of the spectral family we obtain (1.10). ∎

Theorem 1.2 - Let $A_\varepsilon$, A be selfadjoint operators in a Hilbert space H, positively defined in such a way that

$$(A_\varepsilon x, x) \geqslant \alpha \|x\|^2 \quad , \qquad \alpha > 0$$

and an analogous relation for A. Moreover, let us suppose that

(1.10) $$A_\varepsilon^{-1}x \to A^{-1}x \qquad \text{in H strongly} \qquad \forall x \in H$$

and that the real number $\mu$ ($\mu > 0$) is not an eigenvalue of $A_\varepsilon$, A. Then,

(1.11) $$E(A_\varepsilon, \mu)x \to E(A, \mu)x \text{ in H strongly,} \qquad \forall x \in H.$$

Proof - From the hypothesis, $A_\varepsilon^{-1}$, $A^{-1}$ are bounded in norm by $\alpha$. Moreover, $\mu^{-1}$ is not an eigenvalue of $A_\varepsilon^{-1}$, $A^{-1}$, thus by th. 1.1,

$$E(A_\varepsilon^{-1}, \mu^{-1})x \to E(A^{-1}, \mu^{-1})x \quad \text{in H strongly} \quad \forall x \in H$$

Then, by lemma 1.1, taking into account that in $\mu$ the spectral families are continuous and then the left-continuous coïncide with the right-continuous, we have (1.11). ∎

It is noticeable that the preceeding theorem holds without any hypothesis on the domains of $A_\varepsilon$ . Let us now state without proof another theorem own to Rellich [ 1 ].

Theorem 1.3 - Let H be a Hilbert space, and $A_\varepsilon$, A selfadjoint operators in H such that

a) The intersection $\mathcal{D}$ of the domains $D(A_\varepsilon)$ is such that, for any $x \in H$, there exists the sequence $x_i \in \mathcal{D}$ for which

$$x_i \to x \qquad \text{in } H \text{ strongly}$$

$$Ax_i \to Ax \qquad \text{in } H \text{ strongly}$$

b) For any $x \in \mathcal{D}$,

$$A_\varepsilon x \to Ax \qquad \text{in } H \text{ strongly}$$

c) $\mu$ is not an eigenvalue of A.

Then, $\quad E(A_\varepsilon , \mu)x \to E(A , \mu)x$ in H strongly $\forall x \in H$ .

2.- Application to singular perturbations - We are going to study some spectral properties of the singular perturbations introduced in chap. 9, sect. 1 and 3 in the self-adjoint case, for compact and non compact operators.

Let V, W, H be three Hilbert spaces and $H' = H$, $V'$ , $W'$ their duals, satisfying

(2.1) $\qquad V \subset W \subset H = H' \subset W' \subset V'$

with dense and continuous (not necessarily compact) injections.

Moreover, let a(u , v) (resp. b(u , v)) be a sesquilinear continuous form on V (resp. W) satisfying :

(2.2) $\qquad a(v , v) \geqslant \alpha\|v\|_V^2 \quad ; \quad \alpha > 0 \quad ; \quad \forall v \in V$

(2.3) $\qquad b(v , v) \geqslant \beta\|v\|_W^2 \quad ; \quad \beta > 0 \quad ; \forall v \in W$

Moreover, let $B_\varepsilon$ , $B_0$ the selfadjoint operators of H associated to the forms $b + \varepsilon a$ and b according to the first representation theorem ( $\varepsilon$ is a small positive parameter) :

(2.4) $\qquad (B_\varepsilon u , v)_H = b(u , v) + \varepsilon\, a(u , v)$

(2.5) $\qquad (B_0 u , v)_H = b(u , v)$

Let $E_\varepsilon(\lambda)$, $E_0(\lambda)$ be the spectral families associated with the operators $B_\varepsilon$ and $B_0$ respectively.

Theorem 2.1 - Under the hypothesis (2.1) - (2.5), we have :

$E_\varepsilon(\lambda)v \to E_0(\lambda)v$ in H strongly, for any $v \in H$ and any $\lambda$ which is not an eigenvalue of $B_0$ and $B_\varepsilon$ .

Proof - For any $f \in H$, from chap. 9, theorem 1.1, we have :

$$B_\varepsilon^{-1} f \to B_0^{-1} f \qquad \text{in } H \text{ strongly}$$

Moreover, from (2.2) - (2.5), for some $\gamma > 0$ :

$$(B_\varepsilon v, v) \geqslant \beta \| v \|_W^2 \geqslant \gamma \| v \|_H^2 ,$$

and theorem 2.1 follows from theorem 1.2. ■

Remark 2.1 - The preceeding theorem applies in particular to operators with continuous spectrum (without eigenvalues). Examples of such operators for boundary value problems in unbounded domains will be seen in part IV. ■

More precise results are obtained if the embedding $W \subset H$ is compact. In this case, the operators $B_\varepsilon^{-1}$ and $B_0^{-1}$ are compact, and for fixed $\varepsilon$, we are in the framework of chap. 2, sect. 6 (Note that for fixed $\varepsilon > 0$ and $\varepsilon = 0$, the forms on the right hand side of (2.4) and (2.5) may be taken as scalar products on V and W, respectively). The eigenvalues may be written

(2.6) $$0 < \lambda_1^\varepsilon \leqslant \lambda_2^\varepsilon \leqslant \dots \leqslant \lambda_n^\varepsilon \leqslant \dots \to +\infty$$

with the corresponding eigenvectors

(2.7) $$w_1^\varepsilon , w_2^\varepsilon \dots w_n^\varepsilon \dots \quad ; \quad \| w_n^\varepsilon \|_H = 1$$

and analogous sequences for $\varepsilon = 0$. Moreover, we shall write (2.6) under the form

(2.8) $$0 < \lambda_{(1)}^\varepsilon < \lambda_{(2)}^\varepsilon < \dots < \lambda_{(n)}^\varepsilon < \dots \to +\infty$$

if each eigenvalue $\lambda_{(i)}$ is written once (its multiplicity will be m(i)). We then have the following theorem (which was also proved in chap. 11, sect. 3 in a slightly different form, even for not selfadjoint operators).

Theorem 2.2 - Under the hypothesis (2.1) - (2.5), if the embedding of W into H is compact, if $\lambda_{(i)}^0$ is an eigenvalue of $B_0$ with multiplicity m, and $]\lambda_{(i)}^0 - \delta , \lambda_{(i)}^0 + \delta[$ is an interval containing no other eigenvalue of $B_0$, then, for sufficiently small $\varepsilon$, the operator $B_\varepsilon$ has eigenvalues in that interval with total multiplicity m . Moreover, the corresponding eigenvectors may be chosen in such a way that they tend (in the norm of H) to the eigenvectors of $B_0$ corresponding to $\lambda_{(i)}^0$.

To prove this theorem, we shall establish several lemmas.

Lemma 2.1 - If $\lambda$ is not an eigenvalue of $B_0$, then, for sufficiently small $\varepsilon$, it is not an eigenvalue of $B_\varepsilon$ .

<u>Proof</u> - By contradiction. If $\lambda$ is an eigenvalue of $B_\varepsilon$ for a sequence $\varepsilon \searrow 0$, there exists $f^\varepsilon$ such that

$$B^\varepsilon f^\varepsilon = \lambda f^\varepsilon \quad ; \quad \| f^\varepsilon \|_H = 1 \tag{2.9}$$

and, by weak compactness we may assume

$$f^\varepsilon \to f^0 \quad \text{in } H \text{ weakly} \Rightarrow \text{in } W' \text{ weakly} \tag{2.10}$$

Then, from chapter 9, theorem 1.1,

$$f^\varepsilon = \lambda B_\varepsilon^{-1} f \longrightarrow \lambda B_0^{-1} f^0 \text{ in } W \text{ weakly} \tag{2.11}$$

and thus <u>in H strongly</u>. By comparaison with (2.10) we see that

$$f_o = \lambda B_o^{-1} f^o \tag{2.12}$$

and the convergence in (2.10) is in H strongly. This implies

$$\| f^0 \| = 1 \quad f^0 \neq 0 \quad ; \quad B_0^{-1} f_0 = \lambda f_0$$

i.e. $\lambda$ is an eigenvalue of $B_0$, which is in contradiction with the hypothesis. ∎

Then, by applying the theorem 2.1, we obtain

<u>Lemma 2.2</u> - If $\lambda$ is not an eigenvalue of $B_0$, we have

$$E_\varepsilon(\lambda)v \to E_0(\lambda)v \qquad \text{in } H \text{ strongly}, \quad \forall v \in H.$$

This implies :

<u>Lemma 2.3</u> - If $\lambda$ is not an eigenvalue of $B_0$, for sufficiently small $\varepsilon$, we have

$$\dim E_\varepsilon(\lambda)H \geqslant \dim E_0(\lambda)H$$

<u>Proof</u> - If not, there exists a sequence $\varepsilon \searrow 0$ such that

$$\dim E_\varepsilon(\lambda)H < \dim E_0(\lambda)H$$

and then there exists

$$x_\varepsilon \in E_0(\lambda)H \quad ; \quad \| x_\varepsilon \|_H = 1$$

orthogonal to $E_\varepsilon(\lambda)H$ ; moreover, $E_0(\lambda)H$ has finite dimension, and we may assume

$$x_\varepsilon \to x \qquad \text{in } H \text{ strongly}, \quad \| x \| = 1$$

By lemma 2.2, we have

$$E_\varepsilon(\lambda)x - E_0(\lambda)x \to 0 \quad \text{in } H \text{ strongly} \tag{2.13}$$

Moreover, the left hand side of (2.13) writes

$$E_\varepsilon(x - x_\varepsilon) + (E_\varepsilon - E_0)x_\varepsilon + E_0(x_\varepsilon - x)$$

the first and third terms tend to zero. As for the second,

it is equal to $-E_0 x_\varepsilon$ , which tends to $-E_0 x = -x$ and we have a contradiction. ■

<u>Lemma 2.4</u> - If $\lambda$ is not an eigenvalue of $B_0$, we have

$$\dim E_\varepsilon(\lambda)H \leqslant \dim E_0(\lambda)H$$

for all $\varepsilon > 0$.

<u>Proof</u> - If not, for certain $\lambda$ and $\varepsilon$, there exists an eigenvector $\tilde{w}^\varepsilon$ of $B_\varepsilon$ corresponding to the eigenvalue $\tilde{\lambda}^\varepsilon < \lambda$ which is orthogonal in H to all the eigenvectors $w_i^0$ with $\lambda_i^0 < \lambda$ . We then have (all the eigenvectors are normalized in H) :

$$B_\varepsilon^{1/2}\, \tilde{w}^\varepsilon = (\tilde{\lambda}^\varepsilon)^{1/2}\, \tilde{w}^\varepsilon \Rightarrow \|B_\varepsilon^{1/2}\, \tilde{w}^\varepsilon\|_H^2 = \tilde{\lambda}^\varepsilon < \lambda\ , \quad \text{i.e. :}$$

$$b(\tilde{w}^\varepsilon , \tilde{w}^\varepsilon) + \varepsilon\, a(\tilde{w}^\varepsilon , \tilde{w}^\varepsilon) < \lambda \tag{2.14}$$

On the other hand,

$$B_0^{1/2}\, \tilde{w}^\varepsilon = \Sigma(\lambda_i^0)^{1/2}(\tilde{w}^\varepsilon , w_i^0)\, w_i^0$$

and by taking into account that the terms of the preceeding sum are zero for $\lambda_i^0 \leqslant \lambda$ ,

$$\|B_0^{1/2}\, \tilde{w}^\varepsilon\|^2 \geqslant \lambda\, \|\tilde{w}^\varepsilon\|^2 = \lambda\ , \quad \text{i.e. :}$$

$$b(\tilde{w}^\varepsilon , \tilde{w}^\varepsilon) \geqslant \lambda \tag{2.15}$$

which is in contradiction to (2.14). ■

From lemmas 2.3 and 2.4, we see that for small $\varepsilon$ , the dimensions of $E_\varepsilon(\lambda)H$ and $E_0(\lambda)H$ are equal, and by applying this result to the points $\lambda^0_{(i)} - \delta$ , $\lambda^0_{(i)} + \delta$ , we see that the projectors $P_\varepsilon$ and $P_0$ associated to this interval have the same dimension. If $w^0_{(i)1}, \ldots\ w^0_{(i)m}$ are the eigenvectors of $B_0$ corresponding to $\lambda^0_{(i)}$ we may take

$$w^\varepsilon_{(i)j} = P_\varepsilon\, w^0_{(i)j}$$

as the eigenvectors of $B_\varepsilon$ corresponding to the interval. They are linearly independent and converge as $\varepsilon \to 0$ to $w^0_{(i)j}$. Theorem 2.2 is proved. ■

## 3.- <u>Remarks about hyperbolic equations and Fourier transform</u>. <u>Application to a problem of homogenization</u> -

The aim of this section is to give a representation of the solution of a class of hyperbolic problems and its Fourier transform which will be used in the sequel.

We consider hyperbolic problems in the framework of chap. 4, sect. 5. More exactly, we have with the standard hypothesis and notations : ( $\|.\|$ and ( , ) are the norm and scalar product in H) :

(3.1) $$V \subset H \subset V'$$

and let a(u , v) be a symmetric continuous form on V, such that there exists a $\gamma > 0$ for which

(3.2) $$a(v , v) \geqslant \gamma \|v\|_V^2$$

Let A be the selfadjoint operator of H associated to the form a according to the second representation theorem. We consider the problem :

(3.3) $$\frac{d^2u(t)}{dt^2} + Au(t) = 0$$

(3.4) $$u(0) = 0 \quad ; \quad u'(0) = u_1 \in H$$

(Note that u(0) is taken to be zero : this is sufficient in the sequel). We know (chap. 4, sect. 5) that (3.3), (3.4) has a unique solution in the framework of semigroup theory (Stone's group !) in V $\times$ H. We then have the following characterization of the solution u :

Proposition 3.1 - The solution u of (3.3), (3.4) is the unique function such that, for any fixed T (either positive or negative) satisfies

(3.5) $$u \in L^\infty(0 , T ; V) \quad ; \quad u' \in L^\infty(0 , T ; H)$$

(3.6) $$u(0) = 0$$

(3.7) $$\int_0^T [a(u(t) , \psi(t)) - (u'(t) , \psi'(t))] \, dt = (u_1 , \psi(0))$$

for any test function $\psi$ of the form $\psi(t) = \phi(t)v$ where v is any element in a dense set of V and $\phi$ is any function of the class

(3.8) $$\phi \in \{ \phi \ ; \ \phi \in C^1([0 , T]) \ ; \ \phi(T) = 0 \}$$

Proof - First, let u be the solution of (3.3), (3.4). Then, it satisfies (3.5), (3.6). Moreover, in the particular case where $u_1 \in V$, the initial values belong to the domain of the operator $\mathcal{A}$ (see chap. 4, sect. 5) and consequently (3.3) makes sense for any t. By multiplying it by $\psi$ , we have

(3.9) $$(u'' , \psi) + a(u , \psi) = 0$$

and of course

(3.10) $$\frac{d}{dt}(u' , \psi) = (u'' , \psi) + (u' , \psi')$$

Then, if we integrate (3.10) by taking into account (3.9) and (3.4), we obtain

(3.7). If $u_1 \in H$, according to semigroup theory, we consider u as the limit in the spaces (3.5) of the solutions $u^i$ corresponding to $u_1^i \to u_1$ in H, with $u_1^i \in V$. We then have (3.7).

Conversely, if u satisfies (3.5) - (3.7), to see that u is the solution of (3.3), (3.4), it suffices to prove that u is unique. First, we see that u satisfies (3.7) for any $\psi$ of the class

$$(3.11) \qquad \begin{cases} \psi \in L^2(0\,,T\,;V) \quad ; \quad \psi' \in L^2(0\,,T\,;H) \\ \psi(T) = 0 \end{cases}$$

because the functions which are finite sums of functions of the type $\phi v$ with $\phi \in$ (3.8), and v in a dense set in V are dense in (3.11). Now, we prove the uniqueness by using the classical device of Ladyzhenskaya. Let u satisfy (3.5), (3.6) and (3.7) with $u_1 = 0$ with $\psi$ in the class (3.11). Then, for any chosen $s \in ]\,0,T\,[$, we define

$$\psi(t) = \begin{cases} \int_s^t u(\sigma)\, d\sigma & \text{if } t < s \\ 0 & \text{if } t \geq s \end{cases}$$

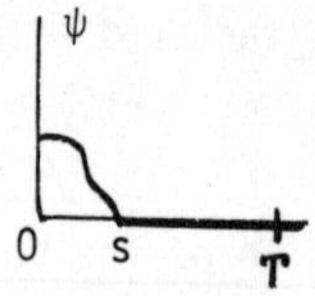


$$\psi'(t) = \begin{cases} u(t) & \text{if } t < s \\ 0 & \text{if } t \geq s\,. \end{cases}$$

which belongs to the class (3.11). Thus, (3.7) becomes

$$\int_0^s [\, a(\psi'\,,\psi) - (u'\,,u)\,]\, dt = 0$$

and taking the real part,

$$0 = \int_0^s \frac{d}{dt}[\, a(\psi\,,\psi) - (u\,,u)\,]\, dt = -\, a(\psi(0)\,,\psi(0)) - \|u(s)\|^2$$

which implies, by (3.2) that the two terms in the right hand side are zero. In particular $u(s) = 0$ (for arbitrary s !). The proposition is proved. ■

We now consider the Fourier transform (from t into $\lambda$) of $u(t)$. The solution of (3.3) (3.4) may be written

$$(3.12) \qquad u(t) = A^{-1/2} \sin(A^{1/2} t)\, u_1$$

$$(3.13) \qquad u'(t) = \cos(A^{1/2} t) u_1$$

which is easily checked (see Milkhlin [1], sect. 24.8). According to (3.2), there exists $\beta^2 > 0$ such that

$$(Av , v) \geqslant \beta^2 \|v\|^2$$

and thus the spectral family $E(A , \lambda)$ is zero for $\lambda < \beta^2$ (c.f. chap. 2, remark 4.7). We have

$$A = \int_{-\infty}^{+\infty} \lambda dE(A , \lambda) = \int_{\beta^2}^{\infty} \lambda \, dE(A , \lambda)$$

$$A^{1/2} = \int_{-\infty}^{+\infty} \mu^{1/2} \, dE(A , \mu) = \int_{\beta}^{\infty} \lambda \, dE(A^{1/2} , \lambda) \Longrightarrow$$

(3.14)
$$E(A^{1/2} , \lambda) = \begin{cases} E(A , \lambda^2) & \text{for } \lambda > \beta \\ 0 & \text{for } \lambda < \beta \end{cases}$$

Consequently, $A^{1/2}t$ writes

(3.15)
$$\cos A^{1/2}t = \frac{1}{2} \int_{\beta}^{\infty} (e^{i\lambda t} - e^{-i\lambda t}) dE(B^{1/2} , \lambda) =$$

$$= \frac{1}{2} \int_{|\lambda| > \beta} e^{it\lambda} \, d[E(B^{1/2} , \lambda) - E(B^{1/2}, -\lambda) ] = \frac{1}{2} \int_{-\infty}^{+\infty} e^{it\lambda} \, d[ \ ]$$

where

(3.16)
$$[ \ ] \equiv [E(B^{1/2} , \lambda) - E(B^{1/2} , -\lambda)]$$

Now, we multiply (3.13) by any test element $v \in H$ ; by virtue of (3.15) we have :

(3.17)
$$(u'(t), v) = \frac{1}{2} \int_{-\infty}^{+\infty} e^{it\lambda} \, d([ \ ] u_1 , v)$$

and we see that (3.17) is formally the inverse Fourier transform of

(3.18)
$$\frac{de(\lambda)}{d\lambda} \quad \text{where } e(\lambda) \equiv ( [ \ ] \, u_1 , v)$$

In fact, (3.17) is the inverse Fourier transform of (3.18) in the sense of tempe- rated distributions. For, $e(\lambda)$ is a function of bounded variation (this is easily seen from the estimate of chap. 2, remark 4.4, see also Kato [ 2 ] , chap. 6, sect. 5), the total variation is $\leqslant 2 \|u_1\| \, \|v\|$ and thus $e(\lambda)$ is the difference of two increasing functions $e_1$, $e_2$, such that

$$\int_{-\infty}^{+\infty} de_i < +\infty \qquad ; \qquad i = 1 , 2$$

The Fourier transform of temperated distributions generalizes that of such functions (see Schwartz [ 1 ], chap. 7, sect. 2 and 6). Consequently, we have

Proposition 3.2 - If $u(t)$ is the solution of (3.3), (3.4), for any $v \in H$ we have

(3.19)
$$(u'(t) , v)^\wedge = \sqrt{\frac{\pi}{2}} \frac{d}{d\lambda} ( [E(B^{1/2}, \lambda) - E(B^{1/2}, -\lambda)] \, u_1 , v)$$

where ^ and $d/d\lambda$ denote the Fourier transform and the derivation in the sense of tempered distributions.

Remark 3.1 - Under the supplementary hypothesis that the embedding $V \subset H$ is compact, the operator A has eigenvalues and eigenvectors as in (2.6) (2.7). The relation (3.19) becomes

$$(3.20)\quad \begin{cases} (u'(t)\ ,\ v)^{\wedge} = \sqrt{\frac{\pi}{2}}\ \sum_{1}^{\infty} (u_1, w_i)(w_i\ ,\ v)\ [\ \delta(\lambda - \alpha_i) + \delta(\lambda + \alpha_i)] \\ \alpha_i^2 = \lambda_i \qquad ; \qquad \delta(\lambda \mp \alpha_i) = \text{Dirac's distribution} \end{cases}$$

We see that the Fourier transform is a sum of Dirac's distributions at the points ± the square roots of the eigenvalues. This is also easily seen from the expression at the end of chap. 4, remark 5.1. ■

Remark 3.2 - Proposition 3.2 furnishes a method to obtain spectral properties of problems depending on a parameter if the properties for the corresponding hyperbolic problems are known. It suffices to apply the Fourier transform bearing in mind that it is continuous in the topology of tempered distributions. ■

As an example, let us study an eigenvalue problem in homogenization. We consider the operator $A_\varepsilon$ and the function $\rho(\frac{x}{\varepsilon})$ in the framework of chap. 5, def. 6.2. We want to study properties of the limit process $\varepsilon \to 0$ for the spectral problem

$$(3.21)\qquad A_\varepsilon\, w_i^\varepsilon = \lambda_i^\varepsilon\, \rho(\tfrac{x}{\varepsilon})\, w_i^\varepsilon$$

It is clear that this problem is not the classical eigenvalue problem for the operator $A_\varepsilon$ (which is studied in chap. 11, sect. 3), but of the operator $A_\varepsilon$ with respect to the operator product by $\rho(x/\varepsilon)$ (which depends on $\varepsilon$!)

For fixed $\varepsilon$, this problem is in the framework of propositions 3.1 and 3.2, if we take $H = L^2_\varepsilon$ , $V = H^1_0(\Omega)$ where $L^2_\varepsilon$ is the space $L^2(\Omega)$ equipped with the scalar product

$$(3.22)\qquad (u\ ,\ v)_{L^2_\varepsilon} = \int_\Omega \rho(\tfrac{x}{\varepsilon}) uv\, dx$$

Analogous considerations hold for the homogenized problem

$$(3.23)\qquad A_h\, w_i^h = \lambda_i^h\, \tilde{\rho}\, w_i^h$$

where $A_h$ is the classical homogenized operator and $\tilde{\rho}$ is the mean value of $\rho(y)$.

Let us fix $u_1$, $v \in L^2(\Omega)$. Then, if we take $u_o = 0$, $u_1 = u_1$ in theorem 6.3 of chap. 5, we have

(3.24) $$u^{\varepsilon} \to u^{h} \quad \text{in } L^{\infty}(-\infty\ ,\ -\infty\ ;\ L^2(\Omega)) \text{ weakly } *$$

(note that that theorem also holds for negative t). Then, after multiplication by v in $L^2(\Omega)$ we have :

(3.25) $$(u^{\varepsilon}\ ,\ v)_{L^2(\Omega)} \to (u^{h}\ ,\ v)_{L^2(\Omega)} \quad \text{in } L^{\infty}(-\infty\ ,\ +\infty) \text{ weakly } * .$$

But the weak convergence in $L^{\infty}$ implies convergence in the space of tempera-ted distributions S', and thus the Fourier transform of (3.25) converge in S' :

(3.26) $$\widehat{(u^{\varepsilon}\ ,\ v)}_{L^2(\Omega)} \to \widehat{(u^{h}\ ,\ v)}_{L^2(\Omega)} \quad \text{in } S'$$

We now write (3.26) by using (3.20). To this end, we bear in mind that, according to (3.22)), the scalar products in (3.20) are in fact scalar products in $L^2_{\varepsilon}$ (and in the analogous space $L^2_h$ for the homogenized problem) according to the following equivalence :

(3.27) $$(f\ ,\ g)_{L^2_{\varepsilon}} = \int_{\Omega} \rho^{\varepsilon}\ f\ g\ dx = (f\ ,\ \rho^{\varepsilon}\ g)_{L^2} \quad ; \quad \text{where } \rho^{\varepsilon}(x) = \rho(\frac{x}{\varepsilon})$$

Then, the left hand side of (3.20) is :

$$\widehat{(u^{\varepsilon}\ ,\ v)}_{L^2} = \widehat{(u^{\varepsilon}, \rho\varepsilon^{-1}\ v)}_{L^2_{\varepsilon}}$$

and with (3.20) (written in $L^2_{\varepsilon}$) we have

(3.28) $$(u^{\varepsilon}\ ,\ v)_{L^2} = \sum_{i=1}^{\infty} (u_1\ ,\ \rho^{\varepsilon}\ w_i^{\varepsilon})_{L^2}\ (w_i^{\varepsilon}\ ,\ v)_{L^2}[\ \delta(\lambda - \alpha_i^{\varepsilon}) + \delta(\lambda + \alpha_i^{\varepsilon})]$$

where $(\alpha_i^{\varepsilon})^2 = \lambda_i^{\varepsilon}$. We evidently have an analogous relation for $u^h$. Then, (3.26) may be expressed as :

<u>Theorem 3.1</u> - If ( , ) is the classical scalar product of $L^2(\Omega)$ and $\rho^{\varepsilon}(x) = \rho(x/\varepsilon)$, the temperated distribution (of the variable $\lambda$)

(3.29) $$\sum_{i=1}^{\infty} (u_1\ ,\ \rho^{\varepsilon}\ w_i^{\varepsilon})(w_i^{\varepsilon}\ ,\ v)[\ \delta\ (\lambda - \alpha_i^{\varepsilon}) + \delta(\lambda + \alpha_i^{\varepsilon})]$$

converges in the sense of the temperated distributions to

(3.30) $$\sum_{i=1}^{\infty} (u_1\ ,\ \tilde{\rho}\ w_i^{h})(w_i^{h}\ ,\ v)\ [\ \delta(\lambda - \alpha_i^{h}) + \delta(\lambda + \alpha_i^{h})]$$

for any $u_1$, $v \in L^2(\Omega)$.

Corollary 3.1 - Each eigenvalue $\lambda_i^h$ of the homogenized problem (3.23) is an accumulation point of eigenvalues $\lambda^\varepsilon$ of the problem (3.21) for $\varepsilon \searrow 0$.

Proof - It suffices to choose $v = w_i^h$, $u_1 = \tilde{\rho}^{-1} w_i^h$ and (3.30) takes the form

$$\delta(\lambda - \alpha_i^h) + \delta(\lambda + \alpha_i^h) \qquad ; \qquad (\alpha_i^h)^2 = \lambda_i^h$$

which is the sum of two Dirac's distributions at the points $\mp \alpha_i^h$. By applying this distribution to a test function $\theta \in \mathcal{D}$ with support in a little neighbourhood of $\alpha_i^h$, the theorem shows that for sufficiently small $\varepsilon$, this neighbourhood contains some $\alpha_j^\varepsilon$. ■

Remark 3.3 - The problem of the acustic vibrations of a inhomogeneous compressible fluid with periodic, piecewise constant structure (a suspension of drops in a homogeneous fluid, for instance) leads to the equation

$$\frac{1}{\rho(x)\, C^2(x)} \frac{\partial^2 p}{\partial t^2} = \operatorname{div}\left(\frac{1}{\rho(x)} \operatorname{\underline{grad}} p\right)$$

where $\rho$, $C$, $p$ are the density, speed of propagation and perturbation of pressure. The study of the eigenfrequencies of that equation leads to a problem of the type (3.21). ■

4.- Acustic vibrations in a domain with very corrugated boundary - Let us consider the acustic vibrations of air in a (connected) vessel $\Omega$ with rigid boundary. If $u$ denotes the perturbation of pressure and we take the physical units in such a way that the speed of propagation is 1, we have the classical Neumann problem for the wave equation (The Neumann condition means that the velocity is tangential to $\partial\Omega$) :

$$(4.1) \qquad \frac{\partial^2 u}{\partial t^2} + B u = 0$$

$$(4.2) \qquad B u = - \Delta u \quad \text{in } \Omega \qquad ; \qquad \frac{\partial u}{\partial n} = 0 \quad \text{on } \partial\Omega .$$

The operator $B$ has the eigenvalue $\mu_1 = 0$ associated to the eigenfunction $u$ = cost. ; the remaining eigenvalues $\mu_2, \mu_3 \ldots$ are strictly positive. As in chap. 4, remark 5.1, the solution of (4.1), (4.2) corresponding to arbitrary initial values is a sum (a series !) of solutions of (4.1) depending on $t$ in the form $\sin \sqrt{\mu_i}\, t$ or $\cos \sqrt{\mu_i}\, t$. The values $\sqrt{\mu_i}$ for $i > 1$ are the eigenfrequencies of $\Omega$ (Note that for $\mu = \mu_1 = 0$ the corresponding solution is constant in time). Each eigenfrequency is associated to a simple sound with a specific pitch.

Moreover, the operator B is associated (see chapter 3, sect. 2) with the form :

$$(4.3) \qquad b(u , v) = \int_\Omega \frac{\partial u}{\partial x_i}\frac{\partial v}{\partial x_i}\, dx \quad ; \quad (Bu , v) = b(u , v)$$

and the classical theory of eigenvalues of symmetric, compact operators applies (see, for instance Courant Hilbert [1] or Necas [1] , chapter 1, sect. 5 ; of course, B has a compact resolvent.

Proposition 4.1 - If $\Omega$ is bounded and connected, there exists an orthonormal basis of $L^2(\Omega)$ formed by eigenvectors $w_1$ , $w_2$, ... associated with the eigenvalues

$$0 = \mu_1 < \mu_2 \leq \mu_3 \leq \ldots \to +\infty$$

of the operator B (defined by (4.2)), and

$$(4.4) \qquad \mu_k = \inf \frac{b(v , v)}{(v , v)_{L^2}}$$

where the inf is calculated for $v \in H^1(\Omega)$ such that

$$(v , w_j)_{L^2} = 0 \quad , \quad j = 1 , 2 \ldots \quad k-1 \quad ,$$

and

$$\mu_i = \frac{b(w_i , w_i)}{\| w_i \|^2_{L^2}}$$

Remark 4.1 - The hypothesis that $\Omega$ is connected ensures that $\mu_2 > \mu_1 = 0$ and so that the eigenvalue $\mu_1$ is simple. ■

Remark 4.2 - In proposition 4.1, it is assumed that $\Omega$ is sufficiently smooth to ensure that the compactness theorem of Rellich holds. $\partial\Omega$ continuous (see Necas Necas [1] chap. 1 for an exact definition) is a sufficient condition. ■

Now, we shall consider the problem 4.1, 4.2 for a sequence of domains $\Omega_\varepsilon$ which converge to $\Omega_0$ in a sense that will be described later. In fact $\partial\Omega_\varepsilon$ is a very corrugated perturbation of $\partial\Omega_0$. We shall show (proposition 4.2) that the eigenvalue $\mu_2^\varepsilon$ tends to zero as $\varepsilon \to 0$, although $\mu_2^0 > 0$. Consequently, the lower eigenfrequency of $\Omega_\varepsilon$ (that is to say $\sqrt{\mu_2^\varepsilon}$ , because $\mu_1^\varepsilon = 0$ is not associated to a vibration) tends to zero, although the lower eigenfrequency of $\Omega_0$ is positive. this situation is very different of that of theorem 2.2, where the eigenvalues of the limit problem were limits of eigenvalues with corresponding multiplicities. On the other hand, we shall prove (proposition 4.3) that each eigenfrequency of the limit problem in $\Omega_0$ is an accumulation point of eigenfrequencies corresponding to $\Omega_\varepsilon$ with $\varepsilon \to 0$.

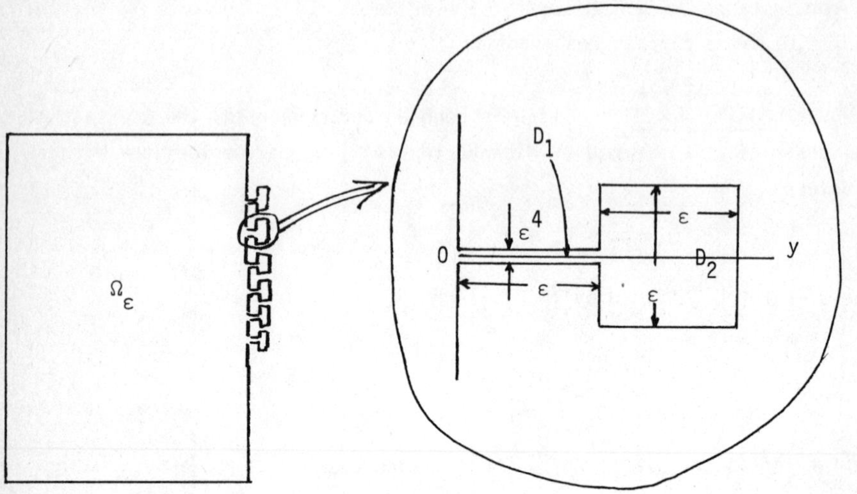


fig. 1 fig. 2

Now we consider the two-dimensional case, but an analogous study may be performed in the N-dimensional case.

Let $\Omega_0$ be a fixed rectangle. $\Omega_\varepsilon$ is the domain enclosed by $\partial\Omega_\varepsilon$ described in fig. 1 and fig. 2. Note that there is an arbitrary number of "protuberences" and this number may tend to infinity as $\varepsilon \searrow 0$, (under the condition that different protuberances must not overlap !). Each protuberance is formed by a "corridor" $D_1$ with length $\varepsilon$ and width $\varepsilon^4$ and by a "room" $D_2$ with length and width $\varepsilon$. (see fig. 1 and fig. 2).

Proposition 4.2 - We consider the domain $\Omega_\varepsilon$(resp. $\Omega_0$) according to the preceeding definition. We consider the second eigenvalue $\mu_2^\varepsilon$ (resp. $\mu_2^0$) of the Neumann problem (4.1), (4.2) in the domain $\Omega_\varepsilon$ (resp. $\Omega_0$). Then :

a) $\mu_2^0 > 0$

b) $\mu_2^\varepsilon \searrow 0$ as $\varepsilon \searrow 0$.

Proof - Part a) is contained in proposition 4.1. To prove part b), we have

$$\mu_2^\varepsilon = \inf \frac{\int_{\Omega_\varepsilon} |\nabla v|^2 \, dx}{\int_{\Omega_\varepsilon} |v|^2 \, dx} \tag{4.5}$$

where the inf is taken in the class of the $H^1$-functions orthogonal to the constant functions :

$$\mathcal{E}^\varepsilon = \{v \; ; \; v \in H^1(\Omega_\varepsilon) \quad ; \quad \int_{\Omega_\varepsilon} v \, dx = 0 \}$$

Then, b) will be proved if we find a function $v^\varepsilon \in \mathcal{E}^\varepsilon$ (depending on $\varepsilon$) such that the quotient of integrals in (4.5) tend to zero as $\varepsilon \to 0$.

To this end, we choose a "protuberance" as in fig. 2,(note that y is the abscissa along $D_1$),and we define

$$v^\varepsilon = \begin{cases} -C(\varepsilon) + y & \text{in } D_1 \\ -C(\varepsilon) + \varepsilon & \text{in } D_2 \\ -C(\varepsilon) & \text{elswhere} \end{cases} \tag{4.6}$$

which is indeed a function of $H^1(\Omega_\varepsilon)$. The condition

$$\int_{\Omega_\varepsilon} v^\varepsilon \, dx = 0 \qquad \text{implies} \quad C(\varepsilon) = O(\varepsilon^3)$$

and we immediately obtain from (4.6) :

$$\int_{\Omega_\varepsilon} |\nabla v^\varepsilon|^2 \, dx = O(\varepsilon^5)$$

$$\int_{\Omega_\varepsilon} |v^\varepsilon|^2 \, dx = O(\varepsilon^4)$$

and the proposition is proved. ∎

Proposition 4.3 - Each eigenfrequency $\sqrt{\mu_i^0}$ of the Neumann problem (4.1), (4.2) in the domain $\Omega_0$ is an accumulation point of eigenfrequencies $\sqrt{\mu_j^\varepsilon}$ corresponding to $\Omega_\varepsilon$ with $\varepsilon \searrow 0$.

To prove this proposition, we first remark that it suffices to prove it for $\mu + 1$ instead of $\mu$, that is to say, we may consider the operator $-\Delta + 1$ instead of $-\Delta$, and we apply the method used in theorem 3.1 and corollary 3.1. We then consider the hyperbolic problem

$$\text{(4.7)} \qquad \frac{\partial^2 u}{\partial t^2} + (-\Delta + 1)u^\varepsilon = 0 \quad \text{in } \Omega_\varepsilon \quad ; \quad \left.\frac{\partial u^\varepsilon}{\partial n}\right|_{\partial\Omega_\varepsilon} = 0$$

$$\text{(4.8)} \qquad u^\varepsilon(0) = 0 \quad ; \quad u^{\varepsilon\prime}(0) = u_1$$

and an analogous problem with $u^0$, $\Omega_0$ instead of $u^\varepsilon$, $\Omega_\varepsilon$. The initial value $u_1$ in (4.8) is an function of $L^2(\Omega_0)$ continuated by zero out of $\Omega_0$ (and thus belonging to $L^2(\Omega_\varepsilon)$).

On the other hand, it is clear that the bilinear form associated to $-\Delta+1$ (for the Neumann boundary condition is

$$\text{(4.9)} \qquad a(u , v) = (u , v)_{H^1}$$

We first establish the following lemma :

Lemma 4.1 - The solution $u^\varepsilon$ of (4.7), (4.8) tends to the solution $u^0$ of the corresponding problem in $\Omega_0$ in the following sense :

$$\text{(4.10)} \qquad u^\varepsilon\big|_{\Omega_0} \to u^0 \quad \text{in } L^\infty(-\infty , +\infty ; H^1(\Omega_0)) \text{ weakly } *$$

$$\text{(4.11)} \qquad u^{\varepsilon\prime}\big|_{\Omega_0} \to u^{0\prime} \quad \text{in } L^\infty(-\infty , +\infty ; L^2(\Omega_0)) \text{ weakly } *$$

Proof - From (4.7), (4.8), (4.9) and the classical property of conservation of the energy norm (see chap. 4, sect. 5), we have :

$$\text{(4.12)} \qquad \|u^{\varepsilon\prime}(t)\|^2_{L^2(\Omega_\varepsilon)} + \|u^\varepsilon(t)\|^2_{H^1(\Omega_\varepsilon)} = \|u_1\|^2_{L^2(\Omega_0)}$$

where the right hand side is independent of $\varepsilon$ and t. By taking the restriction of $u^\varepsilon$ to $\Omega_0$, we may extract a subsequence such that

$$\text{(4.13)} \qquad u^\varepsilon\big|_{\Omega_0} \to u^* \quad \text{in } L^\infty(-\infty , +\infty ; H^1(\Omega_0)) \text{ weakly } *$$

$$\text{(4.14)} \qquad u^{\varepsilon\prime}\big|_{\Omega_0} \to u^{*\prime} \quad \text{in } L^\infty(-\infty , +\infty ; L^2(\Omega_0)) \text{ weakly } *$$

and it suffices to prove that $u^*$ is the solution of the problem (4.7), (4.8) in $\Omega_0$.

To this end, we apply proposition 3.1 taking as v the restriction to $\Omega_\varepsilon$ of a fixed function of $H^1(R^2)$. We have (for any finite T) :

$$\int_0^T [(u^\varepsilon , \psi)_{H^1(\Omega_\varepsilon)} - (u^{\varepsilon\prime} , \psi')_{L^2(\Omega_\varepsilon)}] \, dt = (u_1 , \psi(0))_{L^2(\Omega_0)} \tag{4.15}$$

where we pass to the limit $\varepsilon \to 0$ by writting $\Omega_\varepsilon = \Omega_0 + (\Omega_\varepsilon - \Omega_0)$ ; the integrals over $\Omega_0$ pass to the limit by (4.13), (4.14) ; the integrals over $\Omega_\varepsilon - \Omega_0$ tend to zero by virtue of $|\Omega_\varepsilon - \Omega_0| \to 0$ and estimate (4.12). Then, the limit $u^*$ satisfy a relation analogous to (4.15) in $\Omega_0$ instead of $\Omega_\varepsilon$. Moreover,

$$u^*(0) = 0$$

as a consequence of (4.13), (4.14) and (4.8) (this is a kind of "trace theorem" for t = 0). Consequently, according to proposition 3.1, $u^* = u^0$ and lemma 4.1 is proved. ■

Now, in order to prove proposition 4.3, we take the Fourier transform of (4.11) according to (3.19) and (3.20). The proof is then finished as in corollary 3.1.

<u>Proposition 4.3 is proved</u>.

## CHAPTER 13

## STIFF PROBLEMS IN CONSTANT AND VARIABLE DOMAINS

This chapter is devoted to some examples of "stiff" problems, i.e., boundary value problems for partial differential equations with coefficients depending on a parameter $\varepsilon$ which tends to zero in such a way that the orders of the coefficients are different in two (or several) regions. That regions may be either constant (sect. 1, 2) or dependent on $\varepsilon$ in a singular way (sect. 3, 4). The examples studied are in the framework of the transmission problems of chap. 3, sect. 3. The physical applications are the classical ones in time-independent heat transfert, electrostatics, etc. (c.f. chap. 5, sect. 1), if the physical constants take very different values in different regions. Other questions about stiff problems are studied in chap. 17.

1.- A model stiff problem - Let $\Omega^0, \Omega^1$ be two connected bounded domains of $R^N$ with smooth boundaries $\Gamma$ and S, located as the fig. 1 shows. We also consider

$$\Omega = \Omega^0 \cup S \cup \Omega^1$$

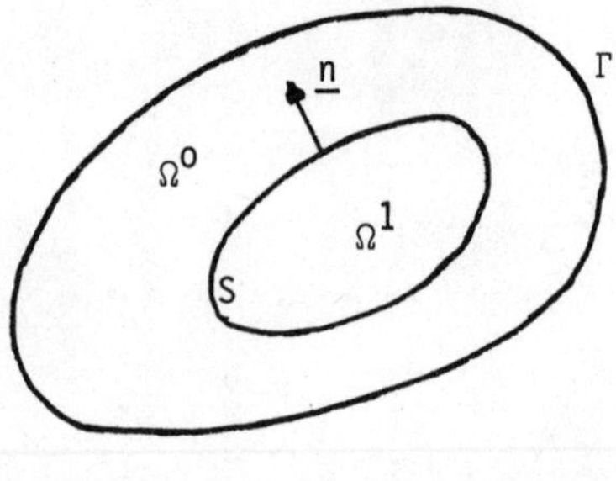


fig. 1

We define the real coefficients

$$a^k_{ij}(x) = a^k_{ji}(x) \quad ; \quad k = 0 , 1 \quad ; \quad i , j = 1 \dots N$$

which are smooth functions on $\Omega^k$ satisfying :

$$(1.1) \qquad a^k_{ij}\, \xi_i\, \xi_j \geqslant \gamma\, |\xi|^2 \qquad \gamma > 0 \qquad \forall \underline{\xi} \in R^N$$

We define the forms :

(1.2) $$\begin{cases} a^k(u\ ,\ v) \equiv \displaystyle\int_{\Omega^k} a^k_{ij} \frac{\partial u}{\partial x_i}\frac{\partial v}{\partial x_j}\, dx\ , \quad k = 0\ ,\ 1 \\ a^\varepsilon(u\ ,\ v) \equiv a^0(u\ ,\ v) + \varepsilon\, a^1(u\ ,\ v) \end{cases}$$

Moreover, let $f \in L^2(\Omega)$ be a given function. It is obvious that, for any $\varepsilon > 0$, there exists a unique $u^\varepsilon$ such that

(1.3) $$u_\varepsilon \in H^1_0(\Omega) \quad ; \quad a^\varepsilon(u_\varepsilon\ ,\ v) = (f\ ,\ v) \qquad \forall v \in H^1_0(\Omega)$$

where ( , ) is for the scalar product in $L^2(\Omega)$.

We define n as the outer unit normal to S. We also write :

(1.4) $$A^k\, v \equiv -\frac{\partial}{\partial x_i}\left(a^k_{ij}\frac{\partial v}{\partial x_j}\right) \quad ; \quad \frac{\partial v}{\partial \nu^k} \equiv a^k_{ij}\frac{\partial v}{\partial x_j}\, n_i \qquad k = 0,1$$

We then have the following formulae of integration by parts for functions which are zero on $\Gamma$ (otherwise, a new term on $\Gamma$ arises) :

(1.5) $$a^0(u\ ,\ v) = \int_{\Omega^0} (A^0 u) v\, dx - \int_S \frac{\partial u}{\partial \nu^0}\, v\, dS$$

(1.6) $$a^1(u\ ,\ v) = \int_{\Omega^1} (A^1 u) v\, dx + \int_S \frac{\partial u}{\partial \nu^1}\, v\, dS$$

Then, the classical formulation of the problem (1.3) (c.f. chap. 3, sect. 3) is :

(1.7) $$A^0\, u^0 = f^0 \ \text{ in } \Omega^0 \quad ; \quad \varepsilon\, A^1\, u^1 = f^1 \ \text{ in } \Omega^1$$

(1.8) $$u_\varepsilon\big|_\Gamma = 0 \quad ; \quad u^0_\varepsilon\big|_S = u^1_\varepsilon\big|_S \quad ; \quad \frac{\partial u^0_\varepsilon}{\partial \nu^0} = \varepsilon\, \frac{\partial u^1_\varepsilon}{\partial \nu^1}$$

where the index $^k$ (k = 0,1) denotes the restriction of the function to $\Omega^k$. It is clear that $u^0_\varepsilon\big|_S$ denotes the trace of $u_\varepsilon$ on S on the side $\Omega^0$, for instance.

Remark 1.1 - It is well known that the necessary and sufficient condition for $v \in H^1(\Omega)$ is :

$$v^k \in H^1(\Omega^k) \qquad (k = 0,1) \quad \text{and} \quad v^0\big|_S = v^1\big|_S\ . \ \blacksquare$$

Remark 1.2 - In the following treatment of the asymptotic problem ( $\varepsilon \searrow 0$), formulae (1.5), (1.6) are used. It follows from standard regularity theory (c.f. Necas [ 1 ] chap. 4, for instance) that they will be applied to functions of $H^2(\Omega^k)$ and consequently they make sense. In particular,

$$A^k u \in L^2(\Omega^k) \quad \text{and} \quad \frac{\partial u}{\partial \nu^k} \in H^{1/2}(S)\ .$$

We now study the asymptotic process $\varepsilon \searrow 0$. It is clear that if we take $\varepsilon = 0$

the problem (1.3) is not coercive ; the solution $u_\varepsilon$ may tend to infinity. It is then natural to search for a formal expansion (which will be justified in the sequel) of the type :

(1.9) $$u_\varepsilon = \varepsilon^{-1} u_{-1} + u_0 + \varepsilon u_1 + \varepsilon^2 u_2 \dots \quad ; \quad u_i \in H_0^1(\Omega)$$

Then, (1.3) becomes

(1.10) $$a^0(\varepsilon^{-1} u_{-1} + u_0 + \varepsilon u_1 + \dots) + \varepsilon a^1(\varepsilon^{-1} u_{-1} + u_0 + \varepsilon u_1 + \dots) = (f,v)$$

and by identifying the powers of $\varepsilon$, we have :

(1.11) $\varepsilon^{-1}$) $\quad a^0(u_{-1}, v) = 0 \qquad \forall v \in H_0^1(\Omega)$

(1.12) $\varepsilon^0$) $\quad a^0(u_0, v) + a^1(u_{-1}, v) = (f, v) \qquad \forall v \in H_0^1(\Omega)$

(1.13) $\varepsilon^j ; j>0$) $\quad a^0(u_j, v) + a^1(u_{j-1}, v) = 0 \qquad \forall v \in H_0^1(\Omega)$

In order to study these problems, we define :

(1.14) $$\begin{cases} X = \{v \; ; \; v \in H^1(\Omega^0) \quad , \quad v|_\Gamma = 0\} \\ Y = \{v \; ; \; v \in H_0^1(\Omega) \quad , \quad v^0 = 0\} \cong \{v \; ; \; v^1 \in H_0^1(\Omega^1), \; v^0=0\} \end{cases}$$

<u>Study of $u_{-1}$</u> - If we take $v = u_{-1}$ in (1.11), it is easily seen that <u>(1.11) is equivalent to</u>

(1.15) $$u_{-1} \in Y \implies u_{-1}^0 = 0$$

Next, we take $v \in Y$ in (1.12), which becomes

$$u_{-1}^1 \in H_0^1(\Omega_1) \quad , \quad a^1(u_{-1}^1, v) = (f^1, v) \quad \forall v \in H_0^1(\Omega^1)$$

which is the variational form of

(1.16) $$A^1 u_{-1}^1 = f^1 \quad ; \quad u_{-1}^1 |_S = 0$$

<u>Then, $u_{-1}$ is given in all $\Omega$ by (1.15), (1.16).</u> ■

<u>Study of $u_0$</u> - Relation (1.16) with (1.6) becomes:

$$a^1(u_{-1}^1, v) = \int_{\Omega_1} f^1 v \, dx + \int_S \frac{\partial u_{-1}^1}{\partial \nu^1} v \, dS$$

and (1.12) becomes :

(1.17) $$u_0^0 \in X \; ; \; a^0(u_0^0, v) = \int_{\Omega_0} f^0 v \, dx - \int_S \frac{\partial u_{-1}^1}{\partial \nu^1} v \, dS \quad \forall v \in X$$

and <u>this gives $u_0^0$ in a unique manner</u> ; in fact, this is the variational formulation of

(1.18) $$A^o u^o_o = f^o \text{ in } \Omega^o \quad ; \quad u^o_o\big|_\Gamma = 0 \quad ; \quad \frac{\partial u^o_o}{\partial \nu^o}\bigg|_S = \frac{\partial u^1_{-1}}{\partial \nu^1}\bigg|_S$$

which is a mixed (Dirichlet - Neuman )non homogeneous problem. Next, to find $u^1_o$, we have $u^1_o$ is equal to $u^0_0$ on S ; on the other hand, (1.13) for j = 1 with $v \in Y$ gives

$$a^1(u^1_o , v) = 0 \qquad \forall v \in H^1_o(\Omega^1)$$

and $u^1_o$ <u>is the unique solution of</u>

(1.19) $$A^1 u^1_o = 0 \qquad ; \qquad u^1_o\big|_S = u^0_0\big|_S$$

Then, $u_0$ <u>is given by (1.18), (1.19)</u>. ∎

<u>Study of $u_m$ $(m > 0)$</u> - This study is analogous to that of $u^0$ but with f = 0. Because $u_{m-1}$ is known, (1.13) with j = m becomes

(1.20) $$a^0(u_m , v) = - \int_S \frac{\partial u^1_{m-1}}{\partial \nu^1} v \, dS \qquad \forall v \in X$$

and of course $u^o_m \in X \Rightarrow u^o_m$ is given by

(1.21) $$A^o u^o_m = 0 \quad \text{in } \Omega^o \quad ; \quad u^o_m\big|_\Gamma = 0 \quad ; \quad \frac{\partial u^o_m}{\partial \nu^o}\bigg|_S = \frac{\partial u^1_{m-1}}{\partial \nu^1}\bigg|_S$$

Then, to find $u^1_m$ we take (1.13) with j = m+1 and $v \in Y$, and we see that $u^1_m$ is given by :

(1.22) $$A^1 u^1_m = 0 \qquad \text{in } \Omega^1 \quad ; \quad u^1_m\big|_S = u^o_m\big|_S$$

Then, $u_m$ <u>is given by (1.21), (1.22)</u>. Note that (1.22) with m = m-1 has been used to write (1.20). ∎

We then have

<u>Proposition 1.1</u> - $u_m$ (m = -1 , 0, 1 ...) are uniquely determined as the solutions of the preceeding boundary value problems.

Moreover, we have :

<u>Theorem 1.1 (Estimate of the error)</u> - Let us define :

$$u^*_m = \varepsilon^{-1} u_{-1} + u_0 + \varepsilon u_1 + \ldots + \varepsilon^m u_m \qquad \text{(approx. order m)}$$

$$w_m = u_\varepsilon - u^*_m \qquad \text{(error of order m)}$$

then, there exists $C_m$ such that

(1.23) $$\| w_m \|_{H_0^1(\Omega)} \| \leqslant C_m \varepsilon^{m+1}$$

Proof - We consider $u_{m+1}$ (one step more than $u_m$). From (1.11), (1.12), and (1.13) with j = 1, 2 ... m+1 we have :

$$a^o(u^*_{m+1}, v) + \varepsilon\, a^1(u^*_m, v) = (f, v) \qquad \forall v \in H_o^1(\Omega)$$

which writes (see (1.2) :

$$a^\varepsilon(u^*_{m+1}, v) = (f, v) + \varepsilon^{m+2}\, a^1(u_{m+1}, v) \qquad \forall v \in H_o^1(\Omega)$$

and taking the difference with (1.3) :

(1.24) $$a^\varepsilon(w_{m+1}, v) = -\,\varepsilon^{m+2}\, a^1(u_{m+1}, v) \qquad \forall v \in H_o^1(\Omega)$$

we then take $v = w_{m+1}$ and we use (1.1)

$$\gamma\varepsilon \|w_{m+1}\|^2_{H_o^1} \leqslant \varepsilon^{m+2} \left| a^1(u_{m+1}, w_{m+1}) \right|$$

(1.25) $$\| w_{m+1} \|_{H_o^1} \leqslant C \|u_{m+1}\|_{H_o^1} \varepsilon^{m+1}$$

But on the other hand,

$$w_m = w_{m+1} + \varepsilon^{m+1} u_{m+1}$$

and we have (1.23) with $C_m = (C+1) \| u_{m+1} \|_{H_o^1}$ . ■

Remark 1.3 - Several variants of the preceeding problem are possible. For instance, if the places of the domains $\Omega^o$ and $\Omega^1$ are exchanged, more involved boundary value problems appear. (see Lions [ 3 ], chapter 1).■

2.- Some spectral properties of stiff problems - We consider the model problem of the preceeding section. In addition, we define

(2.1) $$b^k(u, v) = \int_{\Omega^k} u\, v\, dx \qquad k = 0, 1$$

(2.2) $$b^\varepsilon(u, v) = b^o(u, v) + \varepsilon\, b^1(u, v)$$

Now, for fixed $\varepsilon > 0$, we consider the problem :

(2.3) $$u_\varepsilon \in H_o^1(\Omega) \qquad a^\varepsilon(u, v) - \zeta\, b^\varepsilon(u, v) = (f, v) \quad \forall v \in H_o^1(\Omega)$$

for certain values of the (complex) parameter $\zeta$ . It is clear that for fixed $\varepsilon > 0$, $b^\varepsilon$ is a scalar product on $L^2(\Omega)$ and (2.3) is a classical eigenvalue problem associated with an operator with compact inverse. Moreover, we consider the operators

$$(2.4)\quad \begin{cases} \mathcal{A}^0 \Leftrightarrow A^0 & \text{in } \Omega^0 \text{ with} \quad u\big|_\Gamma = 0 \quad \dfrac{\partial u}{\partial \nu^0}\Big|_S = 0 \\ \mathcal{A}^1 \Leftrightarrow A^1 & \text{in } \Omega^1 \text{ with} \quad u\big|_S = 0 \end{cases}$$

Let $\zeta$ be a point in a compact set K contained in the resolvent sets of $\mathcal{A}^0$, and $\mathcal{A}^1$, i.e. :

$$(2.5)\qquad K \subset\subset \rho(\mathcal{A}^0) \cap \rho(\mathcal{A}^1)$$

we look for a solution $u_\varepsilon$ of (2.3) for small $\varepsilon$ under the form (1.9). We obtain a sequence of problems as (1.11) - (1.13) but with $a^k - \zeta b^k$ instead of $a^k$ $(k = 0,1)$. The solution of these problems is streighforward by virtue of (2.5).

For instance, to study $u_{-1}$, we take $v \in X$ in (1.11) (which is equivalent to $v \in H_0^1(\Omega)$), we obtain

$$u_{-1}^0 \in X \quad , \quad a^0(u_{-1}^0 , v) - \zeta\, b^0(u_{-1}^0 , v) = 0 \qquad \forall v \in X$$

and this implies $u_{-1}^0 = 0$ because $\zeta \in \rho(\mathcal{A}^0)$.

Next, from (1.12) with $v \in Y \Rightarrow$

$$(2.6)\quad u_{-1}^1 \in H_0^1(\mathcal{A}^1) \ ; \ a^1(u_{-1}^1 , v) - \zeta\, b^1(u_{-1}^1, v) = (f^1, v) \qquad \forall v \in H_0^1(\Omega_1)$$

and because $\zeta \in \rho(\mathcal{A}^1)$, $u_{-1}^1$ exists and is unique ; moreover, from standard regularity theory :

$$(2.7)\qquad \| u_{-1}^1 \|_{H^2(\Omega^1)} \leq C \, \| f^1 \|_{L^2(\Omega^1)}$$

where C depends only on the compact K.

We calculate in an analogous way all the terms of the sequence (1.9). Moreover, at each step of resolution, we have an estimate of the type (2.7) in $H^2(\Omega^0)$ or $H^2(\Omega^1)$ (in particular, for inhomogeneous boundary value problems of the type (1.18), we have, from (2.6) :

$$\Big\| \frac{\partial u_{-1}^1}{\partial \nu^1}\Big|_S \Big\|_{H^{1/2}(S)} \Rightarrow \| u_0^0 \|_{H^2(\Omega^0)} \leq C \, \| f \|_{L^2(\Omega)}$$

and so on. Moreover, it is easy to see that the boundary value problems are the same for $j > 0$ (see (1.13) and we then have

$$\| u_j \|_{H^1(\Omega)} \leq C^{j+2} \, \| f \|_{L^2(\Omega)}$$

where C depends only on the compact K but not on j. Consequently, the series (1.9) converges uniformly on K for $|\varepsilon| < C^{-1}$. It is also clear that the sum of the series satisfies (2.3), and this proves that K is contained in the resolvent set of (2.3).

We have thus proved :

Theorem 2.1 - If K is a compact set satisfying (2.5), then, for sufficiently small $\varepsilon$(i.e. $\varepsilon < C(K)^{-1}$) K is in the resolvent set of (2.3) and the solution $u_\varepsilon(\zeta)$ is given by a series of the type (1.9) which is uniformly convergent (in the norm of $(H^1(\Omega))$ for $\zeta \in K$.

Now we study the eigenvalues of the problem (2.3).

Lemma 2.1 - Let $\gamma$ be a closed simple curve in the plane of the variable $\zeta$. For fixed $\varepsilon > 0$, let $u_\varepsilon(\zeta)$ be the solution of (2.3) (for $\zeta$ in the resolvent set of the problem (2.3)). Then, if

(2.8) $$\int_\gamma u_\varepsilon(\zeta)\, d\zeta \neq 0$$

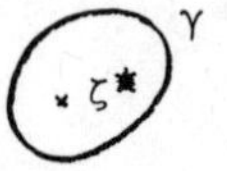


fig. 2

There exists at least a point of the spectrum in the region of the plane enclosed by $\gamma$.

Proof - It is immediate, because $u_\varepsilon(\zeta)$ is a holomorphic function of $\zeta$ on the resolvent set, (2.8) implies that there are at least a singularity of the resolvent at the interior of $\gamma$. ■

Theorem 2.3 - Let $\zeta^*$ be a point of the spectrum of either $\mathfrak{A}^0$ or $\mathfrak{A}^1$ : $\zeta^* \in \sigma(\mathfrak{A}^0) \cup \sigma(\mathfrak{A}^1)$. Moreover, let $\gamma$ be a simple closed curve enclosing $\zeta^*$ contained in $\rho(\mathfrak{A}^0) \cap \rho(\mathfrak{A}^1)$. Then, for sufficiently small $\varepsilon$, there is at least an eigenvalue of the problem (2.3) in the region enclosed by $\gamma$.

Proof - First, let us assume $\zeta^* \in \sigma(\mathfrak{A}^0)$. We take in (2.3) $f^1 = 0$, $f^0$ equal to an eigenvector of $\mathfrak{A}^0$ associated with $\zeta^*$. Then, for $\zeta \in \gamma$, we consider $u_\varepsilon(\zeta)$ given by the theorem 2.1. From the construction of the elements of the series, we see that $u_{-1} = 0$, and we have

$$u_\varepsilon(\zeta) = u_0(\zeta) + \varepsilon\, u_1(\zeta) + \varepsilon^2 \ldots \qquad \zeta \in \gamma$$

which is uniformly convergent for $\zeta \in \gamma$. Then

(2.9) $$\int_\gamma u_\varepsilon(\zeta)\, d\zeta = \int_\gamma u_0(\zeta)\, d\zeta + \varepsilon \int_\gamma u_1(\zeta)\, d\zeta + \ldots$$

By taking the limit $\varepsilon \to 0$, we see that, if

(2.10) $\left( \int_\gamma u_0^0(\zeta)\, d\zeta\, ,\, f^0 \right)_{L^2(\Omega^0)} \neq 0$, then the left hand side of (2.9) is different from zero for small $\varepsilon$ and the conclusion follows from lemma 2.1. But (2.10) is evident, because $u^0(\zeta)$ is the solution of

$$u_0^0 \in X \quad ; \quad a^0(u_0^0 , v) - \zeta b^0(u_0^0 , v) = \int_{\Omega_0} f^0 \, v \, dx \qquad \forall v \in X$$

(which is analogous to (1.17) in our case) and (2.10) is equivalent to

$$\left( \int_\gamma (\mathcal{A}^0 - \zeta)^{-1} \, f^0 \, d\zeta \, , \, f^0 \right)_{L^2(\Omega^0)}$$

which is different from zero by the classical properties of the projector (see chap. 11, sect. 1).

In the case $\zeta^* \in \sigma(\mathcal{A}^1)$ the proof is analogous by taking $f^0 = 0$, $f^1$ an eigenvector of $\mathcal{A}^1$ associated to $\zeta^*$. We evidently have

$$\left( \int_\gamma (\mathcal{A}^1 - \zeta)^{-1} \, f^1 \, d\,\zeta, \, f^1 \right)_{L^2(\Omega^1)} \neq 0$$

which, by (2.6) is equivalent to

$$(2.11) \qquad \left( \int_\gamma u_{-1}^1(\zeta) \, d\,\zeta \, , \, f^1 \right)_{L^2(\Omega^1)} \neq 0$$

and, because the series (1.9) is uniformly convergent on $\gamma$ , we have

$$(2.12) \qquad \int_\gamma u_\varepsilon(\zeta) \, d\,\zeta = \frac{1}{\varepsilon} \int_\gamma u_{-1}(\zeta) \, d\zeta + \int_\gamma u_0(\zeta) \, d\zeta \; + \dots$$

then, for $\varepsilon$ sufficiently small, the right hand side (and then the left !) is $\neq 0$, and the conclusion follows from lemma 2.1. ■

Remark 2.1 - The selfadjointness is not used in section 1 and 2. Consequently, the same method applies to other non-selfadjoint stiff problems. ■

## 3.- Heat transfer through a narrow plate with small conductivity -

We consider a problem analogous to that of sect. 1 but where the domain $\Omega^1$ depends on $\varepsilon$ and tends to a N-1 surface as $\varepsilon \to 0$. In fact, if the thickness of $\Omega^1$ is of order $\varepsilon$ , a balance is established between conductivity and thickness and a limit state exists as $\varepsilon \searrow 0$.

We study the two-dimensional case, but the N-dimensional problem may be handled in a similar way. Moreover, we study the isotropic case $a_{ij} = \delta_{ij}$.

Certain details of the proofs are not given, but they may be find in Sanchez-Palencia [4], where the treatment is a little different.

The domain $\Omega$ is as in section 1, bounded and with smooth boundary. The domains $\Omega^k$ , k = 0,1 depend on $\varepsilon$ and will be noted $\Omega_\varepsilon^k$ . These domains are defined

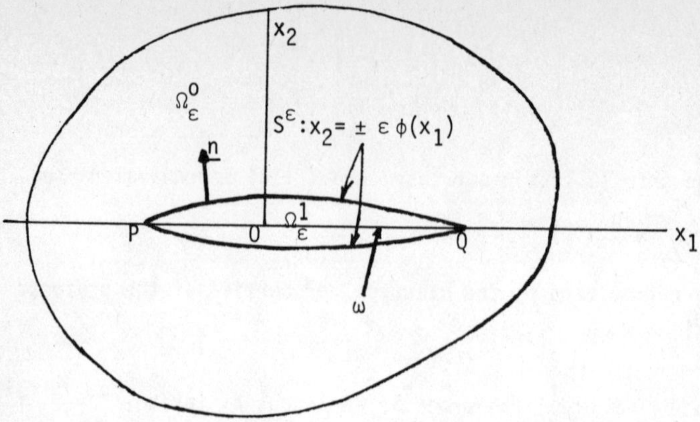


fig. 3

in the following way. Let $\omega = ]P , Q[$ be a segment of the axis $x_1$, strictly contained in $\Omega$. We consider a smooth function $\phi(x_1) \geqslant 0$ defined on $\overline{\omega}$ (the closed segment) and taking the value 0 only at the extremities P, Q, of $\overline{\omega}$ . Moreover, the derivative of $\phi(x_1)$ at P and Q is not zero. Then, $\Omega^1_\varepsilon$ is defined as the "lens-shaped" domain :

(3.1) $$\Omega^1_\varepsilon = \{\underline{x} \; ; \; x_1 \in \omega \; , \; |x_2| < \varepsilon\, \phi(x)\}$$

and of course, $\Omega^0_\varepsilon = \Omega - \overline{\Omega}^1_\varepsilon$ .

We also define :

(3.2) $$a^\varepsilon(u , v) = \int_{\Omega^0_\varepsilon} \frac{\partial u}{\partial x_i} \frac{\partial v}{\partial x_i}\, dx + \varepsilon \int_{\Omega^1_\varepsilon} \frac{\partial u}{\partial x_i} \frac{\partial v}{\partial x_i}\, dx$$

We also define $\underline{V = H^1_0(\Omega) \; ; \; H = L^2(\Omega)}$ and we give a function $f \in H$. Then, $u_\varepsilon$ is defined by :

(3.3) $$u_\varepsilon \in V \quad ; \quad a^\varepsilon(u , v) = (f , v) \qquad \forall v \in V$$

which is of course equivalent to :

(3.4) $$\begin{cases} -\Delta u_\varepsilon = 0 \quad \text{in } \Omega^k_\varepsilon \; , \; k = 0,1 \; . \\ u_\varepsilon\big|_\Gamma = 0 \quad ; \quad u^1_\varepsilon\big|_{S^\varepsilon} = u^2_\varepsilon\big|_{S^\varepsilon} \quad ; \quad \dfrac{\partial u^0_\varepsilon}{\partial n}\Big|_{S^\varepsilon} = \varepsilon\, \dfrac{\partial u^1_\varepsilon}{\partial n}\Big|_{S^\varepsilon} \end{cases}$$

where n is the outer unit normal to $S^\varepsilon$.

In order to define the limit problem as $\varepsilon \searrow 0$, we consider the space

(3.5) $$W = \{v \in H^1(\Omega - \overline{\omega}) \quad ; \quad v\,|_\Gamma = 0\,\}$$

equipped with the scalar product (which is associated to a norm equivalent to that of $H^1(\Omega-\overline{\omega})$)

(3.6) $$(u\ ,\ v)_W = \int_\Omega \frac{\partial u}{\partial x_i}\frac{\partial v}{\partial x_i}\,dx$$

Remark 3.1 - It is clear that W is a space of distributions on the domain $\Omega - \overline{\omega}$ ($\Omega$ minus the closed segment $\overline{\omega}$). Consequently, if $v \in W$, its traces on $\omega$ on the two sides $x^2 \gtrless 0$ are in general different. W is a space larger than V and V is not dense in W. ■

For functions $v \in W$, we define

(3.7) $$[v] = v\big|_{\omega^+} - v\big|_{\omega^-}$$

where $v|_{\omega^+}$, $v|_{\omega^-}$ are the traces of v on $\omega$ on the sides $x_2 > 0$, $x_2 < 0$ respectively. It is clear that $v \in W \Rightarrow [v] \in L^2(\omega)$. We shall see later a sharper result.

Now, the limit problem (we shall prove later that $u^\varepsilon$ tends to the solution of this problem) is :

(3.8) $$u \in W\ ;\ \int_\Omega \frac{\partial u}{\partial x_i}\frac{\partial v}{\partial x_i}\,dx + \int_\omega \frac{1}{2\phi(x_1)}[u]\,[v]\,dx_1 = (f\ ,\ v) \qquad \forall v \in W.$$

Remark 3.2 - It is noticeable that $\phi(x_1)$ tends to zero as $x_1 \to \partial\omega$, and it is not not evident that the left hand side of (3.8) is a bounded bilinear form on W. We shall prove this in the next lemma ; then, u exists and is unique by virtue of the Lax-Milgram theorem. ■

Lemma 3.1 - The application $v \longrightarrow \frac{1}{\sqrt{\phi}}[v]$ is continuous from W into $L^2(\omega)$.

Proof - We have to prove that there exists a constant C such that

(3.9) $$\int_\omega \frac{1}{\phi}\,[v]^2\,dx_1 \leqslant C\,\|v\|_W^2$$

In fact, by virtue of the trace theorem, it is sufficient to prove (3.9) with the integral on a neighbourhood (independent of v) of $\partial\omega$ (see remark 3.2). Let us take polar coordinates $r,\theta$ centered at P (or Q). Moreover, from the hypothesis that $\phi$ is smooth and $\phi'(P) > 0$, we see that in a neighbourhood of $r = 0$ (for instance $r < R$) we have $\phi > ar$. We then have (see figure 4) :

$$\frac{1}{\phi(r)}[v(r)]^2 \leqslant \frac{1}{ar}|v(M^+)-v(M^-)|^2 = \frac{1}{ar}\left(\int_0^{2\pi}\frac{1}{r}\frac{\partial v}{\partial\theta}\,r\,d\theta\right)^2 \leqslant$$
$$= \frac{1}{ar}\left(\int_0^{2\pi}|\underline{\text{grad}}\,v|\,r^{1/2}\,r^{1/2}\,d\theta\right)^2 \leqslant \frac{1}{ar}\int_0^{2\pi}|\underline{\text{grad}}\,v|^2\,r\,d\theta\int_0^{2\pi}r\,d\theta =$$
$$= \frac{2\pi}{a}\int_0^{2\pi}|\text{grad}\,v|^2\,r\,d\theta$$

and by integrating in r from 0 to R :

$$\int_0^R \frac{1}{\phi(r)} [v(r)]^2 \, dr \leqslant \frac{2\pi}{a} \int_0^R \int_0^{2\pi} |\text{grad } v|^2 \, r \, d\theta \, dr = \frac{2\pi}{a} \|v\|_W^2 \,. \blacksquare$$

The classical formulation of the limit problem (3.8) in terms of equations and boundary conditions is easily obtained from (3.8). By taking $v \in \mathcal{D}(\Omega - \overline{\omega})$ we see that

(3.10) $$-\Delta u = f \quad \text{in } \Omega - \overline{\omega}$$

We of course have

(3.11) $$u\Big|_\Gamma = 0$$

Then, by integrating by parts (3.8) in $\Omega - \overline{\omega}$ we obtain :

$$\int_\omega \left( - \frac{\partial u}{\partial x_2}\Big|_{\omega^+} v\Big|_{\omega^+} + \frac{\partial u}{\partial x_2}\Big|_{\omega^-} v\Big|_{\omega^-} + \frac{1}{2\phi}[u][v] \right) dx_2 = 0 \qquad \forall v \in W$$

we first consider v taking the same values on the two sides of $\omega$ , but otherwise arbitrary $\Rightarrow$

$$\frac{\partial u}{\partial x_2}\Big|_{\omega^+} = \frac{\partial u}{\partial x_2}\Big|_{\omega^-}$$

and then $[v]$ arbitrary gives :

(3.12) $$\frac{\partial u}{\partial x_2}\Big|_{\omega^+} = \frac{1}{2\phi}[u] = \frac{\partial u}{\partial x_1}\Big|_{\omega^-}$$

Conversely, it is also possible to obtain (3.8) from (3.10) - (3.12). Then, the classical formulation of the limit problem (3.8) is (3.10) - (3.12).

Remark 3.3 - There is a heuristic reasonning to obtain the limit problem (3.10)-(3.12). If we assume that in a neighbourhood of the plate the variations in $x_1$ may

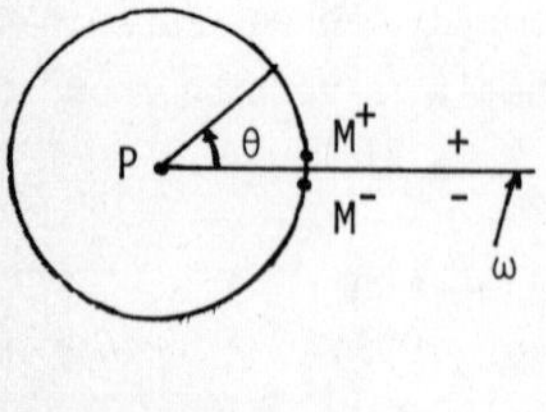


fig. 4

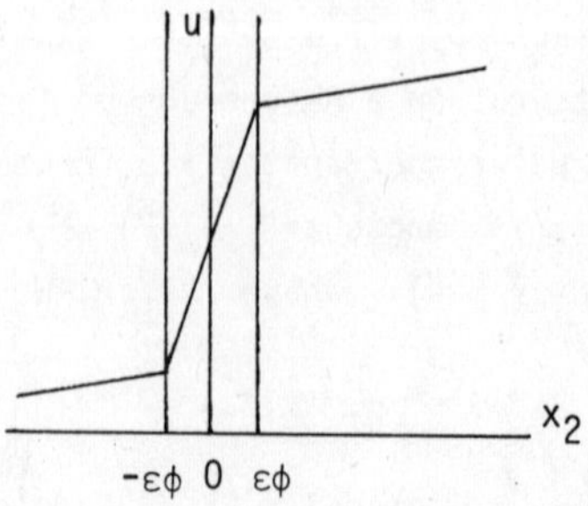


fig. 5

be neglected with respect to the variations in $x_2$, we have the one-dimensional problem associated to (3.4), and we immediately obtain (3.12) (see figure 5).■

Now, we prove the convergence of $u_\varepsilon$ to the solution u of the limit problem. This result will be given in the theorems 3.1 and 3.2. The second is a sharper form of the first.

<u>Theorem 3.1</u> - If $u_\varepsilon$ and u are defined by (3.3) and (3.8) we have

(3.13) $$u_\varepsilon \to u \text{ in } L^2(\Omega) \text{ weakly.}$$

Moreover, if $K_\alpha$ is the cylinder

(3.14) $$K_\alpha = \{x \; ; \; x_1 \in \omega \, , \quad |x_2| < \alpha\} \quad \text{then,}$$

then,

(3.15) $$u_\varepsilon \to u \quad \text{in } H^1(\Omega - \overline{K}_\alpha) \text{ weakly, for fixed } \alpha \, .$$

The proof will be given in several lemmas.

<u>Lemma 3.2</u> - There exists $C > 0$ independent of $\varepsilon$ such that

(3.16) $$\int_\Omega |v|^2 \, dx \leq C \, a^\varepsilon(v \, , \, v) \qquad \forall v \in V$$

<u>Proof</u> - Let us fix $\alpha$ in such a way that $\varepsilon \phi < \alpha$ for $x \in \omega$ and all the $\varepsilon$ considered (i.e., $\Omega_\varepsilon^1 \subset K_\alpha$). We consider $x_2 > 0$ ; for $x_2 < 0$ analogous inequalities hold. For $0 < x_2 < \varepsilon \alpha$, we have :

$$|v(x_1 \, , \alpha) - v(x_1 \, , x_2)| = \left| \int_{x_2}^{\alpha} \frac{\partial v}{\partial x_2} \, d\xi \right| \leq \left| \int_{x_2}^{\varepsilon\phi} \right| + \left| \int_{\varepsilon f}^{\alpha} \right| \Rightarrow$$

$$|v(x_1 \, , \alpha) - v(x_1 \, , x_2)|^2 \leq 2\left( \int_{x_2}^{\varepsilon\phi} \right)^2 + 2\left( \int_{\varepsilon\phi}^{\alpha} \right)^2 \leq$$

$$\leq 2 \, \varepsilon\phi \int_0^{\varepsilon\phi} \left| \frac{\partial v}{\partial x_2} \right|^2 dx_2 + 2 \, \alpha \int_{\varepsilon\phi}^{\alpha} \left| \frac{\partial v}{\partial x_2} \right|^2 dx_2 \Rightarrow$$

$$\int_0^{\varepsilon\phi} |v(x_1 \, , \alpha) - v(x_1 \, , x_2)|^2 \, dx_2 \leq 2 \, \varepsilon^2 \, \phi^2 \int_0^{\varepsilon\phi} |\text{grad } v|^2 \, dx_2 + 2 \, \alpha\varepsilon\phi \int_{\varepsilon\phi}^{\alpha} |\text{grad } v|^2 dx_2$$

and for $$\varepsilon\phi < x_2 < \alpha$$

$$\int_{\varepsilon\phi}^{\alpha} |v(x_1 \, , \alpha) - v(x_1 \, , x_2)|^2 \, dx_2 \leq 2\alpha^2 \int_{\varepsilon\phi}^{\alpha} |\text{grad } v|^2 \, dx_2$$

and by additionating and integrating on $\omega$ :

(3.17) $$\iint_{\omega \, 0}^{\alpha} |v(x_1 \, , \alpha) - v(x_1, x_2)|^2 dx \leq C \, \varepsilon^2 \int_{\Omega_\varepsilon^1} |\text{grad } v|^2 dx + 2\alpha(\alpha+\varepsilon) \int_{\Omega_\varepsilon^0} |\text{grad } v|^2 \, dx$$

We consider v continuated out of $\Omega$ with value 0. Then, by taking $\alpha$ sufficiently large, $v(x_1, \alpha) = 0$ and inequality (3.16) follows (note that for $x_1 \notin \omega$ it is trivial).■

Lemma 3.3 - We have the apriori estimates :

$$\int_{\Omega_\varepsilon^0} |\text{grad } u_\varepsilon|^2 \, dx + \varepsilon \int_{\Omega_\varepsilon^1} |\text{grad } u_\varepsilon|^2 \, dx \leqslant C \tag{3.18}$$

$$\int_{\Omega} | u_\varepsilon|^2 \, dx \leqslant C \tag{3.19}$$

Proof - We take $v = u_\varepsilon$ in (3.3) ; by virtue of (3.16) we have

$$a^\varepsilon(u_\varepsilon , u_\varepsilon) \leqslant C' \, a^\varepsilon(u_\varepsilon , u_\varepsilon)^{1/2}$$

which is equivalent to (3.18). (3.19) then follows from (3.16).■

Lemma 3.4 - We consider the set $\mathcal{E}$ of the functions of W which are of class $C^\infty$ in the closed sets

$$\bar{\Omega} \cap \{x_2 \geqslant 0 \} \quad \text{and} \quad \bar{\Omega} \cap \{ x_2 \leqslant 0 \}$$

and are continuous in a neighbourhood of P and Q (and consequently, the only discontinuity is strictly contained in $\omega$). Then, $\mathcal{E}$ is dense in W.

Proof - The proof is easily obtained by using standard techniques in density theorems (cf. Nečas [1 ] , chap. 2). First, under a $C^\infty$ -diffeomorphism,

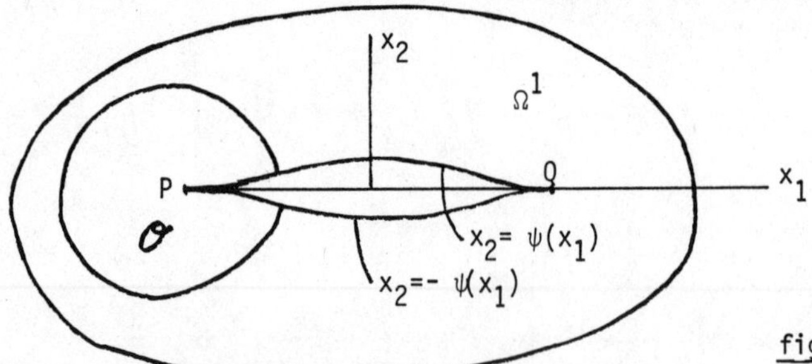


fig. 6

we may consider, instead of $\omega$, a domain

$$|x_2| < \psi(x_1)$$

with $\psi \in C^\infty(-\infty, +\infty)$, $\psi(x)$ equal to zero for $x \notin \omega$ and $\psi(x) > 0$ for $x \in \omega$ .

Next, we consider a partition of unity. The only non trivial case is that of functions with support in a domain $\mathcal{O}$ enclosing the point P (or Q). We may assume that $\psi(x_1)$ is increasing in the region interior to $\mathcal{O}$. Next, we consider a translation of the functions in the direction $x_1$ :

$$u(x_1 , x_2) \Rightarrow u(x_1 - a , x_2) \quad , \quad a > 0 .$$

consequently the translated function is defined in $\mathcal{O} + a$ , and then in a part of the domain

$$|x_2| < \psi(x_1)$$

Finally, we apply a regularization process to the translated functions. The classical properties of translation and regularization (see Necas [ 1 ] chap. 2) show that at each step, the functions form a dense set in W. Of course, at the last step, the functions belong to $\mathcal{E}$ and the lemma is proved. ■

Lemma 3.5 - After extraction of a subsequence,

$$u_\varepsilon \to u^* \tag{3.20}$$

in $L^2(\Omega)$ weakly and in $H^1(\Omega - \overline{K}_\alpha)$ weakly for any $\alpha$. Moreover, $u^* \in W$.

Proof - We consider a sequence $\alpha_i \to 0$. For $\alpha_1$ we extract a sequence of $u_\varepsilon$ such that (3.20) holds in $L^2(\Omega)$ and $H^1(\Omega - \overline{K}_{\alpha_1})$. Then, for $\alpha_2$ we extract a subsequence, and so on. Finally, by a diagonalization process, we have (3.20). Moreover, from (3.18) :

$$\| u^* |_{\Omega - \overline{K}_\alpha} \|^2_{H^1} < C \tag{3.21}$$

independent of $\alpha$. It is clear that $u^* \in \mathcal{D}'(\Omega - \overline{\omega})$ is associated to a function defined almost everywhere and from (3.21) we have $u^* \in H^1(\Omega - \overline{\omega})$, Q.E.D. ■

Now we shall prove that $u^*$ is the solution of (3.8). To this end, we shall construct test-functions of a special kind.

Lemma 3.6 - Let w be a fixed function of $\mathcal{E}$ (see lemma 3.4). We shall define $w_\varepsilon$ by :

$$\begin{cases} w_\varepsilon(x_1 , x_2) = w(x_1 , x_2) & \text{if } (x_1 , x_2) \in \Omega^0_\varepsilon \\ w_\varepsilon(x_1 , x_2) = \frac{1}{2}[w(x_1 , \varepsilon\phi(x_1)) + w(x_1 , -\varepsilon\phi(x_1))] + & \\ \quad + \frac{1}{2}[w(x_1, \varepsilon\phi(x_1)) - w(x_1 , -\varepsilon\phi(x_1))] \dfrac{x_2}{\varepsilon\phi(x_1)} & \text{if } (x_1 \; x_2) \in \Omega^1_\varepsilon \end{cases} \tag{3.22}$$

Then, $w^\varepsilon \in V$ for any $\varepsilon$ and :

$$\frac{\partial w_\varepsilon}{\partial x_1} \leqslant C \quad ; \quad \frac{\partial w_\varepsilon}{\partial x_2} \leqslant \frac{C}{\varepsilon} \quad ; \quad \int_{\Omega^1_\varepsilon} \left| \frac{\partial w^\varepsilon}{\partial x_1} \right|^2 dx \leqslant C\,\varepsilon \tag{3.23}$$

where C depends on w but not on $\varepsilon$ .

Proof - It is clear that $w_\varepsilon$ is continuous on $\Omega$. It coincides with $w$ on $\Omega_\varepsilon^0$ and the derivative $\partial w^\varepsilon/\partial x_2$ is constant with respect to $x_1$. The estimates (3.23) are evident by virtue of the properties of the functions of $\mathcal{E}$. ■

Lemma 3.7 - As $\varepsilon$ tends to zero, we have

$$\varepsilon \int_{\Omega_\varepsilon^1} \frac{\partial u^\varepsilon}{\partial x_2}\frac{\partial w^\varepsilon}{\partial x_2}\,dx \to \int_\omega \frac{1}{2\phi(x_1)}\,[u^*][w]\,dx_1 \tag{3.24}$$

where the notation (3.7) is used.

Proof - We introduce the notation

$$[u]_{\varepsilon\phi} \equiv u(\varepsilon\phi(x_1)) - u(-\varepsilon\phi(x_1))$$

Then, from (3.22) we have

$$\varepsilon \int_{\Omega_\varepsilon^1} \frac{\partial u_\varepsilon}{\partial x_2}\frac{\partial w_\varepsilon}{\partial x_2}\,dx = \varepsilon \int_\omega \left(\int_{-\varepsilon\phi}^{+\varepsilon\phi} \frac{\partial u_\varepsilon}{\partial x_2}\,dx_2\right)\frac{1}{2\varepsilon\phi}\,[w]_{\varepsilon\phi}\,dx_1 =$$

$$= \int_\omega [u]_{\varepsilon\phi}\,\frac{1}{2\phi}[w]_{\varepsilon\phi}\,dx_1$$

and from the smoothness of $w$, $[w]_{\varepsilon\phi} \to [w]$ uniformly in $\omega$. On the other hand, $[u_\varepsilon]_{\varepsilon\phi} \to [u^*]$ in $L^2(\omega)$ strongly as it is easily seen from the classical properties of the traces. Lemma 3.7 then follows. ■

Lemma 3.8 - As $\varepsilon$ tends to zero :

$$\int_{\Omega_\varepsilon^0} \frac{\partial u_\varepsilon}{\partial x_i}\frac{\partial w_\varepsilon}{\partial x_i}\,dx \to \int_\Omega \frac{\partial u^*}{\partial x_i}\frac{\partial w}{\partial x_i}\,dx \tag{3.25}$$

$$\varepsilon \int_{\Omega_\varepsilon^1} \frac{\partial u_\varepsilon}{\partial x_1}\frac{\partial w_\varepsilon}{\partial x_1}\,dx \to 0 \tag{3.26}$$

$$\int_\Omega f\,w^\varepsilon\,dx \to \int_\Omega f\,w\,dx\quad. \tag{3.27}$$

Proof - (3.25) follows from lemma 3.5 (note that $\int_{K\alpha}$ is as small as desired for little $\alpha$). (3.26) is a consequence of (3.23) and (3.18). Finally, (3.27) follows from the form of $w_\varepsilon$. ■

Then, theorem 3.1 is proved. It suffices to fix $w \in \mathcal{E}$ and to take $v = w^\varepsilon$ in (3.3) ; lemmas 3.7 and 3.8 show that $u^*$ satisfies (3.8) with $v = w$ ; by virtue of the density property (Lemma 3.4), $u^*$ is the (unique !) solution of (3.8).

Theorem 3.2 - If $A_\varepsilon$ and A are the selfadjoint operators of H associated with (3.3) and (3.8) according to the first representation theorem, we consider the equations

$$A_\varepsilon u_\varepsilon = f^\varepsilon \quad ; \quad A u = f$$

If $f^\varepsilon$ tends to f in H weakly, then, $u^\varepsilon$ tends to u in H strongly. Moreover

(3.28) $$\| A_\varepsilon^{-1} - A^{-1} \|_{\mathcal{L}(H,H)} \to 0$$

and the spectral properties of chap. 11, sect. 3, follow.

Proof. - The fact that f depends on $\varepsilon$ is irrelevant (note that $w^\varepsilon$ tends to w in H strongly and consequently the analogous of (3.27) holds. In the same way, the convergence $u_\varepsilon \to u$ holds in the sense of theorem 3.1. Moreover, the convergence holds in H strongly. This is easily seen from (3.17) by fixing $\alpha$ sufficiently small and by using the property that the trace operator is compact from $H^1$ into $L^2$. Consequently, we have

(3.29) $$f^\varepsilon \to f \quad \text{in H weakly} \Rightarrow u_\varepsilon \to u \quad \text{in H strongly}$$

Then, to prove (3.28) we proceed by contradiction, as in chapter 9, sect. 3. If (3.28) is not true, there exists a sequence ($\varepsilon \to 0$) such that

$$f^\varepsilon \to f \quad \text{in H weakly and}$$

(3.30) $$\| (A_\varepsilon^{-1} - A^{-1}) f^\varepsilon \|_H \geqslant \delta \quad \text{for a certain } \delta > 0 \text{ . But}$$

(3.31) $$\| (A_\varepsilon^{-1} - A^{-1}) f^\varepsilon \|_H = \| A_\varepsilon^{-1}(f^\varepsilon - f) + (A_\varepsilon^{-1} - A^{-1}) f + A^{-1}(f - f^\varepsilon) \|_H$$

and the two first terms in the right hand side tend to zero in H strongly by virtue of (3.29). Moreover,

$A^{-1}f$ and $A^{-1}f^\varepsilon$ are the solutions of problem (3.8) with variable f. It is immediately seen that $A^{-1}f \to A^{-1}f^\varepsilon$ in W weakly and then in H strongly (by the Rellich's theorem). Thus, (3.31) implies a contradiction.■

4.- Heat transfer through a narrow plate with high conductivity - Now we study a problem which is in some sense the opposite of that of sect. 3. In fact, we consider the case where the conductivity of the plate tends to infinity ; moreover we consider a cylindrical plate (i.e. the plate is "coin-shaped" instead of "lens-shaped"). As in the preceeding section, certain details are not given. They may be found in Pham Huy and Sanchez-Palencia [ 1 ] .

We consider the bounded, connected domain $\Omega$ and the segment $\omega$ as in the preceeding section. Moreover, for fixed $\lambda > 0$ and $\varepsilon > 0$ (parameter) :

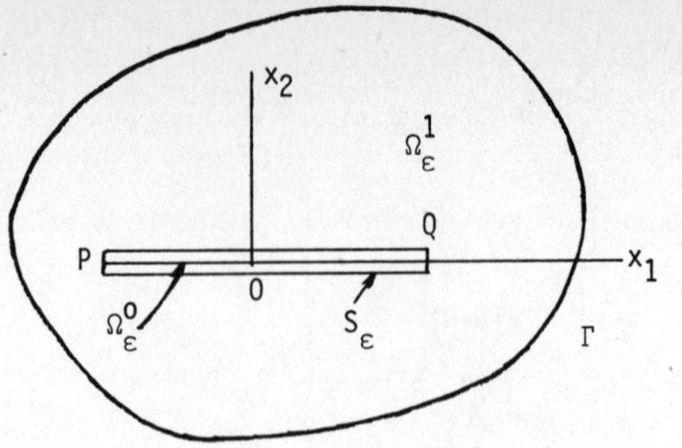


fig. 7

(4.1) $\Omega_\varepsilon^0 = \{(x_1 , x_2) \quad ; \quad x_1 \in \omega \quad ; \quad |x_2| < \lambda\varepsilon\}$

and we denote by $S_\varepsilon$ the boundary of $\Omega_\varepsilon^0$. The outer domain is $\Omega_\varepsilon^1$, and we have

$$\Omega = \Omega_\varepsilon^0 \cup S_\varepsilon \cup \Omega_\varepsilon^1$$

We consider the form

(4.2) $$a^\varepsilon(u , v) = \int_{\Omega_\varepsilon^1} \frac{\partial u}{\partial x_i} \frac{\partial v}{\partial x_i} dx + \frac{1}{\varepsilon}\int_{\Omega_\varepsilon^0} \frac{\partial u}{\partial x_i} \frac{\partial v}{\partial x_i} dx$$

On $V = H_0^1(\Omega)$. We also define $H = L^2(\Omega)$ and

(4.3) $$V_\omega = \{ v ; \ v \in V \quad ; \quad v|_\omega \ H^1(\omega) \}$$

which is a Hilbert space for the scalar product

(4.4) $$(u , v)_{V_\omega} = \int_\Omega \frac{\partial u}{x_i} \frac{\partial v}{x_i} dx + \int_\omega \frac{\partial u}{\partial x_1} \frac{\partial v}{\partial x_1} dx_1$$

We also consider $f^\varepsilon$, $f \in H$. The solution $u^\varepsilon$ of the conduction problem is given by :

(4.5) $$u^\varepsilon \in V \quad ; \quad a^\varepsilon(u_\varepsilon , v) = (f^\varepsilon , v) \qquad \forall v \in V$$

where ( , ) denotes the scalar product in H.

The limit problem is defined as follows :

(4.6) $$u \in V_\omega \quad ; \int_\Omega \frac{\partial u}{\partial x_i} \frac{\partial v}{\partial x_i} dx + 2\lambda \int_\omega \frac{\partial u}{\partial x_1} \frac{\partial v}{\partial x_1} dx_1 = (f , v) \qquad \forall v \in V_\omega$$

and we have the converge theorem :

<u>Theorem 4.1</u> - If $f^\varepsilon \to f$ in H weakly, and $u^\varepsilon$, u are defined by (4.5) and (4.6), then

(4.7) $$u^\varepsilon \to u \quad \text{in } V \text{ weakly}$$

Moreover,

(4.8) $$\| A_\varepsilon^{-1} - A^{-1} \|_{\mathcal{L}(H,H)} \to 0$$

where $A_\varepsilon$ and $A$ are the selfadjoint operators of $H$ associated with the forms in the left hand sides of (4.5), (4.6). The spectral properties of chapter 11, sect. 3, then follow.

Remark 4.1 - It is not difficult to obtain the classical formulation of the limit problem (4.6). The classical techniques give :

(4.9) $$-\Delta u = f \quad \text{in } \Omega - \overline{\omega} \; ; \qquad u \Big|_\Gamma = 0$$

(4.10) $$-\lambda \frac{\partial^2 u}{\partial x_1^2} = \frac{\partial u}{\partial x_2} \Bigg|_{x_2 > 0} - \frac{\partial u}{x_2} \Bigg|_{x_2 > 0} \qquad \text{on } \omega$$

(4.11) $$\frac{\partial u}{\partial x_2}(P) = \frac{\partial u}{\partial x_2}(Q) = 0$$

where (4.10) is in fact an equation of conduction in $\omega$ , with source terms (the right hand side) given by the heat flux across the two faces of $\omega$. (4.11) is a Neumann condition at the extremities of $\omega$(note that the conductivity of $\Omega - \overline{\omega}$ is negligible with respect to that of $\omega$). In fact (4.9) - (4.11) is the coupling of a two-dimensional conduction problem in $\Omega - \overline{\omega}$ and a one-dimensional conduction problem in $\omega$ . ■

Now, we prove theorem 4.1. By taking $v = u_\varepsilon$ in (4.6), we immediately obtain :

(4.12) $$\int_\Omega |\underline{\text{grad}}\, u_\varepsilon|^2 \, dx \leqslant C$$

(4.13) $$\frac{1}{\varepsilon} \int_{\Omega_\varepsilon^0} |\underline{\text{grad}}\, u_\varepsilon|^2 \, dx \leqslant C$$

We define the operator (mean value through $\Omega_\varepsilon^0$) :

(4.14) $$m^\varepsilon v(x_1) = \frac{1}{2\lambda\varepsilon} \int_{-\lambda\varepsilon}^{+\lambda\varepsilon} v(x_1 , x_2) \, dx_2$$

Lemma 4.1 - The operator $m^\varepsilon$ commutes with $\partial/\partial x_1$ when applied to functions of $H^1(\Omega_\varepsilon^0)$. Moreover, it is bounded from $L^2(\Omega_\varepsilon^0)$ into $L^2(\omega)$ and from $H^1(\Omega_\varepsilon)$ into $H^1(\omega)$, with norm $\leqslant (2\lambda\varepsilon)^{-1/2}$ in both cases.

Proof - It suffices to consider the following inequality :

$$\int_\omega |m^\varepsilon v|^2 \, dx_1 = \int_\omega \frac{1}{(2\lambda\varepsilon)^2} \left( \int_{-\lambda\varepsilon}^{+\lambda\varepsilon} v \, dx_2 \right)^2 dx_1 \leqslant$$
$$\leqslant \int_\omega \frac{1}{2\lambda\varepsilon} \int_{-\lambda\varepsilon}^{+\lambda\varepsilon} |v|^2 \, dx_2 = \frac{1}{2\lambda\varepsilon} \int_{\Omega_\varepsilon^0} |v|^2 \, dx \qquad ■$$

Now, from (4.12) and the trace theorem,

(4.15) $$\| u^{\varepsilon}|_{\omega} \|_{L^2(\omega)} \leq C$$

On the other hand, from

$$|v(x_1 , \alpha)- v(x , 0)|^2 = \left| \int_0^{\alpha} \frac{\partial v}{\partial x_2} dx_2 \right|^2 \leq \alpha \int_0^{\alpha} \left| \frac{\partial v}{\partial x_2} \right|^2 dx_2 \Rightarrow$$

(4.16) $$\int_{\omega} |u(x_1 , \alpha)- u(x_1 , 0)|^2 \, dx_1 \leq \alpha \iint_{\omega}\int_0^{\alpha} \left| \frac{\partial u}{\partial x_2} \right|^2 dx_1$$

and from (4.13), (4.15), (4.16) we obtain

(4.17) $$\| m^{\varepsilon} u_{\varepsilon} \|_{L^2(\omega)} \leq C$$

and by applying lemma 4.1, taking into account (4.13) and (4.17),

(4.18) $$\| m^{\varepsilon} u_{\varepsilon} \|_{H^1(\omega)} \leq C$$

Lemma 4.2 - After extracting a subsequence, we have :

(4.19) $$u_{\varepsilon} \rightarrow u^* \quad \text{in } V \text{ weakly}$$

(4.20) $$u_{\varepsilon} |_{\omega} \rightarrow u^* |_{\omega} \quad \text{in } L^2(\omega) \text{ strongly}$$

(4.21) $$m^{\varepsilon} u_{\varepsilon} \rightarrow u^*|_{\omega} \quad \text{in } H^1(\omega) \text{ weakly}$$

and consequently $u^* \in V_{\omega}$ .

Proof - (4.19) is a consequence of (4.12) ; then, (4.20) follows from the trace theorem. Now, by (4.18) we can extract a subsequence such that $m^{\varepsilon} u_{\varepsilon}$ converges in $H^1(\omega)$ weakly. From estimates of the type (4.16) it is easily seen that the limits of $u_{\varepsilon}|_{\omega}$ and $m^{\varepsilon} u_{\varepsilon}$ are the same, and we have (4.21). ∎

Lemma 4.3 - Let $V_{\omega}^d$ the set of the functions of $V_{\omega}$ which are independent of $x_2$ in some domain $x_1 \in \omega$, $|x_2| < \delta$. Then, $V_{\omega}^d$ is dense in $V_{\omega}$.

Proof - By utilisation of a partition of unity, we may consider only functions $v \in V_{\omega}$ with support in a neighbourhood of P (or Q). We have to find a sequence of elements of $V_{\omega}^d$ which tends to v in the topology of $V_{\omega}$.

We consider $v|_{\omega}$ and we continuate it for $x_1 < x_1(P)$ in a function $\tilde{v} \in H^1(\{x_2 = 0\})$. Let $\theta$ be a smooth function of $x_2$ equal to 1 for small $|x_2|$ and with support in a little neighbourhood of $x_2 = 0$. We have $\tilde{v}(x_1)\, \theta(x_2) \in V_{\omega}^d$ , then, we may consider

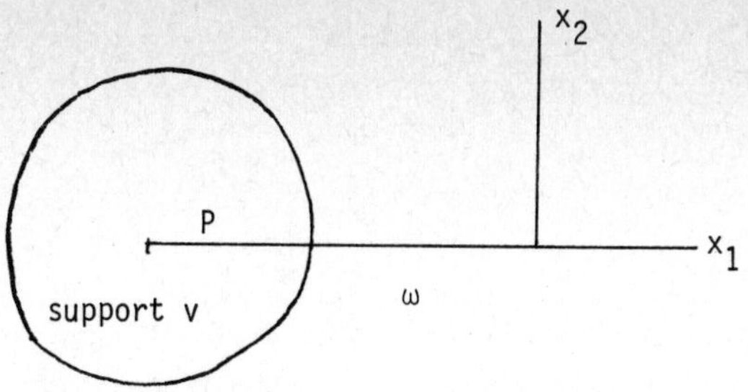


fig. 8

$$w = v - \tilde{v}(x_1)\ \theta(x_2)$$

and the only thing to prove is that a function $w \in H^1(\Omega)$ with trace zero on $\omega$ may be approximated by functions of $V_\omega^d$ in the norm of $H^1(\Omega)$ (which coïncides for such functions with the norm of $V_\omega$). But this is easily proved by using the two following process :

a) translation in the direction $x_1$ negative

b) $w \Rightarrow w^\beta$ defined by

$$w^\beta(x_1, x_2) = \begin{cases} = w(x_1\ , x_2 - \beta\ \phi(x_1)) & \text{if } x_2 > \beta\phi(x_1) \\ = 0 & \text{if } |x_2| < \beta\phi(x_1) \\ = w(x_1\ , x_2 + \beta\phi\ (x_1)) & \text{if } x_2 < -\beta\phi(x_1) \end{cases}$$

where $\phi(x_1)$ is a smooth function, positive for $x_1 > 0$, zero for $x_1 < 0$. The classical properties of mean continuity (see, for instance Necas, [ 1 ] chapter 2) show that such functions approximate w. ■

Now, to prove (4.7), it suffices to show that the function $u^*$ of lemma 4.2 satisfies (4.6) with $v \in V_\omega^d$ (see lemma 4.3). This is immediately obtained by taking $v \in V_\omega^d$ in (4.5). The only non trivial term is

$$\frac{1}{\varepsilon}\int_{\Omega_\varepsilon^0} \frac{\partial u^\varepsilon}{\partial x_1}\frac{\partial v^\varepsilon}{\partial x_1}\, dx = \int_\omega \left(\frac{1}{\varepsilon}\int_{-\lambda\varepsilon}^{+\lambda\varepsilon} \frac{\partial u^\varepsilon}{\partial x_1} dx_2\right) \frac{\partial v}{\partial x_1}\, dx_1 =$$

$$= 2\lambda \int_\omega m^\varepsilon\left(\frac{\partial u^\varepsilon}{\partial x_1}\right) \frac{\partial v}{\partial x_1}\, dx_1$$

and we pass to the limit by using (4.21).

Moreover, the convergence property (4.8) is proved as at the end of sect. 3 (see (3.30), (3.31), by using the fact that weak convergence in V implies strong convergence in H). <u>Theorem 4.1 is proved</u>.

5.- Bibliographical notes - Stiff problems in a more general framework than sect. 1 are studied in Lions [ 3 ]. The study of the spectral properties (sect. 2) in taken from Lobo-Hidalgo and Sanchez-Palencia [ 2 ] .

Stiff problems in variable domains, where the narrowness of the domain "compensates" the high values of the coefficients (sect. 3 and 4) often appears in physics. The results of sect. 4 were obtained almost simultaneously by Pham Huy and Sanchez-Palencia [ 1] and Simonenko [ 3 ], [ 4 ] (see also Boroditskii and Simonenko [ 1 ]). For analogous problems in elasticity theory, see Caillerie [ 1 ], [ 2 ]. These problems have some relation with elasticity problems in "narrow" domains (see Ciarlet and Destuynder [ 1 ], [ 2 ], Ciarlet and Kesavan [ 1 ], Rigolot [ 1 ], [ 2 ]). Sect. 3 is taken from Sanchez-Palencia [ 4 ]. For diffraction stiff problems see chapter 17, sect. 2 and 3. See also Rauch and Taylor [ 1] for other related perturbation problems.

## CHAPTER 14

## AVERAGING AND TWO-SCALE METHODS

The two scale method was widely used in Part II for the study of asymptotic expansions in homogenization theory. In fact, this method was earlier used for the study of perturbation problems in vibration theory and celestial mechanics, where the independent variable is the time and the two scales are in fact time scales. In this framework the two scale method is in connection with the averaging method, which furnishes proofs of the convergence and estimates of the error.

In this chapter we give a short introduction to these questions, limiting ourselves to some very simple situations and results. Bibliographical notes for a deeper study are given in the final section. The averaging method is introduced in sections 1 and 2. The two-scale method and their relations with averaging are given in sect. 3. In sect. 4 an example is handled with the two-scale method and a direct justification (independent of averaging)of the method is given.

1.- Differential equations in a Banach space and integral continuity theorem - Let B be a Banach space. Throughout this chapter only the strong topology (associated with the norm) will be used. Consequently, all limits and convergences will be understood in the norm. Let G be an open domain of B and J a closed interval of the real straight line R. Let $f(x , t)$ be a uniformly continuous function from G x J into B. We then consider the following problem

Find a continuously differenciable function $x(t)$ of t with values in B such that

$$\frac{dx}{dt} = f(x(t) , t) \tag{1.1}$$

$$x(t_0) = x_0 \tag{1.2}$$

where $t_0 \in J$ and $x_0 \in G$ are given.

We then have

Theorem 1.1 - Under the preceeding hypotheses, if in addition $f(x , t)$ satifies a Lipschitz condition in x, i.e., if there exists a constant $\lambda$ such that

$$(1.3) \quad \begin{cases} \|f(x , t) - f(y , t)\| \leqslant \lambda \|x - y\| \\ \text{for any } x , y \in G \quad , \quad t \in J \end{cases}$$

Then, there exists an interval $I \subset J$ with $I \ni t_0$ such that (1.1), (1.2) has a unique solution $x(t)$ defined on I.

Remark 1.1 - If (1.2) is taken as an "initial condition" i.e., $x(t)$ is searched for $t \geqslant t_0$, the interval I may be taken $I = [t_0 , t_0 + h]$

$$(1.4) \qquad h = \inf \{T , \frac{d}{M}\}$$

where T is such that $[t_0 , t_0 + T] \in J$, d is the radius of a ball centered at $x_0$ and contained in G, and M is a bound of $\|f(x , t)\|$ in $G \times J$. An analogous relation holds for $t < t_0$. ■

Remark 1.2 - Condition (1.2) is usually considered as an initial condition, for $t \geqslant 0$, but $x(t)$ is also defined for $t < t_0$. This is an important difference with semigroup theory, where the solution is in general only defined for $t \geqslant 0$ (Note that in semigroup theory $f(x , t) = f(x)$ does not satisfy a Lipschitz condition in general). ■

The proof of theorem 1.1 is classical in differential equations theory. It is based on the iteration method of Picard, after noting that (1.1), (1.2) is equivalent to the integral equation

$$(1.5) \qquad x(t) = x_0 + \int_{t_0}^{t} f(x(s) , s)\, ds$$

Now, we consider differential equations depending on a real positive "small" parameter

$$(1.6) \qquad \varepsilon \in ]0 , \varepsilon_0] = W$$

(where "small" means that our interest is in the limit process $\varepsilon \searrow 0$ ; the value $\varepsilon_0$ is unessential). Note, on the other hand, that $\varepsilon$ cannot take the value 0 (the interval W is open from below).

(1.7) Let $f(x , t , \varepsilon)$ be a continuous function from $G \times J \times W$ into G, uniformly continuous for each $\varepsilon$ and such that there exist positive constants M, $\lambda$ with

(1.8) $\|f(x\ ,\ t\ ,\ \varepsilon)\| \leqslant M \qquad \forall (x\ ,\ t\ ,\ \mu) \in G \times J \times W$

(1.9) $\|f(x\ ,\ t\ ,\ \varepsilon) - f(y\ ,\ t\ ,\ \varepsilon)\| \leqslant \lambda \|x - y\| \quad \forall x\ ,\ y \in G\ ,\ t \in J,\ \varepsilon \in W$

Moreover,

(1.10) Let $F(x\ ,\ t)$ be a uniformly continuous function from $G \times J$ into $G$, such that

(1.11) $\|F(x\ ,\ t)\| \leqslant M \qquad \forall x\ ,\ y \in G \times J$

(1.12) $\left\{\begin{array}{l} \lim\limits_{\varepsilon \searrow 0} \displaystyle\int_{t_0}^{t} f(x\ ,\ s\ ,\ \varepsilon)\, ds \to \int_{t_0}^{t} F(x\ ,\ s)\, ds \\ \text{for any } (x\ ,\ t) \in G \times J. \end{array}\right.$

Remark 1.3 - From (1.12) we have :

$$\left\| \int_{t_1}^{t_2} [F(x\ ,\ s) - F(y\ ,\ s)]\, ds \right\| \leqslant \lambda \|x - y\|\, |t_2 - t_1|$$

for any $x\ ,\ y \in G$, $t_1, t_2 \in J$ and consequently,

(1.13) $\|F(x\ ,\ t) - F(y\ ,\ t)\| \leqslant \lambda \|x - y\| \qquad \forall x,y \in G\ ,\ t \in J$

and $F$ satisfies the hypothesis of theorem 1.1.■

Under the hypothesis (1.7) - (1.12), we have

Theorem 1.2 - Let $x_0 \in G$ be given. If $x(t\ ,\ \varepsilon)$ and $\xi(t)$ are the solutions of

(1.14) $\dfrac{dx}{dt} = f(x\ ,\ t\ ,\ \varepsilon) \quad ; \quad x(t_0) = x_0$

(1.15) $\dfrac{d\xi}{dt} = F(\xi\ ,\ t) \quad ; \quad \xi(t_0) = x_0$

which are uniquely defined on $[t_0\ ,\ t_0+h]$ with $h$ given by (1.4) (independent of $\varepsilon$)

Then, for arbitrarily given $\delta > 0$, there exists $\varepsilon^1(\delta)$ such that

$$\|x(t\ ,\ \varepsilon) - \xi(t)\| \leqslant \delta$$

for $t \in [t_0\ ,\ t_0+h]$ and $\varepsilon \leqslant \varepsilon^1(\delta)$.

Remark 1.4 - It is to be noticed that, under the hypothesis of the theorem,

$$\|f(x\ ,\ t\ ,\ \varepsilon) - F(x\ ,\ t)\|$$

does not tend to zero in general ; we only have the hypothesis of "integral continuity" (1.12). Nevertheless, according to the theorem, the solution $x$ uniforly converges to $\xi$ as $\varepsilon \searrow 0$.■

Proof of theorem 1.2 - From (1.5) we have :

$$x(t\ ,\ \varepsilon) - \xi(t) = \int_{t_0}^{t} [\, f(x(s\ ,\ \varepsilon),\ s\ ,\ \varepsilon) - f(\xi(s)\ ,\ s\ ,\ \varepsilon) + f(\xi(s),\ s\ ,\ \varepsilon) - F(\,\xi(s)\ ,\ s)\,]\ ds$$

and from (1.9),

$$(1.16)\qquad \|x(t\ ,\ \varepsilon) - \xi(t)\| \leq \lambda \int_{t_0}^{t} \|x(s\ ,\ \varepsilon) - \xi(s)\|ds + \Gamma$$

where $\Gamma$ is a constant such that

$$(1.17)\quad \left\| \int_{t_0}^{t} [\, f(\xi(s)\ ,\ s\ ,\ \varepsilon) - F(\xi(s)\ ,\ s)\,]\ ds \right\| \leq \Gamma \quad \text{for } t \in [t_0,\ t_0+h]$$

But it follows from (1.16) that (Gronwall's lemma)

$$\|x(t\ ,\ \varepsilon) - \xi(t)\| \leq \Gamma\ e^{\lambda(t-t_0)}$$

and the theorem will be proved if we show that $\Gamma$ may be taken as small as desired for sufficiently small $\varepsilon$. In order to show this, let us define, for arbitrarily small $\sigma$, a piecewise constant (with a finite number of steps) function $\xi^*(t)$ such that

$$\|\xi(t) - \xi^*(t)\| \leq \frac{\sigma}{\lambda h} \qquad \text{for } t \in [t_0\ ,\ t_0+h\,]$$

(by virtue of the uniform continuity of $\xi(t)$). Then, it follows from (1.8), (1.9), (1.11), (1.13), that

$$(1.18)\quad \begin{cases} \left\| \int_{t_0}^{t} [\, f(\xi(s)\ ,\ s\ ,\ \varepsilon) - f(\xi^*(s)\ ,\ s\ ,\ \varepsilon)\,]\ ds \right\| \leq \sigma \\ \left\| \int_{t_0}^{t} [\, F(\xi(s)\ ,\ s) - F(\xi^*(s)\ ,\ s)\,]\ ds \right\| \leq \sigma \end{cases}$$

Moreover, from (1.12) applied to each interval where $\xi^*(s)$ is constant, we see that, for sufficiently small $\varepsilon$

$$(1.19)\qquad \left\| \int_{t_0}^{t} [\, f(\xi^*(s)\ ,\ s\ ,\ \varepsilon) - F(\xi^*\ ,\ s)\,]\,ds \right\| \leq \sigma$$

and it follows from (1.18), (1.19) that for sufficiently small $\varepsilon$ , we may take $\Gamma = 3\sigma$ (arbitrarily small). The theorem is proved. ∎

2.- The averaging method - The theorem of integral continuity gives a justification of the averaging method for equations describing slowly varying process. Now, we introduce the averaging method.

Let $\varepsilon$ be a "small" positive parameter and $\phi(x , t)$ a uniformly continuous function on $G \times [0 , \infty[$ (where G is an open domain in the Banach space B) such that

$$\|\phi(x , t)\| \leqslant M \qquad \forall (x , t) \in G \times [0 , \infty[ \tag{2.1}$$

$$\|\phi(x , t) - \phi(y , t)\| \leqslant \lambda \|x - y\| \qquad \forall x , y \in G , \quad t > 0 \tag{2.2}$$

Moreover, let $\Phi(x)$ be the average function, defined (if it exists !) by

$$\Phi(x) = \lim_{T\to\infty} \frac{1}{T} \int_0^T \phi(x , t)dt \qquad \forall x \in G \tag{2.3}$$

it follows that

$$\|\Phi(x)\| \leqslant M \qquad \forall x \in G \tag{2.4}$$

$$\|\Phi(x) - \Phi(y)\| \leqslant \lambda \|x - y\| \qquad \forall x , y \in G \tag{2.5}$$

Then, for given $x_0 \in G$, we consider the solutions $x(t , \varepsilon)$ and $\xi(t,\varepsilon)$ of

$$\frac{dx}{dt} = \varepsilon\phi(x , t) \quad ; \qquad x(0 , \varepsilon) = x_0 \tag{2.6}$$

$$\frac{d\xi}{dt} = \varepsilon\, \Phi(\xi) \quad ; \qquad \xi(0 , \varepsilon) = x_0 \tag{2.7}$$

which are defined for $t \in [0 , h/\varepsilon]$ where $h = d/M$ (it suffices to consider theorem 1.1 with the variable $\tau = \varepsilon t$). It is clear that (2.7) is an autonomous equation, easier than (2.6). We shall see that $\xi$ may be used as an approximation of x. To this end, under the re-scaling $\tau = \varepsilon t$ we see that the hypothesis of theorem 1.2 are satisfied : in fact, by using (2.3) we see that

$$\int_0^\tau \phi(x , \frac{\sigma}{\varepsilon})d\sigma = \int_0^{\tau/\varepsilon} \varepsilon\phi(x , s)ds = \tau \frac{\varepsilon}{\tau} \int_0^{\tau/\varepsilon} \phi(x , s)ds$$

converges, as $\varepsilon \searrow 0$ to

$$\tau\, \Phi(x) \equiv \int_0^\tau \Phi(x)\, ds \qquad \forall x \in G , \qquad \tau > 0$$

and consequently (1.12) is satisfied (for $t_0 = 0$, and thus by considering the difference, for any $t_0$). Consequently we have :

Theorem 2.1 - Under the hypothesis of this section, for arbitrarily given $\delta > 0$, there exists $\varepsilon^1(\delta)$ such that

$$\| x(t , \varepsilon) - \xi(t , \varepsilon)\| \leqslant \delta \qquad \text{for } t \in [0 , \frac{h}{\varepsilon}]$$

$$\text{if } \varepsilon < \varepsilon^1(\delta)$$

We then see that $\xi$ (the approximate solution obtained by averaging) approches $x$, and this uniformly on intervals with length which tends to infinity as $\varepsilon$ tends to zero. On the other hand, qualitative properties of differential equations, (such as stability existence of almost periodic solutions ...) are easier to study for the autonomous equation (2.7) than for (2.6). The averaging process is the basis for the study of such properties on infinite intervals of time, $t \in [0 , \infty[$ or $t \in ]-\infty, +\infty[$. (see the bibliographical notes at the end of the chapter).

Remark 2.1 - Theorem 2.1 extends with minor modifications to the case where $\phi = \phi(x , t , \varepsilon)$ depends on the parameter $\varepsilon$ in a continuous manner, under the hypothesis that

$$\|\phi(x , t , \varepsilon) - \phi(x , t , 0)\| \leqslant \gamma(\varepsilon)$$

where $\gamma(\varepsilon)$ tends to zero as $\varepsilon \searrow 0$. In this case, $\Phi(x)$ is defined by a relation analogous to (2.3) with $\phi(x , t , 0)$ instead of $\phi(x , t)$. ■

Remark 2.2 - Theorem 2.1 also extends to the case where $\phi$ is of the form $\phi(x , t , \varepsilon , \varepsilon t)$, i.e., it depends on the "slow time" $\tau = \varepsilon t$ in addition to the "fast time" $t$. In this case, $\Phi = \Phi(x , \tau)$ is defined by

$$\Phi(x , \tau) = \lim_{T \to \infty} \frac{1}{T} \int_0^T \phi(x , t , 0 , \tau)\, dt \tag{2.8}$$

In fact, this case is in the framework of the preceeding remark 2.1 if we consider the system

$$\begin{cases} \dfrac{dx}{dt} = \varepsilon \phi(x , t , \varepsilon , \tau) \quad ; & x(0) = x_0 \\ \dfrac{d\tau}{dt} = \varepsilon \quad ; & \tau(0) = 0 \end{cases} \tag{2.9}$$

(in particular, if $\phi$ does not depend on $\varepsilon$, system (2.9) is of the type (2.6). ■

3.- The two-scale method - It is possible to give a heuristic method (the two-scale method) to obtain the approximation of averaging. The preceeding theorem 2.1 may be considered as a justification of this heuristic method.

To fix ideas, we consider the case of remark 2.2, i.e. the problem

$$(3.1) \qquad \frac{dx}{dt} = \varepsilon\, \phi(x , t , \varepsilon, \varepsilon t) \qquad ; \qquad x(0) = x_0$$

We write $\tau = \varepsilon t$ and we assume that $\phi$ has an expansion of the form

$$(3.2) \qquad \phi(x , t , \varepsilon , \tau) = \phi^0(x , t , \tau) + \varepsilon\, \phi^1(x , t , \tau) + \varepsilon^2 \ldots$$

We search for $x(t , \varepsilon)$ under the form

$$(3.3) \qquad x(t , \varepsilon) = x^0(t , \tau) + \varepsilon x^1(t , \tau) + \varepsilon^2 \ldots ; \qquad \tau = \varepsilon t$$

where it is clear that the different terms must be bounded for

$$(3.4) \qquad t \in [ 0 , \frac{h}{\varepsilon}] \qquad ; \qquad \tau \in [0 , h ]$$

in order that the expansion (3.3) makes sense.

By replacing (3.2) and (3.3) into (3.1) we obtain :

$$\begin{cases} \dfrac{\partial x^0}{\partial t} + \varepsilon(\dfrac{\partial x^0}{\partial \tau} + \dfrac{\partial x^1}{\partial t}) + \varepsilon^2 \ldots = \varepsilon\, \phi^0(x^0 , t , \tau) + \varepsilon^2 \ldots \\ x^0(0 , 0) + \varepsilon x^1(0 , 0) + \varepsilon^2 \ldots = x_0 \end{cases}$$

then, at order $\varepsilon^0$ we have :

$$(3.5) \qquad \frac{\partial x^0(t , \tau)}{\partial t} = 0 \Longrightarrow x^0 = x^0(\tau) \qquad ; \qquad x^0(0) = x_0$$

and at order $\varepsilon^1$ :

$$(3.6) \qquad \frac{dx^0}{d\tau} + \frac{\partial x^1}{\partial t} = \phi^0(x^0(\tau) , t , \tau) \qquad ; \qquad x^1(0 , 0) = 0$$

We integrate (3.6) as an equation for $x^1(t)$ with the parameter $\tau$ :

$$(3.7) \qquad x^1(t , \tau) = t \left[ \frac{1}{t} \int_0^t \phi^0(x^0(\tau), s , \tau)\, ds - \frac{dx^0}{d\tau}(\tau)\right]$$

and the above mentioned condition that $x^1$ must be bounded in (3.4) shows that the function in the brackets of (3.7) must tend to zero as $t \to \infty$ (in fact, $x^1(t , \tau)$ does not depend on $\varepsilon$ and it must be bounded for any $t \in [0 , h/\varepsilon]$ for arbitrarily small $\varepsilon$ , i.e., for arbitrarily large t). This condition may only be satified if

$$(3.8) \qquad \frac{dx^0}{d\tau} = \lim_{t\to\infty} \frac{1}{t} \int_0^t \phi^0(x^0(\tau) , s , \tau)\, ds$$

and consequently we obtain (2.3), (2.7), which appears in a natural, deductive way.

Remark 3.1 - It is clear that condition (3.8) is necessary, but not sufficient for $x^1(t , \tau)$ to be bounded. In fact, (3.8) with appropriate hypotheses ensures the approximation given by theorem 2.1. This is a weaker result than the expansion (3.3) assumed in the two-scale method, which correspond (at least for the first term) to theorem 2.1 with $\varepsilon^1(\delta)$ of the form $c\delta$ . This sharper version of theorem 2.1 is in fact obtained in certain cases, such as $\phi(x , t)$ periodic in t (see Roseau [1 ] , Sanchez-Palencia [8 ] and the following section). ■

4.- Example. Another justification of the two-scale method - We consider the scalar differential equation of second order :

$$\frac{d^2y}{dt^2} + \omega^2 y = \varepsilon f(y , y' , t) \tag{4.1}$$

where $\omega$ is a real constant. This equation is not in the standard form (2.6). We shall write an equivalent equation for slowly varying parameters. We may argue as follows. The solution for $\varepsilon = 0$ depends on two arbitrary constants. For small $\varepsilon > 0$, these parameters will not be constant, but slowly varying functions of time. We then introduce a(t), b(t) by

$$y(t ,\varepsilon) = a(t) e^{i\omega t} + b(t) e^{-i\omega t} \tag{4.2}$$

and we may adjoin another relation in order to define a and b, for instance :

$$\frac{dy(t , \varepsilon)}{dt} = i \omega [a(t) e^{i\omega t} - b(t) e^{-i\omega t} ] \tag{4.3}$$

which implies, with (4.2) :

$$\frac{da}{dt} e^{i\omega t} + \frac{db}{dt} e^{-i\omega t} = 0 \tag{4.4}$$

and by replacing (4.2), (4.3) into (4.1) we obtain an equation which gives, with (4.4) the following system in a, b :

$$\left\{ \begin{aligned} \frac{da}{dt} &= \varepsilon \frac{1}{2\omega i} f(y , y' , t) e^{-i\omega t} \\ \frac{db}{dt} &= \varepsilon \frac{i}{2\omega} f(y , y' , t) e^{i\omega t} \end{aligned} \right. \tag{4.5}$$

where it is understood that y, y' in (4.5) are expressions (4.2), (4.3). This is a system in a, b equivalent to (4.1) ; it has the standard form (2.6). The application of the average method amounts to study the limit (if it exits) (2.3) and the averaged equation (2.7).

But it is also possible to apply the two-scale method directly to equation (4.1). We shall do this now, and we shall give a direct justification (estimate of the error !) of the method.

In order to fix ideas and perform explicit calculations we consider the case where

$$f(y , y' , t) \equiv Ay' + B e^{i\gamma t} \tag{4.6}$$

with A, B, $\gamma$ , real constants, $\gamma \neq \omega$ (case without resonance). We note that the exact solution of (4.1), (4.6) is elementary known ; our study will be of course independent of this fact.

We also consider the initial conditions

(4.7) $$y(0) = 1 \qquad ; \qquad y'(0) = 0$$

We search for a solution of the form :

(4.8) $$y(t\,,\,\varepsilon) = y^0(t\,,\,\tau) + \varepsilon y^1(t\,,\,\tau) + \varepsilon^2 \ldots \qquad ; \qquad \tau = \varepsilon t$$

and consequently :

(4.9) $$\left\{ \begin{aligned} \frac{dy}{dt} &= \frac{\partial y^0}{\partial t} + \varepsilon \left(\frac{\partial y^0}{\partial \tau} + \frac{\partial y^1}{\partial t}\right) + \varepsilon^2 \ldots \\ \frac{d^2y}{dt^2} &= \frac{\partial^2 y^0}{\partial t^2} + \varepsilon \left(2 \frac{\partial^2 y^0}{\partial t\, \partial \tau} + \frac{\partial^2 y^1}{\partial t^2}\right) + \varepsilon^2 \ldots \end{aligned} \right.$$

Then, (4.1) with (4.6) gives at orders $\varepsilon^0$ and $\varepsilon^1$ :

(4.10) $$\frac{\partial^2 y^0}{\partial t^2} + \omega^2 y^0 = 0$$

(4.11) $$2 \frac{\partial^2 y^0}{\partial t\, \partial \tau} + \frac{\partial^2 y^1}{\partial t^2} + \omega^2 y^1 = A \frac{\partial y^0}{\partial t} + B\, e^{i\gamma t}$$

Equation (4.10) gives

(4.12) $$y^0 = a(\tau)\, e^{i\omega t} + b(\tau)\, e^{-i\omega t}$$

and (4.11) becomes

(4.13) $$\frac{\partial^2 y^1}{\partial t^2} + \omega^2 y^1 = e^{i\omega t} \left[ i\,\omega A a - 2\, i\,\omega a' \right] + e^{-i\omega t} \left[ -\, i\,\omega\, A b + 2\, i\,\omega b' \right] + B\, e^{i\gamma t}$$

where $\tau$ plays the role of a parameter. As in (3.4), we search for a solution bounded for $t \in [\,0,h/\varepsilon\,]$ ; this implies that the secular terms must desappear and consequently

(4.14) $$\left\{ \begin{aligned} \frac{da}{d\tau} &= \frac{A}{2}\, a \\ \frac{db}{d\tau} &= \frac{A}{2}\, b \end{aligned} \right\} \Longrightarrow \left\{ \begin{aligned} a(\tau) &= a(0)\, e^{\frac{A}{2}\tau} \\ b(\tau) &= b(0)\, e^{\frac{A}{2}\tau} \end{aligned} \right.$$

and the constants $a(0)$, $b(0)$ are easily obtained in such a way that $y^0(t\,,\,\varepsilon t)$ satisfies the initial conditions (4.7).

The first term, $y^0(t\,,\,\varepsilon t)$, is completely known. Now, in order to study the error, we construct $y^1(t\,,\,\tau)$ in any way (that is to say, we construct any function $y^1(t\,,\,\tau)$ satisfying (4.13), which becomes, with (4.14) :

(4.15) $$\frac{\partial^2 y^1}{\partial t^2} + \omega^2 y^1 = B\, e^{i\gamma t}$$

we choose

(4.16) $$y^1(t\ ,\ \tau) = \frac{B}{\omega^2 - \gamma^2}\, e^{i\gamma t}$$

and we write

(4.17) $$y(\varepsilon\ ,\ t) = y^0(t\ ,\ \varepsilon t) + \varepsilon y^1(t\ ,\ \varepsilon t) + \varepsilon z(t\ ,\ \varepsilon)$$

which is a definition of the function $z(t\ ,\ \varepsilon)$. Because $y^1$ is known, an estimate for z will give an estimate for the error $y - y^0$.

In order to obtain an equation satisfied by z, we write the equations satisfied by $y^0$, $y^1$ with t as only variable, such is to say, with $\tau = \varepsilon t$.

As for $y^1$, because it does not depend on $\tau$ (see (4.16)), we have from (4.15) :

(4.18) $$\varepsilon \frac{d^2 y^1}{dt^2} + \varepsilon\, \omega^2\, y^1 = \varepsilon\, B\, e^{i\gamma t}$$

On the other hand, (4.14) writes

(4.19) $$2\varepsilon \frac{\partial^2 y^0}{\partial t\ \partial\tau} = \varepsilon\, A \frac{\partial y^0}{\partial t}$$

and consequently

(4.20) $$\frac{d^2 y^0}{dt^2} + \omega^2\, y^0 \equiv \left(\frac{\partial^2 y^0}{\partial t^2} + 2\,\varepsilon\, \frac{\partial^2 y^0}{\partial t\ \partial\tau} + \varepsilon^2\, \frac{\partial^2 y^0}{\partial\tau^2}\right) + \omega^2\, y^0 =$$
$$= \varepsilon\, A \frac{\partial y^0}{\partial t} + \varepsilon^2\, \frac{\partial^2 y^0}{\partial \tau^2}$$

Then, if we put (4.17) into (4.1), (4.6) we obtain (with (4.18) and (4.20))

(4.21) $$\left(\frac{d^2}{dt^2} + \omega^2\right)z - \varepsilon\, A\, \frac{dz}{dt} = \varepsilon\left(\frac{\partial^2 y^0}{\partial\tau^2} + \frac{dy^1}{dt}\right) \equiv \varepsilon\, F(t\ ,\ \varepsilon)$$

where it is clear that

(4.22) $$|F(t\ ,\ \varepsilon)\ | \leqslant C \qquad \text{for } t \in [\ 0\ ,\ h/\varepsilon\ ]$$

On the other hand, $y^0(t\ ,\ \varepsilon t)$ satisfies the initial conditions for $y(t\ ,\ \varepsilon)$ and consequently, from (4.17) and (4.16) we see that $z(0)$, $z'(0)$ are well determined constants (independent of $\varepsilon$ ) :

(4.23) $$|z(0)\ |\ ,\ \left|\frac{dz}{dt}(0)\right| \leqslant C$$

and we may obtain an estimate for z from (4.21), (4.23). We multiply (4.21) by dz/dt we have

(4.24) $$\frac{1}{2}\frac{d}{dt}\, E(t) - \varepsilon\, A \left|\frac{dz}{dt}\right|^2 = \varepsilon\, F\, \frac{dz}{dt}$$

were the "energy" E is given by

(4.25) $$E(t) \equiv \left|\frac{dz}{dt}\right|^2 + \omega^2 z^2$$

From (4.24) and (4.22), (C denotes, as usual, several constants independent of ε)

$$\frac{dE(t)}{dt} \leqslant \varepsilon C \left[ C + \left|\frac{dz}{dt}\right|^2 \right] \leqslant \varepsilon C \left[ C + E(t) \right]$$

(4.26) $$E(t) - E(0) \leqslant \varepsilon C \int_0^t [C + E(s)]\, ds$$

and then, for $t \in [0, h/\varepsilon]$ we have (E(0) is bounded by (4.23)) :

$$E(t) \leqslant C + \varepsilon C \int_0^t E(s)\, ds$$

and by the Gronwall's lemma :

(4.27) $$E(t) \leqslant C \exp \left[ \int_0^t \varepsilon C\, ds \right] \leqslant C \quad , \quad \text{for } t \in [0, h/\varepsilon]$$

As a consequence, z(t) and z'(t) are bounded for $t \in [0, h/\varepsilon]$ ; on the other hand, from (4.16), the same holds for $y^1$, $y^{1\prime}$. Consequently, from (4.17) we have :

(4.27) $y(t, \varepsilon) - y^0(t, \varepsilon t)$ and its derivative are bounded by $C\varepsilon$ for $t \in [0, h/\varepsilon]$

This was the desired estimate. We see that we have the analogous of theorem 2.1 with $\varepsilon^1 = C \cdot \delta$ (see remark 3.1).

In the case of linear equations independent of t (antonomous) there is also another justification of the method (which also holds for certain types of equations with unbounded operators in a Hilbert space).

We consider (4.1), (4.6) with B = 0 (antonomous), i.e. :

(4.28) $$\frac{d^2y}{dt^2} + \omega^2 y = \varepsilon A y'$$

and we search for a solution of the form

(4.29) $$y = \alpha e^{\beta(\varepsilon)t} \Longrightarrow$$

$$\beta(\varepsilon)^2 + \omega^2 = \varepsilon A \beta(\varepsilon) \Longrightarrow$$

(4.30) $$\beta(\varepsilon) = \pm i \omega + \frac{\varepsilon A}{2} + O(\varepsilon^2)$$

From (4.14) we remark that the approximation $y^0(t, \varepsilon t)$ for $y(t, \varepsilon)$ amounts to disregarding the term $O(\varepsilon^2)$ in (4.30). But it is clear that, for

$t \in [\,0\,,\,h/\varepsilon\,]$ , the difference

$$e^{\beta(\varepsilon)t} - e^{\pm i\,\omega + \frac{\varepsilon A}{2}}$$

is of order $O(\varepsilon)$. Consequently, we again obtain (4.27). The preceeding reasoning may be performed in a more precise way by taking into account the initial conditions.

5.- Bibliographical notes - Averaging and two-scale are classical methods to obtain approximate solutions of differential equations on long intervals of time. The classical works are associated with the names of Poincaré, Bogolyubov and Mitropolski. For recent results in ordinary differential equations, see Balachandra and Sethna [1] , Banfi [1] , Cerneau [1] , Roseau [1] and Sanchez-Palencia [5] . The case of integro-differential equations is studied in Filatov [1], Filatov and Sharova [1] , Graffi [1]. For differential equations in Banach spaces (including certain types of partial differential equations), see Roseau [3], [5] , and Zabreiko and Fetisov [1] . Formal utilisation of the method for partial differential equations may be seen in Cerneau et Sanchez-Palencia [1], [2] and Mitropolski and Moseenkov [1], Cole [1] (and many others ... !). Justification of the method for certain special types of equations may be seen in Lions [3] sect. VI.6, Sanchez-Palencia [6] and [8] , Simonenko [1] , [2] , and, for integro-partial differential equations, in Turbé [1] .

Our study of the averaging method (sect. 1 and 2) is taken from Roseau [3]. The comparaison of the two methods (sect. 3) is also taken from Roseau [3] (see Morrison [1] for an example). Sect. 4 is a very elementary version of the study of Sanchez-Palencia [8] for partial differential equations.

# PART IV

# DIFFRACTION AND RELATED PROBLEMS

## CHAPTER 15

## GENERALITIES AND POTENTIAL METHOD

1.- Introduction - This chapter deals with the wave equation in unbounded domains of $R^3$, and more exactly in domains $\Omega$ which contain a neighbourhood of infinity, for instance the complement of a bounded region B with smooth boundary S, or the whole space $R^3$.

We shall study solutions of the wave equation

$$(1.1) \qquad \frac{\partial^2 \phi}{\partial t^2} - \Delta\phi = 0 \quad ; \qquad \Delta = \frac{\partial}{\partial x_i} \frac{\partial}{\partial x_i}$$

or of the "reduced wave equation" (or the Helmholtz equation)

$$(1.2) \qquad - \Delta\psi - \omega^2 \psi = 0$$

which is obtained from (1.1) by searching for a solution $\phi$ of the form :

$$(1.3) \qquad \phi(x , t) = \text{Re} [ e^{-i\omega t} \phi(x)]$$

Let us recall that (1.1) is the equation of acoustic waves if the propagation speed is taken to be 1 ; this is also possible after a re-scaling of time. In acoustics, $\phi$ is (for instance !) the velocity potential, and the velocity field is then given by $\underline{v}$ = grad $\phi$.

The parameter $\omega$ in (1.3) is, for the time being, a given constant, either real or complex, but the properties of solutions of (1.2) are very different in the real or complex cases. This point deserves attention in the sequel.

We shall also consider the non-homogeneous equations associated with (1.1) and (1.2) (with given functions at the right hand side instead of zero).

The functions

$$\psi^{\pm} = \frac{-1}{4\pi}\,\frac{e^{\pm i\omega r}}{r} \quad ; \qquad r = |x - y| \tag{1.4}$$

are fundamental solutions of the equation (1.2). This means that

$$(-\Delta - \omega^2)\psi^{\pm} = \delta_y \tag{1.5}$$

where $\delta_y$ is for the Dirac distribution at the point y. The proof of (1.5) is a simple exercise in distribution theory (c.f. for instance Schwartz [2] sect. II. 2.3).

Remark 1.1 - The two fundamental solutions $\psi^{\pm}$ are almost equivalent as for the behaviour for little r because the difference between them is a bounded function. Thus they have the same singularity at the origin. On the other hand, the corresponding solutions of (1.1)

$$\phi^{\pm} = \psi^{\pm} e^{-i\omega t} = \frac{-1}{4\pi}\,\frac{e^{-i\omega(t\mp r)}}{r} \tag{1.6}$$

are constants in the cones $t \mp r = Cst.$ ; in other words, $\phi^+$ (resp. $\phi^-$) is an outgoing (resp. incoming) wave. We shall later see the implications of this property in the construction of solutions. ■

We now study the "radiation conditions" for $\psi^{\pm}$.

Proposition 1.1 - Let us consider $\psi^{\pm}$ for real $\omega$ , with fixed y, as a function of x. If $R = |x|$ is the distance from the origin, $\psi^{\pm}$ satisfies the radiation condition (Sommerfeld condition) :

$$\left|\frac{\partial\psi^{\pm}}{\partial R} \mp i\,\omega\psi^{\pm}\right| = O(R^{-2}) \qquad \text{for } R \to \infty \tag{1.7}$$

and this asymptotic behaviour holds uniformly for y in a bounded region. The same property holds for the derivatives $\frac{\partial\psi^{\pm}}{\partial y_i}$ .

Proof - Let us first consider $y = 0 \Rightarrow r = R$

$$\frac{\partial\psi^{\pm}}{\partial R} = \frac{1}{4\pi}\left(\frac{e^{\pm i\omega r}}{r^2} \pm i\,\omega\psi^{\pm}\right)$$

In the case $y \neq 0$, it suffices to remark that

$$\frac{\partial}{\partial R} = \frac{\partial}{\partial r} \cos\theta \quad \text{and} \quad \cos\theta - 1 = O(R^{-2})$$

for large R

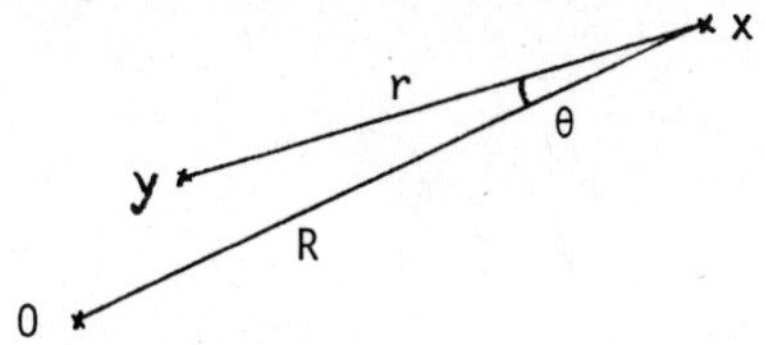


As for $\frac{\partial \psi^{\pm}}{\partial y_i}$ it suffices to remark that it is $= - \frac{\partial \psi^{\pm}}{\partial x_i}$ and the result is obtained by differentiation. ∎

In the sequel, we only consider the + sign (outgoing solutions). Clearly, reversed relations are obtained with the sign - .

Remark 1.2 - It is evident that if $\psi$ is a convergent superposition of solutions of type $\psi^+$ ($\omega$ real), it satisfies

(1.8) $$\left|\frac{\partial \psi}{\partial R} - i\,\omega\,\psi\right| = O(R^{-2}) \quad \text{and} \quad |\psi| = O(R^{-1})$$

Here, convergent superposition means a surface or volume distribution of the type

$$\psi(x) = \int_B \rho(y)\psi^+ \, dy$$

with B bounded and such that $\int_B \rho(y)\, dy$ is finite. By taking into account $\partial\psi^+ / \partial y_i$ instead of $\psi^+$ the result also holds for double layer potentials (see later). ∎

The radiation condition (1.8) plays a fundamental role in the study of the energy flux at large distance. We shall now study this question.

Let D be a bounded region of the domain $\Omega$ . If $\psi$ (and $\phi$, defined by (1.3)) is a solution of the wave equation, we obtain an energy relation by multiplying (1.2) by $\partial\phi/\partial t$ and by integrating it by parts. ($\underline{n}$ is for the unit normal vector to $\partial D$) :

$$0 = \int_D \frac{\partial^2 \phi}{\partial t^2} \frac{\partial \phi}{\partial t}\, dx - \int_D \frac{\partial}{\partial x_i}\left(\frac{\partial \phi}{\partial x_i}\right) \frac{\partial \phi}{\partial t}\, dx = \frac{1}{2}\frac{d}{dt} \int_D \left|\frac{\partial \phi}{\partial t}\right|^2 -$$

$$- \int_{\partial D} \frac{\partial \phi}{\partial n} \frac{\partial \phi}{\partial t}\, dS + \frac{1}{2}\frac{d}{dt} \int_D \sum_i \left|\frac{\partial \phi}{\partial x_i}\right|^2 dx \implies$$

$$(1.9)\qquad \frac{1}{2}\frac{d}{dt}\int_D [(\frac{\partial\phi}{\partial t})^2 + (\underline{\text{grad}}\ \phi)^2]\,dx = \int_{\partial D}\frac{\partial\phi}{\partial n}\frac{\partial\phi}{\partial t}\,dS$$

The volume integral on D may be considered as an energy (the first term is a kinetic energy and the second a potential one) and the surface integral in the right hand side is a "surface energy flux".

$\phi$ given by (1.3) is periodic in time of period $2\pi/\omega$. Let us calculate the energy $\Gamma$ given in a period by the exterior of D to D.

$$\Gamma \equiv \int_0^{2\pi/\omega}\int_{\partial D}\frac{\partial\phi}{\partial n}\frac{\partial\phi}{\partial t}\,dS\,dt \qquad ; \quad \phi = \text{Re}\,[e^{-i\omega t}\,\phi]$$

For real $\omega$, we have : (the bar is for the complex conjugate) :

$$\frac{\partial\phi}{\partial n} = \frac{1}{2}(\frac{\partial\psi}{\partial n}e^{-i\omega t} + \frac{\partial\overline{\psi}}{\partial n}e^{i\omega t})$$

$$\frac{\partial\phi}{\partial t} = \frac{i\omega}{2}(-\ \psi e^{-i\omega t} + \overline{\psi}e^{i\omega t})$$

and we obtain

$$(1.10)\qquad \Gamma = \frac{\pi i}{2}\int_{\partial D}(\frac{\partial\psi}{\partial n}\overline{\psi} - \psi\frac{\partial\overline{\psi}}{\partial n})dS \equiv -\ \pi\ \text{Im}\int_{\partial D}\frac{\partial\psi}{\partial n}\overline{\psi}\,dS$$

If we integrate also the left hand side of (1.9) in a period, we evidently obtain zero,; then, we have

Proposition 1.2 - If $\omega$ is real and we have a solution of the Helmholtz equation (1.2) in a bounded domain D, the energy equation in a period of time takes the form :

$$(1.11)\qquad 0 = \frac{\pi i}{2}\int_{\partial D}(\frac{\partial\psi}{\partial n}\overline{\psi} - \psi\frac{\partial\overline{\psi}}{\partial n})\,dS \equiv -\ \pi\ \text{Im}\int_{\partial D}\frac{\partial\psi}{\partial n}\overline{\psi}\,dS\,.$$

Remark 1.3 - The preceeding deduction shows the physical meaning of the relation (1.11) ; but it is also possible to obtain it directly from (1.2). It suffices to multiply (1.2) by $\overline{\psi}$ and to integrate it by parts :

$$(1.12)\qquad 0 = -\int_D(\Delta\psi + \omega^2\psi)\,\overline{\psi}\,dx = -\int_{\partial D}\frac{\partial\psi}{\partial n}\overline{\psi}\,dS + \int_D(\frac{\partial\psi}{\partial x_i}\frac{\partial\overline{\psi}}{\partial x_i} - \omega^2\psi\,\overline{\psi})dx\,.$$

For $\omega$ real, the integral over D is real, and we obtain (1.11) by taking the imaginary part of (1.12) and by multiplying it by $-\ \pi$. ∎

We are now able to give a new physical interpretation of the Sommerfeld radiation condition (1.8).

Let us consider a solution $\psi$ of (1.2) for real $\omega$ in the region $|x| \geqslant a$. We then apply the relation (1.11) to the region

$$D_R = \{x \; ; \; a < |x| < R\}$$

for large R. For n exterior to $D_R$ we have

$$(1.13) \qquad 0 = \frac{\pi i}{2} \int_{|x|=a} \left(\frac{\partial \psi}{\partial n} \overline{\psi} - \psi \frac{\partial \overline{\psi}}{\partial n}\right) dS + \frac{\pi i}{2} \int_{|x|=R} \left(\frac{\partial \psi}{\partial n} \overline{\psi} - \psi \frac{\partial \overline{\psi}}{\partial n}\right) dS$$

On the other hand, we write the identity

$$\frac{\pi}{2\omega} \int_{|x|=\rho} \left|\frac{\partial \psi}{\partial R} - i\omega \psi\right|^2 dS = \frac{\pi}{2\omega} \int_{|x|=\rho} \left(\left|\frac{\partial \psi}{\partial R}\right|^2 + \omega^2 |\psi|^2\right) dS + \frac{\pi i}{2} \int_{|x|=\rho} \left(\frac{\partial \psi}{\partial R}\overline{\psi} - \psi \frac{\partial \overline{\psi}}{\partial R}\right) dS$$

Then, if $\psi$ satisfies the Sommerfeld condition (1.8), the limit of the left hand side as $\rho \to \infty$ exists and is zero ; the same then holds for the right hand side. By putting this in (1.13) (for $|x| = \rho$, $\partial/\partial R = \partial/\partial n$), we have

$$\frac{\pi i}{2} \int_{|x|=a} \left(\frac{\partial \psi}{\partial n} \overline{\psi} - \psi \frac{\partial \overline{\psi}}{\partial n}\right) dS - \frac{\pi}{2\omega} \int_{|x|=\rho} \left(\left|\frac{\partial \psi}{\partial n}\right|^2 + \omega^2 |\psi|^2\right) dS \to 0 \qquad \rho \to \infty$$

but the integral over $|x| = a$ is independent of $\rho$ ; consequently the integral over $|x| = \rho$ has a limit as $\rho \to \infty$ and this limit is necessarily $\geqslant 0$ because the integrand is $\geqslant 0$. We have then proved :

Proposition 1.3 - Let $\psi$ be a solution of the Helmholtz equation (1.2) with $\omega$ real, in the region $|x| \geqslant a$ satisfying the radiation condition (1.8). Then, there exists a constant $\Phi \geqslant 0$ such that

$$(1.14) \qquad \Phi = \lim_{\rho \to \infty} \frac{\pi}{2\omega} \int_{|x|=\rho} \left(\left|\frac{\partial \psi}{\partial R}\right|^2 + \omega^2 |\psi|^2\right) dS$$

$\Phi$ is equal to the energy flux (from $|x| < a$ towards $|x| > a$) through $|x| = a$ :

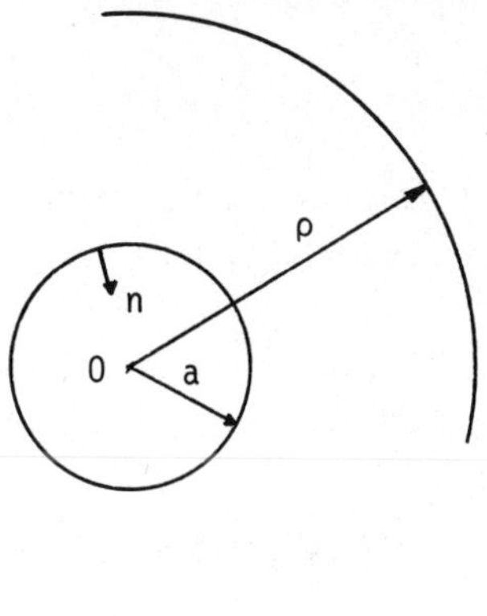

(1.15) $$\Phi = \frac{\pi i}{2} \int_{|x|=a} \left(\frac{\partial\psi}{\partial n} \overline{\psi} - \psi \frac{\partial\overline{\psi}}{\partial n}\right) dS$$

In an analogous manner, if $\psi$ is defined in the exterior of a closed surface $\Sigma$ the same result holds when the integral is taken over $\Sigma$ instead of taking it over $|x| = a$ as in (1.15).

Remark 1.4 - The fact that $\Phi \geqslant 0$ shows that the energy flows towards infinity. This is the energetic meaning of the radiation condition (1.8). Of course if the sign - is taken in (1.7) (and consequently + in (1.8)), the energy flux comes from infinity. This remark may be compared to remark 1.1. ∎

Remark 1.5 - Since the Helmholtz equation is elliptic and has constant coefficients, its solutions are analytic functions in the interior of their domain of definition. But they are not necessarily smooth on the boundaries of their domains of definition. Then, integrals such as (1.15) do not make sense in certain cases. This point shall deserve some attention in the sequel. ∎

2.- Uniqueness theorems - The Helmholtz equation enjoys the property that if

(2.1) $$\lim_{\rho \to \infty} \int_{|x|=\rho} |\psi(x)|^2 \, dS = 0 \qquad \text{then,}$$

$\psi$ is zero. There are several forms of this property, which are in general proved by using an explicit expansion of the solution in special functions. We shall give without proof the following theorem of Rellich [2], which is a suitable form of this property :

Theorem 2.1 - For real $\omega > 0$, let $\psi$ be a solution of the Helmholtz equation for $|x| > a$. Then, for any $b > a$, there exists $M > 0$ and $c > b$ such that, for $R > c$, either

$$\int_{b<|x|<R} |\psi|^2 \, dx \geqslant M R$$

or $\psi$ is identically zero.

Corollary 2.1 - If $\psi$ is a solution of the Helmholtz equation for real $\omega > 0$ and $\psi$ belongs to $L_2$ of an exterior domain, $\psi$ is zero in it. ∎

Corollary 2.2 - If $\psi$ is a solution of the Helmholtz equation for real $\omega > 0$ in an exterior domain, and $\psi$ satisfies (2.1), then, $\psi \equiv 0$. ∎

Corollary 2.3 - In the case of proposition 1.3, $\Phi = 0$ implies $\psi \equiv 0$ (By corollary 2.2). ∎

We can now study the uniqueness of the solution of the Dirichlet, Neumann and transmission problems.

Let B be a bounded domain of $R^3$ with smooth boundary S, and let $\Omega$ be the complement domain (which extends to infinity). $\Omega$ is connected. We then have Defin

Definition 2.1 - Let f be a function defined on S. The Dirichlet (resp. Neumann) problem consists of finding $\psi$ defined in $\Omega$ , satisfying (1.2) for real $\omega > 0$, the radiation condition (1.8) and such that, on S :

(2.2) $$\psi|_S = f \quad \text{(Dirichlet)}$$

(2.3) $$\frac{\partial\psi}{\partial n}\Big|_S = f \quad \text{(Neumann)}$$

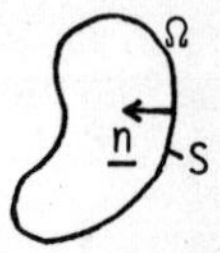


Remark 2.1 - According to remark 1.5, the preceeding definition is only formal. For the time being, we shall suppose that (2.2) and (2.3) are satisfied in the classical sense. More elaborate definitions will be used in the sequel.∎

Theorem 2.2 - The Dirichlet and Neumann problems (defined according to def. 2.1 and remark 2.1) have at most one solution.

Proof - Let $\psi$ be a solution for f = 0. Let $\Omega_\rho$ be the intersection of $\Omega$ with $|x| < \rho$. Let us multiply (1.2) by $\overline{\psi}$, and let us consider also their conjugates. By integrating by parts on $\Omega_\rho$ we have :

$$0 = \int_{\Omega_\rho} [ ( \Delta\psi + \omega^2\psi) \overline{\psi} - (\Delta \overline{\psi} + \omega^2 \overline{\psi}) \psi ] \, dx = \int_{|x|=\rho} (\frac{\partial\psi}{\partial n} \overline{\psi} - \frac{\partial\overline{\psi}}{\partial n} \psi) \, dS$$

where n is the outgoing normal to $|x| = \rho$. (In fact, the integral in the right hand side should be extended also to S, but it is zero there by (2.2) or (2.3) with f = 0). Then, according to the last past of proposition 1.3, we have $\Phi = 0$, and by corollary 2.2, $\psi \equiv 0$ in a neighbourhood of infinity. Moreover, $\psi$ is analitic (see remark 1.5) and $\Omega$ is connected, then by analytic continuation, $\psi$ is zero all over $\Omega$.∎

We let B and $\Omega$ be given as before, and

(2.4) $$b(x) = \begin{cases} 1 & \text{if } x \in \Omega \\ \beta & \text{if } x \in B \end{cases} \quad ; \quad \beta > 0 \quad \text{(real !)}$$

(2.5) $$a(x) = \begin{cases} 1 & \text{if } x \in \Omega \\ \alpha & \text{if } x \in B \end{cases} \quad ; \quad \alpha > 0 \quad \text{(real !)}$$

we then have the following definition of the transmission problem :

Definition 2.2 - Let f be a given function of $L^2(R^3)$ with compact support. The transmission problem consists of finding $\psi$ satisfying the radiation condition (1.8), and the equation (for real $\omega > 0$)

$$(2.6) \qquad -\frac{\partial}{\partial x_i}\left(a(x)\frac{\partial \psi}{\partial x_i}\right) - \omega^2 b(x)\,\psi = f$$

in B and $\Omega$, with the transmission conditions on S associated with (2.6) in the sense of distributions, i.e. :

$$(2.7) \qquad \psi\big|_B = \psi\big|_\Omega \quad ; \quad \alpha \frac{\partial \psi}{\partial n}\Big|_B = \frac{\partial \psi}{\partial n}\Big|_\Omega \quad \text{on } S$$

where n is the unitary normal to S.

Theorem 2.3 - The transmission problem, (defined according to def. 2.2 and the remark analogous to the remark 2.1) has at most one solution.

Proof - We shall prove that any solution with f = 0 is zero. We begin as in the proof of theorem 2.2. We multiply (2.6) by $\overline{\psi}$ and we substract the conjugate of this product. By integrating by parts on $\Omega_\rho$, the integrals over S cancels by vitue of (2.7) and we have :

$$0 = \int_{|x|=\rho} \left(\frac{\partial \psi}{\partial n}\overline{\psi} - \frac{\partial \overline{\psi}}{\partial n}\psi\right) dS$$

By the last part of prop. 1.3, $\Phi = 0$ and by coroll. 2 2, $\psi$ is zero in a neighbourhood of infinity. Moreover, $\psi$ is analytic in $\Omega$ (and in B !). Then, by analytic continuation, $\psi$ is zero on $\Omega$. The study of $\psi$ in B is a little bit delicate. According to (2.6) and (2.7), $\psi$ satisfies

$$(2.8) \qquad -\alpha\,\Delta\,\psi - \omega^2\,\beta\,\psi = 0 \quad \text{in B}$$

$$(2.9) \qquad \psi = 0 \;, \quad \frac{\partial \psi}{\partial n} = 0 \quad \text{on S}$$

Conditions (2.9) are Cauchy data for the elliptic equation (2.8), and (2.8), (2.9) is not a well posed problem. Nevertheless, for equations with analytic coefficients, the Holmgren theorem holds (c.f. Courant-Hilbert [ 1 ], vol. II, chap. III, app. 2). This theorem states the uniqueness of the Cauchy problem in a neighbourhood of the surface S which supports the Cauchy data. Consequently, $\psi$ is zero in the vicinity of S on the side B, and then, by analytic continuation, $\psi = 0$ all over B. ■

## 3.- Representation formula and radiation condition -

Proposition 3.1 - Let D be a bounded domain of smooth boundary $\Gamma$. Let $\psi$ be a two-times differentiable function on $\overline{D}$, satisfying

(3.1) $$- \Delta \psi - \omega^2 \psi = 0$$

( $\omega$ either real or complex) and let P be an interior point (see fig. 4).Then, $\psi$ admits the representation

(3.2) $$\psi(P) = \frac{1}{4\pi} \int_\Gamma [\frac{\partial \psi}{\partial n} \frac{e^{i\omega r}}{r} - \frac{\partial}{\partial n} (\frac{e^{i\omega r}}{r}) \psi ] dS$$

where n is the outer unit normal to $\Gamma$ and r is the distance between P and the running point of $\Gamma$ .

Proof - It is analogous to the classical one for the case $\omega = 0$ (see for instance Mikhlin [1] sect. 11.3). It suffices to apply the Green's formula

$$\int_{D_\rho} [ ( \Delta\psi + \omega^2 \psi) \theta - ( \Delta \theta + \omega^2 \theta) \psi ] dx = \int_{\partial D_\rho} (\frac{\partial \psi}{\partial n} \theta - \frac{\partial \theta}{\partial n} \psi) dS$$

with $\theta = \frac{e^{i\omega r}}{r}$ to the domain $D_\rho$ obtained by removing from D the ball of radius $\rho$ centered at the origin and by taking the limit $\rho \to 0$. ■

We can now establish an analogous formula for solutions defined in an outer domain. It turns out that, if $\psi$ satisfies a radiation condition, it has a representation analogous to (3.2) with the integral extended to the (bounded !) boundary of the outer domain.

Proposition 3.2 - Let D be an outer domain with smooth boundary $\Gamma$ . Let $\psi$ be a two times differentiable function on $\overline{D}$ satisfying

(3.3) $$- \Delta \psi - \omega^2 \psi = 0$$

for real $\omega$ and the radiation condition (1.8).(see fig. 5).

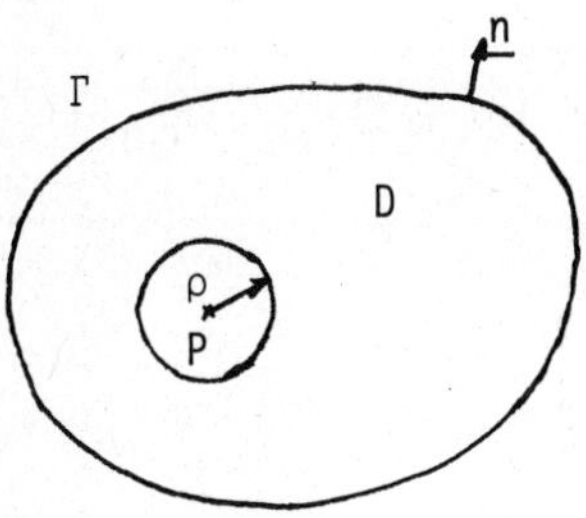


fig. 4

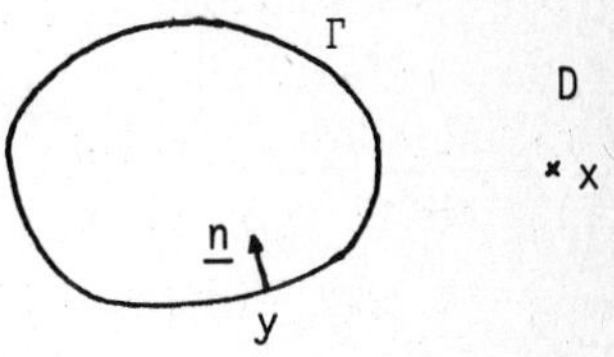


fig. 5

Moreover, let x be an interior point of D. Then, $\psi(x)$ admits the representation (3.2) where n is the outer unit normal to $\Gamma$.

Proof - Let us apply proposition 3.1 to the domain $D_\rho$, the intersection of D with $\{|x| < \rho\}$. We obtain $\psi(x) = \psi^1(x) + \psi^2(x)$ where $\psi^1(x)$ is the desired expression and

$$(3.4) \quad \psi^2(x) = \frac{1}{4\pi} \int_{|y|=\rho} \left[ \frac{\partial \psi}{\partial R}\left(\frac{e^{i\omega r}}{r}\right) - \frac{\partial}{\partial R}\left(\frac{e^{i\omega r}}{r}\right) \psi \right] dS$$

which is independent of $\rho$ (because $\psi$ and $\psi^1$ are so). Moreover, $\psi$ (and $e^{i\omega r}/r$ also !) satisfies (1.8) ; consequently

$$\frac{\partial \psi}{\partial R} = i\,\omega\,\psi + O(\rho^{-2})$$

$$\frac{\partial}{\partial R}\left(\frac{e^{i\omega r}}{r}\right) = i\,\omega\left(\frac{e^{i\omega r}}{r}\right) + O(\rho^{-2})$$

Then, by taking the limit of (3.4) as $\rho \to \infty$, the principal parts cancel and the integrand is $O(\rho^{-3})$ ; because the measure of the sphere is $O(\rho^2)$, the integral tends to zero (it is zero, in fact !), Q.E.D. ∎

Remark 3.1 - The prop. 3.2 is the converse of the evident proposition given in Remark 1.2.

In fact, for real $\omega$, the conditions (1.8) are equivalent to the fact that the solution is a superposition of elementary solutions $\frac{e^{i\omega r}}{r}$. ∎

Remark 3.2 - Let $\psi$ be a solution of (3.3) with real $\omega$ ; if it satisfies (1.8) and is defined in an outer domain $\Omega$ (whose boundary is not necessarily smooth), it is analytic in $\Omega$, and we can take in prop. 3.2 $D = \{x ; |x| > \rho\}$ for sufficiently large $\rho$. Then, $\psi$ has the representation

$$(3.5) \quad \psi(x) = \frac{1}{4\pi} \int_{|y|=\rho} \left[ -\frac{\partial \psi}{\partial R}\left(\frac{e^{i\omega r}}{r}\right) + \frac{\partial}{\partial R}\left(\frac{e^{i\omega r}}{r}\right) \psi \right] dS \quad ; \quad R = |y|$$

for large $|x|$. ∎

It is now possible to give a new definition of the radiation condition.

Definition 3.1 - Let $\psi$ be a solution of $-\Delta\psi - \omega^2\psi$ ($\omega$ either real or complex) in an outer domain. We shall say that it satisfies the outgoing (resp. in coming) radiation condition if it is a superposition (in the sense of remark 1.2) of elementary solutions $\frac{e^{i\omega r}}{r}$ (resp. $\frac{e^{-i\omega r}}{r}$).

For real $\omega$, this definition is analogous to the fact that $\psi$ satisfies

$$(3.6) \quad \left|\frac{\partial \psi}{\partial R} \mp i\,\omega\,\psi\right| = O(R^{-2}) \quad ; \quad |\psi| = O(R^{-1}) \qquad R = |x|$$

with - (resp. + ) for outgoing (resp. incoming) radiation condition.

Remark 3.3 - The definition associated with (3.6) only holds for real $\omega$. In fact, it is easy to see that, for Im $\omega < 0$, the function $\frac{e^{i\omega r}}{r}$ is outgoing but it does not satisfy (3.6) (in fact, it oscillates with exponentially large amplitude for large r.) ∎

4.- Potential theory - This theory deals with some kinds of superpositions (or convolutions) of fundamental solutions

$$\psi^+(x\ ,\ y) \equiv \frac{-1}{4\pi}\frac{e^{i\omega r}}{r} \quad ; \quad r = |x - y| \tag{4.1}$$

This is a classical matter, and it may be read in any classical text book (may be for $\omega = 0$) (see the bibliography given in the last section). Certain results will then be given without proof.

Theorem 4.1 - Let D be a bounded domain of $R^N$ (or of an N-dimensional smooth manifold). Let B(x , y) be a kernel, bounded for $x,y \in \overline{D}$ and continuous for $x \neq y$. Then, the integral operator K defined by

$$(K\ u)\ (x) = \int_D \frac{B(x\ ,\ y)}{r^\lambda}\, u(y)\, dy \qquad r = |x - y| \tag{4.2}$$

is bounded and compact from $L_p(D)$ into $C^0(D)$ provided that $\lambda < N - \frac{N}{p}$.

Theorem 4.2 - Let D and B(x , y) be as in the preceeding theorem.Let N, $\lambda$, p be such that

$$N > \lambda \geqslant N - \frac{N}{p}$$

and the integer $s \leqslant N$ such that

$$s > N - (N - \lambda)p$$

Then, the integral operator K defined by (4.2) is bounded and compact from $L_p(D)$ into $L_q(D_s)$, where $D_s$ is any given s-dimensional section of D (by an s-dimensional smooth manifold) and q satisfies

$$q < \frac{s\,p}{N - (N - \lambda)p}$$

The two preceeding theorems are in fact Lemmas for the proof of the Sobolev embedding theorem. The proofs may be seen in Smirnov [ 1 ] vol. 5, sect. 115.

Remark 4.1 - It is clear that in the theorem 4.2 we may take the section of D by itself (s = N). ∎

Theorem 4.3 - Let D be a bounded domain of $R^3$ of smooth boundary S. Let B(x , y) be a kernel of class $C^1$ for $x \in D$, $y \in S$. Then, the integral operator K defined by

(4.3) $$(K\,u)(x) = \int_S \frac{B(x\,,\,y)}{r}\,u(y)\,dS_y$$

is bounded and compact from $L^2(S)$ into $L^2(D)$.

Proof - Since the limit in norm of a sequence of compact operators is compact, it suffices to prove that K is the sum of a compact operator and of another operator whose norm is as small as desired.

For any $\varepsilon$, let us define a smooth function $\theta$ equal to 0 for $r \geqslant \varepsilon$, equal to 1 for $r \leqslant \varepsilon/2$, and bounded by 1 elsewhere.

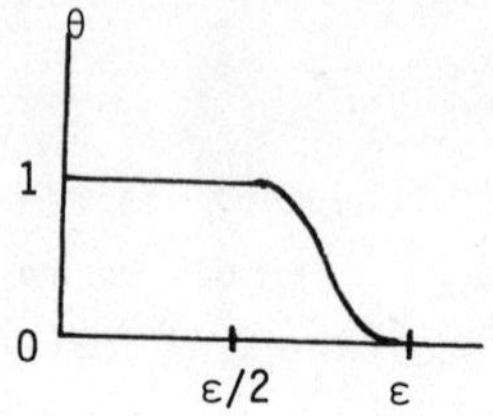


Let us define

$$\left.\begin{aligned} A_\varepsilon(x,\, y) &= \frac{B(x\,,\,y)}{r}\,\theta(r) \\ A'_\varepsilon(x\,,\,y) &= \frac{B(x\,,\,y)}{r}\,(1-\theta(r)) \end{aligned}\right\} \qquad r = |x-y|$$

and let the potentials $K_\varepsilon$ and $K'_\varepsilon$ be defined by

(4.4) $$(K_\varepsilon\,u)\,(x) = \int_S A_\varepsilon(x\,,\,y)\,u(y)\,dS_y$$

and the analogous formula with $A'_\varepsilon$. Thus

$$K = K_\varepsilon + K'_\varepsilon$$

The Kernel $A'_\varepsilon(x\,,\,y)$ as well as $\dfrac{\partial A'_\varepsilon(x\,,\,y)}{\partial x_i}$ is bounded. We then have

$$(K'_\varepsilon\,u)\,(x) = \int_S A'_\varepsilon\,(x\,,\,y)\,u(y)\,dS_y$$

$$\frac{\partial(K'_\varepsilon\,u)}{\partial x_i}(x) = \int_S \frac{\partial A'_\varepsilon}{\partial x_i}\,(x\,,\,y)\,u(y)\,dS_y$$

and we immediately see that $K'_\varepsilon$ is a bounded operator from $L_2(S)$ into $H^1(D)$. Then by the Rellich theorem, $K'_\varepsilon$ is a compact operator from $L_2(S)$ into $L_2(D)$.

It suffices then to prove that for sufficiently small $\varepsilon$, the norm of $K_\varepsilon$ in $\mathcal{L}(L_2(S)\,,\,L_2(D))$, can be made as small as pleased. For, by the Schwarz inequality,

(4.5) $$\left|K_\varepsilon u(x)\right|^2 = \left|\int_{S_y} \frac{B\,\theta}{r^{1/2}} \frac{u(u)}{r^{1/2}}\, dS_y\right|^2 \leqslant C \left(\int_{S_y \cap (r<\varepsilon)} \frac{1}{r}\, dS_y\right)\left(\int_{S_y} \frac{|u(u)|^2}{r} dS_y\right)$$

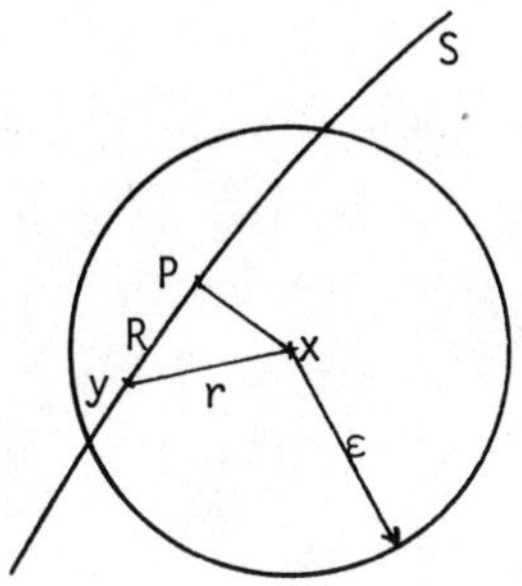


The first integral is bounded above if we put the distance R from y to the projection P of x on S instead of r. The first integral is then

$$\leqslant \int_0^\varepsilon \frac{2\,\pi\, R\, dR}{R}$$

then, (4.5) gives :

$$\| K_\varepsilon\, u \|^2_{L^2(D)} = \int_{\Omega_x} C\,\varepsilon \int_{S_y} \frac{|u(y)|^2}{r}\, dS_y\, dx = C\varepsilon \int_{S_y} |u(y)|^2 \int_{\Omega_x} \frac{dx}{|x-y|} \leqslant$$

$$\leqslant C'\varepsilon\, \| u \|^2_{L_2(S)} \qquad \text{Q.E.D. } \blacksquare$$

We now consider the volume potential defined by the convolution of the fundamental solution $\psi^+$ (c.f. (4.1)) of the Helmholtz equation with mass distributions $\rho$ .

Proposition 4.1 - Let $\rho$ be a distribution of bounded support B. The potential

(4.6) $$u = \psi^+ * \rho$$

which writes (if this makes sense) :

(4.7) $$u(x) = \frac{1}{4\pi} \int_B \rho(y)\, \frac{e^{i\omega r}}{r}\, dy \qquad , \qquad r = |y - x|$$

has the following properties :

a) $- (\Delta + \omega^2)u = \rho$ (in the distributions sense)

b) If $\rho \in L^2(B)$, then u is a continuous function defined on $R^3$ (and hence on B) and the operator from $\rho$ to u is compact from $L^2(B)$ into $C^0(B)$.

c) If $\rho \in L^\infty(B)$, then $u \in C^1(R^3)$ and the first derivatives are given by formal differentiation of (4.7).

d) If $\rho \in L^2(B)$, then $u \in H^2(B)$ and $\| u \|_{H^2} \leq C \|\rho\|_{L^2}$

for a constant C which only depends on B.

Proof - a) is evident by differenciating the convolution (4.6) and using (1.5). Part b) is a consequence of theorem 4.1 with N = 3, $\lambda = 1$, p = 2. For c), see for instance, Muller [1] p. 29-32. To prove d), let us continue u by 0 outside B. u is then a solution of

$$- \Delta u = \rho + \omega^2 u$$

By the standard interior regularity theory for elliptic equations (see.for instance Agmon [1], chapter 6) we have

$$\|u\|_{H^2(B)} \leq C( \| - \Delta u\|_{L^2(\tilde{B})} + \|u \|_{L^2(\tilde{B})} ) = C'(\|u \|_{L^2(\tilde{B})} + \|\rho \|_{L^2(B)})$$

where $\tilde{B}$ is a bounded domain which contains in its interior B. But by part b), the norm in $L_2(\tilde{B})$ is bounded by the product of a constant by $\|\rho\|_{L_2(B)}$. ∎

Now, let S be a closed smooth surface and $\nu$ a function defined on S. The single layer potential is

(4.8) $$u(x) = \int_{S_y} \nu(y) \ \psi^+(r) \, dS \qquad r = |x - y|$$

Theorem 4.4 - If S is a smooth surface and $\nu$ is a continuous function, then

a) The single layer potential u is continuous on $R^3$.

b) Let n be the outer unit normal to S. Then the single layer potential u is of class $C^1$ in the inner and outer regions of S. Moreover, the normal derivatives on the outer and inner sides of S are given by

(4.9) $$\begin{cases} \frac{\partial u}{\partial n}\Big|_a(x) = \frac{-1}{2}\,\nu(x) + (\frac{\partial u}{\partial n})^* \\ \frac{\partial u}{\partial n}\Big|_i(x) = \frac{1}{2}\,\nu(x) + (\frac{\partial u}{\partial n})^* \end{cases}$$ where

(4.10) $$(\frac{\partial u}{\partial n})^* = \int_{S_y} \nu(u) \frac{\partial \psi^+}{\partial n_x} dS$$

is the so called "direct value" of the normal derivative.

Remark 4.2 - It is clear that $\frac{\partial}{\partial n_x}$ is for the derivative in the direction of the normal $\frac{\partial r}{\partial n_x} = \cos(r, n_x)$ and so

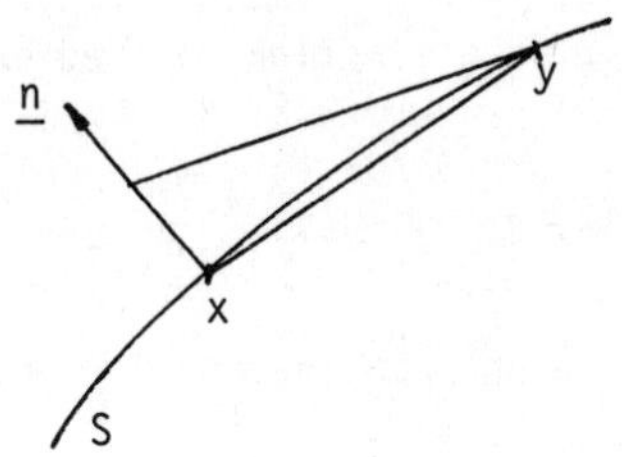


(4.11) $$\frac{\partial}{\partial n_x} \psi^+ = \frac{1}{4\pi} \frac{d}{dr}(\frac{e^{i\omega r}}{r}) \cos(r, n_x)$$

and the expression (4.10) is an integral of the type (4.2) over S instead of D and $\lambda = 1$, because

$$r \to 0 \Rightarrow \begin{cases} \frac{d}{dr}(\frac{e^{i\omega r}}{r}) = O(r^{-2}) \\ \cos(r, n_x) = O(r) \end{cases}$$

Thus, under the hypothesis of theorem 4.4, the expression (4.10) is a continuous function on S. ∎

The proof of theorem 4.4 is the same as for the particular case $\omega = 0$, which may be seen in Mikhlin [1], sect. 18.7. See also Werner [1].

An intuitive idea of (4.9) may be obtained by considering (4.8) as a particular case of (4.6), where $\rho$ is a surface density ; we formally have the formula of proposition 4.1, a), and we integrate it on a flattered domain $\mathcal{D}$ whose intersection with S is a portion of surface $\sigma$ . By integrating by parts we then have :

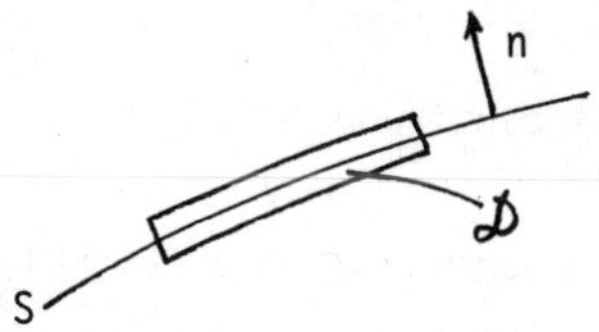

$$- \int_{\partial \mathcal{D}} \frac{\partial u}{\partial n} dS + \int_{\mathcal{D}} \omega^2 u \, dx = \int_{\mathcal{D}} \rho \, dx$$

and in the limit of flat domain,

$$\int_\sigma \left[ - \frac{\partial u}{\partial n}\Big|_a + \frac{\partial u}{\partial n}\Big|_i \right] dS = \int_\sigma \nu \, dS \qquad \forall \sigma \Rightarrow$$

$$\frac{\partial u}{\partial n}\Big|_a - \frac{\partial u}{\partial n}\Big|_i = - \nu \quad .$$

Let us now consider the <u>double layer potential</u>. If as before S is a closed smooth surface and $\mu$ is a function defined on S, the double layer potential is defined by

$$u(x) = \int_{S_y} \mu(y) \frac{\partial \psi^+(r)}{\partial n_y} dS_y \qquad r = |x - y| \tag{4.12}$$

where the derivative $\partial/\partial n_y$ has the meaning (4.11) with $n_y$ instead of $n_x$.

<u>Theorem 4.5</u> - If S is a smooth surface and u is continuous on S, the double layer potential u is continuous in the inner and outer regions of S. Moreover, the outer and inner limits of u on S are given by

$$\begin{cases} u(x)\big|_a = \frac{1}{2} \mu(x) + u^*(x) \\ u(x)\big|_i = \frac{-1}{2} \mu(x) + u^*(x) \end{cases} \tag{4.13}$$

where $u^*(x)$ is the value of u at the point x of S ("direct value").

The proof is the same as in the case $\omega = 0$ (Mikhlin [1], sect. 18.5) ; see also Werner [1].

It is noticeable that the preceeding properties of potentials are given for continuous densities. In the sequel we shall deal with $L_2$ densities. In this connection the following theorem is useful :

<u>Theorem 4.6</u> - Let D be a bounded domain of $R^N$ (or an N-dimensional smooth manifold) and let B(x , y) be a kernel, bounded for x, $y \in \overline{D}$ and continuous for $x \neq y$. Moreover, let $u \in L^2(D)$ be a solution of the integral equation

$$u(x) = \int_D \frac{B(x , y)}{r^\lambda} u(y) \, dy + f(x)$$

with $\lambda < N$. Then, if f is continuous, u is also continuous.

The proof of this theorem (see Mikhlin [1], sect. 8.6) is based on a decomposition of the kernel analogous to that of theorem 4.3 for small $\varepsilon$.

The operator associated to $A_\varepsilon$ is of small norm, and the solution u may be obtained as the uniform sum of a series of continuous terms.

5.- The Neumann problem. Existence and uniqueness for real $\omega$ . - The uniqueness of the solution of the Neumann problem (definition 2.1) was proved in theorem 2.2. The aim of this section is to prove the following result.

Theorem 5.1 - If $f \in C^0(S)$ is given, the Neumann problem (def. 2.1) for real $\omega$ has one and only one classical solution.

Here, "classical" is for analytic functions in $\Omega$ of class $C^1(\overline{\Omega})$ (c.f. remark 1.5). For the sake of convenience we shall take here $\underline{n}$ as the outer unit normal. (2.2) may then be written as

$$\left.\frac{\partial \psi}{\partial n}\right|_S = - f \tag{5.1}$$

We shall search for the solution $\psi$ in the form of a sum of a single layer potential $\psi_1$ and a volume potential on B, $\psi_2$

$$\psi = \psi_1 + \psi_2 \qquad \begin{cases} \psi_1(x) = \int_{S_y} \nu(y)\, \psi^+(r)\, dS_y \\ \psi_2(x) = \int_{B_y} \rho(y)\, \psi^+(r)\, dy \end{cases} \tag{5.2}$$

and the unknowns of the problem are now the densities $\nu$ and $\rho$ of the potentials. We consider $\nu \in C^0(S)$, $\rho \in C^0(B)$. It is then clear that the Helmholtz equation in $\Omega$ and the radiation condition are satisfied. Let us write the boundary condition (5.1). From (4.9), (4.10) we have :

$$\left.\frac{\partial \psi_1}{\partial n}\right|_S (x) = \frac{-\nu(x)}{2} + \int_{S_y} \nu(y) \frac{\partial \psi^+}{\partial n_x} dS_y \qquad (\text{side } \Omega) \tag{5.3}$$

On the other hand, from proposition 4.1, part c), the volume potential is of class $C^1$ on $R^3$, thus

$$\left.\frac{\partial \psi_2}{\partial n}\right|_S = \int_{B_y} \rho(y) \frac{\partial \psi^+}{\partial n_x} dy \tag{5.4}$$

and from (5.3), (5.4), the condition (5.1) becomes :

$$\nu = 2 \int_S \nu \frac{\partial \psi}{\partial n_x} dS + 2 \int_B \rho \frac{\partial \psi^+}{\partial n_x} dy + 2 f \tag{5.5}$$

We then have only one equation to find the two unknowns $\nu$, $\rho$. So it is possible to impose another condition to $\nu, \rho$ . It is clear that $\psi$ is defined on $\Omega$ ; nevertheless, the expression (5.2) is also defined on B. Thus we shall impose that

(5.6) $$- (\Delta + \omega^2)\, \psi = i\, \psi \qquad \text{on } B$$

($i = \sqrt{-1}$). This choice allows to us to overcome a technical difficulty which will be seen at the end of the proof : otherwise, it would be easier to consider $\rho = 0$. By proposition 4.1, part a) we have

$$-(\Delta + \omega^2)\, \psi_1 = 0 \quad ; \quad -(\Delta + \omega^2)\, \psi_2 = \rho$$

and (5.6) thus becomes $\rho = i\, \psi$ , i.e :

(5.7) $$\rho = i \int_S \nu\, \psi^+ \, dS + i \int_B \rho\, \psi^+ \, dy$$

Then, (5.5), (5.7) is a system of two integral equations for $\nu$ and $\rho$. In fact, we shall consider (5.5), (5.7) as an operator equation of the form

(5.8) $$x = Tx + F$$

for the unknown $x = (\nu , \rho)$ in the space $X = L^2(S) \times L^2(B)$.(Note that we are looking for $x \in C^0(S) \times C^0(B)$ !). Here, $F = (2f , 0)$ and $T$ is a matrix with the operator-valued entries which are evident from (5.5), (5.7). We have :

Lemma 5.1 - The operator T is compact in X.

Proof - It suffices to prove that each entry is compact in the corresponding spaces. In fact :

$T_{11} = 2 \int_S \nu \frac{\partial \psi}{\partial n}\, dS$ is compact in $L^2(S)$ by theorem 4.2 with $D = S$, $N = 2$, $s = 2$, $p = q = 2$, $\lambda = 1$ (see remark 4.2).

$T_{12} = 2 \int_B \rho \frac{\partial \psi^+}{\partial n_x}\, dy$ is compact from $L^2(B)$ into $L^2(S)$ by theorem 4.2 with $D = B$, $N = 3$, $s = 2$, $p = q = 2$, $\lambda = 2$.

$T_{21} = i \int_S \nu\, \psi^+ \, dS$ is compact from $L^2(S)$ into $L^2(B)$ by theorem 4.3.

$T_{22} = i \int_B \rho\, \psi^+ \, dy$ is compact in $L^2(B)$ by theorem 4.1 with $D = B$, $N = 3$, $p = 2$ , $\lambda = 1$.

Thus lemma 5.1 is proved. ■

It then appears that uniqueness for equation (5.8) gives existence. This is customarily called "Fredholm alternative" ; in fact, we know (chapter 11, proposition 1.2) that the singularities of the resolvent of the compact operator T are its eigenvalues ; then, if (5.8) have the uniqueness property, 1 is not an eigenvalue and $(T - I)^{-1}$ is a well determined operator of $\mathcal{L}(X , X)$. The existence of the solution then holds. In fact, we have

Lemma 5.2 - If $F = 0$, we have $x = 0$ in (5.8).

Proof - Let $\rho \in L^2(B)$, $\nu \in L^2(S)$ be a solution of (5.5), (5.7) with $f = 0$. Let us consider (5.5) as an integral equation in $\nu$ . The "given term"

$$2 \int_{B_y} \rho(y) \frac{\partial \psi^+}{\partial n_x} dy \tag{5.8 bis}$$

is a continuous function on S (see Remark 9.1 later). We then apply Theorem 4.6 with $D = S$, $\lambda = 1$, $N = 2$ and we find

$$\nu \in C^0(S) \tag{5.9}$$

As for $\rho$ we consider (5.7) as an integral equation in $\rho$ . The "given term"

$$i \int_{S_y} \nu(u) \psi^+ dS_y$$

is a continuous function on B (this results from theorem 4.4 part a) and by theorem 4.6 with $D = B$, $N = 3$, $\lambda = 1$ we find

$$\rho \in C^0(B) \tag{5.10}$$

Bearing in mind (5.9) and (5.10), we consider the corresponding potentials $\psi_1$ and $\psi_2$ given by (5.2). By proposition 4.1 part c and theorem 4.4, part b, we see that $\psi$ is of class $C^1$ in $\overline{\Omega}$ and $\overline{B}$. As a consequence, $\psi$ in $\Omega$ satisfies the homogeneous ($f = 0$) Neumann problem in a classical sense, and by theorem 2.2

$$\psi \equiv 0 \qquad \text{in } \Omega. \tag{5.11}$$

We shall now prove that $\psi$ is also zero in B. By the continuity of $\psi_1$ and $\psi_2$ (proposition 4.1, part b and theorem 4.4 part a) we have

$$\psi|_S = 0 \tag{5.12}$$

In B $\psi$ of course satisfy (5.6). We multiply it by the conjugate $\overline{\psi}$ and we integrate it by parts in B ; by (5.12) the surface integral vanishes and we have

$$\int_B \frac{\partial \psi}{x_i} \frac{\partial \overline{\psi}}{x_i} dx = (\omega^2 + i) \int_B \psi \, \overline{\psi} \, dx \tag{5.13}$$

the two integrals are real and (5.13) is not possible unless

$$\psi \equiv 0 \qquad \text{in B} \tag{5.14}$$

(This is the reason why the imaginary term i was included in (5.6). If not, the conclusion (5.14) would only hold for $\omega^2$ different from the eigenvalues of the

Dirichelet problem in B).

From (5.12), (5.13), by the relation $\rho = i\psi$ we have $\underline{\rho = 0}$. Then, $\psi = \psi_1 = 0$ and by considering the difference of the two equations (4.9) we have $\underline{\nu = 0}$. Lemma 5.2 is proved. ∎

Then, as a consequence of the above considerations on the Fredholm alternative, for any given F there exists $x \in X$ satisfying (5.8). This does not suffice to solve the problem because the obtained densities belong to $L^2(S)$ and $L^2(B)$ ; They are not continuous and we do not know if the corresponding $\psi$ satisfies the boundary conditions. In fact, in theorem 5.1, f is continuous and then, by repeating the process of Lemma 5.2, we see that $\nu \in C^0(S)$, $\rho \in C^0(B)$. Then, the classical properties of the potential hold (see proposition 4.1 and theorem 4.4) and we have a classical solution of the Neumann problem. This achieves the proof of theorem 5.1.

6.- The transmission and Dirichlet problems. Existence and uniqueness for real $\omega$.
We now consider the transmission and Dirichlet problems by using potential methods, as in the preceeding section. We shall see that some difficulties arise in the solution of the Dirichlet problem ; we shall only give a "formal proof" in this case ; since this problem will be handled again in the following chapter by the "limiting obsorption" method.

Theorem 6.1 - If f is a piecewise continuous function with compact support in $R^3$, the transmission problem (definition 2.2, notations (2.4), (2.5)) has one and only one classical solution.

To prove this theorem, we transform the problem by writing the unknown $\psi$ under the form ($\tilde{f}$ is the piecewise continuous function defined later in (6.0)) :

$$\psi = u + v \quad ; \quad v = \psi^+ * \tilde{f} \quad ; \quad \psi^+ = \frac{-1}{4\pi}\frac{e^{i\omega r}}{r}$$

where by virtue of proposition 4.1, v is a $C^1$ class function satisfying the radiation condition and

$$(6.0) \qquad -(\Delta + \omega^2)\, v = \tilde{f} = \begin{cases} f & \text{for } x \in \Omega \\ \frac{1}{\alpha} f & \text{for } x \in B \end{cases}$$

then, the problem for u is : Find u satisfying the radiation condition and

$$(6.1) \qquad -(\Delta + \omega^2)u = 0 \qquad \text{in } \Omega$$

$$(6.2) \qquad -(\Delta + \omega^2)u = (\frac{\beta}{\alpha} - 1)\, \omega^2\, u + \phi \qquad \text{in B}$$

$$(6.3) \qquad u|_B = u|_\Omega \qquad \text{on S}$$

(6.4) $$\left.\frac{\partial u}{\partial n}\right|_{\Omega} = \alpha \left.\frac{\partial u}{\partial n}\right|_{B} + \Phi \qquad \text{on } S$$

where

$$\phi = \omega^2 (\frac{\beta}{\alpha} - 1) v \quad \text{and} \quad \Phi = (\alpha - 1) \left.\frac{\partial v}{\partial n}\right|_{S}$$

are given continuous functions.

We then search for u under the form

(6.5) $$u = u_1 + u_2$$

exactly as in (5.2) (single layer and volume potentials with continuous densities $\nu$ and $\rho$ , respectively). Then, (6.1) and (6.3) are satisfied. As for the Neumann problem, (6.4) and (6.2) give :

(6.6) $$\nu = \frac{2(1 - \alpha)}{(1 + \alpha)} \int_S \nu \frac{\partial \psi^+}{\partial n_x} dS + \frac{2(1 - \alpha)}{(1 + \alpha)} \int_B \rho \frac{\partial \psi^+}{\partial n_x} dy + \frac{2}{1 + \alpha} \Phi$$

(6.7) $$\rho = (\frac{\beta}{\alpha} - 1) \omega^2 \int_S \nu \psi^+ dS + (\frac{\beta}{\alpha} - 1) \omega^2 \int_B \rho \psi^+ dy + \phi$$

Note that here the unknown u is defined in the whole space $R^3$ and we do not have to impose a supplementary equation analogous to (5.6).

The system (6.6), (6.7) is exactely of the same type as (5.5), (5.7), and existence of $(\nu , \rho) \in L^2(S) \times L^2(B)$ follows from uniqueness. Then, for $\phi = \Phi = 0$, we have $u = 0$ by the uniqueness theorem 2.3 (note that $\nu, \rho$ are continuous functions). We then have $\nu = \rho = 0$. The existence of the solution with continuous densities $\nu$ and $\rho$ then follows as for the Neumann problem. Theorem 6.1 is proved.

Theorem 6.2 - If f is a continuous function defined on S, the Dirichlet problem (definition 2.1) has one and only one classical solution.

We search for the solution $\psi$ as a sum of a double layer potential $\psi_1$ and a volume potential $\psi_2$ :

(6.8) $$\psi = \psi_1 + \psi_2 \qquad \begin{cases} \psi_1 = \int_{S_y} \mu(y) \frac{\partial \psi^+}{\partial n_y} dS_y \\ \psi_2 = \int_{B_y} \rho(y) \psi^+ dy \end{cases}$$

with continuous densities $\mu$ and $\rho$. From theorem 4.5 and proposition 4.1, part b,

$$\psi_1 \Big|_S = \frac{\mu}{2} + \int_S \mu \frac{\partial \psi^+}{\partial n_y} dS \qquad \text{(side } \Omega)$$

$$\psi_2 \Big|_S = \int_B \rho \psi^+ dy$$

and the boundary condition (2.2) becomes :

(6.9) $$\mu(x) = -2 \int_{S_y} \mu(y) \frac{\partial\psi^+}{\partial n_y} dS_y - 2 \int_{B_y} \rho(y)\, \psi^+ \, dy + 2 f(x)$$

Moreover, as in the Neumann problem, we impose the supplementary equation (5.6) on B, which writes (as for the Neumann problem)

(6.10) $$\rho(x) = i \int_{S_y} \mu(y) \frac{\partial\psi^+}{\partial n_y} dS_y + i \int_{B_y} \rho(y)\, \psi^+ \, dy$$

and we consider (6.9), (6.10) as an equation of the form (5.8) on $L^2(S) \times L^2(B)$. As in Lemma 5.1, the corresponding operator T is compact in this space, and the Fredholm alternative holds.

We now find some technical difficulties. In order to apply the uniqueness theorem 2.2, we must show that $\psi$ has continuous normal derivatives on S. We proceed as follows. Let $(\rho\ ,\ \nu) \in L^2(S) \times L^2(B)$ be a solution of (6.9), (6.10) for $f = 0$. As in the Neumann problem, we see that $\rho$ and $\nu$ are solutions of an integral equation and they are then continuous ; moreover, they are Holder-continuous (see an analogous proof in Kellog [1] , p. 300) and also Holder-differentiable (see Werner [2] at the end of p. 50). We then have (see Werner [2] , Lemma 7, p. 36) : the double layer potential $\psi_1$ is of class $C^1$ on each side of S and

(6.11) $$\left.\frac{\partial\psi_1}{\partial n}\right|_{\Omega} = \left.\frac{\partial\psi_1}{\partial n}\right|_{B} \qquad \text{on } S$$

In these conditions, we may apply the uniqueness theorem 2.2 and we have

(6.12) $$\psi = 0 \qquad \text{in } \Omega$$

Moreover, from (6.11) we deduce that $\partial\psi/\partial n$ takes the same values on the two sides on S and then

(6.13) $$\left.\frac{\partial\psi}{\partial n}\right|_{S} = 0 \qquad \text{(side B)}$$

and we deduce that $\psi = 0$ on B as in (5.13). But if $\psi = 0$ everywhere, by the relation $\rho = i\psi$ in B we have $\rho = 0 \Longrightarrow \psi = \psi_1$ and the second equation (4.13) gives $\mu = 0$. We then have proved the uniqueness and thus the existence in $L^2(S) \times L^2(B)$. As in the Neumann problem, if f is continuous, so are $\mu$ and $\rho$ and the solution is classical.

## 7.- Analytic continuation of the solutions for complex $\omega$. Scattering frequencies -

We begin this section by an abstract study of the analytical continuation of the inverses of a class of operators.

We already know the concept of bounded-holomorphic operators (chap.11 , def. 4.2). Let us introduce the bounded-meromorphic operators.

Definition 7.1 - Let X be a Banach space and D an open domain of the complex plane of the variable $\mu$ . A function $A(\mu)$ with values in $\mathcal{L}(X , X)$ is said to be bounded-meromorphic in D if it is defined on D except at some isolated points which are poles of $A(\mu)$ (i.e., their Laurent's series have at most a finite number of negative terms).

We then have the following results :

Theorem 7.1 - Let D be an open connected domain of the complex plane of the variable $\mu$ and $T(\mu)$ be a bounded-holomorphic family of compact operators from the Banach space X into itself for $\mu \in D$. Moreover, there exists a point $\mu^* \in D$ such that $(I - T(\mu^*))^{-1} \in \mathcal{L}(X , X)$. Then, $(I - T(\mu))^{-1}$ is a bounded meromorphic function in D, with values in $\mathcal{L}(X , X)$.

It is clear that this theorem means that, if the equation

$$x - T(\mu)\, x = y \tag{7.1}$$

in X is bounded invertible for $\mu = \mu^0$, then it is also invertible for $\mu \in D$ except for certain isolated points, which are poles of the inverse.

Proof of Theorem 7.1 - Let us bear in mind the spectral properties of compact operators (c.f. chapter 11, proposition 1.2). The fact that

$$(I - T(\mu))^{-1} \in \mathcal{L}(X , X)$$

is equivalent to the fact that 1 is not an eigenvalue of $T(\mu)$.

Let Q be the set of the points $\mu$ where $(I - T)$ is not bounded invertible, i.e. :

$$Q = \{\mu \in D \ ; \ 1 \text{ is an eigenvalue of } T(\mu)\}$$

We shall show that Q is a set of isolated points. Let us suppose that

$\mu_n \in Q$

$$\mu_n \to \mu_0 \in D \quad ; \quad n = 1,2, \ldots \tag{7.2}$$

For the time being, we do not know if $\mu_0$ belongs to Q or not. In the complex plane $\lambda$ associated with

$$\lambda - T$$

we consider a small circle $\Gamma$ enclosing the point 1 and no other points of $\sigma(T(\mu_0))$ (the spectrum of $T(\mu_0)$). We then apply the properties of separation of the spectrum given by chapter 11, theorem 4.2. Then, for $|\mu - \mu_0|$ sufficiently small (and thus for large n in (7.2)) we consider the associated projection $P(\mu)$ and the operators $U(\mu)$ and $U^{-1}(\mu)$ (c.f. chap. 11, theorem 4.2) we also define

$$\tilde{T}(\mu) = U(\mu)\ T(\mu)\ U^{-1}(\mu)$$

$$\tilde{B}(\mu) = U(\mu)\ (1 - T(\mu))\ U^{-1}(\mu) = I - \tilde{T}(\mu)$$

which commute with $P(\mu_0)$. Moreover, the pair (independent of $\mu$!):

$$M(\mu_0) = P\ (\mu_0)\ X \quad ; \quad N(\mu_0) = (I-P(\mu_0))\ X$$

decomposes $\tilde{B}(\mu)$. The solvability of

$$\tilde{B}(\mu)x = 0$$

in $M(\mu_0)$ is equivalent to the solvability of

$$(I - T(\mu))x = 0$$

in $M(\mu)$. Moreover, $M(\mu)$ is finite dimensional (because $T(\mu)$ is compact) and consequently $\tilde{B}(\mu)$ is a holomorphic matrix. Its determinant $\Delta(\mu)$ is holomorphic and $\Delta(\mu_n) = 0$ for large n. Then, $\Delta(\mu) = 0$ in the considered neighbourhood of $\mu_0$ ; this means that 1 is an eigenvalue of $T(\mu)$ for $\mu$ in this neighbourhood, and the standard reasoning of continuation shows that (D is connected !) the set Q coincides with D. But this is false because $\mu^*$ is not in D. Then, <u>Q is formed by isolated points</u>.

Elsewhere, the operator $(I - T(\mu))^{-1}$ is bounded holomorphic by chapter 11 proposition 4.5. The only point that we have still to prove is that the singularities are poles. Let $\mu_0$ be such a singularity. We define $P(\mu)$, $U(\mu)$, $U^{-1}(\mu)$ as above. Thus, under the holomorphic transformation defined by $U(\mu)$ , the operator-valued functions $(I - T(\mu))^{-1}$ and $\tilde{B}(\mu)^{-1}$ are equivalent in the vicinity of $\mu_0$. But $\tilde{B}(\mu)^{-1}$ is holomorphic for $N(\mu_0)$ and has a <u>pole</u> for $M(\mu_0)$ (this result is a consequence of the fact that $\tilde{B}(\mu)$ is holomorphic on the finite-dimensional space $M(\mu_0)$, i.e. is a holomorphic matrix ; consequently, by the Cramer's rule, the entries of the matrix $\tilde{B}(\mu)^{-1}$ on $M(\mu_0)$ are given by the quotient of two holomorphic functions). Theorem 7.1 is proved. ■

In physical applications, we often have the case of the preceeding theorem but the operator T depends also on another parameter $\nu$. It is then useful to know the evolution of the poles of $(I - T)^{-1}$ as functions of $\nu$ . In this connection we have

Theorem 7.2 - Let D be an open connected domain of the complex plane $\mu$ . Let $T(\mu , \nu)$ be a family of compact operators (from the Banach space X into itself) defined for $\mu \in D$, $\nu \in [0 , \nu_1]$ (interval of the real line) which is bounded-holomorphic of $\mu \in D$ for each $\nu$, and jointly continuous (in the norm) for $(\mu , \nu) \in D \times [0 , \nu_1]$ . There exists $\mu^* \in D$ such that $(I - T(\mu^* , \nu))^{-1} \in \mathcal{L}(X,X)$ $\forall\ \nu$ , then

a) $(I - T(\mu , \nu))^{-1}$ is, for each $\nu$, a bounded-meromorphic family of $\mu$ on D.

b) If $\mu_0$ is not a pole of $(I - T(\mu , \nu_0))^{-1}$, then $(I - T(\mu , \nu))^{-1}$ is jointly continuous (in the norm) in a neighbourhood of $(\mu_0 , \nu_0)$.

c) The poles of $(I - T(\mu , \nu))^{-1}$ depend continuously on $\nu$ and consequently they can appear and desappear only at the boundary of D. (The sense of "continuously" will be seen in the proof).

Remark 7.1 - The fact that $\nu$ is a real parameter is not essential. The reader can easily obtain analogous results in other cases. ■

Proof of th. 7.2 - Part a) is merely theorem 7.1. As for part b), we recall (chapter 11, sect. 2) that for bounded operators, convergence in the norm amounts to convergence in the gap. Then, chapter 11, theorems 2.3 and 2.2 show that

$$(I - T(\mu , \nu))^{-1} \in \mathcal{L}(X , X)$$

for $(\mu , \nu)$ in a neighbourhood of $(\mu_0 , \nu_0)$ and is continuous in $(\mu , \nu)$.

Let us now prove part c). We shall prove the "continuity" for $\nu \to 0$. More exactly, we shall prove : I and II :

I) The accumulation points of poles for $\nu \to 0$ are poles for $\nu = 0$, (i.e.,if there exists $\nu_i \to 0$ and poles $\mu_i \in D$ of $(I - T(\mu , \nu_i))^{-1}$ such that $\mu_i \to \mu_0 \in D$), then, $\mu_0$ is a pole of $(I - T(\mu , 0))^{-1}$.

II) The poles of $(I - T(\mu , 0))^{-1}$ are accumulation points of poles of $(I - T(\mu , \nu_i))^{-1}$ with $\nu_i \to 0$, i.e., if $\mu_0$ is a pole of $(I - T(\mu , 0))^{-1}$, then, in any neighbourhood of $\mu_0$ there exists atleast one pole $\mu(\nu)$ of $(I - T(\mu , \nu))^{-1}$ for sufficiently small $\nu$.

Proof of I - It is evident from part b, that if $\mu_0$ is not a pole of $(I - T(\mu , 0))^{-1}$, the function $(I - T(\mu , \nu))^{-1}$ cannot be continuous in $(\mu , \nu)$ in a neighbourhood of $(\mu_0, 0)$.

Proof of II - By contradiction : let us suppose that, there exists a certain neighbourhood $\mathcal{V}$ of $\mu_0$ and a sequence $\nu_i \to 0$ such that $(I - T(\mu , \nu_i))^{-1}$ has no poles in $\mathcal{V}$. By part a, these functions are holomorphic of $\mu$ in $\mathcal{V}$. Let us consider a circle $\Gamma$ contained in $\mathcal{V}$, enclosing $\mu_0$ and no other singularities of $(I - T(\mu , 0))^{-1}$. Consequently, this function has a Laurent expansion with non zero principal part. By multiplying it by $(\mu - \mu_0)^m$ with an appropriate integer m, the (operator !) coefficient of $(\mu - \mu_0)^{-1}$ is not zero. Thus :

$$(7.1) \qquad \int_\Gamma (\mu - \mu_0)^m \; [I - T(\mu , 0)]^{-1} \, d\mu \neq 0$$

(i.e., it is an operator different from the zero operator). On the other hand, we have

$$(7.2) \qquad \int_\Gamma (\mu - \mu_0)^m \, [I - T(\mu , \nu_i)]^{-1} \, d\mu = 0 \quad ; \quad i = 1, 2, \ldots$$

We then have a contradiction if we pass to the limit in (7.2). But this is possible, because the integrand is continuous in $(\mu , \nu)$ for $\mu \in \Gamma$ and small $\nu$ (recall part b and the fact that $\Gamma$ is compact).

Theorem 7.2 is proved.■

We may now study the problems of sect. 5 and 6 (Neumann, transmission and <u>Dirichlet) for complex $\omega$</u>. We have seen that, for real $\omega$, the solution of these problems brings back to the study of the equation (5.8), which we write

$$(7.3) \qquad x - T(\omega)\, x = F$$

where T is a compact operator in the space $X = L^2(S) \times L^2(B)$ which depends on the parameter $\omega$. It is clear that the same problems may be considered for complex $\omega$, and that the construction of the solutions with

$$\psi^+ = \frac{e^{i\omega r}}{r}$$

ensures that <u>the radiation condition (definition 3.1) is satisfied</u>. On the other hand, if the given function F in (7.3) is continuous, the potentials $x = (\nu , \rho)$ are continuous on S and B, and we have classical solutions. We shall see that we can use theorem 7.1 for the resolution of these problems. As a preliminary step, we have

<u>Lemma 7.1</u> - The operator $T(\omega)$ is compact in $L^2(S) \times L^2(B)$ for any complex $\omega$ and is bounded-holomorphic on the whole complex plane.

<u>Proof</u> - The compactness is evident as for real $\omega$. To prove that T is holomorphic, it suffices to consider the expansion

(7.4)
$$\frac{e^{i\omega r}}{r} = \frac{e^{i\omega_0 r}}{r} \sum_{0}^{\infty} \frac{[i(\omega - \omega_0)r]^n}{r}$$

for any $\omega_0$. It is then clear that we can formally write

(7.5)
$$T(\omega) = T(\omega_0) \sum_{0}^{\infty} T_n$$

where the operators $T_n$ are less singular than T (in fact for $n \geqslant 1$ or $n \geqslant 2$ for the different entries of the matrix T, the operators $T_n$ are bounded), and from the convergence of (7.4) we obtain the convergence of (2.5). ■

We then have :

<u>Theorem 7.3</u> - Each of the problems (Neumann, Transmission, Dirichlet) with complex $\omega$ has one and only one solution except for a discrete set of values $\omega_i$ <u>called the scattering frequencies</u> of the problem. The imaginary parts of the scattering frequencies are negative. If $\omega_i$ is a scattering frequency, there exits a finite number of solutions of the problem corresponding to F = 0.

<u>Proof</u> - The first part of the problem is immediate : the solution of (7.3) is given by

(7.6)
$$x = [I - T(\omega)]^{-1} F$$

Theorem 7.1 applies because $[I - T(\omega)]$ is invertible for real $\omega$ . Then, the operator in (7.6) is a meromorphic function on the complex plane with values in $\mathcal{L}(X , X)$. For $\omega$ different from the poles $\omega_i$, the solution x is unique and as we have already said, if the given functions are continuous, so are the densities, and the solution $\psi$ constructed with the potentials is classical. Let us now consider a scattering frequency $\omega_1^*$. Then, 1 is an eigenvalue of finite multiplicity of $T(\omega_1^*)$. Let x be an eigenvector.

$$x - T(\omega^*)x = 0$$

it corresponds to F = 0 and by the preceeding assertion, we may construct an associated classical solution $\psi$ , corresponding to zero boundary conditions. If $\mathrm{Im}\, \omega^* > 0$, $\psi$ is a superposition of elementary solutions $\frac{e^{i\omega r}}{r}$ which decay exponentialy to zero as $r \to \infty$. This shows that we have a solution belonging to $L_2(\Omega)$ of the Helmholtz equation

$$- \Delta\psi - \omega^2 \psi = 0$$

with zero boundary conditions. Then, $\psi$ is an eigenfunction of $- \Delta$ ; but this operator is selfadjoint and positive and the corresponding eigenvalue $\omega^2$ must be real $> 0$. We then have $\omega$ real, which is a contradiction. Consequently, $\mathrm{Im}\, \omega^*$ must be $< 0$. The theorem is proved. ■

Remark 7.1 - We have not proved that the scattering frequencies $\omega_i$ really exists ; in fact, it should be possible that the meromorphic function (7.6) was in fact a holomorphic function (entire function). Moreover, it may occur that if $\omega^*$ is a singularity, we have $x \neq 0$ but $\psi = 0$. In fact, we shall see in the examples of chapter 17 that the scattering frequencies really exist . In fact this is also easily seen in the example of the sphere by exact calculations. ■

It is clear that theorem 7.2 immediatly gives the continuity of solutions with respect to a parameter. For instance, in the transmission problem of theorem 6.1 (definition 2.2, we may consider as parameters the coefficients $\alpha$ and $\beta$ of (2.4) and (2.5). The continuity hypothesis of theorem 7.2 are obviously satisfied ; we then obtain, for fixed f in theorem 6.1, the solution $\psi(\omega,\alpha,\beta)$ depends continuously on $\alpha$ and $\beta$ (in the sense that the potentials $(\nu , \rho)$ continuously depend on $\alpha,\beta$ in $L^2(S) \times L^2(B)$) and the scattering frequencies $\omega_i(\alpha,\beta)$ continuously depend on $\alpha$ and $\beta$ .

8.- Solutions in space-time. Interpretation of the scattering frequencies - Let us go back to the study of the wave equation (1.1) and of the functions $\phi$ associated with the solutions $\psi$ of the reduced wave equation (1.2). To fix ideas, we shall confine ourselves to the Neumann problem, but of course the results obviously also hold for the transmission and Dirichlet problems.

Let $\psi(x)$ be, for fixed $\omega$ (real or complex) a solution of the Neumann problem

$$- (\Delta + \omega^2)\, \psi = 0 \qquad \text{in } \Omega \tag{8.1}$$

$$\frac{\partial\psi}{\partial n}\Big|_S = f \qquad \text{in } S \tag{8.2}$$

satisfying the radiation condition (def. 3.1). Moreover, according to the construction of sect. 5, $\psi$ is of the form :

$$\begin{cases} \psi(x) = \displaystyle\int_{S_y} \nu(y)\, \psi^+(r)\, dS_y + \int_{B_y} \rho(y)\, \psi^+(r)\, dy \\ \text{where} \quad \psi^+(r) = \dfrac{-1}{4\pi}\dfrac{e^{i\omega r}}{r} \quad ; \quad r = |x - y| \end{cases} \tag{8.3}$$

If $\omega$ is (resp. is not) a scattering frequency of the Neumann problem, we consider $f = 0$ (resp. $f \neq 0$) in (8.1) - (8.3).

We can now construct according to (1.3) a "naive" solution $\phi(x,t)$ of the wave equation

$$\phi(x,t) = \mathrm{Re}[\, e^{-i\omega t}\, \psi(x)] \tag{8.4}$$

but, as we know, for $\omega$ with imaginary part $\leqslant 0$, this solution has an infinite energy in $\Omega$ (see for instance theorem 2.1 for real $\omega$ ; as for complex $\omega$ with negative imaginary part, it suffices to see that $e^{i\omega r}/r$ is exponentially large for large r).

It is then clear that the solution $\phi(x,t)$ is not appropriate for the study of physical phenomena with finite energy.

We now show <u>how to construct solutions $\phi^*$ of finite energy associated to $\psi$</u>.

Let us suppose that the obstacle B is contained in the ball of diameter d centered at the origin. We then construct a smooth function $\Phi(s)$ equal to $e^{i\omega s}$ for $s \leqslant d$, and zero for $s \geqslant d + a$ (a arbitrarily chosen), i.e. :

$$\Phi(r - t) = \begin{cases} \frac{-1}{4\pi} e^{i\omega(r-t)} & \text{if } r - t \leqslant d \\ & \text{smooth everywhere} \\ 0 & \text{if } r - t \geqslant d + a \end{cases} \tag{8.5}$$

We then construct the function

$$\phi^*(x,t) = \mathrm{Re}\left[ \int_{S_y} \nu(y) \frac{\Phi(r-t)}{r} dS_y + \int_{B_y} \rho(y) \frac{\Phi(r-t)}{r} dy \right] \tag{8.6}$$

Let us study this function for $(x,t) \in \Omega \times [\ 0, +\infty[$ .

<u>Proposition 8.1</u> - The function $\phi^*$ defined by (8.6) is a solution of the wave equation (1.1) in $\Omega \times [0\ , +\infty[$ , moreover, it is zero outside an influence cone and is equal to $\phi(x\ , t)$ (defined by (8.4)) inside an influence cone which contains the body B for $t \geqslant 0$. In particular, $\phi^*$ satisfies the boundary condition

$$\left.\frac{\partial \phi^*}{\partial n}\right|_S = \mathrm{Re}\,[e^{-i\omega t} f] \quad \text{for } (x,t) \in S \times [0, +\infty[. \tag{8.7}$$

<u>Proof</u> - It is well known that

$$\frac{\Phi(r - t)}{r} \qquad r = |x - y|$$

is a solution of the wave equation (x, t variables, y parameter) in three dimensions, for any function $\Phi$ . It is then clear that $\phi^*$ is a superposition (convolution) of solutions of the wave equation and thus a solution.

Moreover, let us consider the two cones

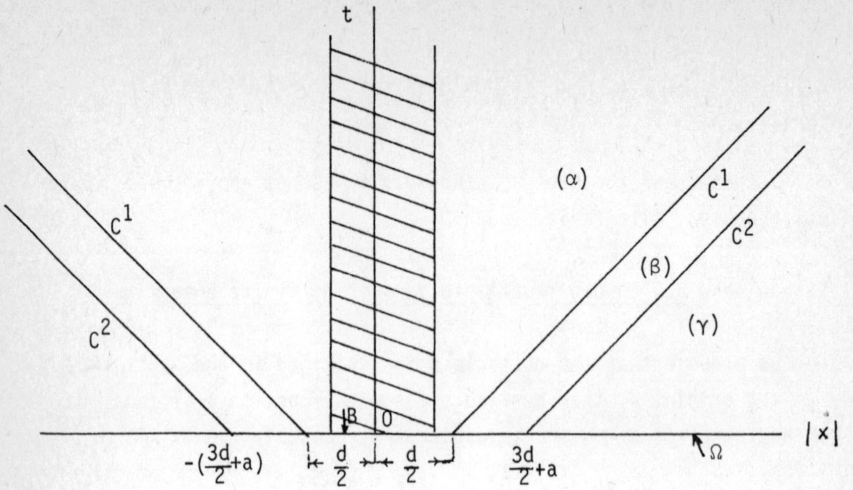


$$\begin{cases} t = |x| - \frac{d}{2} & C^1 \\ t = |x| - (\frac{3d}{2} + a) & C^2 \end{cases}$$

and the three regions (α), (β), (γ) of the figure. It is clear that, if $y \in S$ and $x \in \Omega$ we always have

$$|x| - \frac{d}{2} \leq r \leq |x| + \frac{d}{2} \tag{8.8}$$

Then, in the region (α) :

$$(x , t) \in (\alpha) \Longrightarrow - t \leq -|x| + \frac{d}{2}$$
$$(8.8) \Longrightarrow r \leq |x| + \frac{d}{2}$$

and by addition $r - t \leq d \Longrightarrow \Phi(r - t) = e^{-i\omega t} e^{i\omega r}$ and $\phi^* = \phi$ .

In the region (γ) :

$$(x , t) \in (\gamma) \Longrightarrow - t \geq -|x| + (\frac{3d}{2} + a)$$
$$(8.8) \Longrightarrow r \geq |x| - \frac{d}{2}$$

and by addition, $r - t \geq d + a \Rightarrow \Phi(r - t) = 0$ and $\phi^* = 0$. The proposition is proved. ∎

It is then clear that we may construct $\phi^*$ associated with $\phi$ (or $\psi$) in such a way that the energy is finite ($\phi^*$ has a compact support for fixed t). Moreover, $\phi^*$ is not unique, because the values of the function $\Phi(s)$ for $d < s < d + a$ are arbitrary. If we modify these values, we obtain different functions $\phi^*$, which correspond to different initial values

$$\phi^* \Big|_{t=0} \quad \text{and} \quad \frac{\partial \phi^*}{\partial t}\Big|_{t=0}$$

It is also clear that $\phi^*$ coincides with the "naive" solution $\phi$ (defined by (8.4)) in ($\alpha$), consequently, $\phi^*$ exhibits a wave front in the region ($\beta$) : this wave front matches the solution $\phi$ in ($\alpha$) with 0 in ($\gamma$). It is clear that the construction of this wave front is made possible by the structure of $\phi(x)$ (see (8.3)) with $\psi^+$ elementary solutions instead of $\psi^-$ (see (1.4)). If we wish to construct solutions $\phi$ in the backward time (i.e., $(x, t) \in \Omega \times ]-\infty, t]$) the $\psi^-$ elementary solutions must be cho sen. It is also worthwile to recall remark 1.1.

Now, we may consider, in particular <u>the scattering frequencies $\omega_i$ and the associated scattering functions $\psi_i$</u>. We may construct as before the functions $\phi_i^*(x, t)$. It is clear that $\text{Im}\ \omega_i < 0 \Longrightarrow e^{-i\omega_i t}$ decays exponentially as $t \to +\infty$. Then, for fixed x, $\phi_i^*$ tends to zero as $t \to +\infty$ (any point x belongs, for sufficiently large t, to the region ($\alpha$)). This is the essential difference between a scattering vibration $\phi_i^*$ and an eigenvibration (we know that eigenvibrations of finite energy does not exist !). The scattering vibration decays in time, and the support of $\phi^*(x, t)$ for fixed t grows as t increases. Of course, the energy of the solution :

$$\int_\Omega \left( |\nabla \phi^*|^2 + \left|\frac{\partial \phi}{\partial t}\right|^2 \right) dx$$

is independent of t ; as t increases, the energy spreads out to infinity. Locally (i.e. for fixed x ) the energy decays as t increases. This property is associated to the term scattering vibration. In fact, we shall see in the next chapter that any solution with finite energy of the wave equation in $\Omega$ (with Neumann boundary condition, or another one) locally decays as $t \to +\infty$ .

**9.-** <u>Comments and bibliographical notes</u> - Exterior problems for the Helmholtz equation were initialy studied by potential theory, following the classical schemes developed for the Laplace equation (see Gunther [1] , Kellog [1 ]). The principal difficulty is in the obtainment of uniqueness theorems, because one deals with points belonging to the continuous spectrum of the operator (see chap. 16, sect.2) for real $\omega$ . In fact, one must search for solutions not belonging to the natural $L^2$ space. Uniqueness theorems follow from the Sommerfeld radiation condition which ensures that the energy flows towards infinity and not from infinity (see

Sommerfeld [ 1 ] , Rellich [ 2 ] , as well as Roseau [ 4 ] for other situations). Another interpretation of the radiation condition is given in sect. 8 (possibility of matching with an unperturbed region by a wave front). The potential theory for the Helmholtz equation is given in Kupradse [ 1 ] and Werner [ 1 ] , [ 2 ] , [ 3 ] , Shenk and Thoe [ 1 ]. An analogous theory for the elasticity system is given in Kupradse [ 2 ] , and for the electromagnetism system in Muller [ 1 ] . See also Filippi [ 1 ] for other related problems.

If $\omega$ is considered as a complex parameter, the singularities of the analytic continuation of the solutions are the "scattering frequencies" or "diffusion frequencies". They play the role of eigenvalues for this class of problems (see chapter 17 for examples of problems depending on a parameter showing that the scattering frequencies continuously tend to eigenvalues). See also sect. 8 of the present chapter for interpretation in space-time. Scattering frequencies are customary obtained as the singularities of the scattering matrix (see Lax and Phillips [ 1 ] , Shenk and Thoe [ 2 ] ), but we introduce them here directly from the solution of the boundary value problems for the Helmholtz equation. The main tool to do this is a theorem about the inversion of holomorphic operator-valued functions given in sect. 7 (Steinberg [ 1 ] ; see also Dunford-Schwartz [ 1 ] Lemma 5, p. 586 and Kato [ 2 ] theorem 1.9, p. 370 for less elaborated versions of this theorem).

Remark 9.1 - The proof of Lemma 2.5 is not complete. The fact that the expression (5.8 bis) is a continuous function on S follows from proposition 4.1, part c, if we see that $\rho$ is a function of class $L^\infty(B)$. To see this, we consider (5.7). The first term in the right hand side is a function of $L^\infty(B)$ as it is easily seen as in the proof of theorem 4.3, in particular formula (4.5). Then, we consider (5.7) as an integral equation for $\rho$. We apply an analogous of theorem 4.6 in $L^\infty$ instead of $C^0$ (the proof is the same) and we obtain the desired fact that $\rho \in L^\infty(B)$. ■

# CHAPTER 16

## FUNCTIONAL METHODS

We introduce here new methods for the study of scattering problems. These methods are more general than the potential method of the preceeding chapter and are suited for the study of spectral properties of operator with a continuous spectrum and asymptotic distribution of energy. The method of reduction to a problem in a bounded domain (sect. 4) furnishes in a simple way the scattering frequencies.

1.- Limiting absorption method - We shall introduce this method for a transmission problem more general than that of sect. 6 chap. 15 . A similar treatment for the Dirichlet and Neumann problems is possible with obvious modifications.

As in the preceeding chapter, let us consider the space $R^3$ decomposed in a bounded region B with piecewise smooth boundary and an outer region $\Omega$ . The boundary S of B is supposed to be piecewise smooth.

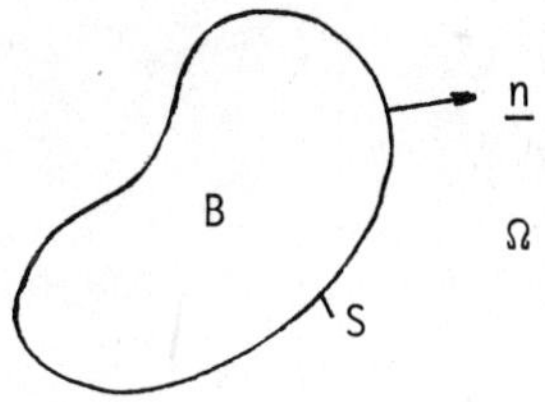


Let us define the real functions of class $L^\infty$ $a_{ij}(x)$, $b(x)$ by

$$(1.1) \qquad b(x) = \begin{cases} 1 & \text{if } x \in \Omega \\ \beta(x) & \text{if } x \in B \end{cases}$$

$$(1.2) \qquad a_{ij}(x) = \begin{cases} \delta_{ij} & \text{if } x \in \Omega \quad ; \quad \delta_{ij} = \text{Kronecker's symbol} \\ \alpha_{ij}(x) & \text{if } x \in B \quad ; \quad \alpha_{ij} = \alpha_{ji} \end{cases}$$

such that

$$a_{ij}(x)\ \xi_i\ \xi_j \geqslant \gamma |\xi|^2 \quad ; \quad b(x) \geqslant \gamma \tag{1.3}$$

for a certain positive constant $\gamma$ .

It is also assumed that $\alpha_{ij}(x)$, $\beta(x)$ are piecewise constant and that the boundaries between the regions where they are constant are piecewise smooth.

Let f be a given function of $L_2(R^3)$ with compact support.

We then consider the problem of searching for a function $\psi$ defined on $R^3$ such that ( $\omega$ is a real given constant) :

$$-\frac{\partial}{\partial x_i}\left(a_{ij}(x)\ \frac{\partial \psi}{\partial x_j}\right) - \omega^2\ b(x)\ \psi = f \tag{1.4}$$

in the sense of distributions and such that $\psi$ satisfies a radiation condition (note that (1.4) for large $|x|$ is the Helmholtz equation)

$$-(\Delta + \omega^2)\psi = 0 \tag{1.5}$$

Now we give the "variational" formulation of this problem. (See chap. 15, sect. 3 for the meaning of the radiation condition (1.6)).

Problem 1.1 - Find $\psi \in H^1_{\ell oc}(R^3)$ such that for $|x|$ sufficiently large it can be written in the form

$$\psi(x) = \int_{|y|=\rho} \left(\frac{\partial \psi(y)}{\partial R}\ \psi^+ - \psi(y)\ \frac{\partial \psi^+}{\partial R}\right) dS_y \tag{1.6}$$

where $R = |y|$, $\psi^+ = \frac{-1}{4\pi}\ \frac{e^{i\omega r}}{r}$ ; $r = |x - y|$

and $\rho$ is any constant such that B and supp f belongs to the ball $\{|x| < \rho\}$.

Moreover, $\psi$ must satisfy (for sufficiently large $\rho$) :

$$\int_{|x|<\rho} a_{ij}(x)\ \frac{\partial \psi}{\partial x_i}\ \frac{\partial \bar{\theta}}{\partial x_j}\ dx - \omega^2 \int_{|x|<\rho} b(x)\ \psi\,\bar{\theta}\,dx - \int_{|x|=\rho} \frac{\partial \psi}{\partial n}\,\bar{\theta}\ dS = \int_{|x|<\rho} f\ \bar{\theta}\ dx \tag{1.7}$$

for any test-function $\theta \in H^1(|x| < \rho)$. n as usual is the outer unit normal to $|x| = \rho$ .

Remark 1.1 - The problem 1.1 makes sense. By taking $\theta \in \mathcal{D}(R^3)$ with support outside B and supp f, it is easily seen that (1.7) shows that $\psi$ is a solution of the helmholtz equation (1.5) for sufficiently large $|x|$ . Then, (1.6) shows that $\psi$ satisfies the radiation condition. It is also clear that the choice of $\rho$ in (1.6) and (1.7) is irrelevant. It is also evident that (1.7) is a generalized formulation of (1.4). ∎

Theorem 1.1 - For real $\omega > 0$, the solution of problem 1.1 is unique.

Proof - Let us consider f = 0 in problem 1.1. By taking the imaginary part of (1.7) with $\theta = \psi$, we have for $\rho$ sufficiently large (note that $a_{ij}$ is symmetric and real) :

$$\text{(1.8)} \qquad \operatorname{Im} \int_{|x|=\rho} \frac{\partial\psi}{\partial n} \bar{\psi}\, dS = 0$$

Moreover, for sufficiently large $|x|$, $\psi$ is a solution of the Helmholtz equation (1.5) ; from proposition 1.3 of chap. 15 (see (1.14) and (1.15) of chapter 15) : we have

$$\Phi = 0 = \lim_{\rho\to\infty} \int_{|x|=\rho} |\psi|^2\, dS$$

and theorem 2.1 of chap.15 shows that $\psi$ is zero in a neighbourhood of infinity. The proof is then achieved as in theorem 2.3 of chap. 15 (note that

$$\psi = 0 \quad , \quad a_{ij} \frac{\partial\psi}{\partial x_j} n_i = 0$$

are homogeneous Cauchy conditions at the boundaries between the regions where the coefficients $\alpha_{ij}$, $\beta$ are constant, and that the equation (1.4) is (piecewise) elliptic with constant coefficients). ∎

Let us consider the Hilbert spaces $L_2(R^3)$ and $H^1(R^3)$ equipped with the scalar products :

$$\text{(1.9)} \qquad (\psi, \theta)_{L_2} = \int_{R^3} b(x)\, \psi \bar{\theta}\, dx$$

$$\text{(1.10)} \qquad (\psi, \theta)_{H^1} = \int_{R^3} \left( \psi\bar{\theta} + \frac{\partial\psi}{\partial x_i} \frac{\partial\bar{\theta}}{\partial x_i}\right) dx$$

(Note that the scalar product of $L_2$ induces a norm equivalent to the classical one ; the scalar product of $H^1$ is classical).

Moreover, let us consider the hermitian form

$$\text{(1.11)} \qquad a(\psi, \theta) = \int_{R^3} a_{ij} \frac{\partial\psi}{\partial x_i} \frac{\partial\bar{\theta}}{\partial x_j}\, dx \quad \text{on } H^1$$

Consequently we are in the framework of the "second representation theorem" (theorem 5.3 of chapter 2) and we consider the positive selfadjoint operator A on $L_2$ associated with the form $a(\psi, \theta)$. It is clear that

$$\text{(1.12)} \qquad D(A) \subset D(A^{1/2}) = H^1$$

The spectrum of A is contained in $R_+$ (the positive part of the real axis) (see proposition 4.1 of chapter 2). Consequently, if $f \in L_2(R^3)$, the equation

$$A\,\psi - \omega^2\,\psi = \frac{1}{b(x)}\,f \tag{1.13}$$

with $\omega^2 \notin R_+$ has a unique solution $\psi \in D(A)$. Let us multiply (1.13) in $L_2$ scalarly by a test function $\theta \in H^1(R^3)$ ; so we have

$$a(\psi,\theta) - \omega^2(\psi,\theta)_{L^2} = \left(\frac{1}{b}\,f\,,\theta\right)_{L^2} \quad \text{i.e.}$$

$$\int_{R^3} a_{ij}\,\frac{\partial\psi}{\partial x_i}\,\frac{\partial\overline{\theta}}{\partial x_j}\,dx - \omega^2\int_{R^3} b(x)\psi\,\overline{\theta}\,dx = \int_{R^3} f\,\overline{\theta}\,dx \tag{1.14}$$

and we have proved the following :

<u>Theorem 1.2</u> - If $\omega^2 \notin R_+$, and $f \in L_2(R^3)$ are given, there exists a unique $\psi \in H^1(R^3)$ such that (1.14) holds for any test element $\theta \in H^1(R^3)$.

<u>Remark 1.2</u> - Uniqueness in theorem 1.2 is not evident because we ask for $\psi \in H^1$ and (1.14) is equivalent to (1.13) only if $\psi \in D(A)$. But it can be easily proved. By taking $f = 0$ , $\theta = \psi$ in (1.14) we have

$$\int_{R^3} a_{ij}\,\frac{\partial\psi}{\partial x_i}\,\frac{\partial\overline{\psi}}{\partial x_j}\,dx - \omega^2\int_{R^3} b\,\psi\,\overline{\psi}\,dx = 0 \tag{1.15}$$

Then, if $\omega^2$ is real $< 0$, (1.15) is the sum of two positive quantities $\Rightarrow \psi = 0$. If the imaginary part of $\omega^2$ is not zero, by taking the imaginary part of (1.15), we have

$$\int_{R^3} b\,|\psi|^2\,dx = 0 \Rightarrow \psi = 0 \quad . \blacksquare$$

Our aim is to obtain the solution of problem 1.1 by taking the limit of solutions of (1.14) with complex $\omega^2$ as the imaginary part tends to zero.

Let us fix $\mu > 0$, and let $\nu$ be a real positive parameter which will tend to zero in the sequel. We then define $\omega = \omega(\nu)$ by

$$\omega(\nu)^2 = \mu^2 + i\nu \qquad ; \qquad \omega(\nu) = +\sqrt{\mu^2 + i\nu} \tag{1.16}$$

where the square root is defined in such a way that the imaginary part of $\omega(\nu)$ is positive.

Let us consider $\omega^2 = \omega(\nu)^2$ in (1.14) with $f \in L^2$ of bounded support and let $\psi_\nu$ be the corresponding solution given by theorem 1.2 (Note that $\psi$ depends in fact on $\omega(\nu)^2$ but not on $\omega(\nu)$).

Let us now study the structure of the solutions $\psi_\nu$ for large $|x|$ .

<u>Lemma 1.1</u> - Let $\rho$ be sufficiently large so that the ball of radius $\rho$ centered at the origin contains the domain B and the support of f. Then, $\psi_\nu$ has the representation

$$(1.17)\qquad \psi_\nu(x) = \int_{|y|=\rho} \left(\psi_\nu(y)\,\frac{\partial\psi^+}{\partial n_y} - \frac{\partial\psi_\nu}{\partial n_y}\,\psi^+\right) dS_y \qquad |x| > \rho$$

$$(1.18)\qquad \psi^+ = \frac{-1}{4\pi}\,\frac{e^{i\omega(\nu)r}}{r} \qquad r = |x - y|$$

where $n_y$ is the outer unit normal to $|y| = \rho$ .

<u>Proof</u> - For $|x| > \rho$ , we take $\rho_2 > |x|$ and we apply proposition 3.1 of chap. 15 to the region $\rho < |x| < \rho_2$. This gives

$$\psi_\nu(x) = \psi_\nu^1(x) - \psi_\nu^2(x)$$

where $\psi_\nu^1$ is for the expression (1.17) and $\psi_\nu^2$ is for an analogous one with $\rho_2$ instead of $\rho$ (Note that the unit outer normal to this region is $n_y$ for $\rho_2$ and $-n_y$ for $\rho$).

It is noticeable that $\psi_\nu^1(x)$ (resp. $\psi_\nu^2(x)$) is defined for $|x| > \rho$ (resp. $|x| < \rho_2$). Moreover, they are solutions of the equation

$$(1.19)\qquad (\Delta + \omega(\nu)^2)\psi_\nu^i = 0 \qquad i = 1\ ,\ 2$$

in their domains of definition. $\psi_\nu$ is also a solution of (1.19) for $\rho < |x| < \rho_2$. But the solutions of (1.19) are holomorphic (as solutions of an elliptic equation with constant coefficients).

We are going to prove that $\psi_\nu^2 \equiv 0$. Let us take $\rho_2$ variable and let it tend to $+\infty$ . The function $\psi_\nu$ is defined in particular for $|x| > \rho$ , and it coincides with its analytic continuation. The same is true for $\psi_\nu^1$. Consequently, the function $\psi_\nu^2(x)$ has an analytic continuation $\tilde{\psi}_\nu^2$ for $|x| > \rho_2$, which is equal to $\psi_\nu - \psi_\nu^1$ . (Note that $\tilde{\psi}_\nu^2(x)$ is in fact an analytic continuation and not the expression

$$\int_{|y|=\rho_2} \left(\psi_\nu(y)\,\frac{\partial\psi^+}{\partial n_y} - \frac{\partial\psi_\nu}{\partial n_y}\,\psi^+\right) dS_y \qquad |x| > \rho_2$$

which has singularities on $|x| = \rho_2$).

Moreover, $\psi_\nu \in L^2(R^3)$ is a solution given by theorem 1.2. On the other hand, $\psi^+$ decays exponentially at infinity (this is a consequence of the choice of the square root in (1.16)) and then $\psi_\nu^1 \in L_2(|x| > \rho)$. Consequently, $\tilde{\psi}_\nu^2 \in L^2(|x| > \rho)$. Moreover, $\psi_\nu^2 = \tilde{\psi}_\nu^2$ is defined and regular for $|x| < \rho$, thus

$\tilde{\psi}_\nu^2 \in L^2(R^3)$.

On the other hand, $\psi_\nu^2$ is, in $\rho < |x| < \rho_2$, a solution of (1.19), and by analytic continuation $\tilde{\psi}_\nu^2$ is a solution of (1.19) in $R^3$. Consequently, $\tilde{\psi}_\nu^2$ is an eigenvector of $-\Delta$ (considered as an operator in $L_2(R^3)$ equipped with its classical norm) corresponding to the eigenvalue $\omega(\nu)^2$ and this is only possible if $\tilde{\psi}_\nu^2 \equiv 0$ because $\omega(\nu)^2$ is not real and $-\Delta$ is selfadjoint. We thus have $\psi_\nu = \psi_\nu^1$ for $|x| > \rho$ and the lemma is proved.■

<u>Lemma 1.2</u> - For sufficiently large $\rho$ (as in lemma 1.1) $\psi_\nu$ satisfies :

$$(1.20) \quad \int_{|x|<\rho} a_{ij}(x) \frac{\partial \psi_\nu}{\partial x_i} \frac{\partial \overline{\psi}_\nu}{\partial x_j} dx - \omega(\nu)^2 \int_{|x|<\rho} b(x)\psi_\nu \overline{\psi}_\nu \, dx - \int_{|x|=\rho} \frac{\partial \psi}{\partial n} \overline{\psi} dS = \int_{|x|<\rho} f \overline{\psi} dx$$

where n is the outer unit normal to $|x| = \rho$.

<u>Proof</u> - Let us take $\omega = \omega(\nu)$, $\theta = \psi$ in (1.14) :

$$(1.21) \quad \int_{R^3} a_{ij} \frac{\partial \psi}{\partial x_i} \frac{\partial \overline{\psi}}{\partial x_j} dx - \omega(\nu)^2 \int_{R^3} b \, \psi \overline{\psi} \, dx = \int_{R^3} f \, \overline{\psi} \, dx \quad .$$

According to lemma 1.1 for $|x| > \rho$, $\psi$ and <u>grad</u> $\psi$ decay exponentially at infinity and satisfy equation (1.19). If we multiply (1.19) by $\overline{\psi}$ and we integrate it by parts, we have :

$$(1.22) \quad \int_{|x|>\rho} \frac{\partial \psi}{\partial x_i} \frac{\partial \overline{\psi}}{\partial x_i} dx - \omega(\nu)^2 \int_{|x|>\rho} \psi \, \overline{\psi} \, dx + \int_{|x|=\rho} \frac{\partial \psi}{\partial n} \overline{\psi} \, dS = 0$$

and by substracting (1.22) from (1.21) we obtain (1.20).■

<u>Lemma 1.3</u> - The solutions $\psi_\nu$ remain bounded as $\nu \to 0$ in the local norm of $L_2$ (i.e., for any $\rho > 0$, the restrictions of $\psi_\nu$ to $\{|x| < \rho\}$ are bounded in $L_2(|x| < \rho)$).

<u>Proof</u> - We shall prove this lemma by contradiction. Let us suppose that it is not true. Then, there exists a $\rho$ and a sequence $\nu_i \to 0$ such that $\psi_i = \psi_{\nu_i}$ tends to infinity in $L_2(|x| < \rho)$. It is clear that we can consider $\rho + 5$ and $\rho$ sufficiently large for Lemmas 1.1 and 1.2 hold for $\rho$. Consequently

$$(1.23) \qquad \|\psi_i\|_{L_2(|x|<\rho+5)} = m_i \to +\infty \quad ; \quad i = 1,2 \ldots$$

Let us consider the corresponding normalized functions

$$(1.24) \qquad \eta_i = \frac{\psi_i}{m_i} \quad ; \quad \|\eta_i\|_{L_2(|x|<\rho+5)} = 1$$

which satisfy the same equations that $\psi_i$ with $f_i/m_i$ instead of f .

The set (1.24) is weakly compact in $L_2(|x|<\rho+5)$ ; thus, after extracting a subsequence, we may consider

(1.25) $$\eta_i \to \eta \qquad \text{weakly in } L_2(|x|<\rho+5)$$

From (1.24) by taking the restrictions to $\rho<|x|<\rho+5$, $\eta_i$ is a solution of the Helmholtz equation (1.19) ; thus we have

(1.26) $$\| \eta_i \|_{L_2(\rho<|x|<\rho+5)} \leqslant 1$$

(1.27) $$\| \Delta \eta_i \|_{L_2(\rho<|x|<\rho+5)} < C$$

where C is a constant independent of i. Then, by standard regularity theory of the solutions of elliptic equations at the interior of their domain of definition (see for instance Agmon [1], chapt. 6) there exists a constant C' which depends only on on $\rho$ and $\rho+5$ such that

(1.27) $$\|\eta_i\|_{H^2(\rho+1<|x|<\rho+4)} \leqslant C'(1+C) = C''$$

and it follows from (1.25) that

(1.28) $$\eta_i \to \eta \quad \text{weakly in } H^2(\rho+1<|x|<\rho+4)$$

It then follows from the trace theorem that

(1.29) $$\begin{cases} \eta_i\Big|_{|x|=\rho+2} \longrightarrow \eta\Big|_{|x|=\rho+2} & \text{weakly in } H^{3/2}(|x|=\rho+2) \\ \text{and the same for } |x|=\rho+3 \end{cases}$$

(1.30) $$\begin{cases} \dfrac{\partial\eta_i}{\partial n}\Big|_{|x|=\rho+2} \longrightarrow \dfrac{\partial\eta}{\partial n}\Big|_{|x|=\rho+2} & \text{weakly in } H^{1/2}(|x|=\rho+2) \\ \text{and the same for } |x|=\rho+3 \end{cases}$$

It is clear that (1.29) and (1.30) hold also in the strong topology of $L_2(|x|=\rho+2)$ and $L_2(|x|=\rho+3)$ (see theorem 3.2 of chapter 1).

Now, we apply Lemma 1.2 with $\eta_i$ and $f/m_i$ instead of $\psi_\nu$ and f, for the region $|x|<\rho+3$ :

(1.31) $$\int_{|x|<\rho+3} a_{kj}(x)\,\frac{\partial\eta_i}{\partial x_k}\,\frac{\partial\bar\eta_i}{\partial x_j}\,dx = \omega(\nu_i)^2 \int_{|x|<\rho+3} b(x)\,\eta_i\,\bar\eta_i\,dx +$$
$$+\int_{|x|=\rho+3} \frac{\partial\eta_i}{\partial n}\,\bar\eta_i\,dS + \int_{|x|<\rho+3} \frac{f}{m_i}\,\bar\eta_i\,dx$$

By virtue of (1.26), (1.29), (1.30), the right hand side of (1.31) is bounded and by taking a lower bound of the left hand side we have

(1.32) $$\|\eta_i\|_{H^1(|x|<\rho+3)} \leq C$$

Thus, by (1.25) and the Rellich theorem :

(1.33) $$\eta_i \to \eta \quad \text{in} \quad H^1(|x| < \rho + 3) \text{ weakly}$$

(1.34) $$\eta_i \to \eta \quad \text{in} \quad L_2(|x| < \rho + 3) \text{ strongly}$$

Moreover, we can use Lemma 1.1 for $\eta_i$ and $\rho + 2$ :

$$\eta_i(x) = \int_{|y| = \rho+2} (\eta_i(y) \frac{\partial\psi^+}{\partial n_y} - \frac{\partial\eta_i}{\partial n_y} \psi^+) \, dS_y$$

From (1.29) and (1.30) we see that $\eta_i$ converges uniformly in $\rho + 3 < |x| < \rho + 5$ to $\eta$, which has also the representation

(1.35) $$\eta(x) = \int_{|y|= \rho+2} (\eta(y) \frac{\partial\psi^+}{\partial n_y} - \frac{\partial\eta}{\partial n_y} \psi^+) \, dS_y \, ; \quad \psi^+ = \frac{-1}{4\pi} \frac{e^{i\mu t}}{r} ,$$

for $\rho + 2 < |x| < \rho + 5$

(the uniform convergence is immediate, as in the proof of theorem 4.3 of chapter 15 for instance). Then, with (1.34) and (1.24), we have :

(1.36) $$\eta_i \to \eta \quad \text{strongly in} \quad L_2(|x| < \rho + 5)$$

(1.37) $$\|\eta\|_{L_2(|x| < \rho + 5)} = 1$$

Moreover, from (1.35) we see that $\eta$ satisfies

(1.38) $$-(\Delta + \mu^2)\eta = 0$$

for $\rho + 2 < |x| < \rho + 5$ (This is also easily seen by passing to the limit $i \to \infty$ in (1.19)), and we can consider the analytic continuation of $\eta$ for $|x| > \rho + 5$, given by (1.35). Consequently, <u>we can consider (1.35) for $|x| > \rho + 2$.</u>

We are going to prove that $\eta = 0$, which will be in contradiction to (1.37). To this end, we shall prove that $\eta$ is a solution of problem 1.1 with $f = 0$.

Let us consider $\theta \in H^1(|x| < \rho + 3)$. This function may be extended to a function $\theta \in H^1(R^3)$ zero for $|x| > \rho + 5$. As in the proof of Lemma 1.2, we see that that $\eta_i$ satisfies

$$(1.39)\quad \int_{|x|<\rho+3} a_{kj}\frac{\partial\eta_i}{\partial x_k}\frac{\partial\overline{\theta}}{\partial x_j}\,dx - \omega(\nu_i)^2\int_{|x|<\rho+3} b\,\eta_i\,\overline{\theta}\,dx - \int_{|x|=\rho+3}\frac{\partial\eta_i}{\partial n}\overline{\theta}dS =$$

$$= \int_{|x|<\rho+3}\frac{f}{m_i}\,\overline{\theta}\,dx$$

Then, we pass to the limit $i\to\infty$ in this expression. By using (1.33), (1.34), as $m_i\to\infty$, we have :

$$(1.40)\quad \int_{|x|<\rho+a} a_{kj}\frac{\partial\eta}{\partial x_k}\frac{\partial\overline{\theta}}{\partial x_j}\,dx - \mu^2\int_{|x|<\rho+3} b\eta\,\overline{\theta}\,dx - \int_{|x|=\rho+3}\frac{\partial\eta}{\partial n}\,\overline{\theta}\,dS = 0$$

and we see that $\eta$ is a solution of problem 1.1 (with $\omega=\mu$ , $f=0$, $\rho+3$ instead of $\rho$) ; and by theorem 1.1, we have $\eta=0$, which is in contradiction to (1.37). Lemma 1.3 is proved. ■

It is now easy to prove the following theorem, which is the principal result of this section :

Theorem 1.2 - The problem 1.1 with $\omega$ real and positive has a solution, which is a limit in $L_{2\,loc}(R^3)$ of the functions $\psi_\nu$.

Proof - It suffices to repeat the proof of lemma 1.2 with $\psi_\nu$, which are bounded by virtue of lemma 1.3, instead of $\eta_i$ and f instead of $f/m_i$. The limit $\psi$ of $\psi_\nu$ is a solution of problem 1.1. ■

We have obtained the existence and uniqueness of the solution of problem 1.1, which is a transmission problem of type more general than that of chapter15, sect. 6, where we only studied the case $a_{ij}=a\,\delta_{ij}$. It is also clear that obvious modifications (in fact, simplifications) allows to study the Dirichlet and Neumann problems, to obtain new proofs of theorems 5.1 and 6.2 of chapter15.

Remark 1.3 - Theorems 1.1 and 1.2 hold also for $\omega=0$. In fact the proofs are in this case easier, but a little different from that of the case $\omega>0$. In fact, for $\omega=0$, $\psi(x)$ given by (1.6) decays at infinity as $|x|^{-1}$ and grad $\psi(x)$ as $|x|^{-2}$. It is then easy to obtain an expression of the type (1.22) ; by addition with (1.7) with $\omega=0$ , $f=0$, we obtain

$$\int_{R^3} a_{ij}(x)\,\frac{\partial\psi}{\partial x_i}\,\frac{\partial\overline{\psi}}{\partial x_j}\,dx = 0$$

and uniqueness follows. ■

Remark 1.4 - If we wish to obtain incoming solutions (instead of outgoing), we shall take

$$\psi^- = \frac{-1}{4\pi} \frac{e^{-i\omega r}}{r}$$

in (1.6) instead of $\psi^+$ and $-\nu$ instead of $\nu$ in (1.16) with the square root of negative imaginary part. The results are the same as that for outgoing solutions. ■

2.- Absolute continuity of the spectrum - In this section we study some abstract properties of the spectrum of the operator A which will be used in the following section for the study of some properties of the solutions of the evolution problem associated with A.

In this section we consider the operator A associated with the transmission problem (see (1.9) - (1.12)) but the results hold for the Dirichlet and Neumann problems as well, which are in fact easier.

Let us recall (Smirnov [1], sect. 74) that a function f of the real variable x is said to be absolutely continuous if for any given $\varepsilon > 0$ there is a corresponding $\eta > 0$ such that if $(a_k , b_k)$ $(k = 1, 2, \ldots n)$ are non-overlapping intervals such that

$$\sum_i^n (b_k - a_k) \leqslant \eta$$

Then,

$$\left| \sum_i^n [f(b_k) - f(a_k)] \right| \leqslant \varepsilon$$

An important property of these functions is the following : If f is absolutely continuous, it is expressible as the integral :

$$f(x) = f(a) + \int_a^x g(y)\, dy$$

for a certain function g of class $L_1$.

It is evident from the definition that the integral of a continuous function is absolutely continuous.

Now, let us consider a selfadjoint positive operator A in a Hilbert space H and let $E(\lambda)$ be its spectral family. Thus (remark 4.7 of chapter 2) :

$$A = \int_0^\infty \lambda \, dE(\lambda) \tag{2.1}$$

We then have :

<u>Definition 2.1</u> - A selfadjoint positive operator is said to have an absolutely continuous spectrum if the function $(E(\lambda)u , u)_H$ is absolutely continuous for any $u \in H$.

The following two propositions will be useful in the sequel. Their proofs may be found in Kato [2], sect. X. 1.2.

<u>Proposition 2.1</u> - If $(E(\lambda) u , u)_H$ is absolutely continuous for any u belonging to a set dense in H, the operator A has an absolutely continuous spectrum.

<u>Proposition 2.2</u> - If A has an absolutely continuous spectrum, $(E(\lambda) u,v)_H$ is absolutely continuous for any $u,v \in H$.

Now, we return to the operator A of the preceeding section and we apply proposition 2.1 to establish the absolute continuity of the spectrum.

First, we note that A has no eigenvalues. For, if $\psi$ is an eigenfunction, it satisfies the Helmholtz equation for sufficiently large $|x|$, with $\omega^2$ real (because A is selfadjoint). Moreover, $\psi \in L_2$ and then $\psi = 0$ by corollary 2.1 of chapter 15.

Consequently, the spectral family $E(\lambda)$ is continuous, and the classical formula ((4.15) of chapter 2) becomes :

$$([E(b) - E(a)] x , y) = \lim_{\sigma\to 0} \frac{1}{2\pi i} \int_a^b ([(A - (\lambda + i\sigma))^{-1} - (A - (\lambda - i\sigma))^{-1}] x,y) d\lambda$$

Let us take a = 0 (and thus E(a) = 0) and the elements x and y are equal to a function $g \in L_2$ with compact support ; then :

$$(2.2) \qquad (E(b)g , g)_{L_2} =$$

$$= \lim_{\nu\to 0} \frac{1}{2\pi i} \int_0^b ([(A - (\lambda + i\nu))^{-1} - (A - (\lambda - i\nu))^{-1}] g , g)_{L_2} \, d\lambda$$

It is easy to see that the expression (2.2) is in fact the integral of a continuous function. Let us consider the "limiting absorption" of the preceeding section. Let $\psi$ be the solution (theorem 1.2) corresponding to a given function f (of compact support, $f \in L_2$). Then,

$$(2.3) \qquad \psi = \lim_{\nu\to 0} \psi_\nu = \lim_{\nu\to 0} (A - (\mu^2 + i\nu))^{-1} \left(\frac{f}{b}\right)$$

where the limit holds in the norm of $L_2(|x| < \rho)$ for any $\rho$. By taking $\rho$ sufficiently large, the support of f is contained in $|x| < \rho$. Thus

$$(2.4)\qquad \lim_{\nu\to 0}\ ([A - (\lambda+i\nu)]^{-1} g , g)_{L_2} = \lim_{\nu\to 0} (\psi_\nu , g )_{L_2} = (\psi , g)_{L_2}$$

where $\psi$, $\psi_\nu$ are the solutions corresponding to the given function $f(x)= b(x)\, g(x)$.

Moreover, the limit in (2.4) is a continuous function of $\lambda$ (to see this it suffices to repeat the proof of theorem 1.2 with $\mu^2$ variable at the same time that $\nu$, which is a somewhat trivial matter). On the other hand, the same considerations hold for negative $\nu$ (see remark 1.4). We can commute integral and limit in (2.2) and the right hand side is an absolutely continuous function (as the integral of a continuous function). Moreover, the set of functions $g \in L_2$ with compact support is dense in $L_2$ ; by virtue of proposition 2.1, we have proved the following :

Theorem 2.1 - The operator A (associated with the transmission, Dirichlet or Neumann problems) has an absolutely continuous spectrum.

The spectral family $E(A^{1/2} , \lambda)$ of $A^{1/2}$ is $E(A^{1/2} , \lambda) = E(A , \lambda^2)$ and it is evident from the definition of absolute continuity that $(E(A^{1/2}, \lambda)u , v)$ is absolutely continuous. In the consequence, $A^{1/2}$ as well as A has an absolutely continuous spectrum. Thus we can write :

$$(2.5)\qquad (E(A^{1/2} , \lambda)\, u , v) = \int_0^\lambda f(\mu)\, d\mu$$

for a certain function f of class $L_1$. Moreover, the left hand side of (2.5) tends to $(u , v)$ as $\lambda \to \infty$ ; we thus have $f \in L^1(0 , \infty)$, and also $f \in L^1(-\infty , +\infty)$, because f is zero for negative $\lambda$.

A classical theorem of Lebesque says that the direct and inverse Fourier transforms of such a function f, i.e.

$$\frac{1}{\sqrt{2\pi}} \int_{-\infty}^{+\infty} e^{\mp i\lambda t}\, f(\lambda)\, d\lambda$$

tend to zero as $|t| \to \infty$. Consequently we have proved the following theorem, which will be used in the next section :

Theorem 2.2 - If $E(A^{1/2} , \lambda)$ is the spectral family of $A^{1/2}$, where A is the operator associated with the transmission, Dirichlet or Neumann problems, the functions

$$\int_{-\infty}^{+\infty} e^{\mp i\lambda t}\, d(E(A^{1/2} , \lambda)\, u , v)_{L_2}$$

tend to zero as $|t| \to \infty$ for any u,v $L_2$.

3.- <u>Local energy decay and limiting amplitude</u> - In this section we study the homogeneous evolution problem

$$(3.1) \qquad \frac{\partial^2 \phi}{\partial t^2} + A\,\phi = 0 \quad ; \quad \phi(0) = \phi_0 \quad ; \quad \frac{\partial \phi}{\partial t}(0) = \phi_1$$

associated with the operator A of section 1 (or to the Dirichlet and Neumann problems, with obvious modifications). The abstract differential equation (3.1) is of course equivalent to the hyperbolic partial differential equation :

$$b(x)\,\frac{\partial^2 \phi}{\partial t^2} = \frac{\partial}{\partial x_i}\left(a_{ij}(x)\,\frac{\partial \phi}{\partial x_j}\right)$$

It is well known that for the wave equation (i.e. $b = 1$, $a_{ij} = \delta_{ij}$) the support of a solution corresponding to initial values concentrated at the point $t = 0$, $|x| = 0$, is the surface of the case $|x| = t$.

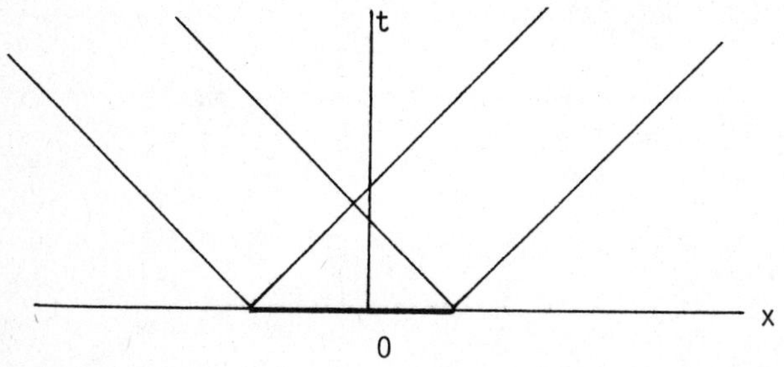


Consequently, if $\phi_0$ and $\phi_1$ are of compact support, the corresponding solution $\phi(x\ ,\ t)$ is identically zero on the domain $|x| < \rho$ for fixed $\rho$ and sufficiently large t.

In the present situation, the obstacle B modifies this simple picture, but in fact, the solution $\phi(x\ ,\ t)$ <u>tends to zero</u> as $t \to +\infty$ in the domain $|x| < \rho$ . This result is given in a rigorous way in theorems 3.1 and 3.2.

The equation (3.1) is in the framework of chapter 4, sect 2, in particular in the framework of remarks 5.1 and 5.2. In fact, the pairs $(\phi\ ,\phi')$ where ' is for the time derivative, form a continuous semigroup in the space $H^1 \times L_2$.

By writting (3.1) under the form :

$$\frac{d}{dt}\begin{pmatrix}\phi\\ \phi'\end{pmatrix} = \mathcal{A}\begin{pmatrix}\phi\\ \phi'\end{pmatrix} \quad ; \quad \mathcal{A} = \begin{pmatrix}0 & I\\ -A & 0\end{pmatrix} \quad ; \quad D(\mathcal{A}) = D(A) \times H^1$$

the semigroup can be written

$$\begin{pmatrix}\phi\\ \phi'\end{pmatrix}(t) = \exp(\mathcal{A}t)\begin{pmatrix}\phi_0\\ \phi_1\end{pmatrix}$$

or (see, for a direct calculation, Mikhlin sect. 2.4.8)

$$(3.2)\quad \begin{cases} \phi(t) = \cos A^{1/2}t \ \ \phi_0 + A^{-1/2} \sin A^{1/2}\, t\, \phi_1 \\ \phi'(t) = -\sin A^{1/2}t\, A^{1/2}\, \phi_0 + \cos A^{1/2}\, t\, \phi_1 \end{cases}$$

We then have :

<u>Theorem 3.1</u> - For any initial values ($\phi_0 \in H^1$, $\phi_1 \in L_2$) and any $\rho$ the solution $\phi(t)$ of (3.1) satisfies :

$$(3.3)\qquad \lim_{t\to+\infty} \phi(t)\big|_{|x|\ \rho} = 0 \quad \text{in } L_2(|x|<\rho) \text{ strongly.}$$

<u>Proof</u> - It suffices to prove that

$$(3.4)\qquad \lim_{t\to+\infty} \phi(t) = 0 \quad \text{in} \quad H^1(R^3) \quad \text{weakly}$$

for, if (3.4) holds, it also holds in $H^1(|x|<\rho)$ weakly, and by the Rellich theorem, we have (3.3).

Let us prove (3.4). Let us remark that the norm of $H^1(R^3)$ is equivalent to

$$(3.5)\qquad \|.\|^2_{H^1} \cong \|.\|^2_{L_2} + a(\ .\ ,\ .\ ) = \|.\|^2_{L_2} + (A^{1/2}\ .\ ,\ A^{1/2}\ .)_{L_2}$$

and consequently, (3.4) is equivalent to

$$(3.6)\quad \begin{cases} (\phi(t)\ ,\ \theta)_{L_2} \xrightarrow[t\to+\infty]{} 0 & \forall\theta \in L_2 \\ (A^{1/2}\,\phi(t)\ ,\ A^{1/2}\,\theta)_{L_2} \xrightarrow[t\to+\infty]{} 0 & \forall\theta \in H^1 \end{cases}$$

But from (3.2)

$$(3.7)\qquad \phi(t) = \cos A^{1/2}t\,(\phi_0) + \sin A^{1/2}t(A^{-1/2}\,\phi_1)$$

where $\phi_0 \in H^1$ , $\phi_1 \in L_2$ and thus the expression in parentheses belong to $L_2$. By writting

$$\cos A^{1/2}t = \frac{1}{2}(e^{iA^{1/2}t} + e^{-iA^{1/2}t})$$

we have

$$(\phi(t)\ ,\theta)_{L_2} = \frac{1}{2}(e^{iA^{1/2}t}\,\phi_0,\ \theta)_L + \dots =$$

$$= \frac{1}{2}\int_{-\infty}^{+\infty} e^{i\lambda t}\, d(E(A^{1/2}\ ,\lambda)\,\phi_0\ ,\ \theta)_{L_2} + \dots$$

where .... is for analogous terms. But this expression tends to zero as $t \to +\infty$ by virtue of theorem 2.2. $(3.6)_1$ is proved. In an analogous way,

$$A^{1/2}\ \phi(t) = \cos A^{1/2}t(A^{1/2}\ \phi_0) + \sin A^{1/2}\ t(\phi_1)$$

where $\phi_0 \in H^1 = D(A^{1/2})$ (by the second representation theorem) and thus the expression in parentheses belong to $L_2$, and we obtain $(3.6)_2$ as before. Theorem 3.1 is proved. ■

It is interesting to improve theorem 3.1 to obtain a stronger convergence, indeed a convergence in the "local energy norm". We shall explain this. We know from chapter 4, sect. 5 that the solution $\phi(t)$ has a constant "energy norm" $\| \ \|_E$ :

(3.8) $$\|\phi\ ,\ \phi'\|_E^2 = a(\phi\ ,\phi) + \|\phi'\|_{L_2}^2 = a(\phi_0\ ,\ \phi_0) + \|\phi_1\|_{L_2}^2$$

$$a(\phi\ ,\phi) = \int_{R^3} a_{ij}\ \frac{\partial \phi}{\partial x_i}\ \frac{\partial \overline{\phi}}{\partial x_j}\ dx$$

We define the local "energy norm for $|x| < \rho$ " by

(3.9) $$\|\phi\ ,\ \phi'\|_{E(|x|<\rho)}^2 = \int_{|x|<\rho} a_{ij}\ \frac{\partial \phi}{\partial x_i}\ \frac{\partial \overline{\phi}}{\partial x_j}\ dx + \int_{|x|<\rho} b\ \phi'\ \overline{\phi}'\ \ dx$$

Then, if we obtain an improved form of theorem 3.1 where (3.3) holds strongly in the local energy norm, <u>the energy of the solution propagates out of any compact set $|x| < \rho$</u>.

In fact, we have :

<u>Theorem 3.2</u> - For any initial values ($\phi_0 \in H^1$, $\phi_1 \in L_2$) and any $\rho$ ,

(3.10) $$\|\phi(t)\ ,\ \phi'(t)\|_{E(|x|<\rho)}^2 \xrightarrow[t \to +\infty]{} 0$$

(see (3.9) for the notation).

<u>Proof</u> - We first prove (3.10) for $\phi_0 \in D(A)$, $\phi_1 \in H^1 = D(A^{1/2})$. Thus from (3.2), we have :

$$\phi'(t) = -\sin A^{1/2}t(A^{1/2}\ \phi_0) + \cos A^{1/2}t(\phi_1)$$

$$A^{1/2}\ \phi'(t) = -\sin A^{1/2}t(A\ \phi_0) + \cos A^{1/2}t(A^{1/2}\ \phi_1)$$

where the expressions in parentheses belong to $L_2$. Thus, as in the proof of theorem 3.1, we have :

(3.11) $$\|\phi'(t)\|_{L_2(|x|<\rho)} \longrightarrow 0 \qquad \text{as } t \to +\infty$$

Moreover,

$$A\ \phi(t) = \cos A^{1/2}t(A\ \phi_0) + \sin A^{1/2}(A^{1/2}\ \phi_1)$$

where the expression in parenthesis belong to $L_2$. Thus, as in the proof at theorem 3.1 :

$$(3.12)\qquad A\phi(t) \xrightarrow[t \to +\infty]{} 0 \qquad \text{in } L^2(R^3) \text{ weakly}$$

and we also have (proof of th. 3.1) :

$$(3.13)\qquad \phi(t) \xrightarrow[t \to +\infty]{} 0 \quad \text{in } L^2(R^3) \text{ weakly}$$

We then apply the classical inequality of local regularity (local in the sense of "up to a neighbourhood of infinity") which can be seen in Lions-Magenes [1] and we obtain

$$(3.14)\qquad \|\phi(t)\|_{H^2(|x|<\rho)} \leqslant C\,(\,\|A\,\phi(t)\|_{L_2} + \|\phi(t)\|_{L_2}\,)$$

and from (3.12) and (3.13) we have

$$\phi(t)\Big|_{|x|<\rho} \longrightarrow 0 \qquad \text{in } H^2 \text{ weakly.}$$

as $t \to +\infty$. Thus, by the Rellich theorem,

$$\phi(t)\Big|_{|x|<\rho} \quad ; \quad \frac{\partial\phi(t)}{\partial x_i}\Big|_{|x|<\rho}$$

tend to zero in $L_2(|x|<\rho)$ strongly, and with (3.11) we obtain (3.10)

It is now easy to obtain (3.10) in the general case where $\phi_o \in H^1 = D(A^{1/2})$, $\phi_1 \in L_2$. In fact, $D(A) \times H^1$ is the domain of $\mathcal{A}$ (see chap. 4, sect. 5) which is dense in $H^1 \times L_2$. Then, for any given $(\phi_o\,,\,\phi_1) \in H^1 \times L_2$, and any positive $\varepsilon$, there exists $(\phi_o^\varepsilon\,,\,\phi_1^\varepsilon) \in D(A) \times H^1$ such that

$$\|(\phi_o - \phi_o^\varepsilon\,,\,\phi_1 - \phi_1^\varepsilon)\|_{H^1 \times L_2} < \varepsilon \qquad \text{and consequently}$$

$$\|(\phi_o - \phi_o^\varepsilon\,,\,\phi_1 - \phi_1^\varepsilon)\|_E < \varepsilon$$

and by virtue of the fact that the norm of the solution in E is independent of t we have

$$\|(\,\phi(t) - \phi^\varepsilon(t)\,,\,\phi'(t) - \phi^{\varepsilon\prime}(t))\|_{E(|x|<\rho)} < \varepsilon \qquad \forall t$$

and by using the preliminary result for $\phi^\varepsilon$, $\phi^{\varepsilon\prime}$,

$$\overline{\lim_{t\to\infty}}\ \|\phi(t)\,,\,\phi'(t)\,\|_{E(|x|<\rho)} < \varepsilon$$

for any $\varepsilon$, and (3.10) holds. ∎

<u>Remark 3.1</u> - The role of the inequality (3.14) is fundamental to prove theorem 3.2, but it is not used in the proof of theorem 3.1. Consequently, in more complicated problems, we shall obtain the local energy decay (3.10) only if we have a

regularity theory, but the weaker form (3.3) of the local decay holds without regularity theory. ■

The preceeding results hold for the evolution problem when the equation (3.1) is homogeneous (i.e. equal to zero). It is easy to obtain analogous results for non homogeneous equations.

Let us consider again the solutions in the space-time $\phi^*$ associated with a stationary solution $\psi$ of the Helmholtz equation. We have seen that several functions $\phi^*(x , t)$ are associated with a fixed stationary solution $\psi(x)$. Let us fix a certain $\phi^*(x , t)$. It is determined by its initial values $\phi^*(x,0)$, $\partial\phi^*/\partial t(x,0)$. According to the construction of $\phi^*$, these initial values are of compact support. In fact, if we take other initial values belonging to $H^1 \times L_2$, the difference of the corresponding solutions is a solution of the homogeneous equation with initial values in $H^1 \times L_2$. Theorem 3.2 holds and we obtain the following property :

Proposition 3.1 (Limiting amplitude property) - Consider the Neumann problem (analogous results hold for the transmission and Dirichlet problems). find $\phi(x , t)$ satisfying

$$\frac{\partial^2\phi}{\partial t^2} = \Delta\phi \qquad \text{in } \Omega \times [0,\infty[$$

$$\frac{\partial\phi}{\partial n} = \operatorname{Re}\{f(x)\, e^{-i\omega t}\} \qquad \text{on } S$$

and having initial values $\phi(0) = \phi_0 \in H^1$, $\phi'(0) = \phi_1 \in L_2$ .

Then, we consider the solution $\psi(x)$ of the Helmholtz equation

$$-(\Delta + \omega^2)\psi = 0 \qquad \text{in } \Omega$$

$$\frac{\partial\psi}{\partial n} = f(x) \qquad \text{on } S$$

which satisfies the radiation condition. Moreover, we construct the function

$$\tilde{\phi}(x , t) = \operatorname{Re}\{\psi(x)\, e^{-i\omega t}\}$$

Then, $\phi(x , t) - \tilde{\phi}(x , t)$ tends to zero as $t \to +\infty$ in the local energy norm for $|x| < \rho$ (see (3.9)) for any $\rho$ .

4.- Reduction to a problem in a bounded domain - We now consider a new method to study the existence of the solution of the exterior problem, as well as the scattering frequencies. We first present the method for the Dirichlet problem, and afterwards for the transmission problem.

Let B and $\Omega$ be given as in section 1. Let f be a function of $L_2(\Omega)$ with compact support. We search for a function u defined on $\Omega$ which satisfies

(4.1) $$-(\Delta + \omega^2)u = f \qquad \text{in } \Omega$$

(4.2) $$u|_S = 0 \qquad \text{on } S$$

(4.3) u satisfies the outgoing radiation condition.

More exactly, we consider :

Problem 4.1 (Dirichlet) - Study of the existence and uniqueness of u satisfying (4.1) - (4.3) for real positive $\omega$ and the complex values of $\omega$ for which $u \neq 0$ exists with f = 0.

Let us begin by the study of the case where $\omega$ is real and positive. The uniqueness is known. (Theorem 2.2 of chap.15. Note that the local regularity theory for elliptic problems allows to us to apply the uniqueness theorem).

We now study the existence. Let $\rho$ be a real number such that the ball $|x| < \rho$ contains B and the support of f. Let g be a function of $L_2(|x| < \rho + 1)$ which we consider known for the time being. We construct the function

(4.4) $$w = g * \psi^+ \quad ; \quad \psi^+ = \frac{-1}{4\pi} \frac{e^{i\omega r}}{r}$$

for $\omega$ either real or complex, which satisfies (according to chapt. 15, proposition 4.1)

(4.5) $$-(\Delta + \omega^2)\, w = g \qquad \text{on } R^3$$

(4.6) w satisfies the outgoing radiation condition.

Next, let us construct a function $v \in H^1(\Omega_{\rho+1})$ (where $\Omega_{\rho+1} = \Omega \cap \{x\ ; |x| < \rho + 1\}$) such that

(4.7) $$(-\Delta + \lambda_0)v = (-\Delta + \lambda_0)\, w \quad \text{in } \Omega_{\rho+1}$$

(4.8) $$v\big|_{|x| = \rho+1} = w\big|_{|x| = \rho+1}$$

(4.9) $$v|_S = 0$$

where $\lambda_o$ is any chosen complex with $\mathrm{Im}\,\lambda_o > 0$. Note that by virtue of (4.5), the right hand side of (4.7) is a function of $L_2(\Omega_{\rho+1})$ ; (4.7) - (4.9) constitute a non homogeneous Dirichlet problem which has a unique solution because $\mathrm{Im}\lambda_o > 0$ and consequently, $\lambda_o$ is not in the spectrum.

Now, we consider a function $\gamma(|x|)$ of class $C^\infty$ which is equal to 1 (resp. 0) for $|x| \leqslant \rho + 1/3$ (resp. $|x| \geqslant \rho + 2/3$), and we construct the function (defined in $\Omega$)

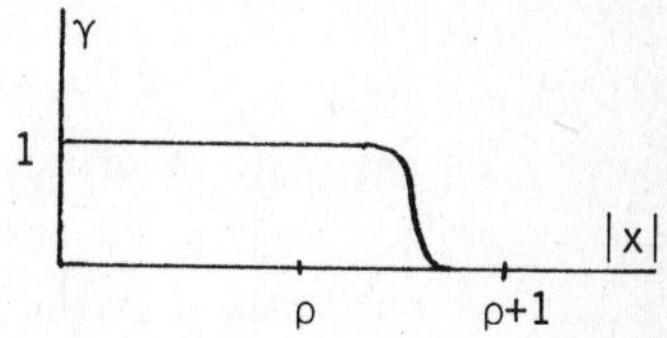


(4.10) $$u = w - \gamma(w - v)$$

Of course, for $|x| \geqslant \rho + 2/3$, u is equal to w and thus it satisfies the outgoing radiation condition. For $x \in S$, we have $|x| < \rho$ and $u|_S = v|_S = 0$. In consequence, u defined by (4.10) satisfies (4.2) and (4.3). In order to satisfy (4.1), the following relation between f and g must hold in $\Omega$

$$f = -(\Delta + \omega^2)u = -(\Delta + \omega^2)w + \Delta[\gamma(w - v)] + \omega^2\gamma(w - v) =$$

$$= -(\Delta + \omega^2)w + \Delta\gamma(w - v) + \frac{\partial\gamma}{\partial x_i}\frac{\partial(w - v)}{\partial x_i} + \gamma\Delta(w - v) + \omega^2\gamma(w - v)$$

which by virtue of (4.5) and (4.7) takes the form :

(4.11) $$f = g + (\Delta\gamma + \omega^2\gamma)(w - v) + \frac{\partial\gamma}{\partial x_i}\frac{\partial(w - v)}{\partial x_i} + \gamma\lambda_o(w - v)$$

Note that the right hand side is identically zero for $|x| > \rho + 1$. It is then possible to consider (4.11) as a relation between elements of $L_2(\Omega_{\rho+1})$. It is clear that the right hand side of (4.11) is determinated by the function g. If we write (4.11) under the form

(4.12) $$f = g + T(\omega)g$$

we have :

Proposition 4.1 - The operator $T(\omega)$ is compact in $L_2(\Omega_{\rho+1})$ and it depends holomorphically on $\omega \in C$ (i.e., it is a bounded holomorphic operator, compact for each $\omega \in C$).

Proof - The operator which takes $g \mapsto w$ is bounded from $L_2(\Omega_{\rho+1})$ into $H^2(\Omega_{\rho+1})$ (by chapter 15 proposition 4.1, d) and is bounded - holomorphic (the proof is analogous to that of chapter 15 , lemma 7.1). The right hand side of (4.7) is a holomorphic function of $\omega$ with values in $L_2(\Omega_{\rho+1})$ ; the right hand side of (4.8) is a holomorphic function of $\omega$ with values in $H^{3/2}(|x| = \rho+1)$ and the solution v of (4.7) - (4.9) is a holomorphic function of $\omega$ with values in $H^1(\Omega_{\rho+1})$. Moreover, by standard interior regularity theory of the solutions of elliptic equations (see for instance Agmon, [1], chapter 6) v is holomorphic with values in $H^2$ for $\rho + 1/3 < |x| < \rho + 2/3$. It is then clear by the Rellich compactness theorem that the operator $g \mapsto w - v$ is bounded holomorphic and compact from $L_2(\Omega_{\rho+1})$ into $L_2(\Omega_{\rho+1})$ and the operator which takes $g \to \frac{\partial(w-v)}{\partial x_i}$ is bounded holomorphic and comptact from $L_2(\Omega_{\rho+1})$ into $L_2(\rho + 1/3 < |x| < \rho + 2/3)$. The proposition follows immediately (note that the support of $\partial\gamma/\partial x_i$ is contained in $\rho + 1/3 < |x| < \rho + 2/3$) ∎

<u>Remark 4.1</u> - We have not assumed that the surface S is smooth. If it is so, by regularity theory, the operator which takes $g \to v$ is bounded - holomorphic from $L_2(\Omega_{\rho+1})$ into $H^2(\Omega_{\rho+1})$ and the preceeding proof is simplified ∎

It is then clear that the Fredholm alternative holds for equation (4.12). We have

<u>Proposition 4.2</u> - For real positive $\omega$, $f = 0$ implies $g = 0$. Consequently, for $f \in L_2(\Omega_\rho)$ (and thus $\in L_2(\Omega_{\rho+1})$), there exists a unique $g \in L_2(\Omega_{\rho+1})$ which solves problem 4.1.

<u>Proof</u> - If $f = 0$ by the uniqueness theorem (chap.15, theorem 2.2) we have $u = 0$ and (4.10) becomes

(4.13) $$0 = w - \gamma(w - v) \qquad \text{in } \Omega_{\rho+1}$$

consequently

$$v = 0 \qquad \text{for } |x| < \rho + 1/3$$

and thus

(4.14) relation (4.12) holds for $|x| < \rho + 1$. (even in B).

Let us study the function $\tilde{w} = w - v$, defined in $|x| < \rho + 1$. By (4.7) and (4.8),

(4.15)$_1$ $$(-\Delta + \lambda_0)\tilde{w} = 0 \qquad \text{in } \Omega_{\rho+1}$$

(4.15)$_2$ $$\tilde{w}\big|_{|x| = \rho+1} = 0 \qquad \text{on } |x| = \rho + 1$$

Let us multiply (4.15)$_1$ by $\bar{\tilde{w}}$; by integrating it by parts we have

(4.16) $$\lambda_0 \int_{\Omega_{\rho+1}} \tilde{w}\,\bar{\tilde{w}}\,dx + \int_{\Omega_{\rho+1}} \frac{\partial\tilde{w}}{\partial x_i}\,\frac{\partial\bar{\tilde{w}}}{\partial x_i}\,dx = -\int_S n_i\,\frac{\partial\tilde{w}}{\partial x_i}\,\bar{\tilde{w}}\,dS$$

where n is the unit outer normal to S. Moreover, in B, by (4.13), $\tilde{w} = w$ and by (4.5), taking into account that the support of g is in $\Omega_\rho$, we have

$$(4.17) \qquad -(\Delta + \omega^2)\tilde{w} = 0 \qquad \text{in } B$$

Again, let us multiply (4.17) by $\bar{\tilde{w}}$, by integrating it by parts, we have :

$$(4.18) \qquad -\omega^2 \int_B \tilde{w}\,\bar{\tilde{w}}\,dx + \int_B \frac{\partial \tilde{w}}{\partial x_i}\frac{\partial \bar{\tilde{w}}}{\partial x_i}\,dx = \int_S n_i \frac{\partial \tilde{w}}{\partial x_i}\bar{\tilde{w}}\,dS$$

Moreover, note that, by (4.13), in a neighbourhood of S, $\tilde{w} = w$ is a function of class $H^2$ and thus the traces $\tilde{w}_{|S}$, $\frac{\partial \tilde{w}}{\partial x_i}\Big|_S$ are well determined elements of $H^{3/2}(S)$ and $H^{1/2}(S)$ respectively, and (4.16) and (4.18) make sense. By adding them, the right hand sides cancel, and by taking the imaginary part we have :

$$\operatorname{Im} \lambda_0 \int_{\Omega_{\rho+1}} |\tilde{w}|^2\,dx = 0 \Rightarrow \tilde{w} = 0 \quad \text{in } \Omega_{\rho+1}$$

Then, by (4.12), we have $w = 0$ in $\Omega_{\rho+1}$ and (4.5) gives $g = 0$ (note that the support of g is in $\Omega_\rho$). Q.E.D.■

The proposition 4.2 furnishes the solution of the problem 4.1 for real $\omega$. This shows in particular that $(I + T(\omega))$ is invertible for $\omega$ real. By applying chap. 15 , theorem 7.1, we see that $(I + T(\omega))^{-1}$ is a meromorphic function on the complex plane with values in $\mathcal{L}(L_2(\Omega_{\rho+1}), L_2(\Omega_{\rho+1}))$. The poles of this function are such that there exists a $g \neq 0$ for $f = 0$, and we can construct solutions of problem 4.1 by using w and v as in the proof at proposition 4.2. The corresponding values of $\omega$ are the scattering frequencies. Problem 4.1 is completely solved.

If we consider $\partial u/\partial n_{|S} = 0$ instead of (4.2), we have the Neumann problem, and the preceeding considerations hold without modification.

The transmission problem is a little different. We shall study it now. We consider again the piece wise constant functions b(x) and $a_{ij}(x)$ given by (1.1), (1.2). Let f be a function of $L_2(R^3)$ with compact support. We search for a function u defined on $R^3$ which satisfies :

$$(4.19) \qquad -\frac{\partial}{\partial x_i}\left(a_{ij}\frac{\partial u}{\partial x_j}\right) - \omega^2 b\,u = f$$

(in the regions where $a_{ij}$, b are constant)

$$(4.20) \qquad \left[n_i a_{ij}\frac{\partial u}{\partial x_j}\right] = 0 \quad ; \quad [u] = 0 \quad \text{on } \Gamma$$

(where $\Gamma$ are the smooth surfaces of discontinuity of $a_{ij}$, b)

(4.21) u satisfies the outgoing radiation condition.

We then consider

Problem 4.2 (Transmission) - Study of the existence and uniqueness of u satisfying (4.19) - (4.21) for real positive $\omega$ and the complex values of $\omega$ for which $u \neq 0$ exists with $f = 0$.

We first consider the case where $\omega$ is real and positive. The uniqueness is easily proved as in chapter 15, theorem 2.3 (note that u and $n_i \, a_{ij} \frac{\partial u}{\partial x_i}$ are Cauchy's data for the elliptic equation (4.19)).

To study the existence, we choose $\rho$ in such a way that the ball $|x| < \rho$ contains B and the support of f. Then, for $g \in L_2(|x| < \rho + 1)$ we construct w as in (4.4). Consequently, we have (4.5) and (4.6).

Next, we construct $v \in H^1(|x| < \rho + 1)$ satisfying the following conditions:

(4.22) $$v|_{|x| = \rho+1} = w|_{|x| = \rho+1}$$

(4.23) $$\int_{|x| < \rho+1} a_{ij} \frac{\partial v}{\partial x_i} \frac{\partial \theta}{\partial x_j} \, dx + \lambda_0 \int_{|x| < \rho+1} v \, \theta \, dx =$$
$$= \int_{|x| < \rho+1} - \Delta w \, \theta \, dx + \lambda_0 \int_{|x| < \rho+1} w \, \theta \, dx$$

for any $\theta \in H^1_o(|x| < \rho + 1)$

Note that (4.22), (4.23) is a non homogeneous boundary value problem ; because $w \in H^2(|x| < \rho + 1)$ it is easily transformed into a homogeneous problem with a modified right-hand side in (4.23). By taking as before $\operatorname{Im} \lambda_0 > 0$, the solution v exists and is unique. Moreover, by interior regularity theory, $v \in H^2(\rho + 1/3 < |x| < \rho + 2/3)$. Note also that v satisfies the transmission conditions

(4.24) $$[\, n_i \, a_{ij} \frac{\partial v}{\partial x_j} \,] = 0 \quad ; \quad [\, v \,] = 0 \quad \text{on } \Gamma \, .$$

Let us now search for the solution u under the form (4.10) in $R^3$, with $\gamma$ defined as before. For $|x| > \rho + 1$ we have $u = w$ and then (4.21) is satisfied. Moreover, for $|x| < \rho$, $u = v$ and (4.24) shows that (4.20) is satisfied. Consequently, we only have to impose (4.19) in order to obtain the solution u, i.e. :

(4.25) $$f = - a_{ij} \frac{\partial^2 u}{\partial x_i \, \partial x_j} - \omega^2 \, b \, u$$

We shall calculate (4.25) in B and in $\Omega_{\rho+1} = \{x \; ; \; |x| < \rho + 1\} - B$

separately (note that for $|x| \geqslant \rho + 1$ this relation is automatically satisfied).

In B, in the regions where $a_{ij}$, b are constant, (4.23) gives

$$- a_{ij} \frac{\partial^2 v}{\partial x_i \, \partial x_j} + \lambda_o \, v = - \Delta \, w + \lambda_o \, w \tag{4.26}$$

Moreover, u = v and (4.25), gives (taking into account (4.26) and (4.5) :

$$f|_B = g + \omega^2 \, w + \lambda_o (w - v) - \omega^2 \, b \, v \tag{4.27}$$

In $\Omega_{\rho+1}$, $a_{ij} = \delta_{ij}$ ; b = 1, and (4.25) becomes

$$f_{|\Omega_{\rho+1}} = - \Delta u - \omega^2 u \tag{4.28}$$

and we obtain exactely the expression (4.11) in $\Omega_{\rho+1}$

We then consider f in $|x| < \rho + 1$ ; (4.27) in B and (4.11) in $\Omega_{\rho+1}$, become

$$f = g + (\Delta \, \gamma + \omega^2 \gamma)(w - v) + \frac{\partial \gamma}{\partial x_i} \frac{\partial (w - v)}{\partial x_i} + \gamma \lambda_o (w - v) - \omega^2 (b - 1) v \tag{4.29}$$

$$\text{in } |x| < \rho + 1$$

which we write in the form :

$$f = g + T^1(\omega) g \tag{4.30}$$

and we have

<u>Proposition 4.3</u> - The operator $T^1(\omega)$ is compact in $L_2(|x| < \rho + 1)$, and holomorphic for $\omega \in C$.

The proof at this proposition is entirely analogous to that of proposition 4.1.

<u>Proposition 4.4</u> - For real positive $\omega$ , f = 0 implies g = 0. Consequently, for $f \in L_2(|x| < \rho)$ (and thus $\in L_2(|x| < \rho + 1)$) there exists a unique $g \in L_2(|x| < \rho + 1)$ which solves the problem 4.2.

<u>Proof</u> - f = 0 implies, by uniqueness, u = 0 and (4.10) becomes

$$0 = w - \gamma(w - v) \tag{4.31}$$

and for $|x| < \rho + 1/3$, 0 = v ; but in the region where v is not zero, we have $a_{ij} = \delta_{ij}$ , b = 1, and (4.22) may be written :

$$(- \Delta + \lambda_o) \, (v - w) = 0 \qquad \text{in } |x| < \rho + 1$$

and this with Im $\lambda_o > 0$ and (4.22) gives

$$v - w = 0 \qquad \text{in } |x| < \rho + 1$$

Consequently, (4.31) gives $w = 0$ for $|x| < \rho + 1$ and (4.5) shows that $g = 0$, Q.E.D.■

The scattering frequencies are obtained as in the Dirichlet and Neumann problems (see the underlined considerations after the proof of proposition 4.2).

5.- Bibliographical notes - The limiting absorption method may be found in Eidus [1], Guillot and Wilcox [1], Wilcox [2] and others. It is the starting point to study the spectral properties of the boundary value problems, as in sect. 2. The energy decay (sect. 3) may be seen in Lax and Phillips [1], Morawetz, Ralston and Strauss [1] and Wilcox [1]. For the associated property of limiting amplitude see Ladyzhenskaya [1]. Sect. 4 is inspired from Majda [1] and furnishes in a simple way the solution of the diffraction problem as well as the scattering frequencies (by using analytic continuation, as in chapter 15, sect. 7) without studying the scattering matrix (as in Lax and Phillips [1], Shenk and Thoe [2]).

## CHAPTER 17

# SCATTERING PROBLEMS DEPENDING ON A PARAMETER

This chapter deals with scattering problems in the general framework of the preceeding chapters (15 and 16) which depend on a parameter $\varepsilon$ in a singular manner, i.e., some qualitative property of the problem is different in the cases $\varepsilon > 0$ and $\varepsilon = 0$.

In sect. 1 we study the problem of the acoustic resonator. Mathematically, we deal with a domain $\Omega_\varepsilon$ which is connected and contains a neighbourhood of infinity for $\varepsilon > 0$ ; it has scattering frequencies ; for $\varepsilon = 0$ the domain is not connected and is formed by a bounded domain and an unbounded domain, it has eigenfrequencies and scattering frequencies. As $\varepsilon \searrow 0$, the eigenfrequencies of the limit problem appear as limits (in some sense) of scattering frequencies corresponding to $\varepsilon > 0$.

Analogous questions are considered in sect. 2 for the problem of the vibrations of an elastic, bounded body immersed in a gas (air) with small density $\varepsilon$. For $\varepsilon > 0$ the problem has scattering frequencies which tend in some sense to the eigenfrequencies of the problem for $\varepsilon = 0$ (body in the vacuum).

In sect. 3 we consider diffraction problems with narrow screens with either small or large rigidity, in the framework of the stiff problems considered in chap. 13, sect. 3 and 4. It appears that the limiting process of chapter 13 also applies to the diffraction case.

Section 4 contains some comments and bibliographical notes.

## 1.- Scattering frequencies for acoustic resonators.

We consider the Dirichlet problem in an exterior domain $\Omega^\varepsilon$ ($\varepsilon > 0$) of $R^3$ as $\varepsilon \searrow 0$. The domain tends to a non-connected domain $\Omega^0$ formed by an exterior region $\Omega^0_e$ and an interior one $\Omega^0_i$. The scattering frequencies for $\varepsilon > 0$ tend in a sense to be defined in the sequel, to the set formed by the eigenfrequencies of

$\Omega_i^0$ and the scattering frequencies of $\Omega_e^0$ .

In acoustics, the unknown function u is the velocity potential (or the pressure) and the Dirichlet boundary condition is associated with a soft wall. The more realistic case of a rigid wall leads to a Neumann condition ; the corresponding problem may be seen in Beale [1] . Roughly speaking, $\Omega^\varepsilon$ is formed by a bounded cavity, an exterior region and a small "hole" joining them (see figure 1 later). This is in fact the "Helmholtz resonator". The result of the following study amounts to saying that the scattering frequencies of $\Omega^\varepsilon$ are nearly the scattering frequencies of $\Omega_e^0$ and the eigenfrequencies of $\Omega_i^0$. The last are real, and consequently $\Omega^\varepsilon$ has scattering frequencies $\omega$ with very small imaginary part. The corresponding solutions of the wave equation in $\exp(-i\omega t)$ are thus very slowly damped and resonance phenomena may arise. In particular, the corresponding vibrations may be heared.

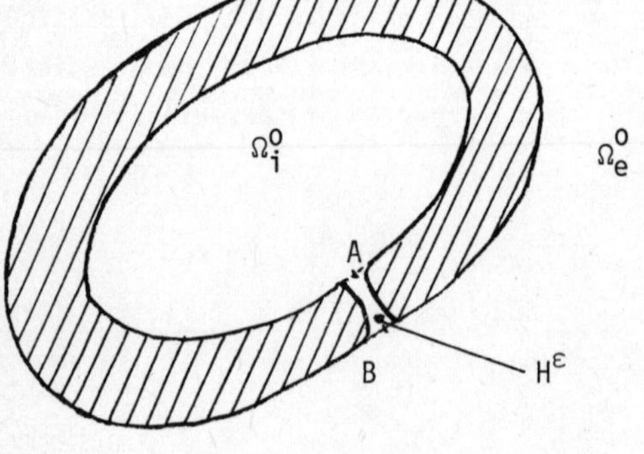


fig. 1

We consider an outer domain $\Omega_e^0$ of $R^3$ with smooth boundary $\partial\Omega_e^0$, moreover, let $\Omega_i^0$ be a bounded connected domain with smooth boundary $\partial\Omega_i^0$ contained in the complement of $\Omega_e^0$ (see fig. 1). Then, let $\Omega^\varepsilon$ be the outer connected domain (non-shaded in fig. 1) formed by $\Omega_e^0$ , $\Omega_i^0$ and a region $H^\varepsilon$ (the hole !) contained in a cylinder with radius $\varepsilon$ and axis A, B, where A and B are points of $\partial\Omega_i^0$ and $\partial\Omega_e^0$ resp. The whole boundary $\partial\Omega^\varepsilon$ of $\Omega^\varepsilon$ is assumed to be smooth. We then consider the problem :

(1.1) $$-\Delta u^\varepsilon - \omega^2 u^\varepsilon = f \quad \text{in } \Omega^\varepsilon$$

(1.2) $$u^\varepsilon = 0 \quad \text{on } \partial\Omega^\varepsilon$$

(1.3) $u^\varepsilon$ satisfies the radiation condition.

It is clear that, for fixed $\varepsilon > 0$, if $\omega$ is not a scattering frequency of the problem, the solution of (1.1) - (1.3) exists and is unique for given

$f \in L^2(\Omega^\varepsilon)$ with compact support. If $\omega$ is a scattering frequency, a non-zero solution $u^\varepsilon$ exists with $f = 0$. It is useful to give a formulation of this problem of the type of Problem 1.1 of chap. 16.

Dirichlet problem in $\Omega^\varepsilon$ - Find $u^\varepsilon \in H^1_{loc}(\Omega^\varepsilon)$ with $u^\varepsilon\big|_{\partial\Omega^\varepsilon} = 0$ such that for $|x|$ sufficiently large, it may be written in the form

$$(1.4)\qquad u^\varepsilon(x) = \int_{|y|=\rho} \left(\frac{\partial u^\varepsilon(y)}{\partial R}\,\psi^+ - u^\varepsilon(y)\,\frac{\partial \psi^+}{\partial R}\right) dS_y$$

where $R = |y|$, $\psi^+ = \frac{-1}{4\pi}\frac{e^{i\omega r}}{\gamma}$ ; $r = |x - y|$

and $\rho$ is any constant such that $\partial\Omega^\varepsilon$ and the support of f are contained in the ball $\{\,|x| < \rho\}$. Moreover, $u^\varepsilon$ must satisfy, for sufficiently large $\rho$ :

$$(1.5)\qquad \int_{\Omega^\varepsilon_\rho} \frac{\partial u^\varepsilon}{\partial x_i}\frac{\partial \bar v}{\partial x_i}\,dx - \omega^2 \int_{\Omega^\varepsilon_\rho} u^\varepsilon\,\bar v\,dx - \int_{|x|=\rho} \frac{\partial u^\varepsilon}{\partial n}\,\bar v\,dS = \int_{\Omega^\varepsilon_\rho} f\,\bar v\,dx$$

for any test function $v \in H^1_{loc}(\Omega^\varepsilon_\rho)$ with $v\,\big|_{\partial\Omega^\varepsilon} = 0$. Here, the domain $\Omega^\varepsilon_\rho$ denotes $\Omega^\varepsilon \cap \{|x| < \rho\}$.

Lemma 1.1 - In the hole $H^\varepsilon$ the following Dirichlet-like inequality holds for any $v \in H^1_{loc}(\Omega^\varepsilon)$ with $v\big|_{\partial\Omega^\varepsilon} = 0$ :

$$(1.6)\qquad \int_{H^\varepsilon} |v|^2\,dx \leq C\,\varepsilon^2 \int_{H^\varepsilon} |\mathrm{grad}\, v|^2\,dx$$

Proof - Under a diffeomorphism $x \to y$ which only modifies the constant C, the domain $H^\varepsilon$ becomes (see fig. 2) a domain contained in the region $|y_3| < b$, $0 < y_1$ , $y_2 < a\varepsilon$

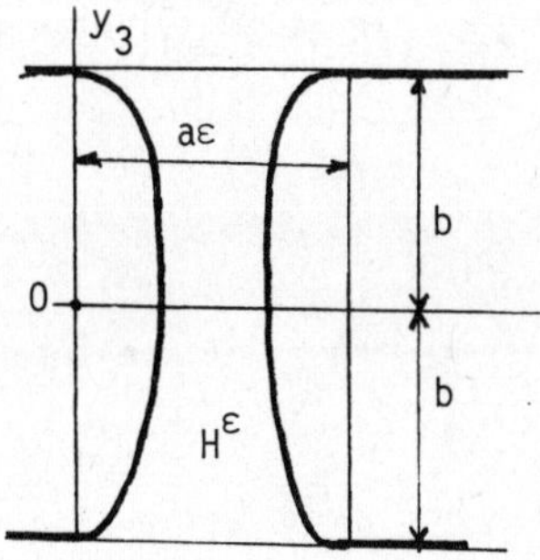


fig. 2

We continuate v with value zero out of $\Omega^\varepsilon$, and then, $v(0\,,\,y_2,\,y_3) = 0 \implies$

$$|v(y_1,y_2,y_3)|^2 = \left|\int_0^{y_1} \frac{\partial v}{\partial y_1}(\alpha , y_2 , y_3)\, d\alpha \right|^2 \leqslant$$

$$\leqslant a\,\varepsilon \int_0^{a\varepsilon} |\text{grad } v|^2 \, dy_1$$

and by integrating with respect to $y_1$ :

$$\int_0^{a\varepsilon} |v|^2 \, dy_1 \leqslant a^2 \varepsilon^2 \int_0^{a\varepsilon} |\text{grad } v|^2 \, dy_1$$

and (1.6) is obtained by integrating with respect to $y_2$, $y_3$. ■

Theorem 1.1 - Let $\omega^*$ be an accumulation point of scattering frequencies $\omega(\varepsilon)$ of the Dirichlet problem in $\Omega^\varepsilon$ as $\varepsilon \searrow 0$. Then, $\omega^*$ is either an eigenfrequency of the Dirichlet problem in $\Omega_i^0$ (if $\omega^*$ is real) or a scattering frequency of the Dirichlet problem in $\Omega_e^0$ (if $\omega^*$ is not real).

Proof - The reasoning that follows recalls that of chapter 16, lemma 1.3. We fix $\rho$ in such a way that $|x| < \rho + 1$ contains $\partial\Omega^\varepsilon$ . We write (1.4), (1.5) for $\omega = \omega(\varepsilon)$, $f = 0$. The corresponding scattering functions are $u^\varepsilon$ , and we normalize them in $L^2(\Omega_\rho^\varepsilon + 1)$

(1.7)
$$\|u^\varepsilon\|_{L^2(\Omega_\rho^\varepsilon+1)} = 1$$

(1.8)
$$\int_{\Omega_\rho^\varepsilon} \frac{\partial u^\varepsilon}{\partial x_i}\frac{\partial \bar v}{\partial x_i}\, dx - \omega(\varepsilon)^2 \int_{\Omega_\rho^\varepsilon} u^\varepsilon \bar v \, dx - \int_{|x|=\rho} \frac{\partial u^\varepsilon}{\partial \eta}\, \bar v \, ds = 0$$

From (1.7), after extracting a subsequence (again noted $\varepsilon$ ),

(1.9)
$$u^\varepsilon\Big|_{\Omega^0} \longrightarrow u^* \quad \text{in } L^2(\Omega^0_{\rho+1}) \text{ weakly.}$$

Moreover, from (1.8), $u^\varepsilon$ satisfies :

$$-\Delta u^\varepsilon = \omega(\varepsilon)^2 u^\varepsilon$$

and by standard elliptic regularity theory at the interior of the domain :

$$\|u^\varepsilon\|_{H^2(\rho-1/2<|x|<\rho+1/2)} \leqslant C'[\, \|\Delta u^\varepsilon\|_{L^2(\Omega^\varepsilon_{\rho+1})} + \|u^\varepsilon\|_{L^2(\Omega^\varepsilon_{\rho+1})}] \leqslant C$$

and by the trace theorem (after extracting subsequences).

(1.10)
$$u^\varepsilon \longrightarrow u^* \quad \text{in } H^2(\rho - 1/2 < |x| < \rho + 1/2) \text{ weakly}$$

(1.11)
$$u^\varepsilon\Big|_{|x|=\rho} \longrightarrow u^*\Big|_{|x|=\rho} \quad \text{in } H^{3/2}(|x| = \rho) \text{ weakly, } L^2 \text{ strongly}$$

(1.12) $\left.\frac{\partial u^{\varepsilon}}{\partial \eta}\right|_{|x|=\rho} \to \left.\frac{\partial x^{*}}{\partial \eta}\right|_{|x|=\rho}$ in $H^{1/2}(|x|=\rho)$ weakly, $L^2$ strongly

and by passing to the limit $\varepsilon \to 0$ in (1.4) we see that $u^*$ satisfies (1.4) and

(1.13) $u^{\varepsilon} \to u^{*}$ in $C^0(\rho < |x| < \rho + 1)$

On the other hand, by taking $v = u^{\varepsilon}$ in (1.8), with (1.11), (1.12), we obtain :

(1.14) $\|u^{\varepsilon}\|_{H^1(\Omega_\rho^\varepsilon)} \leqslant C$

(1.15) $\left.u^{\varepsilon}\right|_{\Omega^0} \to u^{*}$ in $H^1(\Omega_\rho^0)$ weakly

Moreover, from lemma 1.1 and (1.14) :

(1.16) $\|u^{\varepsilon}\|_{L^2(H^\varepsilon)} \leqslant C\,\varepsilon$

and from (1.13), (1.15)

(1.17) $\left.u^{\varepsilon}\right|_{\Omega^0} \to u^{*}$ in $L^2(\Omega_{\rho+1}^0)$ strongly

but (1.16), (1.7) and (1.17) show that $u^*$ has a norm in $L^2(\Omega_\rho^0)$ equal to one, and consequently

(1.18) $u^{*} \neq 0$

On the other hand, from the fact that $u^{\varepsilon}$ is zero on $\partial\Omega^{\varepsilon}$ and (1.15), we obtain

(1.19) $u^{*} \in H^1_{loc}(\Omega^0) \quad ; \quad \left.u^{*}\right|_{\partial\Omega^0} = 0$

Now, for fixed $v \in H^1_{loc}(\Omega_\rho^0)$ (with value zero on $\partial\Omega^0$), we pass to the limit $\varepsilon \searrow 0$ in (1.5) ($v$ is continuated by zero on $H^{\varepsilon}$). We obtain

$$\int_{\Omega_\rho^\varepsilon} = \int_{\Omega_\rho^0} + \int_{H^\varepsilon}$$

and using (1.14), (1.16) and the fact that the measure of $H^{\varepsilon}$ tends to zero, we obtain

(1.20) $$\int_{\Omega_\rho^0} \frac{\partial u^*}{\partial x_i} \frac{\partial \bar{v}}{\partial x_i}\, dx - \omega^{*2} \int_{\Omega_\rho^0} u^* \bar{v}\, dx - \int_{|x|=\rho} \frac{\partial u^*}{\partial \eta} \bar{v}\, dx = 0$$

We then have two possibilities :

a) $\left.u^{*}\right|_{\Omega_i^0} \not\equiv 0$ and $\left.u^{*}\right|_{\Omega_e^0} \equiv 0$ . Then, $\omega^*$ is an eigenfrequency of $\Omega_i^0$ (and $\omega^*$ is real).

b) $u^* |_{\Omega_i^0} \equiv 0$ and $u^* |_{\Omega_e^0} \not\equiv 0$ . Then, $\omega^*$ is a scattering frequency of $\Omega_e^0$ (and $\omega^*$ is not real).

Of course, $u^*$ non zero in $\Omega_i^0$ and $\Omega_e^0$ is impossible ($\omega^*$ should be both real and non-real !). Theorem 1.1 is proved. ■

Now we prove a result which is in some sense the converse of theorem 1.1.

Theorem 1.2 - Let $\omega^*$ be either an eigenfrequency of the Dirichlet problem in $\Omega_i^0$ or a scattering frequency of the Dirichlet problem in $\Omega_e^0$. Moreover, let $\gamma$ be a simple closed curve enclosing $\omega^*$ and no other eigenfrequency or scattering frequency.

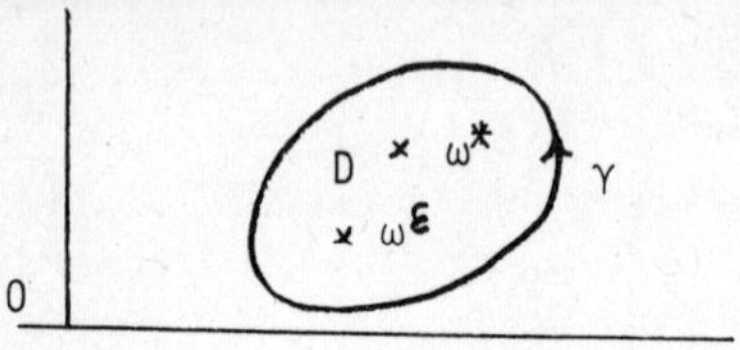


fig. 3

Let D be the region of the plane enclosed by $\gamma$ . Then, for sufficiently small $\varepsilon$ , there exists at least a scattering frequency of $\Omega^\varepsilon$ contained in $\overline{D} \equiv D \cup \gamma$.

Before proving this theorem, we note that the eigenfrequencies of $\Omega_i^0$ as well as the scattering frequencies of $\Omega_e^0$ are isolated points of the complex plane ; consequently $\gamma$ always exists and may be taken as small as desired. The theorem shows that $\omega^*$ is an accumulation point of scattering frequencies $\omega^\varepsilon$ of $\Omega^\varepsilon$ with $\varepsilon \searrow 0$.

Now we are going to prove theorem 1.2. We prove it in the case where $\omega^*$ is a scattering frequency of $\Omega_e^0$ (the case $\omega^*$ = eigenfrequency of $\Omega_i^0$ is of course analogous).

Let us choose $f \in L^2(\Omega_e^0)$ with bounded support such that the problem

$$\begin{cases} (-\Delta \ -\sigma^2)v = f & \text{in } \Omega_e^0 \\ v \text{ satisfies the radiation condition} & \\ v = 0 & \text{on } \partial\Omega_e^0 \end{cases}$$

has a singularity for $\sigma = \omega^*$ in the standard way (see chapter 16, sect. 4). Then, for $\sigma \in \overline{D}$ but different from $\omega^*$, we construct $v(\sigma)$ and we prolongate v and f with zero values to $\Omega_i^0$. It then satisfies

$$(1.21) \qquad \begin{cases} (-\Delta \ -\sigma^2)v = f & \text{in } \Omega^0 \\ v \text{ satisfies the radiation condition} & \\ v = 0 & \text{on } \partial\Omega^0 \end{cases}$$

Of course $\omega^*$ is the only singularity (pole) of $v(\sigma)$ for $\sigma \in \overline{D}$. Now we prove the theorem by contradiction. If the conclusion does not hold, there exists a sequence

$$\varepsilon_i \searrow 0 \quad ; \quad i = 1, 2, 3 \dots$$

such that the problem

$$(1.22) \quad \begin{cases} (-\Delta - \sigma^2)\, u(\sigma , \varepsilon_i) = f & \text{in } \Omega^\varepsilon \\ u(\sigma , \varepsilon_i) \text{ satisfies the radiation condition} \\ u(\sigma , \varepsilon_i) = 0 & \text{on } \partial\Omega^\varepsilon \end{cases}$$

has no singularity for $\sigma \in \overline{D}$ and consequently $u(\sigma , \varepsilon_i)$ is well defined for $\sigma \in \overline{D}$, holomorphic with values in $H^1_{loc}(\Omega^\varepsilon)$ (for fixed $\varepsilon_i$).

Now we fix $\rho$ sufficiently large (this will be described in a more precise manner later), and we have

Lemma 1.2 - In the preceeding conditions, there exists a constant C such that, for $\sigma \in \gamma$ and i = 1, 2, ...

$$(1.23) \qquad \| u(\sigma , \varepsilon_i) \|_{L^2(\Omega^\varepsilon_\rho)} \leq C$$

Proof - If (1.23) does not hold, for a subsequence (again noted i = 1, 2, 3, ...) and some $\sigma_i$ we will have

$$\| u(\sigma_i , \varepsilon_i) \|_{L^2(\Omega^i_\rho)} = m_i \nearrow \infty$$

and after normalizing them in the standard way (we consider $u/m_i$ instead of u) we have

$$(1.24) \qquad \| u(\sigma_i , \varepsilon^i) \|_{L^2(\Omega^i_\rho)} = 1$$

which satisfies (1.22) with $f/m_i$ instead of f. Moreover, we may suppose

$$\sigma_i \to \sigma^* \in \gamma \quad (\gamma \text{ is compact !})$$

and we pass to the limit $i \to \infty$ in (1.22) (with $f/m_i$ instead of f) exactly as in theorem 1.1 (the only difference is the supplementary term

$$\int_{\Omega^\varepsilon} \frac{f}{m_i} \overline{w}\, dx$$

at the right hand side, which tends to zero as $m_i \nearrow \infty$). We then see that $\sigma^* \in \gamma$ is either a scattering frequency of $\Omega^0_e$ or a eigenfrequency of $\Omega^0_i$, and this is impossible by construction of $\gamma$ . We then have (1.23).■

Lemma 1.3 - For fixed $\sigma \in \gamma$, the solution $u(\sigma , \varepsilon_i)$ of (1.22) converges as $i \to \infty$ to the solution $v(\sigma)$ of (1.21) in the following sense :

$$u(\sigma, \varepsilon_i)\Big|_{\Omega^0_\rho} \to v(\sigma)\Big|_{\Omega^0_\rho} \quad \text{in } L^2(\Omega^0_{\rho+1}) \text{ strongly.}$$

Proof - It is again analogous to that of theorem 1.1 but with the supplementary right hand side

$$\int_{\Omega^\varepsilon} f\, w\, dx \quad . \blacksquare$$

Now, we finish the proof of the theorem 1.2. Because $v(\sigma)$, the solution of (1.21), has an isolated singularity for $\sigma = \omega^*$, the corresponding Laurent series has a non-zero term in $(\sigma - \omega^*)^{-m}$ for some $m \geq 1$. Consequently, the function $(\sigma - \omega^*)^{m-1}\, v(\sigma)$ has a non-zero term in $(\sigma - \omega^*)^{-1}$ and the integral

$$\int_\gamma (\sigma - \omega^*)^{m-1}\, v(\sigma)\, d\sigma = a \neq 0$$

where $a \in L^2_{loc}(\Omega^0)$. In particular, for sufficiently large $\rho$,

$$\int_\gamma (\sigma - \omega^*)^{m-1}\, v(\sigma)\Big|_{\Omega^0_\rho} d\sigma = a\Big|_{\Omega^0_\rho} \neq 0 \tag{1.25}$$

On the other hand, $v(\sigma, \varepsilon_i)$ has no singularity on $\overline{D}$, and consequently

$$\int_\gamma (\sigma - \omega^*)^{m-1}\, v(\sigma, \varepsilon_i)\, d\sigma = 0 \tag{1.26}$$

We take the restriction of (1.26) to $\Omega^0_\rho$ and we pass to the limit $i \to \infty$ (this is possible by dominated convergence, by virtue of lemmas 1.2 and 1.3) and we have a contradiction with (1.25). Theorem 1.2 is proved.

Remark 1.1 - In this section, the radiation condition was written under the form (1.4). On the other hand, if we study the scattering frequencies in the framework of the "reduction to a problem in a bounded domain" (chap. 16, sect. 4) the radiation condition was written under the form

$$u^\varepsilon(x) = \int_{\text{support } g} g(y)\, \psi^+\, dy \tag{1.27}$$

where g is a function with bounded support (this is true even for the scattering frequencies). The two forms (1.4) and (1.27) are equivalent for real $\omega$ (see chap. 15, proposition 3.2). Then, by taking g fixed (as well as $\varepsilon$, of course), we see that $u^\varepsilon$, is a holomorphic function of $\omega$ with values in $H^2_{loc}$. It is then easily seen that $u^\varepsilon$ has the two representations (1.4) and (1.27). ■

## 2. - Vibration of an elastic body surrounded by a gas of small density.

We consider an elastic bounded body $\Omega_1$ which is immersed into a gas (air) the density of which is small ($\varepsilon$) with respect to that of the body. If the gas does not exist (i.e., $\varepsilon = 0$, body in the vacuum), the vibration problem is associated with an elliptic system in a bounded domain, and the classical eigenfrequencies and eigenfunctions of the vibration exist. On the other hand, for $\varepsilon > 0$, the vibration of the body is coupled with that of the surrounding air, and we deal with elliptic problems in exterior domains and eigenfrequencies does not exist. Instead of this, for $\varepsilon > 0$ we have scattering frequencies.

We study here the limit process $\varepsilon \to 0$. This problem presents some analogy with that of the preceeding section. As $\varepsilon$ tends to zero, the set of scattering frequencies tends to the eigenfrequencies of the body (and also to the scattering frequencies corresponding to the vibration of air with the solid at rest).

A study of the coupled vibration problem in terms of the displacement vector may be seen in Hachem [1]. Here we study directly the scattering frequencies following, with some modifications, the study of Sanchez-Palencia [6].

We consider a bounded connected domain $\Omega_1$ of $R^3$ with smooth boundary $\Gamma \cup S$ (see figure 4) where S is an outer surface

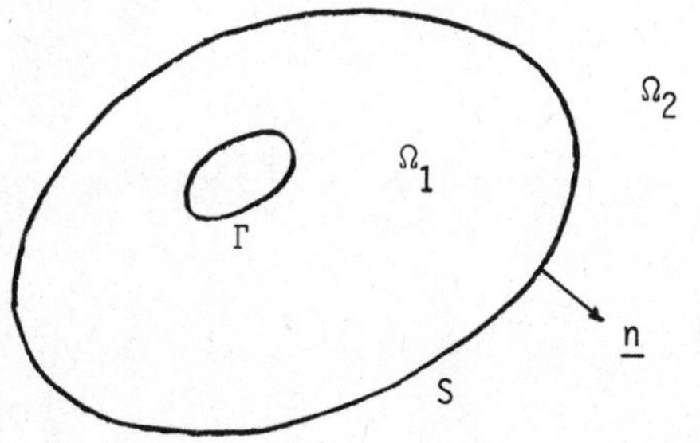


fig. 4

and the "air" fills the outer domain $\Omega_2$ (which contains a neighbourhood of infinity). The body $\Omega_1$ is supposed to be fixed by the inner boundary $\Gamma$ (in order to disregard solid displacements of the body ; but the role of $\Gamma$ is not essential).

We consider a rest state for the body and for the air, and we then study the vibrations of the body and the air, linearized with respect to this rest position. We write the equations and boundary conditions in a non-dimensional form. The displacement vector in the body $\Omega_1$ is denoted by $\underline{u}$ and the velocity potential in the gas $\Omega_2$ is $\phi$ ; $\underline{\text{grad}}\ \phi$ is the velocity vector in $\Omega_2$. The equations to be studied are ($a_{ijkh}$ are the classical elasticity coefficients, see chap. 6)

(2.1) $$\frac{\partial^2 u_i}{\partial t^2} = \frac{\partial \sigma_{ij}}{\partial x_j} \qquad \text{in } \Omega_1 \ , \text{ where}$$

$$\sigma_{ij} = a_{ijkh}\, e_{kh}(\underline{u}) \qquad ; \qquad e_{kh}(\underline{u}) \quad \frac{1}{2}\left(\frac{\partial u_k}{\partial x_h} + \frac{\partial u_h}{\partial x_k}\right)$$

(2.2) $$\frac{\partial^2 \phi}{\partial t^2} = \Delta \phi$$

The pressure (perturbation of) in $\Omega_2$ is given by

(2.3) $$p = -\varepsilon \frac{\partial^2 \phi}{\partial t^2}$$

where $\varepsilon$ is for the (small) density of the gas. The boundary conditions at $\Gamma$ and S are

(2.4) $$\underline{u} = 0 \qquad \text{on } \Gamma$$

(2.5) $$u_i \, n_i \Big|_{\Omega_1} = \frac{\partial \phi}{\partial x_i} n_i \Big|_{\Omega_2} \qquad \text{on S}$$

(2.6) $$\boxed{\sigma_{ij} \, n_j \Big|_{\Omega_1} = - p \, n_i = \varepsilon \frac{\partial^2 \phi}{\partial t^2} n_i \Big|_{\Omega_2}} \qquad \text{on S}$$

where (2.4) express that the solid is fixed on $\Gamma$ ; (2.5) is the continuity of the normal component of velocity across S (note that the tengential component of velocity may not be continuous) ; (2.6) is the continuity of stress.

Now, we search for solutions of the form (there will be no ambiguity between the two meanings of $\underline{u}$ and $\phi$ ) :

(2.7) $$\underline{u}(x\ ,\ t) = e^{-i\omega t}\, \underline{u}(x) \quad ; \quad \phi(x\ ,\ t) = e^{-i\omega t}\, \phi(x)$$

<u>where $\phi$ satisfy a radiation condition</u>. The system of equations and boundary conditions become :

(2.8) $$-\omega^2 \, u_i = \frac{\partial \sigma_{ij}}{\partial x_j} \qquad \text{in } \Omega_1$$

(2.9) $$-\omega^2 \, \phi = \Delta \, \phi \qquad \text{in } \Omega_2$$

(2.10) $\phi$ satisfies a radiation condition

(2.11) $$\underline{u} = 0 \qquad \text{on } \Gamma$$

(2.12) $$\boxed{u_i \, n_i \Big|_{\Omega_1} = \frac{\partial \phi}{\partial x_i} n_i \Big|_{\Omega_2} = \frac{\partial \phi}{\partial n}} \qquad \text{on S}$$

(2.13) $$\sigma_{ij} \, n_j \Big|_{\Omega_1} = - \omega^2 \, \varepsilon \, \phi \, n_i \Big|_{\Omega_2} \qquad \text{on S}$$

Now, we reduce this problem to a problem in $\Omega_1$. In a first step, we consider $\phi$ known, and we establish a weak formulation of the problem. We define the space

(2.13) $$V = \{\underline{u} \ ;\ \underline{u} \in \underline{H}^1(\Omega_1) \ ;\ \underline{u}\,\big|_\Gamma = 0\ \}$$

Then, by integration by parts as in chap. 6, sect. 1, (2.8), (2.11), (2.13) amounts to find $\underline{u} \in V$ such that

$$(2.14)\begin{cases} -\omega^2(\underline{u}\ ,\ \underline{v})_{\underline{L}^2(\Omega_1)} + a(\underline{u}\ ,\ \underline{v}) = \varepsilon\omega^2 \int_S \phi\ \overline{v}_i\ n_i\ dS & \forall\ \underline{v} \in V \\ a(\underline{u}\ ,\ \underline{v}) \equiv \int_{\Omega_1} a_{ijkh}\ e_{kh}(\underline{u})\ e_{ij}(\overline{\underline{v}})dx \end{cases}$$

On the other hand, if (2.9), (2.10), (2.12) are satisfied, $\phi|_S$ is the trace of the solution of the Neumann problem in $\Omega_2$ with given data $u_i\ n_i$. It is easily seen from the resolution of such problem (chapter 15, sect. 5 and 7, in particular theorem 7.3) that the operator

$$\frac{\partial\phi}{\partial n}\bigg|_S \longmapsto \phi\bigg|_S = T(\omega)\ \frac{\partial\phi}{\partial n}\bigg|_S$$

is holomorphic of $\omega$ (for $\omega$ is a neighbourhood of each real value) with values in $\mathcal{L}(L^2(S),\ L^2(S))$. Consequently, by taking into account (2.12), the right hand side of (2.14) may be written

$$(2.15) \qquad \varepsilon\ \omega^2 \int_S T(\omega)\ (u_i\ n_i)(\overline{v}_i\ n_i)dS$$

which is a holomorphic continuous form on V by the trace theorem.

Consequently, (2.14) writes : find $\underline{u} \in V$ such that

$$(2.16) \qquad -\ \omega^2(\underline{u}\ ,\ \underline{v})_{\underline{L}^2} + b(\varepsilon\ ,\omega\ ;\ \underline{u}\ ,\ \underline{v}) = 0 \qquad \forall\ \underline{v} \in V$$

where the form b is given by

$$b(\varepsilon,\omega\ ;\ \underline{u}\ ,\ \underline{v}) \equiv a(\underline{u}\ ,\ \underline{v}) + \varepsilon\omega^2 \int_S T(\omega)(u_i\ n_i)(\overline{v}_i\ n_i)dS$$

which is a holomorphic form of $\omega$ (defined in a neighbourhood of each real value) continuous on V. Moreover, for sufficiently small $\varepsilon$ it is coercive on V (because a does, see chapter 6, sect. 1). The scattering frequencies of the problem (2.8)-(2.13) for fixed $\varepsilon > 0$ are the values $\omega$ such that there exists a non-zero solution $\underline{u}$ of (2.16). For $\varepsilon = 0$, the eigenfrequencies of the vibration of the body in vacuum are the values $\omega$ such that there exists a non-zero solution of (2.16).

The preceeding framework is suitable to study the limits of the scattering frequencies which are real (see theorems 2.1 and 2.2 hereafter). Another framework will be needed to study the non real limits (see theorem 2.3).

<u>Theorem 2.1</u> - Let $\varepsilon_i \searrow 0$ be a sequence such that there exists the corresponding scattering frequencies $\omega_i$ and scattering functions $\underline{u}^i$ such that

$$(2.17) \qquad \omega_i \longrightarrow \omega^* \quad ; \quad \omega^* \text{ real}$$

Then, $\omega^*$ is an eigenfrequency of the vibration of the body for $\varepsilon = 0$.

Proof - After normalization, we have

$$\| \underline{u}^i \|_{\underline{L}^2(\Omega_1)} = 1$$

and by using (2.16) and the coerciveness of b we see that the $\underline{u}^i$ have bounded norms in V. Then, after extracting a subsequence, we have :

(2.18) $$\underline{u}^i \longrightarrow \underline{u}^* \quad \text{in V weakly, and } \underline{L}^2(\Omega_1) \text{ strongly}$$

and in particular,

(2.19) $$\| \underline{u}^* \|_{L^2} = 1 \Longrightarrow \underline{u}^* \neq 0$$

and by passing to the limit $\varepsilon_i \searrow 0$ in (2.16) we see that $\omega^*$ is an eigenfrequency of the limit problem (and $\underline{u}^*$ is the corresponding eigenfunction). ■

Theorem 2.2 - Let $B(\varepsilon, \omega)$ and A be the operators of $\underline{L}^2(\Omega_1)$ associated with the forms $b(\varepsilon, \omega)$ and a defined by (2.16), (2.14). (They are operators with compact resolvent). Let $\omega^*$ be an eigenfrequency of A. We construct a simple closed curve $\gamma$ enclosing $\omega^*$ and no other eigenfrequency of A (moreover, $\gamma$ and its interior D are contained in the region where we form $b(\varepsilon, \omega)$ is defined for small $\varepsilon$). Then, for sufficiently small $\varepsilon$, there is at least a scattering frequency $\omega(\varepsilon)$ of (2.16) contained in $\overline{D} = D \cup \gamma$ .

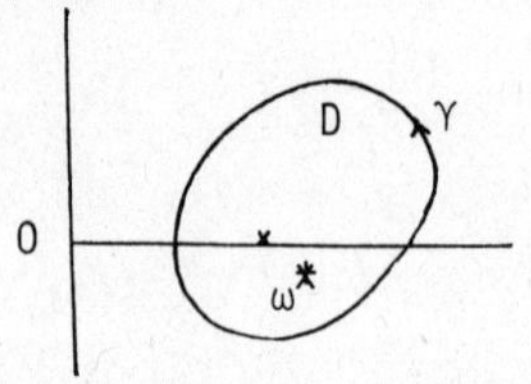


fig. 5

Proof - For any given $\underline{f} \in \underline{L}^2(\Omega_1)$, the problem

(2.20) $$- \sigma^2 \, v(\sigma) = A \, v(\sigma) + f \quad ; \quad \sigma \in \gamma$$

has a unique solution, which is a holomorphic function of $\sigma$ with values in $\underline{L}^2(\Omega_1)$. Moreover, it is holomorphic in $\overline{D}$ excepted the point $\omega^*$. We choose f such that $v(\sigma)$ has a singularity (for instance, f = an eigenfunction associated with $\omega^*$). The Laurent series of $v(\sigma)$ has at least a term (in $(\sigma - \omega^*)^{-m}$) different from zero, and consequently

(2.21) $$\int_\gamma (\sigma - \omega^*)^{m-1} \, v(\sigma) \, d\sigma = \theta \neq 0$$

where $\theta$ is a non-zero element of $\underline{L}^2(\Omega_1)$.

Now, we prove the theorem by contradiction. If the conclusion does not hold, let $v(\varepsilon , \sigma) \in V$ be the solution of

$$(2.22) \qquad - \sigma^2 v(\varepsilon , \sigma) = B(\varepsilon , \sigma) v(\varepsilon , \sigma) + f \qquad \sigma \in \gamma$$

which exists for $\varepsilon$ small (if not, there exists, for $\varepsilon_i \searrow 0$, singularities on $\gamma$, i.e. on $\overline{D}$). Moreover, for fixed $\varepsilon$, $B(\varepsilon , \sigma)$ is holomorphic of $\sigma$, and

$$v(\varepsilon , \sigma) = -\left(\sigma^2 + B(\varepsilon , \sigma)\right)^{-1} f$$

is holomorphic of $\sigma$ on $\overline{D}$. Consequently, we can write

$$(2.23) \qquad \int_\gamma (\sigma - \omega^*)^{m-1} v(\varepsilon , \sigma) d\sigma = 0$$

Then, to obtain a contradiction between (2.21) and (2.23) it suffices to prove that we can pass to the limit $\varepsilon \searrow 0$ in the integral of (2.23). This is easily seen by dominated convergence, as in sect. 1. First, $\| v(\varepsilon , \sigma)\|_{L^2}$ for $\sigma \in \gamma$, $\varepsilon \searrow 0$ remains bounded, if not, after normalizing,

$$w(\varepsilon_i , \sigma_i) = \frac{v(\varepsilon_i, \sigma_i)}{m_i} \qquad m_i = \|v(\varepsilon_i, \sigma_i)\|_{L^2} \nearrow \infty$$

are solutions of (2.22) with $(f/m_i) \to 0$ instead of f. Moreover,

$$\varepsilon_i \to 0 \quad ; \quad \sigma_i \to \sigma^* \in \gamma$$

and we have a contradiction by passing to the limit as in theorem 2.1. Secondly, for fixed $\sigma \in \gamma$ the integrand in (2.23) pass to the limit in the same way. Then we can pass to the limit in (2.23) and we have a contradiction with (2.21).■

Now, we consider another functional framework for the problem (2.8)-(2.13), suitable for the study of the scattering frequencies which tend to non-real values.

We consider again the resolution of the exterior Neumann problem for the Helmholtz equation in $\Omega_2$,

$$(2.24) \qquad \frac{\partial\phi}{\partial n} = f \qquad \text{on } S$$

where $f \in L^2(S)$ is given. According to chapter 15, theorem 7.3, using the potential method to search $\phi$ amounts to search the potential $x \in L^2(S) \times L^2(\Omega_1)$ satisfying the equation

$$(2.25) \qquad x - T^1(\omega) x = F \qquad \text{where} \qquad F = (2f , 0)$$

In problem (2.8) - (2.13) we search for a solution such that

$$f = \frac{\partial\phi}{\partial n} = u_i n_i$$

where $\underline{u} \in V$ is the solution of the problem (2.8), (2.11), (2.13) for, say, $\omega$ non real ; This boundary value problem is selfadjoint and consequently the solution for $\omega$ non-real is holomorphic with values in V ; more exactly, if $\phi \in L^2(S)$ is given, the resolution operator

$$\phi \to \underline{u}$$

of (2.8), (2.11), (2.13) is holomorphic of $\omega$ (non real) with values in $\mathcal{L}(L^2(S),V)$ and by the trace theorem, the operator

$$\phi \Longrightarrow \frac{\partial\phi}{\partial n} = u_i \; n_i \;\Big|_S$$

writes $\omega^2\ \varepsilon\ T^2(\omega)$ and is holomorphic with values in $\mathcal{L}(L^2(S),\ L^2(S))$, compact.

Then, the equation (2.25) becomes

$$x - T^1(\omega)x = (2\ \omega^2\ \varepsilon\ T^2(\omega)\ \phi\ ,\ 0) \tag{2.26}$$

On the other hand, if the potential x is given, by virtue of chapter 15, theorems 4.1, 4.2, and 7.3,

$$\phi = T^3(\omega)x \tag{2.27}$$

where $T^3(\omega)$ is holomorphic with values in $\mathcal{L}(L^2(S) \times L^2(\Omega_1),\ L^2(S))$. Then, by writting

$$T^4(\omega)x = (2\ \omega^2\ T^2(\omega)\ T^3(\omega)\ x\ ,\ 0)$$

which is holomorphic and compact from $L^2(S) \times L^2(\Omega_1)$ into itself, it follows that (2.25) writes :

$$x - [T^1(\omega) + \varepsilon\ T^4(\omega)\ ]\ x = 0 \tag{2.28}$$

and we of course search for $x \neq 0$ satisfying this equation. This problem is in the framework of chapter 15, theorem 7.2 : we take $\omega = \mu$, $\varepsilon = \nu$, and D any region of the complex plane not intersecting the real axis. We immediately obtain :

Theorem 2.3 - The scattering frequencies $\omega(\varepsilon)$ of the problem (2.8)-(2.13) depend continuously on $\varepsilon$ , and appear and desappear only at the boundary of any complex domain D not intersecting the real axis. Consequently, if $\omega^*$ is a scattering frequency of the exterior Neumann problem (i.e., a pole of (2.28) for $\varepsilon = 0$), it is an accumulation point of scattering frequencies $\omega(\varepsilon_i)$ with $\varepsilon_i \searrow 0$. Conversely, any such accumulation point is a scattering frequency of the exterior Neumann problem.

## 3. - Diffraction problems with narrow obstacles.

We consider here stiff problems of the type of chapter 13, sect. 3 and 4: transmission problems through narrow plates with either small or large conductivity in unbounded domains. (In chapter 13, sect. 3, 4 only the two-dimensional case was explicitely handled, but natural extensions to three-dimensional problems hold with minor modifications). Moreover, we study some spectral properties of the limit problem and the limiting process for diffraction problems in the framework of chapter 16, sect. 1.

The "small conductivity" case - We consider the space $R^3$ refered to the axis $Ox_1, x_2, x_3$. Let $\omega$ be a bounded domain of the plane $Ox_1, x_2$, with smooth boundary $\partial\omega$. We then define the lens-shaped domains $\Omega_\varepsilon^1$:

$$\Omega_\varepsilon^1 \equiv \{ x \; ; \; (x_1 , x_2) \in \omega \quad , \quad |x_3| \leqslant \varepsilon \, \phi(x_1 , x_2) \} \tag{3.1}$$

where $\phi(x_1 , x_2)$ is a smooth function defined on $\overline{\omega}$ with non-zero gradient on $\partial\omega$. The outer region $R^3 - \overline{\Omega}_\varepsilon{}^1$ is noted $\Omega_\varepsilon^0$.

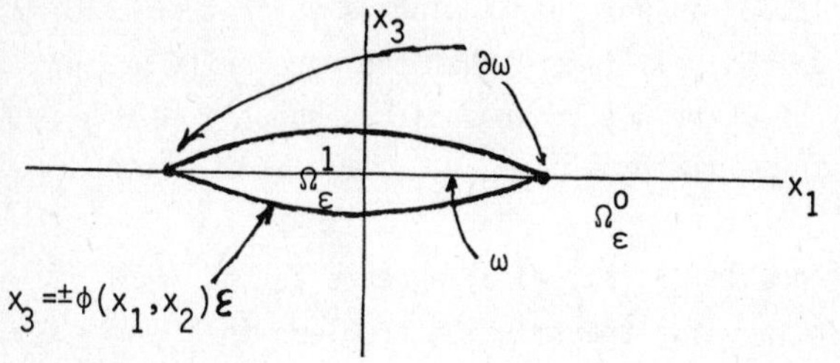


fig. 6

As in chapter 13, sect. 3, we consider the "lens" $\Omega_\varepsilon^1$ with "conductivity" $\varepsilon$ (see chapter 3 for details). Some properties of the limit problem are given by :

Theorem 3.1 - Let $f \in L^2(R^3)$ with compact support be given. We then search for $u \in H^1_{loc}(R^3 - \overline{\omega})$ such that for $|x|$ sufficiently large it can be written

$$u(x) = \int_{|y|=\rho} \left( \frac{\partial u(y)}{\partial R} \psi^+ - u(y) \frac{\partial \psi^+}{\partial R} \right) dS_y \tag{3.2}$$

where $R = |y|$ ; $\psi^+ = \frac{-1}{4\pi} \frac{e^{i\omega r}}{r}$ : $r = |x - y|$ and $\rho$ is any constant such that $\omega$ and support f are contained in the ball $\{|x| < \rho\}$. Moreover, u must satisfy

$$\int_{|x|<\rho} \frac{\partial u}{\partial x_i} \frac{\partial \overline{v}}{\partial x_i} dx + \int_\omega \frac{1}{2\phi(x_1,x_2)} [u] [\overline{v}] \, dx_1 \, dx_2 - \tag{3.3}$$

$$- c^2 \int_{|x|<\rho} u \, \overline{v} \, dx - \int_{|x|=\rho} \frac{\partial u}{\partial n} \overline{v} \, dS = \int_{|x|<\rho} f \, \overline{v} \, dx \quad ; \; c > 0$$

where $[u] = u\big|_{\omega^+} - u\big|_{\omega^-}$, for any $v \in H^1(\{ |x| < \rho \} - \overline{\omega})$.

(3.4) Then, u exists and is unique.

Let A be the selfadjoint operator of $L^2(R^3)$ associated with the form

$$\int_{R^3} \frac{\partial u}{\partial x_i} \frac{\partial \overline{v}}{\partial x_i} dx + \int_\omega \frac{1}{2\phi(x_1,x_2)} [u] [\overline{v}] \, dx_1 \, dx_2$$

(3.5) Then, A has an absolutely continuous spectrum

Moreover, if u(t) is the solution of the evolution problem

(3.6)
$$\begin{cases} \dfrac{d^2 u(t)}{dt^2} + A\, u(t) = 0 \\ u(0) = u_o \in H^1(R^3 - \overline{\omega}) \quad ; \quad u'(0) = u_1 \in L^2(R^3) \end{cases}$$

Then, the local energy decay holds, i.e.

(3.7)
$$u(t)\big|_{|x|<\rho} \xrightarrow[t\to+\infty]{} 0 \quad \text{in } L^2(|x| < \rho) \text{ strongly}$$

Proof - The proof is enterely analogous to that of theorems 1.1, 1.2, 2.1 and 3.1 of chapter 16. We only remark that the regularity properties for operator A in the vicinity of $\omega$ are not known and consequently we do not have the energy decay in the local energy norm (i.e., of the type of theorem 3.2 of chap.16)■

Of course the problem (3.2), (3.3) appears as the limit ($\varepsilon \to 0$) of the solution $u^\varepsilon$ of the corresponding transmission problems in $\Omega_\varepsilon^1$, $\Omega_\varepsilon^0$.

Theorem 3.2 - Let $u^\varepsilon \in H^1_{loc}(R^3)$ be such that it has a representation of the form (3.2) for sufficiently large $|x|$ and such that ($c > 0$)

(3.8)
$$\varepsilon \int_{\Omega_\varepsilon^1} \frac{\partial u^\varepsilon}{\partial x_i}\frac{\partial \overline{v}}{\partial x_i}\,dx + \int_{\Omega_\varepsilon^0 \cap \{|x|<\rho\}} \frac{\partial u^\varepsilon}{\partial x_i}\frac{\partial \overline{v}}{\partial x_i}\,dx - c^2\int_{|x|<\rho} u^\varepsilon\,\overline{v}\,dx -$$
$$-\int_{|x|=\rho} \frac{\partial u^\varepsilon}{\partial n}\,\overline{v}\,dS = \int_{|x|<\rho} f\,\overline{v}\,dx \qquad \forall\, v \in H^1(|x| < \rho)$$

Then, $u^\varepsilon$ converges as $\varepsilon \searrow 0$ to the solution u of (3.2), (3.3) in

$$H^1([R^3 - \overline{K}_\alpha] \cap \{\,|x| < \rho\,\}) \text{ weakly}$$

and in $L^2(|x|< \rho)$ weakly for any fixed $\rho,\alpha$ where $K_\alpha$ is the domain

(3.9)
$$K_\alpha = \{x \,;\, (x_1, x_2) \in \omega \;,\; |x_3| < \alpha\}$$

Proof - This theorem is easily proved by using techniques analogous to that of the limiting absorption (chapter 16, sect. 1) and the limit process for the proof of theorem 3.1 of chapter 13. Of course, $u^\varepsilon$ exists and is unique (see chapter 16, sect. 1).

In a first step, we prove the analogous of lemma 1.3 of chapter 16, i.e. for any fixed $\rho$ ,

(3.10)
$$\|u^\varepsilon\|_{L^2(|x|<\rho)} \leqslant C$$

holds for some C independent of $\varepsilon$. This is proved by contradiction. If not, the normalized functions

$$v^\varepsilon = \frac{u^\varepsilon}{m^\varepsilon} \quad ; \qquad m^\varepsilon \equiv \| u^\varepsilon \|_{L^2(|x| < \rho+5)} \nearrow \infty$$

have norm in $L^2(|x| < \rho+5)$ equal to one, and after extracting a subsequence.

(3.11) $\qquad v^\varepsilon \longrightarrow v \qquad$ in $L^2(|x| < \rho + 5)$ weakly

Then, by interior elliptic regularity, we see, as in chap. 16 that the convergence (3.11) holds in $H^2(\rho + 1 < |x| < \rho + 4)$ weakly and consequently the traces of $v^\varepsilon$ and $\partial v^\varepsilon/\partial n$ for $|x| = \rho + 2$ and $\rho + 3$ converge in $L^2$ strongly. From this we see that the limit v satisfies a radiation condition of the type (3.2), and convergence (3.11) holds in $L^2(\rho+1 < |x| < \rho+5)$ strongly. Moreover, from (3.8) with $v^\varepsilon$ and $f/m^\varepsilon$ instead of $u^\varepsilon$ and f) we obtain

(3.12) $$\varepsilon \int_{\Omega_\varepsilon^1} \sum_i \left|\frac{\partial v^\varepsilon}{\partial x_i}\right|^2 dx + \int_{\Omega_\varepsilon^0 \cap \{|x| < \rho+1\}} \sum_i \left|\frac{\partial v^\varepsilon}{\partial x_i}\right|^2 dx \leqslant C$$

and we pass to the limit in (3.8) (with $v^\varepsilon$ , $f/m^\varepsilon$ instead of $u^\varepsilon$, f) as in chap.**13**, lemma 3.5 and 3.6. We then see that v is a solution of the limit problem (3.2), (3.3) with f = 0 ; consequently, <u>v = 0</u> . On the other hand, from the preceeding properties and (3.12) it follows that

(3.13) $\qquad v^\varepsilon \longrightarrow v = 0 \quad$ in $L^2(\{|x| < \rho + 5\} - \overline{K}_\alpha)$

for arbitrarily small $\alpha$. Moreover, the traces of $v^\varepsilon\big|_{|x_3| = \alpha}$ converge in $L^2(\omega)$ strongly ; then, by taking $\alpha$ and $\varepsilon$ sufficiently small in estimates (3.17) of chap. 13, we see that (3.13) holds in $L^2(|x| < \rho+5)$ strongly, and we have a contradiction with $\| v^\varepsilon \| = 1$ in $L^2(|x| < \rho + 5)$. Then, (3.10) is proved. In a second step, starting with (3.10), we repeat the limit process with $u^\varepsilon$ and f (instead of $v^\varepsilon$ and $f/m^\varepsilon$) and theorem 3.2 is proved. ■

<u>The high conductivity case</u> - Now, we consider a problem analogous to that of chap. 16, sect. 4.

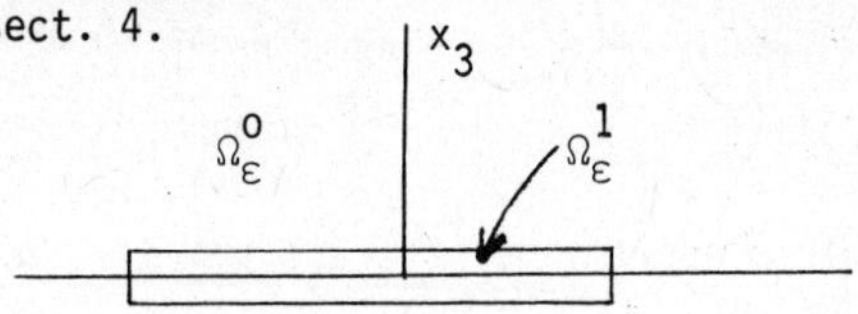


<u>fig. 7</u>

We consider the bounded domain $\omega$ of the plane $Ox_1, x_2$ with smooth boundary as in the preceeding problem. We then define the "coin-shaped" domain

$$\Omega_\varepsilon^1 = \{x \ ; \ (x_1 , x_2) \in \omega \ ; \ |x_3| < \varepsilon \lambda \}$$

where $\lambda$ is a given positive constant. The complement of $\overline{\Omega}_\varepsilon^1$ is noted $\Omega_\varepsilon^0$.

We define the space $V_{\omega\,loc}(R^3)$ as the space of the functions which are locally in $H^1$ and such that the traces on $\omega$ belong to $H^1(\omega)$. Then, we have the following theorem about the properties of the "limit problem" :

Theorem 3.3 - Let $f \in L^2(R^3)$ with compact support be given. We search for $u \in V_{\omega\,loc}(R^3)$ such that for sufficiently large $|x|$ it has a representation of the type (3.2) and satisfies

(3.14) $$\int_{|x|<\rho} \frac{\partial u}{\partial x_i}\frac{\partial \overline{v}}{\partial x_i}\,dx + 2\lambda\int_\omega \sum_1^2 \frac{\partial u}{\partial x_i}\frac{\partial \overline{v}}{\partial x_i}\,dx_1\,dx_2 - c^2\int_{|x|<\rho} u\,\overline{v}\,dx -$$
$$-\int_{|x|<\rho} \frac{\partial u}{\partial n}\overline{v}\,dS = \int_{|x|=\rho} f\,\overline{v}\,dx \qquad \forall v \in V_{\omega loc}(R^3) \quad .$$

(3.15) Then, u exists and is unique.

Moreover, let A be the selfadjoint operator of $L^2(R^3)$ associated with the form

$$\int_{R^3} \frac{\partial u}{\partial x_i}\frac{\partial \overline{v}}{\partial x_i}\,dx + 2\lambda\int_\omega \sum_1^2 \frac{\partial u}{\partial x_i}\frac{\partial \overline{v}}{\partial x_i}\,dx_1\,dx_2$$

Then, A has an absolutely continuous spectrum. Moreover, if u(t) is the solution of the evolution problem

(3.16) $$\begin{cases} \dfrac{d^2u(t)}{dt^2} + A\,u(t) = 0 \\ u(0) = u_0 \in V_\omega(R^3) \quad ; \quad \dfrac{du}{dt}(0) = u_1 \in L^2(R^3) \end{cases}$$

the local energy decay holds, i.e.

(3.17) $$u(t)\Big|_{|x|<\rho} \xrightarrow[t\to+\infty]{} 0 \qquad \text{in } L^2(|x|<\rho) \text{ strongly}$$

The proof of this theorem is analogous to that of theorem 3.1 (and then analogous to theorems 1.1, 1.2, 2.1, and 3.1 of chapter 16).

As for the convergence properties of the solutions of the transmission problems in $\Omega_\varepsilon^1$ , $\Omega_\varepsilon^0$, we have :

Theorem 3.4 - Let $u^\varepsilon \in H^1_{loc}(R^3)$ be such that it has a representation of the form (3.2) for sufficiently large $|x|$ and such that

(3.18) $$\frac{1}{\varepsilon}\int_{\Omega_\varepsilon^1} \frac{\partial u^\varepsilon}{\partial x_i}\frac{\partial \bar{v}}{\partial x_i}\,dx + \int_{\Omega_\varepsilon^0 \cap \{|x|<\rho\}} \frac{\partial u^\varepsilon}{\partial x_i}\frac{\partial \bar{v}}{\partial x_i}\,dx - c^2\int_{|x|<\rho} u^\varepsilon \bar{v}\,dx -$$
$$- \int_{|x|=\rho} \frac{\partial u^\varepsilon}{\partial n}\,\bar{v}\,dS = \int_{|x|<\rho} f\,\bar{v}\,dx \qquad \forall v \in H^1_{loc}(R^3)$$

Then, $u^\varepsilon$ converges as $\varepsilon \to 0$ to the solution u of (3.2), (3.14) in $H^1_{loc}(R^3)$ weakly.

The proof is analogous to that of theorem 3.2.

4.- Bibliographical notes - The study of sect. 1 (Helmoltz resonator) is taken from Beale [ 1 ] . This problem is also studied in Arsen'ev [ 1 ] . The problem of sect. 2 (vibration of an elastic body surrounded by a gas) have some relation with the preceeding one. For experimental results concerning this problem, see Gottlieb [ 1 ] . Formal study using the two-scale method may be find in Cerneau et Sanchez-Palencia [ 2 ] ; the study of sect. 2 is a modified version of Sanchez-Palencia [ 6 ]. Convergence of the spectral families in the framework of chapter 12, sect. 3, 4, may also be obtained in the present problem. (see some indications in Lobo-Hidalgo and Sanchez-Palencia [ 2 ] ). The results of sect. 3, concerning diffraction problems by narrow screens with either small or large rigidity (in the framework of chapter 13, sect. 3, 4) are new. For other physically interesting scattering problems, see Beale [ 2 ] , Guillot and Wilcox [ 1 ] , Rauch and Taylor [ 1 ] , Wilcox [ 2 ] and Codegone [ 1 ] , [ 2 ] .

# APPENDIX

## INCOMPRESSIBLE FLUID FLOW IN A POROUS MEDIUM -

## CONVERGENCE OF THE HOMOGENIZATION PROCESS

by Luc TARTAR

Dept. of Mathematics
University of Paris - Sud, Orsay. France

(This appendix was written from an original draft given by L. Tartar on june 1979).

In this appendix we prove the convergence of the formal asymptotic expansion of chapter 7, sect. 2. In fact, we consider the problem in slightly more restricted conditions, when the solid part of each period is strictly contained into the period. This model is not realist in three dimensions, (the solid parts do not form a connected body) but it is acceptable in two dimensions : flow through an arrangement of columns fixed by their (far located !) extremities.

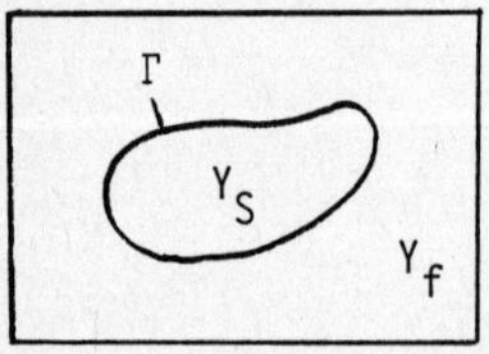


fig. 1

As usual, the regions $Y_f$, $Y_S$ are repeated by periodicity and fill the space $R^N$ (N = 2 or 3). Moreover, we consider a bounded open domain $\Omega$ of $R^N$, and we define $\Omega_\varepsilon$ as the domain obtained from $\Omega$ if we pick out the domains $\varepsilon Y_S$ which do not intersect $\partial\Omega$ (see fig.)

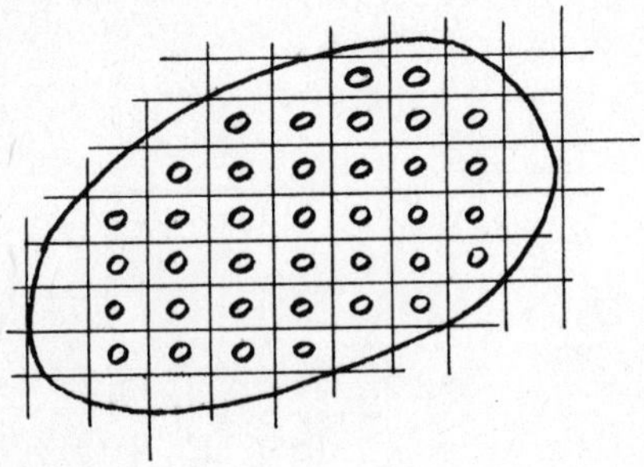

fig. 2

Let f be a smooth function independent of $\varepsilon$ defined on $\Omega$ .

We consider the solution $\underline{u}^\varepsilon \in \underline{H}_0^1(\Omega_\varepsilon)$, $p^\varepsilon \in L^2(\Omega_\varepsilon)$ of the Stokes problem in $\Omega_\varepsilon$ :

(1) $$0 = - \frac{\partial p^\varepsilon}{\partial x_i} + \Delta u_i^\varepsilon + f_i \qquad \text{in } \Omega_\varepsilon$$

(2) $$\text{div } \underline{u}^\varepsilon = 0 \qquad \text{in } \Omega_\varepsilon$$

We know from theorem 1.5 of chapter 7 that $\underline{u}^\varepsilon$ , $p^\varepsilon$ exists and are unique ($p^\varepsilon$ is unique up to an additive constant, it is unique if we consider the corresponding equivalence class : $p^\varepsilon \in L^2(\Omega_\varepsilon)/R$).

Moreover, we consider the local periodic problem (as in chapter 7, (2.10) - (2.12)) in $Y_f$ : if $\underline{e}_k$ is the unitary vector in the direction of the axis k $\underline{v}^k$ and $q^k$ are defined ($p^k$ up to an additive constant) by

(3) $$0 = \Delta_y \underline{v}^k - \underline{\text{grad}}_y\, q^k + \underline{e}_k$$

(4) $$\text{div}_y \underline{v}^k = 0 \quad : \quad \underline{v}^k\big|_\Gamma = 0$$

(5) $$\underline{v}^k ,\ q^k \qquad Y\text{-periodic.}$$

We continue $\underline{v}^k$ with value zero to $Y_S$. (The notation is not changed after continuation). The mean value $\tilde{\underline{v}}^k$ is defined by

$$\tilde{\underline{v}}^k = \frac{1}{|Y|} \int_Y \underline{v}^k \, dy$$

and the permeability tensor $K_{ij}$ (which is symmetric and positive definite, by chapter 7, proposition 2.2) is defined by

(6) $$K_{ij} = \tilde{v}^i_j \qquad \text{(component } j \text{ of } \underline{\tilde{v}}^i)$$

Moreover, we define the asymptotic velocity $\underline{u}^o$ and pressure $p^o$ by

(7) $$\operatorname{div} \underline{u}^o = 0 \qquad \text{in } \Omega$$

(8) $$u^o_i = K_{ij}(f_i - \frac{\partial p^o}{\partial x_j}) \qquad \text{in } \Omega$$

(9) $$\underline{u}^o \,.\, \underline{n} = 0 \qquad \text{on } \partial\Omega$$

Remark 1 - By replacing (8) into (7), we see that (7) - (8) is a Neumann problem (for the elliptic equation $K_{ij} \ \partial^2 p^o/_{\partial x_i \, \partial x_j} = F_i$) and $p^o$ is well defined up to an additive constant ; we consider it as an element of $L^2(\Omega)/R$. ∎

The main result of this appendix is the following :

Theorem 1 - Let $\underline{u}^\varepsilon$, $p^\varepsilon$ (resp. $u^o$, $p^o$) be defined by (1), (2) [resp. (7) - (9) ]. There exists a continuation of $\underline{u}^\varepsilon$ and $p^\varepsilon$ to $\Omega$ (in fact $\underline{u}^\varepsilon$ is continuated by zero) such that

(10) $$\frac{\underline{u}^\varepsilon}{\varepsilon^2} \to \underline{u}^o \qquad \text{in } L^2(\Omega) \text{ weakly}$$

(11) $$p^\varepsilon \to p^o \qquad \text{in } L^2(\Omega) \text{ strongly}$$

The proof of this theorem will be given after several lemmas.

Lemma 1 - The constant of the Friedrichs inequality in $\Omega_\varepsilon$ is of the form $C\,\varepsilon^2$, i.e. :

(12) $$\int_{\Omega_\varepsilon} |u|^2 \, dx \leqslant C\,\varepsilon^2 \int_{\Omega_\varepsilon} |\underline{\operatorname{grad}}_x \, u|^2 \, dx$$

Proof - We consider a period $Y_f$. We have a Friedrichs inequality of the form

$$\int_{Y_f} |u|^2 \, dy \leqslant C(Y_f) \int_{Y_f} |\underline{\operatorname{grad}}_y \, u|^2 \, dy$$

for any $u \in H^1(Y)$ with $u|_\Gamma = 0$. Then, by considering

$$x = \varepsilon y \quad ; \quad dx = \varepsilon^N dy \quad ; \quad \frac{\partial}{\partial y} = \varepsilon \frac{\partial}{\partial x} \implies$$

$$\int_{\varepsilon Y_f} |u|^2 \, dx \leqslant C(Y_f)\, \varepsilon^2 \int_{\varepsilon Y_f} |\underline{\operatorname{grad}}_x \, u|^2 \, dx$$

and by taking the sum for all the periods $\varepsilon Y$ we obtain (12) (the constant $C(Y_f)$ is the same for all them !). In fact, we must consider separately the periods containing a portion of $\partial\Omega$ ; but they yield at a distance $O(\varepsilon)$ of $\partial\Omega$ ; where $u^\varepsilon$ are zero;

the corresponding inequality is immediately obtained. ∎

Lemma 2 - We continue $\underline{u}^\varepsilon$ (initially defined on $\Omega_\varepsilon$) to $\Omega$ with value zero out of $\Omega_\varepsilon$. Then, after extraction of a subsequence (it will be proved to be the whole sequence by uniqueness of its limit):

(13) $$\frac{\underline{u}^\varepsilon}{\varepsilon^2} \to \underline{u}^* \qquad \text{in } L^2(\Omega) \text{ weakly}$$

where

(14) $$\operatorname{div} \underline{u}^* = 0 \quad \text{on } \Omega \quad ; \quad \underline{u}^* \cdot \underline{n} = 0 \quad \text{on } \partial\Omega$$

moreover

(15) $$\| \operatorname{grad} \underline{u}^\varepsilon \|_{L^2(\Omega)} \leqslant C\,\varepsilon \quad ; \quad \| u^\varepsilon \|_{L^2(\Omega)} \leqslant C\,\varepsilon^2$$

Proof - (1) is an identity between elements of $\underline{H}^{-1}(\Omega_\varepsilon)$. We take the duality product by $\underline{u}^\varepsilon \in \underline{H}_0^1(\Omega_\varepsilon)$. We obtain (see chap. 7, theorem 1.4) :

(16) $$\int_{\Omega_\varepsilon} |\operatorname{grad} \underline{u}^\varepsilon|^2\, dx = \int_{\Omega_\varepsilon} f_i\, u_i^\varepsilon\, dx \leqslant C\, \|u^\varepsilon\|_{\underline{L}^2(\Omega_\varepsilon)} \leqslant$$
$$\leqslant [\text{ by } (12)] \leqslant C'\,\varepsilon \left( \int_{\Omega_\varepsilon} |\operatorname{grad} \underline{u}^\varepsilon|^2\, dx\right)^{1/2}$$

and we have $(15)_1$ and then, by using again (16), we obtain $(15)_2$. Then, we continue $\underline{u}^\varepsilon$ with value 0 out of $\Omega_\varepsilon$. It is clear that, after continuation, (15) holds ; moreover,

(17) $$\underline{u}^\varepsilon \in H_0^1(\Omega) \quad ; \quad \operatorname{div} \underline{u}^\varepsilon = 0 \quad \text{in } \Omega$$

By weak compactness of the balls in a Hilbert space, we can extract a subsequence (13) ; moreover, from (17) we see that the convergence (13) holds in the weak topology of $H(\Omega\, ,\, \operatorname{div})$ (see chap. 7, sect. 1) and (14) follows from chapter 7, theorem 1.1. ∎

Now we prove a lemma which will be fundamental for the continuation of the pressure

Lemma 3 - We consider a period Y. We consider a smooth surface $\gamma$ strictly contained in Y, enclosing $Y_S$, and we note $Y_M$ the domain contained between $\gamma$ and $\Gamma$ (See fig. 3). Then

For given $\underline{u} \in \underline{H}^1(Y)$, there exists $\underline{v} \in H^1(Y_M)$, $q \in L^2(Y_M)/R$ satisfying

(18) $$-\Delta \underline{v} = -\Delta \underline{u} + \underline{\operatorname{grad}}\, q \qquad \text{in } Y_M$$

(19) $$\operatorname{div} \underline{v} = g \equiv \operatorname{div} \underline{u} + \frac{1}{|Y_M|} \int_{Y_S} \operatorname{div} \underline{u}\, dy \qquad \text{in } Y_M$$

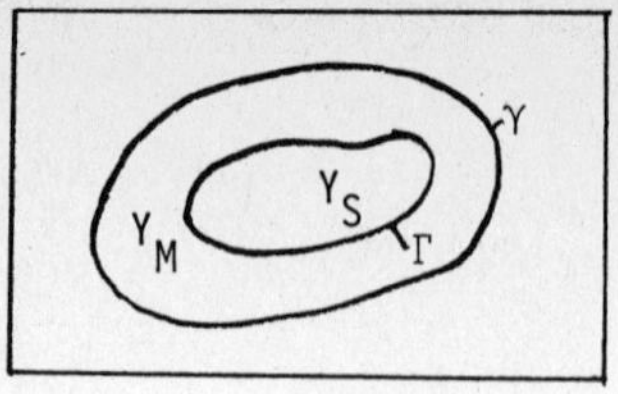


fig. 3

(20) $$\underline{v}\Big|_{\gamma} = \underline{u}\Big|_{\gamma} \quad ; \quad \underline{v}\Big|_{\Gamma} = 0$$

Moreover, there exists C independent of $\underline{u}$ such that

(21) $$\| \underline{v} \|_{\underline{H}^1(Y_M)} \leqslant C \| \underline{u} \|_{H^1(Y)}$$

Proof - We transform the problem in another with (19) and (20) homogeneous by considering a new unknown function $\underline{w}$ (instead of $\underline{v}$) :

(22) $$\underline{v} = \underline{\alpha} + \underline{\beta} + \underline{w}$$

where $\underline{\alpha}$ and $\underline{\beta}$ are defined in the following way.

We construct $\underline{\alpha} \in H^1(Y_M)$ with

(23) $$\underline{\alpha}\Big|_{\gamma} = \underline{u}\Big|_{\gamma} \quad ; \quad \underline{\alpha}\Big|_{\Gamma} = 0 \quad ; \quad \| \underline{\alpha} \|_{\underline{H}^1(Y_M)} \leqslant C \| \underline{u} \|_{\underline{H}^1(Y)}$$

which exists by standard trace and lift properties.

We construct $\underline{\beta} \in H^1_o(Y_M)$ such that

(24) $$\text{div } \underline{\beta} = F \equiv - \text{div } \underline{\alpha} + \text{div } \underline{u} + \frac{1}{|Y_M|} \int_{Y_S} \text{div } \underline{u} \, dy \qquad \text{in } Y_M$$

(25) $$\| \beta \|_{\underline{H}^1_o(Y_M)} \leqslant C \| F \|_{L^2(Y_M)}$$

which exists (see Tartar [ 5 ] p. 30, Lemmas, perhaps also Temam [ 1 ] chap. 1, Lemma 2.4) because the compatibility condition

$$\int_{Y_M} F \, dy = 0$$ is satisfied. In fact, from (24) we have :

$$\int_{Y_M} F\,dy = \int_{Y_M} -\operatorname{div}\alpha\,dy + \int_{Y_M} \operatorname{div}\underline{u}\,dy + \frac{|Y_M|}{|Y_M|}\int_{Y_S} \operatorname{div} u\,dy =$$

$$= \int_{\gamma} -\underline{\alpha}\,.\,\underline{n}\,dS + \int_{\gamma} \underline{u}\,.\,\underline{n}\,dS = 0$$

Then, the problem for $\underline{w}$ is : $\underline{w} \in H_0^1(Y_M)$

$$-\Delta\underline{w} = -\Delta(\underline{u} - \underline{\alpha} - \underline{\beta}) + \underline{\operatorname{grad}}\, q \quad \text{in} \quad Y_M$$

$$\operatorname{div}\underline{w} = 0$$

which exists, with $\|\underline{w}\|_{\underline{H}_0^1} \leqslant C\|u\|_{\underline{H}_0^1}$ by standard theory (see chapter 7, theorem 1.5) and the estimates (23) and (25). ∎

**Lemma 4** - There exists a (restriction) operator

(26) $\quad R_\varepsilon \;:\; H_0^1(\Omega) \longmapsto H_0^1(\Omega_\varepsilon)$ such that

(27) $\quad \underline{w} \in H_0^1(\Omega_\varepsilon) \Longrightarrow R_\varepsilon\,\underline{w} = \underline{w}$

(elements of $H_0^1(\Omega_\varepsilon)$ are continuated by 0 to $\Omega$)

(28) $\quad \operatorname{div}\underline{w} = 0 \Longrightarrow \operatorname{div} R_\varepsilon\,\underline{w} = 0$

(29) $\quad \|R_\varepsilon\,\underline{w}\|_{L^2(\Omega_\varepsilon)} \leqslant C\|\underline{w}\|_{L^2(\Omega)} + C\varepsilon\|\underline{\operatorname{grad}}\,\underline{w}\|_{L^2(\Omega)}$

(30) $\quad \|\underline{\operatorname{grad}}\, R_\varepsilon\,\underline{w}\|_{L^2(\Omega_\varepsilon)} \leqslant \frac{C}{\varepsilon}\|\underline{w}\|_{L^2(\Omega)} + C\,\|\underline{\operatorname{grad}}\,\underline{w}\|_{L^2(\Omega)}$

*Proof* - For a period Y, we consider again the functions $\underline{u}$ and $\underline{v}$ of Lemma 3. We define

(31)
$$R\,\underline{u}(y) = \begin{cases} u(y) & \text{if} \quad y \in Y \setminus (Y_S \cup Y_M) \\ v(y) & \text{if} \quad y \in Y_M \\ 0 & \text{if} \quad y \in Y_S \end{cases}$$

we evidently have

(32)
$$\|R\,\underline{u}\|_{H^1(Y)} \leqslant C\,\|\underline{u}\|_{H^1(Y)}$$

and $R\underline{u}$ coincides with $\underline{u}$ if $\underline{u}$ is zero on $Y_S$ (i.e., if $\underline{u} \in H_0^1(Y_f)$) and $\operatorname{div}\underline{u} = 0 \Rightarrow \operatorname{div} R\,\underline{u} = 0$.

Then, $R_\varepsilon$ is defined by applying R to each $\varepsilon Y$ period. Consequently, (26), (27), (28) are satisfied. Moreover, from (32), as in Lemma 1, we have

$$\varepsilon^2 \int_{\Omega_\varepsilon} |\underline{\text{grad}}_x R_\varepsilon \underline{w}|^2 dx + \int_{\Omega_\varepsilon} |R_\varepsilon \underline{w}|^2 dx \leq C\left[ \varepsilon^2 \int_{\Omega} |\underline{\text{grad}}_x \underline{w}|^2 dx + \int_{\Omega} |\underline{w}|^2 dx\right]$$

and (29), (30) follow. ∎

Proof of theorem 1 - We see from (1) that $\underline{\text{grad}}\ p^\varepsilon \in \underline{H}^{-1}(\Omega_\varepsilon)$. We continue it to $\Omega$ in the following way. Let us define $\underline{F}^\varepsilon \in \underline{H}^{-1}(\Omega)$ by (brackets are for the duality products between $H^{-1}$ and $H_0^1$) :

$$(33) \qquad < \underline{F}^\varepsilon, \underline{w} >_\Omega = < \underline{\text{grad}}\ p^\varepsilon , R_\varepsilon \underline{w} >_{\Omega_\varepsilon} \qquad \forall\ \underline{w} \in \underline{H}_0^1(\Omega)$$

where $R_\varepsilon$ is defined in Lemma 4. We calculate the right hand side of (33) by using (1) and we have

$$(34) \qquad < \underline{F}^\varepsilon, \underline{w} >_\Omega = -\int_{\Omega_\varepsilon} \frac{\partial u_i^\varepsilon}{\partial x_j} \frac{\partial (R_\varepsilon w)_i}{\partial x_j} dx + \int_{\Omega_\varepsilon} f_i (R_\varepsilon w)_i \, dx$$

and by using (29),(30) for fixed $\varepsilon$ we see that it is a bounded functional on $\underline{H}_0^1(\Omega)$, and in fact $\underline{F}^\varepsilon \in \underline{H}^{-1}(\Omega)$. Moreover, if $\underline{w} \in \underline{H}_0^1(\Omega_\varepsilon)$ and we continue it by zero out of $\Omega_\varepsilon$ , we see from (33) and (27) that

$$(35) \qquad \underline{F}^\varepsilon|_{\Omega_\varepsilon} = \underline{\text{grad}}\ p^\varepsilon \qquad \text{(on } \Omega_\varepsilon \text{ !)}$$

Moreover, if div $\underline{w} = 0$, by (28), (33) and the classical orthogonality property (see chapter 7, theorem 1.4), $< \underline{F}^\varepsilon , \underline{w} >_\Omega = 0$ and this implies (again by the orthogonality property) that $\underline{F}^\varepsilon$ is the gradient of a function of $L^2(\Omega)$. This means that $\underline{F}^\varepsilon$ is a continuation of $\underline{\text{grad}}\ p^\varepsilon$ to $\Omega$ , and that this continuation is a gradient. We also may say that $p^\varepsilon$ has been continuated to $\Omega$(we use the same notation for a function and its continuation) and

$$(36) \qquad \underline{F}^\varepsilon \equiv \underline{\text{grad}}\ p^\varepsilon \quad ; \qquad p^\varepsilon \in L^2(\Omega)/R$$

(it is clear that $\underline{F}^\varepsilon$ is well determined, $p^\varepsilon$ is determined up to an additive constant).

Let us estimate $p^\varepsilon$ and its gradient. From (34), with $(15)_1$, (29) and (30) we have

$$(37) \qquad | < \underline{\text{grad}}\ p^\varepsilon , \underline{w} >_\Omega | \leq C [ \ \| \underline{w} \|_{L^2(\Omega)} + \varepsilon \| \underline{\text{grad}}\ \underline{w} \|_{L^2(\Omega)} \ ]$$

Then, as $\varepsilon < 1$, we see that the right hand side of (37) is bounded by $C \| \underline{w} \|_{\underline{H}_0^1}$ and consequently

$$(38) \qquad \| \underline{\text{grad}}\ p^\varepsilon \|_{\underline{H}^{-1}(\Omega)} \leq C$$

and from the inequality (see for instance Tartar [ 5 ] , or Temam [ 1 ] , chap. 1, (1.32) and Necas [ 1 ]).

(39) $$\| p \|_{L^2(\Omega)/R} \leqslant C(\Omega) \| \underline{grad}\, p \|_{\underline{H}^{-1}(\Omega)}$$

we see that $p^\varepsilon$ remains bounded in $L^2(\Omega)/R$ and consequently, after extraction of a subsequence

$$p^\varepsilon \to p^* \qquad \text{in } L^2(\Omega)/R \text{ weakly}$$

(40) $$\underline{grad}\, p^\varepsilon \to \underline{grad}\, p^* \qquad \text{in } \underline{H}^{-1}(\Omega) \text{ weakly}$$

Moreover, let $\underline{w}^\varepsilon$ be a sequence of elements of $\underline{H}_0^1(\Omega)$ such that

(41) $$\underline{w}^\varepsilon \to \underline{w}^* \qquad \text{in } H_0^1(\Omega) \text{ weakly.}$$

From (37) applied to $\underline{w}^\varepsilon - \underline{w}^*$ we have :

$$|< \underline{grad}\, p^\varepsilon, \underline{w}^\varepsilon > - < \underline{grad}\, p^*, \underline{w}^* > | \leqslant$$

$$\leqslant | < \underline{grad}\, p^\varepsilon, \underline{w}^\varepsilon - \underline{w}^* > | + | < \underline{grad}\, p^\varepsilon - \underline{grad}\, p^*, w^* > | \leqslant$$

$$\leqslant C \left[ \| \underline{w}^\varepsilon - \underline{w}^* \|_{\underline{L}^2} + \varepsilon \| \underline{w}^\varepsilon - \underline{w}^* \|_{\underline{H}_0^1} \right] + (\text{term which} \to 0)$$

which tends to zero by virtue of (41) and the Rellich theorem. This implies

(42) $$\underline{grad}\, p^\varepsilon \to \underline{grad}\, p^* \qquad \text{in } \underline{H}^{-1}(\Omega) \text{ strongly}$$

(see, if necessary, chapter 8, considerations after (3.23)). And from (39) we have

(43) $$p^\varepsilon \to p^* \qquad \text{in } L^2(\Omega)/R \text{ strongly}$$

From (13), (14) and (43) it only remains to prove that $\underline{u}^*$ and $p^*$ satisfy equation (8), then $\underline{u}^*$ and $p^*$ are the solution $\underline{u}^0$ and $p^0$ of (7) - (9).

We write the local problem (3) - (5) in terms of $x = \varepsilon y$ :

(44) $$\underline{v}_\varepsilon^k(x) \equiv \underline{v}^k(x/\varepsilon) \quad ; \quad q_\varepsilon^k(x) \equiv q^k(x/\varepsilon)$$

(45) $$0 = \varepsilon^2 \Delta_x \underline{v}_\varepsilon^k - \varepsilon\, \underline{grad}_x\, q_\varepsilon^k + \underline{e}_k$$

(46) $$\text{div}_x\, \underline{v}_\varepsilon^k = 0$$

and because $\underline{v}^k(y)$ , $q^k(y)$ are independent of $\varepsilon$, we have

(47) $$\| q_\varepsilon^k \|_{L^2(\Omega_\varepsilon)} \leqslant C \quad ; \quad \| \underline{v}_\varepsilon^k \|_{L^2(\Omega)} \leqslant C$$

(48) $$\| \underline{grad}_x\, \underline{v}_\varepsilon^k \|_{L^2(\Omega)} \leqslant \frac{C}{\varepsilon}$$

Now, we apply the standard method to prove convergence in homogenization problems. We take $\phi \in \mathcal{D}(\Omega)$ and we multiply (45) by $\phi u_i^\varepsilon$ and integrate on $\Omega$

(in fact, we consider the duality product) (note that $\phi\underline{u}^\varepsilon$ is not divergence-free !):

$$(49) \int_\Omega \frac{\partial v^k_{\varepsilon i}}{\partial x_j} \frac{\partial(\phi u^\varepsilon_i)}{\partial x_j} dx = \frac{1}{\varepsilon} \int_\Omega q^k_\varepsilon \, \mathrm{div}(\phi \, \underline{u}^\varepsilon) \, dx + \frac{1}{\varepsilon^2} \int_\Omega \underline{e}_k \phi \, \underline{u}^\varepsilon \, dx$$

Then, by $(47)_1$, (2), $(15)_2$ :

$$\left| \frac{1}{\varepsilon} \int_\Omega q^k_\varepsilon \, \mathrm{div}(\phi \, \underline{u}^\varepsilon) dx \right| = \left| \frac{1}{\varepsilon} \int_\Omega q^k_\varepsilon \frac{\partial \phi}{\partial x_i} u^\varepsilon_i \, dx \right| \leq C \, \varepsilon \to 0$$

then, passing to the limit in the right hand side of (49) by (13) we have :

$$(50) \qquad \int_\Omega \frac{\partial v^k_{\varepsilon i}}{\partial x_j} \frac{\partial(\phi u^\varepsilon_i)}{\partial x_j} dx \twoheadrightarrow \int_\Omega \phi \, u^*_k \, dx \qquad \text{as} \quad \varepsilon \to 0$$

On the other hand, we take the duality product of (1) by $\phi \, \underline{v}^k_\varepsilon$ :

$$(51) \int_{\Omega_\varepsilon} \frac{\partial u^\varepsilon_i}{\partial x_j} \frac{\partial(\phi v^k_{\varepsilon i})}{\partial x_j} dx = \langle \underline{f} \, , \, \phi \, \underline{v}^k_\varepsilon \rangle_{\Omega_\varepsilon} + \langle p^\varepsilon \, , \, \mathrm{div}(\phi \, \underline{v}^k_\varepsilon) \rangle_{\Omega_\varepsilon} =$$

$$= \int_\Omega f_i \, \phi \, v^k_{\varepsilon i} \, dx + \int_\Omega p^\varepsilon \frac{\partial \phi}{\partial x_i} v^k_{\varepsilon i} \, dx$$

(note that $\underline{v}$ and $\underline{u}$ are zero out of $\Omega_\varepsilon$ , and we may write the integrals either on $\Omega_\varepsilon$ or $\Omega$). Moreover, by the classical lemma on Y-periodic functions (chapter 5, lemma ) :

$$(52) \qquad v^k_{\varepsilon i} \rightharpoonup \tilde{v}^k_i = K_{ki} \qquad \text{in} \quad L^2(\Omega) \text{ weakly}$$

where $K_{ki}$ are the components of the permeability tensor (6). We pass to the limit in (51) by using (43) and (52) :

$$(53) \int_\Omega \frac{\partial u^\varepsilon_i}{\partial x_j} \frac{\partial(\phi v^k_{\varepsilon i})}{\partial x_j} dx \to K_{ki} \int_\Omega f_i \, \phi \, dx + K_{ki} \int_\Omega p^* \frac{\partial \phi}{\partial x_i} dx$$

(it is to be noticed that

$$\int_\Omega \frac{\partial \phi}{\partial x_i} v^k_{\varepsilon i} \, dx = \int_\Omega \mathrm{div}(\phi \, v^k_\varepsilon) \, dx = \int_{\partial\Omega} \phi \, \underline{v}^k_\varepsilon \, . \, \underline{n} \, dS = 0$$

and then $p^\varepsilon$ is defined up to an additive constant in (51) ; consequently the convergence in $L^2(\Omega)/R$ suffices to pass to the limit).

Now, we compare the left hand sides of (50) and (53). Their difference is

$$\left| \int_\Omega \frac{\partial v^k_{\varepsilon i}}{\partial x_j} u^\varepsilon_i \frac{\partial \phi}{\partial x_j} dx - \int_\Omega \frac{\partial u^\varepsilon_i}{\partial x_j} v^k_{\varepsilon i} \frac{\partial \phi}{\partial x_j} dx \right| \leq \left| \int \right| + \left| \int \right| \leq C \, \varepsilon \to 0$$

where (15), (47) and (48) have been used. Consequently, the right hand sides of (50) and (53) are equal and by writing them as distributions we have :

$$< u_k^* , \phi > = K_{ki} < f_i - \frac{\partial p^*}{\partial x_i} , \phi >$$

which is satisfied for any $\phi \in \mathcal{D}(\Omega)$

$$u_k^* = K_{ki}(f_i - \frac{\partial p^*}{\partial x_i})$$

which is (8) for $\underline{u}^*$, $p^*$ and theorem 1 is proved. ■

## REFERENCES


Agmon, S.

[1] Lectures on elliptic boundary value problems. Van Nostrand, New York (1965).

Arsen'ev, A.A.

[1] On the singularities of the analytic continuation of the resonance properties of the solution of the scattering problem for the Helmholtz equation. U.S.S.R. Compt. Math. Phys. 12 (1972) p. 139-173 (= Zh. Vychis. Mat. Mat. Fiz 12, 1972, p. 112-138).

Artola, M. et Duvaut, G.

[1] Homogénéisation d'une plaque renforcée par un système périodique de barres curvilignes. Compt. Rend. Acad. Sci.Paris, sér. A 286 (1978) p. 659-662.

[2] Homogénéisation d'une classe de problèmes non linéaires. Compt. Rend. Acad. Sci. Paris, sér. A 288 (1979) p. 775-778.

Auriault, J.L. et Sanchez-Palencia, E.

[1] Etude du comportement macroscopique d'un milieu poreux saturé déformable. Jour. Mécanique, 16 (1977) p. 575-603.

Babuska, I.

[1] Solution of interface problems by homogenization. Siam Jour. Appl. Math 7 (1976) part I, p. 603-634 ; part II, p. 635-645.

Bakhvalov, N.S.

[1] Averaging of nonlinear partial differential equations with rapidly oscillating coefficients. Dokl. Akad. Nauk 225 (1975) n° 2, p. 249-52.

[2] Homogenization and perturbation. Proc. Fourth Intern. Symp. Compt. Methods in Science and Engineering, Versailles, France (1979)

Balachandra, M. and Sethna, P.R.

[1] A generalization of the method of averaging for systems with two time scales. Arch. Rat. Mech. Anal. 58 (1975) p. 261-283.

Banfi, C.

[1] Sull'approssimazione di processi non stazionari in meccanica non lineare. Boll. Un. Mat. Ital. 22 p. 442-450 (1967).

Bardos, C., Brezis, D. and Brezis, H.

[1] Perturbations singulières et prolongements maximaux d'opérateurs positifs. Arch. Rat. Mech. Anal. 53, p. 69-100 (1973).

Beale, J.T.

[1] Scattering frequencies of resonators. Comm. Pure App. Math. 36 (1973) p. 549-563.

[2] Eigenfunction expansions for objects floating in an open sea. Comm. Pure Appl. Math. 30 (1977) p. 283-313.

Bensoussan, A., Lions, J.L. et Papanicolaou, G.

[1] Sur quelques phénomènes asymptotiques d'évolution. Compt. Rend. Acad. Sci. Paris sér. A 281 (1975) p. 317-322.

[2] Asymptotic Analysis for Periodic Structures. North Holland, Amsterdam (1978).

[3] Perturbations et augmentation des conditions initiales in "Singular perturbations and boundary layer theory. (p. 10-29). Springer, Berlin (1977).Lect. Notes in Math. 594.

Biroli, M.

[1] G-convergence for elliptic variational and quasi-variational inequalities in Proc. Intern. Meeting on Recent Methods in Non Linear Anal., ed. De Giorgi Magenes, Mosco, ed. Pitagora, Bologne (1979), p. 361-383.

Boccardo, L. et Marcellini, P.

[1] Sulla convergenza delle soluzioni di disequazioni variazionali. Ann. Mat. Pura Appl. 110 (1976) p. 137-159.

Boroditskii, M.P. and Simonenko, I.B.

[1] Different dimensions variational problem. Funct. Anal. Appl. 9, p. 325-326 (1975) (Transl. Funk. Anal. Prilozh. 9 p. 63-64 (1975).

Bourgat, J.F.

[1] Numerical experiments of the homogenization method for operators with periodic coefficients. Rapport de Recherche n° 277, I.R.I.A. Rocquencourt, France (1978).

Bourgat, J.F. et Dervieux, A.

[1] Méthode d'homogénéisation des opérateurs à coefficients périodiques. Etude des corrections provenant du développement asymptotique. Rapport de Recherche n° 278, I.R.I.A. Rocquencourt, France (1978).

Bourgat, J.F. et Lanchon, H.

[1] Application of the homogenization method to composite materials with periodic structure. Rapport de Recherche n° 208, I.R.I.A., Rocquencourt, France (1976).

Brezis, D.

[1] C.f. Bardos, C., Brezis, D. and Brezis, H.

Brezis, H.

[1] C.f. Bardos, C., Brezis, D. and Brezis, H.

Brizzi, R. et Chalot, J.P.

[1] Homogénéisation dans des ouverts à frontière fortement oscillante. Thèse de Spécialité, Univ. Nice, 1978.

Caillerie, D.

[1] Sur le comportement limite d'une inclusion mince de grande rigidité dans un corps élastique. Compt. Rend. Acad. Sci. Paris sér. A 287 (1978), p. 675-678.

[2] The effect of a thin inclusion of high rididity in an elastic body. Math. Meth. Appl. Sci. (to be published 1980).

Carbone, L.

[1] Sur un problème d'homogénéisation avec des contraintes sur le gradient. Jour. Math. Pures Appl. 58 (1979) p. 275-297.

Cerneau, S.

[1] Méthode de centrage, estimation de l'erreur dans un cas de stabilité partielle. Comp. Rend. Acad. Sci., A., p. 875-877 (1976).

Cerneau, S. et Sanchez-Palencia, E.

[1] Sur les oscillations libres des corps élastiques légèrement viscoélastiques. Jour. Mécan. 15, p. 237-263 (1976).

[2] Sur les vibrations libres des corps élastiques plongés dans les fluides. Jour. Mécan. 15, p. 399-425 (1976).

Chalot, J.P.

[1] C.f. Brizzi, R. et Chalot, J.P.

Ciarlet, P.G. and Destuynder, P.

[1] A justification of the two-dimensional linear plate model. Jour. Mecan. 18 (1979), p. 315-344.

[2] A justification of the non linear model in plate theory. Compt. Meth. Appl. Mech. Engin. 17/18 (1979) p. 227-258.

Ciarlet, P. et Kesavan, S.

[1] Les équations des plaques élastiques comme limite des équations de l'élasticité tridimensionnelle. Problèmes aux valeurs propres. Proc. Fourth Intern. Symp. Compt. Methods in Science and Engineering, Versailles, France (1979).

Cioranescu, D. and Saint Jean Paulin, J.

[1] Homogenization in open sets with holes. Jour. Math. Anal. Appl. 71 (1979) p. 590-607.

Codegone, M.

[1] Problèmes d'homogénéisation en théorie de la diffraction. Compt. Rend. Acad. Sci. Paris, sér. A. 288 (1979) p. 387-389.

[2] Scattering of elastic waves through a heterogeneous medium. Math. Meth. Appl. Sci., (to be published, 1980).

Cole, J.D.

[1] Perturbation methods in applied Mathematics. Blaisdell, Toronto (1968).

Courant, R. and Hilbert, D.

[1] Methods of Mathematical Physics I and II. Interscience, New York 1953

Damlamian, A.

[1] Homogénéisation du problème de Stefan. Compt. Rend. Acad. Sci. Paris, sér. A. 289 (1979) p. 9-11.

De Giorgi, E.

[1] Convergence problems for functionals and operators in Proceed Intern. Meeting on Recent Methods in Non Linear Analysis, ed. De Giorgi, Magenes, Mosco, ed. Pitagora, Bologne (1979) p. 131-188.

De Giorgi, E. et Sagnolo, S.

[1] Sulla convergenza delli integrali dell energia per operatori ellittici del secondo ordine. Boll. Unione Mat. Ital. 8, p. 391-411 (1973).

Dervieux, A.

[1] Bourgat, J.F. et Dervieux, A.

Desgraupes, B.

[1] Comportement asymptotique de la solution d'un problème elliptique d'ordre 4. Compt. Rend. Acad. Sci. Paris sér. A 287 (1978), p. 647-9.

Destuynder, P.

[1] C.f. Ciarlet, P. et Destuynder, P.

Dunford, N. and Schwartz, J.T.

[1] Linear operators, I. Interscience, New York 1967.

Duvaut, G.

[1] Etude de matériaux composites élastiques à structure périodique. Homogénéisation. Proc. Congress of theoretical and Applied Mechanics. Delft (1976), Ed. Koiter, North Holland, Amsterdam.

[2] Comportement macroscopique d'une plaque perforée périodiquement in "Singular perturbations and boundary layer theory" (p. 131-145), Lecture Notes in Mathematics vol. 594, Springer, Berlin 1977.

[3] Homogénéisation des plaques à structure périodique en théorie non linéaire de Von Karman. Lect. Notes Math. n° 665 (Jour. Analyse Non linéaire, Besançon (1977) p. 56-69).

[4] C.f. Artola, M. et Duvaut, G., Metellus, A.M. et Duvaut, G.

Duvaut, G. et Lions, J.L.

[1] Les inéquations en mécanique et en physique. Dunod, Paris (1972).

Ekeland, I. et Temam, R.

[1] Analyse convexe et problèmes variationnels. Dunod-Gauthier Villars, Paris (1974).

Ene, H.I. et Sanchez-Palencia, E.

[1] Equations et phénomènes de surface pour l'écoulement dans un modèle de milieu poreux. Jour. Mécan. 14 p. 73-108 (1975).

Eydus, D.M.

[1] The principle of limiting obsorption. Mat. Sbornik 57(99) p. 13-44 (1962)

Fetisov. Yu. I.

[1] C.f. Zabreiko, P.P. and Fetisov Yu. I.

Filatov, A.N.

[1] Asymptotic methods in the theory of differential and integrodifferential equations. (in russian) Fan, Tachkent (1974).

Filatov, A.N. and Sharova, L.V.

[1] Integral inequalities and theory of nonlinear oscillations. (in russian) Nauka, Moscow (1976).

Filippi, P.

[1] Problème de transmission pour l'équation de Helmholtz scalaire et problèmes aux limites équivalents : application à la transmission gaz parfait - milieux poreux. Jour. Mécan. 18 (1979), p. 565-591.

Fleury, F.

[1] Sédimentation de particules solides dans un fluide visqueux incompressible. Jour. Mécan. 18 (1979), p. 345-354.

[2] Propagation des ondes dans une suspension de particules solides. Compt. Rend. Acad. Sci. Paris, sér. A, 288 (1979), p. 77-80.

[3] Propagation of waves in a suspension of solid particles. Wave Motion (to be published, 1980).

Fleury, F., Paşa, G. et Polişevshi, D.

[1] Elasticité. Homogénéisation de corps composites sous l'action de forces de grande fréquence spatiale. Compt. Rend. Acad. Sci. Paris. Sér.B, 289 (1979), p. 241-244.

Garnir, H.G.

[1] Les problèmes aux limites de la Physique Mathématique. Birkhauser, Basel (1958).

Germain, P.

[1] Cours de Mécanique des milieux continus I. Masson, Paris (1973).

Geymonat, G. and Sanchez-Palencia, E.

[1] On the vanishing viscosity limit for acoustic phenomena in a bounded region. (To be published).

Gottlieb, H.P.W.

[1] Acoustical radiation damping of vibrating solids. Jour. Sound. Vibr. 40 (1975) p. 521-533.

Graffi, D.

[1] Oscillazioni non lineari in sistemi con hereditarietà. Symposia Mathematica Bologna, 6, p. 177-197 (1971).

Greenlee, W. M.

[1] Singular perturbations of eigenvalues. Arch. Rat. Mech. Anal. 34, p. 143-164 (1969).

Groen, P.P.N. de

[1] Spectral properties of second order singularly perturbed boundary value problems with turning point. Jour. Math. Anal. Appl. 57 (1977) p. 119-149.

[2] Singular perturbations of spectra in Asymptotic Analysis, Lect. Notes Math. 717 (1979) p. 9-32.

Guillot, J.C. and Wilcox, C.H.

[1] Steady-State wave propagation in simple and compound acoutic wave guides. Math. Zeit. 160 (1978) p. 89-102.

Gunther, N.M. [1] Théorie du potentiel, Gauthier-Villars, Paris (1934)

Hachem, G.

[1] Sur les fréquences de diffusion (scattering) d'un corps élastique couplé avec l'air. Anna. Fac. Sc. Toulouse (à paraître 1980).

Harris, W.A.

[1] Singular perturbations of eigenvalue problems. Arch. Rat. Mech. Anal. 7 (1961), p. 224-241).

Hille, E. and Phillips, R.S.

[1] Functional Analysis and semigroups. Amer. Math. Soc. Coll. Publ. 31 (1948).

Huet, D.

[1] Phénomènes de perturbation singulière dans les problèmes aux limites. Ann. Inst. Fourier, 10 p. 61-150 (1960).

[2] Approximation d'un espace de Banach et perturbations singulières. Compt. Rend. Acad. Sci. Paris, sér. A, 289 (1979) p. 69-70.

Kato, T.

[1] Perturbation theory of semibounded operators. Math. Ann. 125, p. 435-447 (1953).

[2] Perturbation theory for linear operators. Springer, Berlin (1966)

[3] Singular perturbation and semigroup theory in Turbulence and Navier Stokes equation, p. 104-112. Lecture Notes in Mathematics vol. 565, Berlin (1976).

Kellog, O.D. [1] Foundations of Pontential Theory, Springer, Berlin (1929).

Kesavan, S.

[1] Homogenization of elliptic eigenvalue problems. Appl. Math. Optim. Part I, 5 (1979), p. 153-167, Part II, 5 (1979), p. 197-216.

[2] C.f. Ciarlet, P. et Kesavan, S.

Krein, S.G.

[1] On the oscillations of a viscous fluid in a vessel. Sov. Math. Dokl. 5 (1964) p. 1467-1471 (= Dokl. Akad. Nauk 159, 1964, p. 262-265).

Kupradse, W.D.

[1] Randwertaufgaben der Schwingungstheorie und Integralgleichungen. Verlag Wissenschaften, Berlin (1959).

[2] Potential methods in the theory of elasticity. Israel program for scientific translation Jerusalem (1965).

Ladyzhenskaya, O.A.

[1] The principle of limiting amplitude. Usp. Mat. Nauk 11, 3(75) p. 161-164(1957)

[2] The mathematical theory of viscous incompressible flow. Gordon and Breach. New York (1963) (= Nauka, Moscow 1961).

Ladyzhenskaya, O.A. et Uralceva, N.N.

[1] Equations aux dérivées partielles de type elliptique. Dunod, Paris (1968) (= Nauka, Moscou, 1964).

Lanchon, H.

[1] C.f. Bourgat, J.F. et Lanchon, H.

Lax, P. and Phillips, R.S.

[1] Scattering theory. Acad. Press, New York, 1967.

Lené, F.

[1] Comportement macroscopique de matériaux élastiques comportant des inclusions rigides ou des trous répartis périodiquement. Compt. Rend. Acad. Sci. Paris, A, 286, p. 75-78 (1978).

Levy, Th.

[1] Equations et conditions d'interface pour des phénomènes acoustiques dans des milieux poreux in Singular perturbations and boundary layer theory, p. 301-311, Lecture Notes in Mathematics vol. 594, Springer, Berlin, 1977.

[2] Acoustic phenomena in elastic porous media. Mech. Res. Comm. 4 p. 253-257 (1977)

[3] Propagation of waves in a fluid-saturated porous elastic solid. Intern. J. Engin. Sci., 17 (1979) p. 1005-1014.

[4] Propagation of waves in a mixture of fluids. Intern. J. Engin. Sci. (To be published, 1980).

Levy, Th. and Sanchez-Palencia, E.

[1] On boundary conditions for fluid flow in porous media. Intern. J. Engin. Sci., 13, p. 923-940 (1975).

[2] Equations and interface conditions for acoustic phenomena in porous media. Jour. Math. Anal. Appl. 61 p. 813-834 (1977).

Lions, J.L.

[1] Problèmes aux limites en théorie des distributions. Acta Math. 94 (1955) p. 13-153.

[2] Equations differentielles opérationnelles et problèmes aux limites. Springer, Berlin (1961)

[3] Perturbations singulières dans les problèmes aux limites et en contrôle optimal. Lecture Notes in Mathematics, vol. 323, Springer Berlin (1973).

[4] Quelques méthodes de résolution des problèmes aux limites non linéaires. Dunod, Paris (1969).

[5] Sur quelques questions d'analyse de mécanique et de contrôle optimal. Presses Univ. Montréal (1976).

[6] Remarks on non local phenomena in composite materials. Lecture at the Ninth Annual Iranian Math. Conference, Isfahan (1978).

[7] Remarks on some asymptotic problems in composite materials and in perforated materials. Proc. of the Iutam symposium Northwestern Univ., Ed. Nemat-Nasser, North Holland (1979).

[8] Homogénéisation non locale in Proceed Intern. Meeting on Recent Methods in Non-Linear Analysis ed. De Giorgi, Magenes, Mosco, ed. Pitagora, Bologne (1979), p. 189-203.

[9] C.f. Bensoussan, A., Lions, J.L. and Papanicolaou, G., Duvaut, G. et Lions, Lions, J.L. et Magenes, E.

Lions, J.L. et Magenes, E.

[1] Problèmes aux limites non homogènes et applications, I. Dunod, Paris (1968).

Lobo-Hidalgo, M. et Sanchez-Palencia, E.

[1] Sur certaines propriétés spectrales des perturbations du domaine dans les problèmes aux limites. Comm. Part. Diff. Eq. 4, p. 1085-1098 (1979)

[2] Perturbation of spectral properties for a class of stiff problems. Proc. Fourth Intern. Symp. Compt. Methods in Science and Engineering, Versailles, France (1979).

Majda, A.

[1] Outgoing solutions for perturbation of $-\Delta$ with applications to spectral and scattering theory. J. Diff. Equat. 16 (1974) p. 515-547.

Marcellini, P.

[1] C.f. Boccardo, L. et Marcellini, P.

Marcellini, P. and Sbordone, C.

[1] An approach of the asymptotic behaviour of elliptic-parabolic operators. Jour. Math. Pures Appl. 56 (1977) p. 157-182.

Metellus, A.M. et Duvaut, G.

[1] Homogénéisation d'une plaque mince en flexion de structure périodique et symétrique. Compt. Rend. Acad. Sci. Paris. A, 283 p. 947-950 (1977).

Mignot, F., Puel, J.P. et Suquet, P.

[1] Homogénéisation d'un problème de bifurcation. Proc. Int. Meeting on Recent Methods in Non-Linear Analysis, ed. De Giorgi, Magenes, Mosco, ed. Pitagora, Bologne, (1979), p. 281-307.

Mikhlin, S.G.

[1] Mathematical physics, an advanced course. North Holland, Amsterdam 1970.

Mitropolskii, Y.A. and Moseenkov, B.I.

[1] Lectures on the applicationsof the asymptotic method to the resolution of partial differential equations. (in russian). Inst. Akad. Nauk Ukr. S.S.R., Kiev (1968).

Morawetz, C.S., Ralston, J.V. and Strauss, W.A.

[1] Decay of solutions of the wave equation outside nontrapping obstacles. Commun. Pure Appl. Math. 30 (1977) p. 447-508.

Morrison, J.A.

[1] Comparaison of the modified method of averaging and the two-variable expansion procedure. SIAM Review, 8, 1966, p. 66-85.

Moseenkov, B.I.

[1] C.f. Mitropolski, Y.A. and Moseenkov, B.I.

Muller, C.

[1] Foundations of the mathematical theory of electromagnetic waves. Springer, Berlin (1969).

Murat, F.

[1] Compacité par compensation. Partie I, Ann. Scuala Norm. Sup. Pisa 5 (1978) p. 489-507 ; Partie II, Proc. Intern. Meeting on Recent Methods in Non Linear Analysis, ed. De Giorgi, Magenes, Mosco, ed. Pitagora, Bologna (1979), p. 245-256.

[2] H-convergence. Rapport du Séminaire d'Analyse Fonctionnelle et numérique de l'Université d'Alger (1978).

Nagy, B.Sz.

[1] C.f. Riesz, F. et Nagy,B.Sz.

Nečas, J.

[1] Les méthodes directes en théorie des équations elliptiques. Masson-Academia, Paris-Prague (1967).

Ohayon, R.

[1] Homogénéisation par développements asymptotiques mixtes. Vibrations harmoniques d'un anneau hétérogène à structure périodique. Compt. Rend. Acad. Sci. Paris, sér. A. 288 (1979), p. 173-176.

Papanicolaou, G.

[1] C.f. Bensoussan, A., Lions, J.L. and Papanicolaou, G.

Paşa, G.

[1] C.f. Fleury, F., Paşa, G. et Polişevshi, D.

Pham Huy, H. et Sanchez-Palencia, E.

[1] Phénomènes de transmission à travers des couches minces de conductivité élevée. Jour. Math. Anal. Appl. 47 p. 284-309 (1974).

Phillips, R.S.

[1] Dissipative operators and hyperbolic systems of partial differential equations. Trans. Amer. Math. Soc. 90 p. 193-254 (1959)

[2] C.f. Hille, E. and Phillips, R.S., Lax, P. and Phillips, R.S.

Polişevshi, D.

[1] C.f. Fleury, F., Paşa, G. et Polişevshi, D.

Puel, J.P.

[1] C.f. Mignot, F., Puel, J.P. et Suquet, P.

Ralston, J.V.

[1] C.f. Morawetz, C.S., Ralston, J.V. and Strauss, W.A.

Rauch, J. and Taylor, M.

[1] Potential and scattering theory on wildly perturbed domains. Jour. Func. Anal. 18 (1975) p. 27-59.

Rellich, F.

[1] Störungstheorie der Spektralzerlegung, II, stetige Abhängigkeit der Spektralschar von einem Parameter. Math. Ann. 113 p. 677-685 (1937).

[2] Uber das asymptotische Verhalten der Lösungen von $\Delta u + ku = 0$ in unendlichen Gebieten. Jber. Deutsch. Math. Verein 53, 57-64 (1943).

Riesz, F. et Nagy, B. SZ.

[1] Leçons d'analyse fonctionnelle. Gauthier Villars - Akademiai Kiado, Paris-Budapest, 1952.

Rigolot, A.

[1] Sur une théorie asymptotique des poutres. Jour. Mécan. 11 p. 674-703 (1972).

[2] Approximation asymptotique des vibrations de flexion des poutres droites élastiques. Jour. Mécan. 16 p. 493-529 (1977).

Roseau, M.

[1] Vibrations non linéaires. Springer, Berlin (1966).

[2] Asymptotic wave theory. North Holland Amsterdam (1976).

[3] Equations différentielles. Masson, Paris, (1976)

[4] Ondes internes de première classe et condition de radiation des Sommerfeld: résultat de non-unicité. Compt. Rend. Acad. Sci. Paris, sér. A., 284 (1977) p. 629-632.

[5] Méthode asymptotique et oscillations non linéaires de systèmes physiques dont l'espace de configuration est de dimension infinie, in Mélanges Th. Vogel, p. 349-369, ed. Rybak, Janssens, Jessel, Presses Univ. de Bruxelles, (1978).

Saint Jean Paulin, J.

[1] C.f. Cioranescu, D. et Saint Jean Paulin, J.

Sanchez-Hubert, J.

[1]] Etude de certaines équations intégro-différentielles issues de la théorie de l'homogénéisation. Boll. Unione Mat. Ital. 16 - B (1979), p. 857-875.

[2] Asymptotic study of the macroscopic behaviour of a solid-liquid mixture. Math. Meth. Appl. Sci. (To be published 1979-80).

Sanchez-Palencia, E.

[1] Equations aux dérivées partielles. Solutions périodiques par rapport aux variables d'espace et applications. Compt. Rend. Acad. Sci. Paris A, p. 1129-1132 (1970).

[2] Equations aux dérivées partielles dans un type de milieux hétérogènes. Compt. Rend. Acad. Sci. Paris A, 272 p. 1410-1413 (1971).

[3] Comportements local et macroscopique d'un type de milieux physiques hétérogènes. Intern. Jour. Engin. Sci. 12 p. 331-351 (1974).

[4] Problèmes de perturbations liés aux phénomènes de conduction à travers des couches minces de grande résistivité. Jour. Math. Pures Appl. 53 p.251-270 (1974).

[5] Méthode de centrage. Estimation de l'erreur et comportement des trajectoires dans l'espace des phases. Intern. Jour. Non Linear Mech. 11 p. 251-263 (1976).

[6] Perturbations spectrales liées à la vibration d'un corps élastique dans l'air in Singular perturbations and boundary layer theory, (p. 437-455). Lecture Notes in Mathematics vol. 594. Springer, Berlin (1977).

[7] Phénomènes de relaxation dans les écoulements des fluides visqueux dans les milieux poreux. Compt. Rend. Acad. Sci. Paris A. 286, p. 185-188 (1978).

[8] Justification de la méthode des échelles multiples pour une classe d'équations aux dérivées partielles. Ann. Mat. Pura Appl. 116 (1978) p.159-176.

[9] Méthode d'homogénéisation pour l'étude de matériaux hétérogènes. Phénomènes de mémoire. Rendiconti Sem. Mat. Univ. Polit. Torino 36 (1978), p. 15-25.

[10] C.f. Auriault, J.L., et Sanchez-Palencia, E., Cerneau,S. et Sanchez-Palencia, Geymonat, G. et Sanchez-Palencia, E., Levy, Th. et Sanchez-Palencia, E., Lobo-Hidalgo, M. et Sanchez-Palencia, E., Pham Huy, H. et Sanchez-Palencia, E., Sanchez-Palencia, E. et Sanchez-Hubert, J.

Sanchez-Palencia, E. et Sanchez-Hubert, J.

[1] Sur certains problèmes physiques d'homogénéisation donnant lieu à des phénomènes de relaxation. Compt. Rend. Acad. Sci. Paris, A. 286, p. 903-906 (1978).

Sbordone, C.

[1] C.f. Marcellini, P. et Sbordone, C.

Schwartz, L.

[1] Théorie des distributions. Hermann, Paris, 1966.

[2] Méthodes mathématiques pour les sciences physiques. Hermann, Paris 1965.

Schwartz, J.T.

[1] C.f. Dunford, N. and Schwartz, J.T.

Sendeckyj, G.P.

[1] Mechanics of composite materials. (Vol. 2 of the series "Composite materials" edited by Broutman and Kroch) Academic Press, New York (1974).

Sethna, P.R.

[1] C.f. Balachandra, M. and Sethna, P.R.

Sharova, L.V.

[1] C.f. Filatov, A.N. and Sharova, L.V.

Shenk, N. and Thoe, D.

[1] Outgoing solutions of $(-\Delta + q - k^2)u = f$ in an exterior domain. Jour. Math. Anal. Appl. 31 (1970) p. 81-116.

[2] Resonant stakes and poles of the scattering matrix for perturbations of $-\Delta$. Jour. Math. Anal. Appl. 37 (1972) p. 467-491.

Simonenko, I.B.

[1] A justification of the averaging method for abstract parabolic equations. Math. U.S.S.R. Sbornik 10 p. 51-59 (1970) (Transl. of Mat. Sbornik 81, 1970).

[2] A justification of the averaging method for a problem of convection in a field of rapidly oscillating forces and for other parabolic equations. Math U.S.S.R. Sbornik 16 p. 245-263 (1972) (Transl. of Mat. Sbornik 87 n° 2 (1972)).

[3] Electrostatic problems for non uniformal media. The case of a thin dielectric with a high dielectric constant, I and II, Differ. Equat., I, 10 p. 223-229 (1974) ; II 11 p. 1398-1404 (1975).(Transl. of Differ. Uravn., I, 10 p. 301-309 (1974), II 11 p. 1870-1878 (1975)).

[4] Limit problem in thermal conductivity in a homogeneous medium. Siberian Math. J. 16 p. 991-998 (1975) (Transl. of Sibirs. Mat. Zh. 16 p. 1291-1300 (1975)).

[5] C.f. Boroditskii, M.P. and Simonenko, I.B.

Smirnov, V.I.

[1] A course of higher mathematics, V. Pergamon, Oxford (1964).

Sommerfeld, A.

[1] Partial differential equations. Academic Press, New York (1952).

Spagnolo, S.

[1] Sul limite delle soluzioni di problemi di Cauchy relativi all'equazione del calore. Ann. Scuola Norm. Sup. Pisa 21 (1967) p. 657-699.

[2] Sulla convergenza di soluzioni di equazioni paraboliche ed ellittiche. Ann. Scuola Norm. Sup. Pisa 22 (1968) p. 577-597.

[3] C.f. De Giorgi, E. et Spagnolo, S.

Steinberg, S.

[1] Meromorphic families of compact operators. Arch. Rat. Mech. Anal. 31 p. 372-379 (1968).

Strauss, W.A.

[1] C.f. Morawetz, C.S., Ralston, J.V. and Strauss, W.A.

Stummel, F.

[1] Diskrete Konvergenz linearer Operatoren. Part I, Math. Ann. 190, p. 45-92 (1970), part II, Math. Zeit, 120, p. 231-264 (1971).

[2] Singular perturbations of elliptic sesquilinear forms. Proceedings of the conference on ordinary and partial differential equations at Dundee. Lecture Notes in Mathematics vol. 208, Springer, Berlin (1972).

[3] Perturbation of domains in elliptic boundary value problems.in Applications of methods of functional analysis to problems in mechanics. Lecture Notes in Mathematics 503, Springer, Berlin (1976), p. 110-136.

Suquet, P.

[1] C.f. Mignot, F., Puel, J.P. et Suquet, P.

Tartar, L.

[1] Une nouvelle méthode de résolution d'équations aux dérivées partielles non linéaires. Lect. Notes Math. n° 665, (Journées Analyse Non Linéaire, Besançon 1977) p. 228-241.

[2] Homogénéisation en hydrodynamique in Singular Perturbations and Boundary Layer Theory. Lect. Notes Math. 594 (1977) p. 474-481.

[3] Cours Peccot (Collège de France, 1977).

[4] Quelques remarques sur l'homogénéisation. Functional Anal. and Num. Anal, Proc. Japan-France seminar 1976. Ed. Fujita, Jap. Soc. for the Promotion of Sci. (1978) p. 469-482.

[5] Topics in Non Linear Analysis. Report of the University of Orsay, France (1978) (Course at the University of Wisconsin, Madison, U.S.A. 1974-5).

[6] Compensated compactness and applications to P.D.E. in Non Linear Analysis and Mechanics, Heriot-Watt Symposium, vol. IV, ed. Knops, Research Notes in Math. Pitmann, London (1979) p. 136-212.

Taylor, M.

[1] Rauch, J. and Taylor, M.

Temam, R.

[1] Navier - Stokes equations. North Holland Amsterdam (1977).

Thoe, D.

[1] C.f. Shenk, N. and Thoe, D.

Trenogin, V.A.

[1] Development and applications of the asymptotic method of Lyusternik and Vishik. Rus. Math. Surveys (1969) p. 119-136.

Turbe, N.

[1] On the two-scales method for a class of integro-differential equations appearing in viscoelasticity. Int. Jour. Engin. Sci. 17 (1979) p.857-868.

Uralceva, N.N.

[1] C.f. Ladyzhenskaya, O.A. et Uralceva, N.N.

Van Dyke, M.

[1] Perturbation methods in fluid mechanics. Academic Press, New York (1964).

Vanninathan, M.

[1] Homogénéisation des valeurs propres dans les milieux perforés. Compt. Rend. Acad. Sci. Paris sér. A. 287 (1978) p. 403-406.

[2] Homogénéisation des problèmes de valeurs propres dans les milieux perforés. Problème de Dirichlet. Compt. Rend. Acad. Sci. Paris 287 (1978) p. 823-5.

Višik, M.I. and Lyusternik, L.A.

[1] Regular degeneration and boundary layer for linear differential equations with small parameter. Uspekhi Mat. Nauk 12 (1957) p. 3-177 (= Amer. Math. Sov. Transl. ser. 2, 20 p. 239-364).

Vogel, T.

[1] Sur les conditions aux limites de l'équation du son. Acustica, 2 p.281-286 (1952).

Werner, P.

[1] Zur mathematischen Theorie akustischer Wellenfelder. Arch. Rat. Mech. Anal. 6 (1960) p. 231-260.

[2] Randwertprobleme der mathematischen Akustik. Arch. Rat. Mech. Anal. 10 (1962) p. 29-66.

[3] Beugunsprobleme der mathematischen Akustik. Arch. Rat. Mech. Anal. 12 (1963) p. 155-184.

Wilcox, C.H.

[1] Scattering states and wave operators in the obstract theory of scattering. Jour. Funct. Anal. 12 p. 257-274 (1973).

[2] Scattering theory for the d'Alembert equation in exterior domains. Lectures Notes in Mathematics vol. 442, Springer Berlin (1975).

[3] Spectral analysis of the Pekeris operator in the theory of acoustic wave propagation in shallow water. Arch. Rat. Mech. Anal. 60 (1976) p. 259-300.

[4] C.f. Guillot, J.C. and Wilcox, C.H.

Yosida, K.

[1] Functional Analysis. Springer Berlin (1971).

Zabreiko, P.P. and Fetisov, Yu. I.

[1] The small-parameter method for hyperbolic equations. Diff. Equat. 1975, p. 626-634 (Transl. of Diff. Uravn. 8 p. 823-834, 1972).

# INDEX

## NOTATIONS

We use the convention of repeated indexes :

$$a_i\, b_i = \sum_1^N a_i\, b_i$$

$\underline{x}$ or $x = \{ x_1 , x_2 \dots x_N \}$ is the current point of $R^N$

$\underline{x}$ vector in $R^N$ (or $C^N$)

$\overline{x}$ complex conjugate

<u>If A is an operator</u> :

D(A) = domain of A

R(A) = range of A

R(z) = resolvent = $(A - z)^{-1}$

AX = image of X by A

$\sigma(A)$ = spectrum

$\rho(A)$ = resolvent set

Re = real part

Im = imaginary part

o,O are the classical notations for orders :

$b = o(a)$ means $\frac{b}{a} \to 0$

$b = O(a)$ means $\frac{b}{a}$ is bounded

$\Omega$ , $\partial\Omega$, $\overline{\Omega}$ domain $\Omega$ , its boundary and its closure

$\mathcal{D}(\Omega)$ space of functions $C^\infty$ with compact support in $\Omega$

$\mathcal{D}'(\Omega)$ distributions on $\Omega$

$L^2(\Omega)$ , $H^m(\Omega)$ , $H_0^m(\Omega)$ are the classical Sobolev spaces

$\underline{L}^2 = (L^2)^N$ , $\underline{H}^m = (H^m)^N$

$\mathcal{L}(X , Y)$ = space of the linear bounded operators from X into Y.

$u|_\Omega$ is the restriction of u to $\Omega$

$u|_{\partial\Omega}$ = trace on $\partial\Omega$

$H^1_{loc}(\Omega)$ (if $\Omega$ is unbounded, $u \in H^1_{loc}(\Omega)$ means that $u|_{\{|x|<\rho\}} \in H^1(\Omega \cap \{|x|<\rho\})$ for any $\rho$

| | |
|---|---|
| $\hat{a}$ | Laplace or Fourier transform of a |
| $\underline{a} \wedge \underline{b}$ | vector product |
| $\mathcal{L}(B)$ | class of distributions with values in B having a Laplace transform in the framework of chapt. 4, sect. 6 |
| $*$ | convolution product |
| $\sim$ | average on a period Y : $\tilde{\cdot} = \frac{1}{\vert Y \vert} \int_Y \cdot \, dy$ |
| [ , ] ⟨ , ⟩ ( , ) | duality and scalar products |
| $\Longrightarrow$ | implies |
| $\Longleftrightarrow$ | equivalent |
| $\rightarrow$ | tends |
| $\mapsto$ | sends |
| $\overline{\lim}$ , $\underline{\lim}$ | upper and lower limits |
| $\mathrm{grad}_x, \mathrm{grad}_y$ | = gradient with respect to the variable x or y |
| supp f | = support of f |
| $\Delta$ | = Laplacian / domain in the complex plane |
| V' | = dual of V |
| $\delta_{ij}$ | = Kronecker symbol |
| u' | $= \frac{du}{dt}$ |
| [u] on $\Gamma$ | means the difference of values of the function u on the two sides of the surface $\Gamma$ |
| $a^{\varepsilon}(x)$ | is sometimes written for $a(\frac{x}{\varepsilon})$ |
| $u_i^0$ , $u_i^1$ | (in chapter 13) restrictions of $u_i$ to $\Omega^0$ and $\Omega^1$ |
| $\Omega_\rho$ | (in chapter 17) = $\Omega \cap \{ x ; \vert x \vert < \rho \}$ |
| $\pi$ | (in chapters 7 and 8) = porosity = $\vert Y_f \vert / \vert Y \vert$ |
| C | denotes constants, it may take several values in the same formula |

All the Hilbert spaces considered are separable

The numbering of formulae, lemmas, propositions, remarks and theorems is independent for each chapter. Moreover, (3.5), for instance, is the 5th formula of sect. 3.

Selected Issues from

# Lecture Notes in Mathematics

Vol. 662: Akin, The Metric Theory of Banach Manifolds. XIX, 306 pages. 1978.

Vol. 665: Journées d'Analyse Non Linéaire. Proceedings, 1977. Edité par P. Bénilan et J. Robert. VIII, 256 pages. 1978.

Vol. 667: J. Gilewicz, Approximants de Padé. XIV, 511 pages. 1978.

Vol. 668: The Structure of Attractors in Dynamical Systems. Proceedings, 1977. Edited by J. C. Martin, N. G. Markley and W. Perrizo. VI, 264 pages. 1978.

Vol. 675: J. Galambos and S. Kotz, Characterizations of Probability Distributions. VIII, 169 pages. 1978.

Vol. 676: Differential Geometrical Methods in Mathematical Physics II, Proceedings, 1977. Edited by K. Bleuler, H. R. Petry and A. Reetz. VI, 626 pages. 1978.

Vol. 678: D. Dacunha-Castelle, H. Heyer et B. Roynette. Ecole d'Eté de Probabilités de Saint-Flour. VII-1977. Edité par P. L. Hennequin. IX, 379 pages. 1978.

Vol. 679: Numerical Treatment of Differential Equations in Applications, Proceedings, 1977. Edited by R. Ansorge and W. Törnig. IX, 163 pages. 1978.

Vol. 681: Séminaire de Théorie du Potentiel Paris, No. 3, Directeurs: M. Brelot, G. Choquet et J. Deny. Rédacteurs: F. Hirsch et G. Mokobodzki. VII, 294 pages. 1978.

Vol. 682: G. D. James, The Representation Theory of the Symmetric Groups. V, 156 pages. 1978.

Vol. 684: E. E. Rosinger, Distributions and Nonlinear Partial Differential Equations. XI, 146 pages. 1978.

Vol. 690: W. J. J. Rey, Robust Statistical Methods. VI, 128 pages. 1978.

Vol. 691: G. Viennot, Algèbres de Lie Libres et Monoïdes Libres. III, 124 pages. 1978.

Vol. 693: Hilbert Space Operators, Proceedings, 1977. Edited by J. M. Bachar Jr. and D. W. Hadwin. VIII, 184 pages. 1978.

Vol. 696: P. J. Feinsilver, Special Functions, Probability Semigroups, and Hamiltonian Flows. VI, 112 pages. 1978.

Vol. 702: Yuri N. Bibikov, Local Theory of Nonlinear Analytic Ordinary Differential Equations. IX, 147 pages. 1979.

Vol. 704: Computing Methods in Applied Sciences and Engineering, 1977, I. Proceedings, 1977. Edited by R. Glowinski and J. L. Lions. VI, 391 pages. 1979.

Vol. 710: Séminaire Bourbaki vol. 1977/78, Exposés 507–524. IV, 328 pages. 1979.

Vol. 711: Asymptotic Analysis. Edited by F. Verhulst. V, 240 pages. 1979.

Vol. 712: Equations Différentielles et Systèmes de Pfaff dans le Champ Complexe. Edité par R. Gérard et J.-P. Ramis. V, 364 pages. 1979.

Vol. 716: M. A. Scheunert, The Theory of Lie Superalgebras. X, 271 pages. 1979.

Vol. 720: E. Dubinsky, The Structure of Nuclear Fréchet Spaces. V, 187 pages. 1979.

Vol. 724: D. Griffeath, Additive and Cancellative Interacting Particle Systems. V, 108 pages. 1979.

Vol. 725: Algèbres d'Opérateurs. Proceedings, 1978. Edité par P. de la Harpe. VII, 309 pages. 1979.

Vol. 726: Y.-C. Wong, Schwartz Spaces, Nuclear Spaces and Tensor Products. VI, 418 pages. 1979.

Vol. 727: Y. Saito, Spectral Representations for Schrödinger Operators With Long-Range Potentials. V, 149 pages. 1979.

Vol. 728: Non-Commutative Harmonic Analysis. Proceedings, 1978. Edited by J. Carmona and M. Vergne. V, 244 pages. 1979.

Vol. 729: Ergodic Theory. Proceedings 1978. Edited by M. Denker and K. Jacobs. XII, 209 pages. 1979.

Vol. 730: Functional Differential Equations and Approximation of Fixed Points. Proceedings, 1978. Edited by H.-O. Peitgen and H.-O. Walther. XV, 503 pages. 1979.

Vol. 731: Y. Nakagami and M. Takesaki, Duality for Crossed Products of von Neumann Algebras. IX, 139 pages. 1979.

Vol. 733: F. Bloom, Modern Differential Geometric Techniques in the Theory of Continuous Distributions of Dislocations. XII, 206 pages. 1979.

Vol. 735: B. Aupetit, Propriétés Spectrales des Algèbres de Banach. XII, 192 pages. 1979.

Vol. 738: P. E. Conner, Differentiable Periodic Maps. 2nd edition, IV, 181 pages. 1979.

Vol. 742: K. Clancey, Seminormal Operators. VII, 125 pages. 1979.

Vol. 755: Global Analysis. Proceedings, 1978. Edited by M. Grmela and J. E. Marsden. VII, 377 pages. 1979.

Vol. 756: H. O. Cordes, Elliptic Pseudo-Differential Operators – An Abstract Theory. IX, 331 pages. 1979.

Vol. 760: H.-O. Georgii, Canonical Gibbs Measures. VIII, 190 pages. 1979.

Vol. 762: D. H. Sattinger, Group Theoretic Methods in Bifurcation Theory. V, 241 pages. 1979.

Vol. 765: Padé Approximation and its Applications. Proceedings, 1979. Edited by L. Wuytack. VI, 392 pages. 1979.

Vol. 766: T. tom Dieck, Transformation Groups and Representation Theory. VIII, 309 pages. 1979.

Vol. 771: Approximation Methods for Navier-Stokes Problems. Proceedings, 1979. Edited by R. Rautmann. XVI, 581 pages. 1980.

Vol. 773: Numerical Analysis. Proceedings, 1979. Edited by G. A. Watson. X, 184 pages. 1980.

Vol. 775: Geometric Methods in Mathematical Physics. Proceedings, 1979. Edited by G. Kaiser and J. E. Marsden. VII, 257 pages. 1980.